Grundlehren der mathematischen Wissenschaften 238

A Series of Comprehensive Studies in Mathematics

Colin C. Graham
O. Carruth McGehee

Essays in Commutative Harmonic Analysis

Springer-Verlag
New York Heidelberg Berlin

Colin C. Graham

Department of Mathematics
Northwestern University
Evanston, Illinois 60201
USA

O. Carruth McGehee

Department of Mathematics
Louisiana State University
Baton Rouge, Louisiana 70803
USA

AMS Subject Classifications 43A25, 43A45, 43A46, 43A70, 42A45, 42A55, 42A63, 43A10

With 1 Figure

Library of Congress Cataloging in Publication Data

Graham, Colin C.
Essays in commutative harmonic analysis.

(Grundlehren der mathematischen Wissenschaften; 238)
Bibliography: p.
Includes index.
1. Harmonic analysis. 2. Locally compact abelian groups. 3. Fourier transformations.
I. McGehee, O. Carruth, joint author. II. Title.
III. Series.
QA403.G7 515′.2433 79-13096

Printed in the United States of America.

9 8 7 6 5 4 3 2 1

ISBN 0-387-90426-3 New York Heidelberg Berlin
ISBN 3-540-90426-3 Berlin Heidelberg New York

To my wife, Jill Wescott Graham

To my father and mother,
Oscar M. McGehee and Louise Blanche Carruth McGehee

Preface

This book considers various spaces and algebras made up of functions, measures, and other objects—situated always on one or another locally compact abelian group, and studied in the light of the Fourier transform. The emphasis is on the objects themselves, and on the structure-in-detail of the spaces and algebras.

A mathematician needs to know only a little about Fourier analysis on the commutative groups, and then may go many ways within the large subject of harmonic analysis—into the beautiful theory of Lie group representations, for example. But this book represents the tendency to linger on the line, and the other abelian groups, and to keep asking questions about the structures thereupon. That tendency, pursued since the early days of analysis, has defined a field of study that can boast of some impressive results, and in which there still remain unanswered questions of compelling interest.

We were influenced early in our careers by the mathematicians Jean-Pierre Kahane, Yitzhak Katznelson, Paul Malliavin, Yves Meyer, Joseph Taylor, and Nicholas Varopoulos. They are among the many who have made the field a productive meeting ground of probabilistic methods, number theory, diophantine approximation, and functional analysis. Since the academic year 1967–1968, when we were visitors in Paris and Orsay, the field has continued to see interesting developments. Let us name a few. Sam Drury and Nicholas Varopoulos solved the union problem for Helson sets, by proving a remarkable theorem (2.1.3) which has surely not seen its last use. Gavin Brown and William Moran and others fleshed out the framework that Joseph Taylor had provided for the study of convolution algebras, and Thomas Körner's construction techniques made child's play of problems once thought intractable.

The book is for those who work in commutative harmonic analysis, for those who wish to do so, and for those who merely want to look into it. In the areas that we have chosen to treat, we have tried to make more accessible than before not only the results for their own sakes, but also the techniques, the points of view, and the sources of intuition by which the subject lives.

We have had repeatedly to choose whether to present material in the abstract setting of an arbitrary locally compact abelian group G, or on, say, the circle group T. As often as not, restricting the discussion to a concrete setting makes the essential ideas more vivid, and one loses nothing but technical clutter. But sometimes one must concede the greater usefulness and aesthetic appeal of a general treatment. So we have made sometimes the one choice, and sometimes the other. But let us emphasize that the subject is truly the union, not the intersection, of the studies on the various abelian groups.

The order of the chapters does not have the usual significance, even though we did choose it with care. One reviewer suggests that 12 and 11 should appear between 4 and 5. In any event, whenever the material of one chapter depends on some part of another one, the reader is provided with a specific reference. Therefore one who is not discouraged by the Prerequisites, and who is familiar with our Symbols, Conventions, and Terminology, may begin reading at any one of the chapters.

We thank our home departments, at Northwestern and Louisiana State, for their support over the years. We thank also the several other mathematics departments where one or both of us have visited and found pleasant conditions for work: in Paris, Jerusalem, Urbana, Eugene, and Honolulu.

We thank the many colleagues and friends who have given us encouragement and help. In particular, for their extensive and critical attention to drafts of various parts of the book, we thank Aharon Atzmon, John Fournier, Yitzhak Katznelson, Thomas Ramsey, and George Shapiro. Especially do we thank Sadahiro Saeki, who read over half the book with care and made many valuable suggestions.

Evanston, Illinois Colin C. Graham

Baton Rouge, Louisiana O. Carruth McGehee

March, 1979

Contents

Chapter 4

Sets of Uniqueness, Sets of Multiplicity 91

Chapter 5

A Brief Introduction to Convolution Measure Algebras 122

Chapter 6

Independent Power Measures 138

Chapter 7

Riesz Products 196

Prerequisites

The areas in which it is most important for the reader to have both knowledge and facility are as follows.

1. Basic functional analysis, as in Dunford and Schwartz [1, Chapter II and Sections V.1–V.6] or Rudin [3].
2. The theory of measure and integration, as in Royden [1, Parts 1 and 3].
3. Commutative Banach algebra theory, as in Rudin [3, Chapter 11].
4. Fourier analysis on the line and the circle, as in Katznelson [1, Chapters I, IV, and VI; also Sections II.1 and V.1].
5. Fourier analysis on locally compact abelian groups, as in Rudin [1, Chapters 1 and 2]. In particular, we shall use the structure theorem: every locally compact abelian group G has an open subgroup of the form $R^n \times H$, where $n \geq 0$ and H is compact. For another treatment of that theorem, see Hewitt and Ross [1, Section 24].

In addition, the reader will find it helpful to have sampled the theory of exceptional subsets ("thin sets") of groups, as for example in Lindahl and Poulsen [1, Chapter 1] and Kahane [1, Chapters III and IV].

Some of the elementary material is treated in the Appendix. For example, the results of Section 2.6 in Rudin [1] are given a different treatment in A.5.

Besides the works that we have recommended here, there are of course other excellent sources from which to acquire the same background knowledge.

There are isolated places in the book where we use other, more advanced and specialized material, and at such points we give specific references.

Symbols, Conventions, and Terminology

Before beginning any of the chapters, the reader should study this list of symbols and terms that are used most frequently. Each item is attended by a brief definition, and perhaps also a remark or two about relevant conventions and basic facts. Some of the definitions make use of others on the list. The order is alphabetical, with the Greek entries grouped all together after the Latin ones; except that we single out several items to explain at the outset.

The symbol G stands for an arbitrary locally compact abelian group, except when some other meaning is specified. The same is true for the symbol Γ. When G and Γ appear in the same context, each denotes the dual group of the other; and then for $x \in G$ and $\gamma \in \Gamma$, the value of γ at x is denoted by $\langle x, \gamma \rangle$. Thus if Γ is considered as an additive group, $\langle x, \gamma_1 + \gamma_2 \rangle = \langle x, \gamma_1 \rangle \cdot \langle x, \gamma_2 \rangle$. If f is an element of a Banach space and S an element of the dual space, then too, $\langle f, S \rangle$ means the value of S at f.

The symbol E nearly always stands for a closed subset of Γ. Whenever $X = X(\Gamma)$ is a Banach algebra of functions on Γ (such as A, AP, B, B_0, or M_p), the symbol $X(E)$ (or $X(E, \Gamma)$) stands for the Banach algebra of restrictions to E of functions in X with norm

$$\|f\|_{X(E)} = \inf\{\|g\|_X : g = f \quad \text{on} \quad E\}.$$

Equivalently, $X(E)$ may be defined as the quotient algebra X/I, where I is the ideal $\{f \in X : f^{-1}(0) \supseteq E\}$. But when X is a space of distributions on G (such as M, M_1, M_c, M_d, PF, or PM), then the symbol $X(E)$ stands for the subspace of X consisting of the elements with support contained in E.

$A(\Gamma)$	—the Fourier representation of the convolution algebra $L^1(G)$; that is, the Banach algebra of Fourier transforms $\hat{f}$ of elements f of $L^1(G)$. The operators are the usual pointwise ones, and the norm, denoted by $\|\hat{f}\|_{A(\Gamma)}$ or $\|\hat{f}\|_A$, is defined to equal the $L^1(G)$-norm of f. Note the natural norm-decreasing inclusion: $A(\Gamma) \subseteq C_o(\Gamma)$.
$AP(\Gamma)$	—the algebra of almost periodic functions on Γ, with pointwise operations and the supremum norm. It is realizable as $C(b\Gamma)$.

$B(\Gamma)$	—the Fourier representation of the convolution algebra $M(G)$; that is, the Banach algebra of Fourier transforms $\hat{\mu}$ of measures $\mu \in M(G)$. The elements of $B(\Gamma)$ are also called the Fourier–Stieltjes transforms on Γ. The operations are the usual pointwise ones, and the norm, denoted by $\|\hat{\mu}\|_{B(\Gamma)}$ or $\|\hat{\mu}\|_B$, is the total variation of μ. Note the natural isometric and isomorphic inclusions: $L^1(G) \subseteq M(G)$, $A(\Gamma) \subseteq B(\Gamma)$.
$B_o(\Gamma)$	$= B(\Gamma) \cap C_o(\Gamma)$.
BV	—the space of functions of bounded variation
$b\Gamma$	—the Bohr compactification of Γ; equivalently, the dual group of G_d.
$\mathbb{C}$	—the complex number system.
$C(X)$	—where X is a topological space: the Banach algebra of bounded continuous complex-valued functions on X, with the usual pointwise operations and the supremum norm.
$C_c(X)$	— $\{f \in C(X)$: the support of f is compact$\}$.
$C_o(X)$	—the subalgebra of $C(X)$ consisting of the functions that vanish at infinity.
countable	—in one-to-one correspondence with some subset of the positive integers.
D	—the Cantor group; that is, the product group $\prod_{j=1}^{\infty} G_j$, where each G_j is the two-element group.
$\# E$ *or* Card E	—the cardinality of the set E.
E-polynomial, E-function, E-measure	Let E be a subset of Γ. An E-polynomial, E-function, or E-measure is a trigonometric polynomial, a function, or a measure, respectively, whose Fourier transform vanishes on the complement of E.
$\hat{f}$	—(1) the Fourier transform of f, where f is a function, bounded Borel measure, or distribution defined on (say) G. Thus if $f \in L^1(G)$, $$\hat{f}(\gamma) = \int_G \langle x, -\gamma\rangle dm_G(x) \qquad \text{for } \gamma \in \Gamma.$$ More generally, for $\mu \in M(G)$, $$\hat{\mu}(\gamma) = \int_G \langle x, -\gamma\rangle d\mu(x) \qquad \text{for } \gamma \in \Gamma.$$ The Fourier transform provides isometric isomorphisms $L^1(G) \triangleq A(\Gamma)$, $M(G) \triangleq B(\Gamma)$, since $(\mu_1 * \mu_2)^\wedge = \hat{\mu}_1\hat{\mu}_2$.

	—(2) the Gelfand transform of an element f of a Banach algebra.
F_σ	—the class of all sets that are countable unions of closed sets.
G_δ	—the class of all sets that are countable intersections of open sets.
$\hat{G}$	—the dual group of G.
G_d	—G, but with its topology replaced by the discrete topology.
G^n	—the product group $\prod_{j=1}^n G_j$, where each G_j is G.
$Gp\ H$	—where $H \subseteq G$: the group generated algebraically by H.
Helson set	See $\alpha(E)$ below.
Hermitian	—$\mu \in M(G)$ is Hermitian if $\tilde{\mu} = \mu$.
$h(I)$	—where I is an ideal in a Banach algebra of functions on a set X: the *hull* of I, $h(I) = \{x \in X : f(0) = 0 \text{ for all } f \in I\}$.

$I(E)$, $I_o(E)$, $J(E)$ — When $E \subseteq \Gamma$ and $A = A(\Gamma)$ is the algebra under discussion, these symbols denote respectively the largest ideal whose hull is E:

$$I(E) = \{f \in A : f^{-1}(0) \supseteq E\},$$

the smallest closed ideal whose hull is E, and the smallest ideal whose hull is E: $J(E) = \{f \in A : f^{-1}(0) \text{ is a neighborhood of } E\}$. Note that $I_o(E)$ is the closure in A of $J(E)$.

independent — A set $E \subseteq G$ is independent if whenever $x_1, \ldots, x_n \in E$ and $u_1, \ldots, u_n \in Z$, and $\sum_{j=1}^n u_j x_j = 0$, then

$$(1) \qquad u_j x_j = 0 \qquad \text{for } 1 \le j \le n.$$

But when $G = T$, we replace (1) in the definition by

$$(2) \qquad u_j = 0 \qquad \text{for } 1 \le j \le n.$$

Kronecker set — A set $E \subseteq G$ is a Kronecker set if for every $f \in C(E)$ and every $\varepsilon > 0$ there exists $\gamma \in \Gamma$ such that $|f(x) - \langle x, \gamma \rangle| < \varepsilon$ for all $x \in E$. A Kronecker set is evidently a Helson set.

K_p-set — A set $E \subseteq G = \prod_{j=1}^\infty G_j$, where each G_j is T_p, is a K_p-set if for every continuous $f: E \to T_p$ there exists

$\gamma \in \Gamma$ such that $f(x) = \langle x, \gamma \rangle$ for all $x \in E$. A K_p-set is evidently a Helson set.

$L^1(G)$ —the convolution algebra of Haar-integrable complex-valued functions (or rather, equivalence classes thereof) on G. Convolution is given by:

$$f * g(x) = \int_G f(x - y)g(y)dm_G(y).$$

The norm, under which $L^1(G)$ is a Banach algebra, is given by:

$$\|f\|_1 = \|f\|_{L^1(G)} = \int_G |f(x)|dm_G(x).$$

$L^p(G)$ $(1 \leq p < \infty)$ —the Banach space of equivalence classes of measurable functions f on G such that $|f|^p$ is integrable, with norm $\|f\|_p = \|f\|_{L^p(G)} = \int_G |f(x)|^p\, dm_G(x)$.

$L^\infty(G)$ —the Banach space of equivalence classes of essentially bounded measurable functions f on G, with norm

$$\|f\|_\infty = \|f\|_{L^\infty(G)} = \inf\{c : |f(x)| \leq c \quad \text{l.a.e.-}m_G\}.$$

$L_E^p(G)$ $= \{f \in L^p(G) : \hat{f}(\gamma) = 0 \text{ for } \gamma \notin E\}$.

m_G —a Haar measure on G, normalized so that $m_G(\{0\}) = 1$ if G is discrete and infinite; or so that $m_G(G) = 1$ if G is compact. We often write dx for $dm_G(x)$, $d\gamma$ for $dm_\Gamma(\gamma)$, and so forth. As for the real line, $dm_R(x)$ is alternatively Lebesgue measure dx, or $dx/2\pi$; thus for $f \in L^1(R)$, $\hat{f}(y) = \int_{-\infty}^{\infty} f(x)e^{-iyx}\, dx$, and if $\hat{f} \in L^1(R)$, $f(x) = (1/2\pi)\int_{-\infty}^{\infty} \hat{f}(y)e^{iyx}\, dy$.

$M(G)$ —the convolution algebra of bounded complex-valued Borel measures on G. Convolution is given by: $\mu * \nu(E) = \int_G \mu(E - x)d\nu(x)$ for every Borel set E. The norm, under which $M(G)$ is a Banach algebra, is given by

$$\|\mu\|_M = \|\mu\|_{M(G)} = \int_G |d\mu(x)|.$$

$M(G)$ is the Banach space dual of $C_o(G)$, with the pairing given by:

$$\langle f, \mu \rangle = \int f(x)\overline{d\mu(x)}.$$

$M(E)$ —where $E \subseteq G$: the subspace of $M(G)$ consisting of the measures with support contained in E.

$M_1(E)$ $= \{\mu \in M(E)\colon \operatorname{supp} \mu \text{ is a finite set}\}$.

$M_c(E)$ $= \{\mu \in M(E)\colon \mu$ is continuous, that is, $\mu(\{x\}) = 0$ for each $x \in E\}$.

$M_d(E)$ $= \{\mu \in M(E)\colon \mu \text{ is discrete}\} \simeq l^1(E)$.

$M_o(E)$ $= \{\mu \in M(E)\colon \hat{\mu} \in C_o(\Gamma)\}$.

M-set A set E is an M-set if $PF(E) \neq \{0\}$.

M_o-set A set E is an M_o-set if $PF(E) \cap M(E) \neq \{0\}$.

$N(E)$ —the annihilator of $I(E)$ in $PM(\Gamma)$; equivalently, the Banach space dual of the quotient algebra $A(E) = A(G)/I(E)$.

Parseval relation —an identity of a certain kind, of which the following are examples. (1) For $f \in A(G)$ and $\mu \in M(G)$, $\int_G f(x)\overline{d\mu(x)} = \int_\Gamma \hat{f}(\gamma)\overline{\hat{\mu}(\gamma)}d\gamma$. (2) For f, $g \in L^2(G)$, $\int_G f(x)g(x)dx = \int_\Gamma \hat{f}(\gamma)\hat{g}(\gamma)d\gamma$.

Pisot number —an algebraic integer $\theta > 1$ with conjugates $\theta, x_1, \ldots, x_{n-1}$ such that $|x_j| < 1$ for each j.

portion If U is an open interval on the line or circle, and if $U \cap E$ is closed and nonempty, then $U \cap E$ is called a portion of E.

$PF(\Gamma)$ —the subspace of $PM(\Gamma)$ consisting of those pseudomeasures S such that $\hat{S} \in C_o(G)$. Its elements are called pseudofunctions. Its dual space is $B(\Gamma)$, which of course equals $A(\Gamma)$ when G is discrete.

$PM(\Gamma)$ —the Banach space dual of $A(\Gamma)$. If $S \in PM(\Gamma)$, S is called a pseudomeasure, and there exists $\hat{S} \in L^\infty(G)$ such that for every $\hat{f} \in A(\Gamma)$,

$$\langle \hat{f}, S \rangle = \int_G f(x)\overline{\hat{S}(x)}dx.$$

Conversely, every element $S \in L^\infty(G)$ gives rise to a pseudomeasure, and $\|S\|_{PM} = \|S\|_{PM(\Gamma)} = \|\hat{S}\|_{L^\infty(\Gamma)}$. Note the natural norm-decreasing inclusion: $A(\Gamma) \subseteq C_o(\Gamma)$ and its adjoint: $M(\Gamma) \subseteq PM(\Gamma)$. As usual with a Banach space and its pre-dual, $PM(\Gamma)$ is a module over $A(\Gamma)$; for $S \in PM$ and $f \in A$, we define $fS \in PM$ by: $\langle g, fS \rangle = \langle gf, S \rangle$ for $g \in A$. Note that $\|fS\|_{PM} \leq \|f\|_A \|S\|_{PM}$; and that $(fS)^\wedge = \hat{f} * \hat{S}$.

R —the real number system.

R^n —n-dimensional Euclidean space.

Rad I —the radical of an ideal I in a Banach algebra B; that is, the intersection of the maximal modular ideals of B that contain I. An example when $B = M(G)$: Rad $L^1(G) = \{\mu \in M(G): \hat{\mu}(\psi) = 0 \text{ for all } \psi \in \Delta\setminus\Gamma\}$.

Sidon set See $\alpha(E)$ below.

supp S —the support of the distribution S.

T —the circle group, realized additively as R mod 2π, or multiplicatively as $\{z \in C: |z| = 1\}$.

T_p —the subgroup $\{z \in T: z^p = 1\}$ (where p is a positive integer).

weak∗ topology —the topology $\sigma(X, X^*)$, the weakest topology on the dual space X^* of a Banach space X with respect to which the mapping $x^* \to \langle x, x^* \rangle$ is continuous for every $x \in X$.

Z —the integer group.

Z_p —Z mod p (where p is a positive integer).

$\alpha(E)$ —the Helson constant of a closed set $E \subseteq G$, called also the Sidon constant when G is discrete. The set E is a *Helson set* (called a Sidon set when G is discrete) if it is a set of interpolation for the algebra $A(G)$—that is, if $A(E) = C_o(E)$. For arbitrary E, the inclusion map: $A(E) \subseteq C_o(E)$ is one-to-one and norm-decreasing, and $A(E)$ is dense in $C_o(E)$, so that E is a Helson set if and only if the quantity

$$\alpha(E) = \sup\left\{\frac{\|f\|_{A(E)}}{\|f\|_{C_o(E)}} : f \in A(E), f \neq 0\right\}$$

is finite. Evidently $1 \leq \alpha(E) \leq \infty$, and by duality

$$\alpha(E) = \sup\left\{\frac{\|\mu\|_M}{\|\mu\|_{PM}} : \mu \in M(E), \mu \neq 0\right\}.$$

When G is discrete, E is a Helson set (a Sidon set) if and only if $B(E) = l^\infty(E)$.

δ_x or $\delta(x)$ —the measure μ such that $\mu(E) = \begin{cases} 1 \text{ if } x \in E, \\ 0 \text{ if } x \in E. \end{cases}$

ΔB —the maximal ideal space of the commutative Banach algebra B.

∂B —the Šilov boundary of ΔB, where B is a commutative Banach algebra; that is, the smallest closed set $S \subseteq \Delta B$ such that for every $f \in B$, $\sup_{h \in \Delta B} |\hat{f}(h)|$ is attained at some $h \in S$.

$\tilde{\mu}$	—the conjugate of the measure $\mu \in M(G)$, defined by the condition that $\tilde{\mu}(E) = \overline{\mu(-E)}$ for every Borel set $E \subseteq G$. Note that $\hat{\tilde{\mu}} = \bar{\hat{\mu}}$ on Γ.
$\mu\|_E$	—the restriction to the set E of the measure μ; that is, the measure ν such that $\nu(F) = \mu(E \cap F)$.
μ_c	—the continuous part of the measure μ.
μ_d	—the discrete (or atomic) part of μ.
μ_s	—the singular part of the measure μ.
$\mu \perp \nu$	—means that μ and ν are mutually singular.
$\mu \ll \nu$	—means that μ is absolutely continuous with respect to ν.
$\mu \approx \nu$	—means that $\mu \ll \nu$ and $\nu \ll \mu$.
ΣB	—the set of symmetric maximal ideals in the commutative Banach algebra B, that is, B has an involution $f \mapsto \tilde{f}$ and $\psi \in \Sigma B$ if and only if $\tilde{f}^\wedge(\psi) = \hat{f}(\psi)^-$ for all $f \in B$. In the case of $M(G)$, it is clear that $\Gamma \subseteq \Sigma M(G)$.
χ_E	—the indicator function (sometimes called the characteristic function) of the set E: $\chi_E(x) = 1$ if $x \in E$, 0 otherwise.

Chapter 1

The Behavior of Transforms

1.1. Introduction

A recurring theme of our subject is the effort to say which functions on Γ belong to the Fourier–Stieltjes transform space $B(\Gamma)$, and which do not—what the transforms can do, and what they cannot do. The question is a deep one. The known characterizations are neither subtle nor powerful. Indeed, it appears that the property of being a transform is not truly reducible. This Chapter treats three aspects of transform behavior. Our three topics, which may be read independently, are as follows.

(1) We devote Sections 1.2 through 1.5, as well as most of the present Section, to the identification of the idempotents in $B(\Gamma)$, and to a discussion of what may be called the idempotent principle of transform behavior. The Idempotent Theorem 1.1.1 is the most important result in the Chapter, and the reader who wishes to restrict attention to 1.1.1 and its proof (in Section 1.2) will find it convenient to do so.

(2) It is an elementary fact that unless Γ is finite, the inclusion mapping of $B(\Gamma)$ into $C(\Gamma)$ is not surjective. The most efficient and most concrete way to prove it is to point out examples of measures $\mu \in M(G)$ such that the ratio of $\|\mu\|_{M(G)} \equiv \|\hat{\mu}\|_{B(\Gamma)}$ to $\|\mu\|_{PM(G)} \equiv \|\hat{\mu}\|_{\infty}$ is arbitrarily large. Section 1.6 provides a supply of such examples; they are very simple objects, very useful in constructions. Section 1.7 explains the use of such objects to construct a Banach space that has no basis.

(3) If $f \in B(\Gamma)$ and $c = \|f\|_{B(\Gamma)}$, then of course

$$(*) \qquad \left| \int f \, d\mu \right| \leq c\|\mu\|_{PM} \text{ for every } \mu \in M(\Gamma).$$

It turns out that condition (*) characterizes the functions $f \in C(\Gamma)$ that belong to the ball of radius c in $B(\Gamma)$. In fact, if f is merely defined and measurable on a measurable set $E \subseteq \Gamma$, and if $|\int f \, d\mu| \leq c\|\mu\|_{PM}$ for every discrete measure with support contained in E, then f agrees a.e. on E with some Fourier–Stieltjes transform whose $B(\Gamma)$-norm is no greater than c. Section 1.8 is devoted to such results.

We turn now to the matter of idempotents, and idempotent-like behavior.

Let E be a subset of Γ that is both open and closed. Then we may ask how to tell whether its indicator function χ_E belongs to $B(\Gamma)$. For $\mu \in M(G)$, $\mu * \mu = \mu$ if and only if $\hat{\mu}^2 = \hat{\mu}$, which is to say that $\hat{\mu}$ takes on no values other than 0 or 1. Thus to identify the idempotent measures $\mu \in M(G)$ is to identify the sets $E \subset \Gamma$ such that $\chi_E \in B(\Gamma)$. The problem is trivial if Γ is connected; $\chi_E \in B(\Gamma)$ if and only if $E = \Gamma$ or $\varnothing$. The problem is more interesting if, for example, $\Gamma = Z$. If F is a finite subset of Z, then χ_F is the transform of the measure $\sum_{n \in F} e^{inx}\, dm_T(x)$. If E is a coset of an infinite subgroup of Z, say $E = nZ + k$, then χ_E is the transform of the measure $e^{ikx}\, dm_{T_n}(x)$, where m_{T_n} is the Haar measure of the subgroup $T_n = \{z \in T: z^n = 1\}$. Those examples, and suitable finite combinations of them, turn out to be the only ones.

There is a complete and satisfactory characterization of the idempotent measures in $M(G)$ in terms of the *coset ring* of Γ, which is the smallest family of subsets of Γ that contains all cosets of open subgroups and is closed under complements and finite unions and intersections. The result is as follows; we shall prove it in Section 1.2, and present another way to approach it in Section 5.3.

1.1.1. The Idempotent Theorem. *Let $E \subset \Gamma$. Then $\chi_E \in B(\Gamma)$ if and only if E belongs to the coset ring of Γ.*

Considering sets E in the coset ring, we go beyond the Idempotent Theorem and try to understand when $\|\chi_E\|_B$ is large. We shall see that $\|\chi_E\|_B = 1$ if and only if E is, very simply, a coset of a subgroup. Looking at examples, one may reasonably make the vague conjecture that the more complicated a combination of cosets E is, the larger $\|\chi_E\|_B$ is. The following result accords with that conjecture.

1.1.2. Theorem. *There exists a sequence, $C_N \to \infty$, such that $\|\chi_E\|_{B(Z)} \geq C_N$ whenever E is an N-element subset of Z.*

Theorem 1.1.2, with $C_N = C \log N$, is known as the Littlewood Conjecture. It was proposed in Hardy and Littlewood [1, p. 168]. In Section 1.3 we shall explain why it is plausible, and prove the best result yet obtained, in which $C_N = C(\log N)^{1/2}$. We shall also obtain a result that is a bit more general and useful. Leaving aside its quantitative aspects—which are of secondary significance, however interesting they may be to those who have tried to improve on them—the result may be formulated as follows. Let $\# S$ denote the cardinality of the set S.

1.1.3. Theorem. *There exist sequences, $0 < \varepsilon_N \to 0$ and $C_N \to \infty$, with the following properties. Let G be a compact abelian group whose dual group Γ has only K torsion elements, where K is a finite number. Let $f \in B(\Gamma)$ and suppose that for every $n \in \Gamma$, either $|f(n)| \leq \varepsilon_N$ or $|f(n)| \geq 1$. If $NK \leq \#\{n \in \Gamma: |f(n)| \geq 1\} < \infty$, then $\|f\|_{B(\Gamma)} \geq C_N$.*

That Theorem, and the principle manifested in it, are useful in understanding the behavior of transforms even on non-discrete groups, such as R—as we shall see (1.3.3, 1.3.14).

Applied to $M(T)$, the Idempotent Theorem implies that if a *continuous* measure $\mu \in M(T)$ is idempotent, and $\hat{\mu} = \chi_E$, then E is but a finite set. That fact has an interesting generalization, in which the continuity condition must be replaced by a stronger one for groups other than T.

Definition. A measure $\mu \in M(G)$ is *strongly continuous* if $|\mu|$ assigns no mass to any coset of any closed subgroup whose index is infinite.

1.1.4. Theorem. *There exist sequences, $\varepsilon_j \to 0$ and $b_j \to \infty$, with the following property. Let G be a compact abelian group. Let $\mu \in M(G)$ be a strongly continuous measure such that $\|\mu\| \leq j$. Let*

$$Q = Q(\mu) = \{n \in \Gamma : |\hat{\mu}(n)| \geq 1\},$$

and suppose that $|\hat{\mu}(n)| \leq \varepsilon_j$ for all $n \notin Q$. Then Q is a finite set. Furthermore, if the dual group Γ has only a finite number K of torsion elements, then $\# Q \leq Kb_j$.

Note that if Λ is a finite subgroup of Γ, then $\|\chi_\Lambda\|_{B(\Gamma)} = 1$ and χ_Λ is the transform of a trigonometric polynomial times Haar measure. It follows that in the last sentence of the Theorem, the hypothesis of only finitely many torsion elements is essential.

We shall prove Theorem 1.1.4 in Section 1.4. In Section 1.5 we shall prove the following result, and others related to it.

1.1.5. Theorem. *For every $\varepsilon > 0$ there exists $\delta > 0$ such that if $\mu \in M(T)$, $\|\mu\| \leq 1$, and* $\limsup_{n \to -\infty} |\hat{\mu}(n)| < \delta$, *then* $\limsup_{n \to +\infty} |\hat{\mu}(n)| < \varepsilon$.

Remarks. Other aspects of Fourier transform behavior are treated by Kahane [2, Chapter II]. In particular, he discusses the relation between various smoothness conditions and membership in $A(T)$. We recommend also the short paper by deLeeuw, Kahane, and Katznelson [1], which provides a striking counterpoint to the fact that $C(T)^\wedge \subseteq l^2$, to wit: for every non-negative sequence $\{c_n\} \in l^2$, there exists $f \in C(T)$ such that $|\hat{f}(n)| \geq c_n$.

1.2. The Idempotents in the Measure Algebra

Proof of 1.1.1. It is easy to show that $\chi_E \in B(\Gamma)$ whenever E is in the coset ring of Γ, as follows. Let Λ be an open subgroup of Γ. Then the quotient group Γ/Λ is discrete and its dual group, $H = \Lambda^\perp = \{x \in G: \langle x, \gamma \rangle = 1 \text{ for all } \gamma \in \Lambda\}$, is compact. Then $m_H \in M(G)$ and $\hat{m}_H = \chi_\Lambda$. For $\gamma_0 \in \Gamma$, the measure

$dv(x) = \langle x, \gamma_0 \rangle dm_H(x)$ is idempotent also, and $\hat{v} = \chi_{\Lambda - \gamma_0}$. Now note that if μ and v are idempotent elements of $M(G)$, with $\hat{\mu} = \chi_E$ and $\hat{v} = \chi_F$, then $\chi_{E \cap F} = (\mu * v)^\wedge$, $\chi_{E \cup F} = (\mu + v - \mu * v)^\wedge$, and $\chi_{\Gamma \setminus E} = (\delta_0 - \mu)^\wedge$. The result follows from those remarks. It remains to prove the "only if" part of the Theorem. We need three lemmas.

The *support group of a measure* μ is the smallest closed subgroup of G that contains the support of μ.

1.2.1. Lemma. *If $\mu \in M(G)$ and the range of $\hat{\mu}$ is a finite set, and if H is the support group of μ, then H is compact.*

Proof. Let $s = \min\{|a - b| : a, b \in \text{range}(\hat{\mu}), a \neq b\}$, $\varepsilon = s/2(\|\mu\| + 2)$. Let K be a compact subset of H such that $|\mu|(G \setminus K) < \varepsilon$. Let V be a neighborhood of 0 in $\hat{H}$ such that $|1 - \langle x, \gamma \rangle| < \varepsilon$ for all $x \in K$ and $\gamma \in V$. Then for all $\gamma \in V$ and $\lambda \in \Gamma$, $|\hat{\mu}(\lambda - \gamma) - \hat{\mu}(\lambda)| = |(\gamma\mu - \mu)^\wedge(\lambda)| \leq \varepsilon\|\mu\| + 2\varepsilon < s/2$. Therefore $\gamma\mu = \mu$, so that γ must equal 1 on the support of μ and hence on H. Thus the only element of V is the identity character. Therefore $\hat{H}$ is discrete and H is compact. □

1.2.2. Lemma. *Let G be compact, $\mu \in M(G)$, and $\mu = \mu_a + \mu_s$, where μ_a is absolutely continuous and μ_s is singular with respect to m_G. Let $\{\gamma_\alpha\}$ be a net in Γ such that $\gamma_\alpha \to \infty$. Suppose that the net $\{\gamma_\alpha \mu\}$ converges weak* to $v \in M(G)$. Then $|v|(E) \leq |\mu_s|(E)$ for every Borel set $E \subset G$; in particular, $v \perp m_G$.*

Proof. We may suppose that $\gamma_\alpha \mu_a$ converges weak* to some measure v_0. It suffices to prove that $v_0 = 0$. And in fact, if μ_0 is *any* measure such that $\hat{\mu}_0 \in C_0(\Gamma)$, and if $\gamma_\alpha \mu_0 \to v_0$ weak* where $\gamma_\alpha \to \infty$, then $v_0 = 0$. For if $f \in C(G)$ and $\varepsilon > 0$, let p be a trigonometric polynomial such that $\|f - p\|_{C(G)} \|v_0\| < \varepsilon$. Then $|\int f\, dv_0| \leq |\int p\, dv_0| + \varepsilon = \lim_\alpha |\int p\gamma_\alpha\, d\mu_0| + \varepsilon = \varepsilon$. The Lemma follows. □

1.2.3. Lemma. (i) *Let $0 < a \leq b\sqrt{2}$. Let X be a set in a Hilbert space such that $\|x\| = b$ for every $x \in X$, and such that $\|x - x'\| \geq a$ whenever x and x' are two distinct elements of X. Let y belong to the weak closure of X but not to X. Then $\|y\| \leq b(1 - (a^2/2b^2))^{1/2}$.*

(ii) *Let $v \in M(G)$, where $v \neq 0$ and $\hat{v}$ is integer-valued on Γ. Let μ be a weak* cluster point in $M(G)$ of the set $\{\gamma v : \gamma \in \Gamma\}$. Then $\|\mu\| \leq \|v\|(1 - (1/2\|v\|))^{1/2}$.*

Proof. (i) The weak closure of the convex set is the same as its norm closure (see Rudin [3, 3.12]). It follows by an exercise that for every $\varepsilon > 0$ there are distinct elements $x_1, \ldots, x_n \in X$ and numbers $c_1, \ldots, c_n$ such that $\sum c_k = 1$, $\|y - \sum c_k x_k\| < \varepsilon$, and $0 < c_k < \varepsilon$ for each k. For $j \neq k$, $a^2 \leq \langle x_j - x_k, x_j - x_k \rangle = 2b^2 - 2\,\text{Re}\langle x_j, x_k \rangle$, so that $2\,\text{Re}\langle x_j, x_k \rangle \leq 2b^2 - a^2$; and $\|\sum c_k x_k\|^2 = 2\sum_{j<k} c_j c_k \,\text{Re}\langle x_j, x_k \rangle + b^2 \sum c_k^2 \leq b^2 - (a^2/2) + b^2\varepsilon$. It follows that $\|y\|^2 \leq b^2 - (a^2/2)$, and Part (i) is proved.

(ii) Let $\rho = |\nu|$, $b = \|\rho\|^{1/2} = \|\nu\|^{1/2}$, $a = \|\rho\|^{-1/2}$. We may write μ in the form $y\rho$ and each $\gamma\nu$ in the form $x_\gamma \rho$, where y and x_γ belong to the Hilbert space $L^2(\rho)$. Let $X = \{x_\gamma : \gamma \in \Gamma \text{ and } \gamma\nu \neq \mu\}$. Note that $\|x\|_2 = b$ for each $x \in X$, and that $\|x_\gamma - x_{\gamma'}\|_2 \geq a\|(\gamma\nu - \gamma'\nu)^\wedge\|_\infty \geq a$ if $\gamma\nu \neq \gamma'\nu$. It is easy to see that y is in the weak closure of X in $L^2(\rho)$. Therefore by Part (i), and the fact that $\|\mu\| = \|y\|_1 \leq b\|y\|_2$, Part (ii) is proved. □

Proof of 1.1.1, completed. It remains to show that if $\mu \in M(G)$ and $\hat{\mu} = \chi_E$, then E is in the coset ring of Γ.

When the result is established for compact G, the general result may be deduced easily, as follows. By 1.2.1, the support group H of μ is compact. If π is the canonical projection from Γ onto the discrete group $\hat{H} = \Gamma/H^\perp$, then $\hat{\mu}(\gamma)$ depends only on $\pi\gamma$, $E = \pi^{-1}(\pi E)$, and $\chi_{\pi E} \in B(\hat{H})$. Therefore πE is in the coset ring of $\hat{H}$. It follows that E is in the coset ring of Γ.

We may suppose, then, that G is compact.

It suffices to show that if $\mu \in M(G)$, $\mu \neq 0$, and $\hat{\mu}$ is integer-valued, then there exists a measure μ_1 of the form

(1) $$\left(\sum_{i=1}^{m} n_i \gamma_i\right) m_H \neq 0,$$

where $n_i \in Z$, $\gamma_i \in \Gamma$, and H is a closed subgroup of G; and such that

(2) $$\|\mu - \mu_1\| \leq \|\mu\| - 1$$

and $(\mu - \mu_1)^\wedge$ is integer-valued. For then either $\mu = \mu_1$, or the statement just made can be applied again, with $\mu - \mu_1$ in the role of μ; thus μ is a finite sum of measures each of the form (1), and it follows that the support of $\hat{\mu}$ is in the coset ring of Γ. In particular, if $\hat{\mu} = \chi_E$ then E is in the coset ring.

For $\mu \in M(G)$, $\mu \neq 0$, $\hat{\mu}$ integer-valued, let $S(\mu)$ be the weak∗ closure of the set $\{\gamma\mu : \gamma \in \Gamma, (\gamma\mu)^\wedge(0) \neq 0\}$. Then $S(\mu)$ contains an element ν of minimal norm, and $\|\nu\| \geq |\hat{\nu}(0)| \geq 1$. Let H be the support group of ν. If the set $\{\gamma\nu : \gamma \in \hat{H}, (\gamma\nu)^\wedge(0) \neq 0\}$ were infinite, then it would have a cluster point, which would of course belong to $S(\mu)$, and whose norm, according to 1.2.3(ii), would be smaller than $\|\nu\|$. By the choice of ν, that cannot happen. Therefore ν equals a finite sum, of the form (1).

The measure ν belongs to the weak∗ closure of the set

$$\{\gamma\mu|_H : \gamma \in \hat{H}, (\gamma\mu|_H)^\wedge(0) \neq 0\}.$$

If ν did not belong o the set itself, then it would be the limit of a net $\{\gamma_\alpha \mu|_H\}$ where $\gamma_\alpha \to \infty$ in $\hat{H}$, so that according to 1.2.2, $\nu \perp m_H$, which cannot be the case for a measure of the form (1). Therefore $\nu = \gamma\mu|_H$ for some $\gamma \in \Gamma$. Let $\mu_1 = \bar{\gamma}\nu$. Then μ_1 is of the form (1), and $\|\mu\| = \|\mu - \mu_1\| + \|\mu_1\| \geq \|\mu - \mu_1\| + 1$, so that (2) is satisfied. The Idempotent Theorem is proved. □

Remarks and Credits. Explicit descriptions are known for all idempotent measures μ with norm less than $(1 + \sqrt{17})/4$. If $\|\mu\| < (1 + \sqrt{2})/2$, then $\|\mu\| = 1$ and $d\mu(x) = \langle x, y\rangle \, dm_H(x)$ for some $\gamma \in \Gamma$ and some subgroup H. If $(1 + \sqrt{2})/2 \le \|\mu\| < (1 + \sqrt{17})/4$, then μ is of the form $(\langle x, \gamma_1\rangle + \langle x, \gamma_2\rangle) \, dm_H(x)$. For details, see Saeki [1 and 2].

The characterization of norm-one idempotents in $M(G)$ was done by Ito and Kawada [1] (1940). Theorem 1.1.1 was proved by Helson [1] in the case when $G = T$ using function theory; see also Helson [4]. Lemma 1.2.2 is due to him. Rudin [9] extended the Theorem to T^n, and to discrete G, and pointed out Lemma 1.2.1. The general Idempotent Theorem was proved in Cohen [1], a paper that won the Bôcher Memorial Prize for the period 1960–64. The same proof, essentially, appears in the presentation of Rudin [1, Chapter 3]. The short proof we have presented, whose main point is the use of Lemma 1.2.3(ii) is due to Amemiya and Ito [1]. In Section 5.3, we shall present a different approach to the Theorem.

The Idempotent Theorem may be applied to obtain a characterization of the homomorphisms from $L^1(G_1)$ into $M(G_2)$, where G_1 and G_2 are two locally compact abelian groups; and a characterization of the countable strong Ditkin subsets of an arbitrary G. For expositions of those two matters, see Rudin [1, Chapter 4] and Saeki [10], respectively. For an application to the representation theory of nilpotent Lie groups, see Richardson [1].

1.3. Paul Cohen's Theorem on the Norms of Idempotents

The main theorem of this Section, Theorem 1.3.6, implies a quantitatively explicit version of Theorem 1.1.3. Of course, it gives information about the norms of a great many elements other than idempotents. Both this theorem and its proof-procedure will be used in Section 1.4.

We shall also prove (Theorem 1.3.11) that $\|\chi_E\|_{B(Z)} \ge C(\log N)^{1/2}$ whenever $E \subseteq Z$ and $\#E = N$, using a new approach due to S. K. Pichorides.

The proof of 1.1.3 reduces easily to the case when $K = 1$, when Γ is "torsion-free." For, if we suppose the Theorem known for torsion-free groups, and if Γ has K torsion elements, then we may argue as follows. Let Λ be a maximal subgroup containing only 0 and elements of infinite order. Then Γ is the union of K cosets of Λ; and one of the cosets, $\Lambda + \lambda$, say, contains at least N elements of the set $\{\gamma \in \Gamma : |f(\gamma)| \ge 1\}$. We may suppose that $\lambda = 0$, since otherwise we may replace $f(\cdot)$ by $f(\cdot - \lambda)$. Since Λ is torsion-free, the known case of the Theorem implies that $C_N < \|f\chi_\Lambda\|_{B(\Lambda)}$. But $\|f\chi_\Lambda\|_{B(\Lambda)} = \|f\chi_\Lambda\|_{B(\Gamma)} \le \|f\|_{B(\Gamma)}$, and the general case is proved.

In this Section, then, we consider only the special case of 1.1.3 in which Γ is torsion-free. It would be equivalent to say that G is connected; or, that Γ may be ordered (see Rudin [1, 8.1.2]). We select an ordering and denote it by $<$.

Before we arrive at Theorem 1.3.6, we wish to present some much simpler results that indicate where the Littlewood Conjecture comes from, and whose proofs may make it easier to understand that of 1.3.6. For suitable $f \in B(\Gamma)$ and $g \in L^\infty(G)$, it makes sense to write: $|\sum_{n \in \Gamma} f(n)\hat{g}(n)| \leq \|f\|_{B(\Gamma)}\|g\|_{L^\infty(G)}$. The method of proving that $\|f\|_{B(\Gamma)}$ is large will be to construct a suitable g such that $\|g\|_\infty$ is small and $\sum f\hat{g}$ is large. In the proofs of the next two propositions, the "g" is easy to describe.

1.3.1. Proposition. *Let $E = \{n_j: 1 \leq j \leq N\}$ be a set of integers satisfying a lacunarity condition: $n_{j+1} \geq 3n_j$ for $1 \leq j < N$. If $|f(n)| \geq 1$ for $n \in E$ and $f(n) = 0$ for $n \notin E$, then $\|f\|_{A(Z)} \geq (N/4e)^{1/2}$.*

Proof. Apply A.1.2. The "g" is a Riesz product. Incidentally, the result is about the best possible, since $N^{1/2} = \|\chi_E\|_2 = \|\hat{\chi}_E\|_2 \geq \|\hat{\chi}_E\| = \|\chi_E\|_{A(Z)}$. □

1.3.2. Proposition. *Let $E = \{a + jb: 1 \leq j \leq N\}$, where $a, b \in Z$ and $b \neq 0$. Then $\|\chi_E\|_{A(Z)} \geq (1/\pi)\log N$.*

Remark. In fact, $\|\chi_E\|_{A(Z)} = (4/\pi^2)\log N + 0(1)$ as $N \to \infty$, as shown in Zygmund [1, Section II.12]; the Proposition, which gives only a lower estimate, is presented for the sake of exhibiting a technique of proof.

Proof. We may suppose that $a = 0$ and $b = 1$. Let $g(x) = \pi - x$ for $0 \leq x < 2\pi$. Then $\hat{g}(0) = 0$, $\hat{g}(n) = 1/in$ for $n \neq 0$, $\|g\|_\infty \leq \pi$, $|\sum_{n \in Z} \chi_E(n)\hat{g}(n)| = \sum_{n=1}^{N} (1/n) > \log N$. The result is proved. □

It is easy to establish, as a corollary of the fact that $\|\chi_{[-n,n]}\|_{A(Z)} \to \infty$ as $n \to \infty$, that there exists a continuous function whose Fourier series does not converge pointwise; see Katznelson [1, Section II.2].

The Littlewood Conjecture, that $\|\chi_E\|_{A(Z)} \geq c \log(\#E)$ for some $c > 0$ and all finite E, appears to be supported by the two propositions above, since, considering all the possible ways that E can lie in Z, one is inclined to suspect that the lacunary sequence of 1.3.1 and the arithmetic progression of 1.3.2 represent the two extreme cases.

Although most of this discussion concerns functions on discrete groups, the principles are relevant also to the behavior of transforms on non-discrete groups. The next result gives an example of this relevance.

1.3.3. Proposition. *Let $f \in A(R)$ and suppose that f is real and increasing on the interval $(x_0 - h, x_0 + h)$. Then*

$$f(x) - f(x_0) = 0(|\log|x - x_0||^{-1}) \quad \textit{as} \quad x \to x_0.$$

Proof. It is easy to show that if $g \in L^\infty(T)$ and $|\hat{g}(n)| \leq K/|n|$, then $|\sum_{|n| \leq N} \hat{g}(n)e^{int}| \leq \|g\|_\infty + 2K$ for all N and t (see Katznelson [1, problem

I.3.3]). Applying that fact to the function g of the previous proof, we find that

$$\left|\sum_{n=1}^{N} \frac{2}{n} \sin nt\right| \leq \pi + 2 \quad \text{for all} \quad N, t.$$

It suffices to consider the case when $x_0 = 0 = f(x_0)$, and to show that for some $C > 0$ and some N,

$$f(x) \leq C/|\log x| \quad \text{for} \quad 0 < x < h/N.$$

Suppose, to the contrary, that for every C and N there exists $x \in (0, h/N)$ such that $f(x) > C/|\log x|$. For such C, N, and x, the discrete measure $\mu = \sum_{n=1}^{N} (1/n)(\delta_{nx} - \delta_{-nx})$ is supported within $(-h, h)$, and $\int f\, d\mu \geq \sum_{n=1}^{N} (f(nx)/n) > f(x)\log N > C \log N/|\log(h/N)|$, which tends to C as $N \to \infty$. But $|\int f\, d\mu| \leq \|f\|_A \sup|\hat{\mu}(y)| = \|f\|_A \sup_y|\sum_{n=1}^{N} (2/n)\sin nxy| \leq (\pi + 2)\|f\|_A$. Thus our supposition implies that $\|f\|_A \geq C/(\pi + 2)$ for every C, which cannot be. The Proposition is proved. □

The next two lemmas are needed for proving Theorem 1.3.6.

1.3.4. Lemma. *Let r and N be positive integers, with $r \leq (\log N/4 \log\log N)^{1/2}$. Let Q be a finite subset of a discrete ordered group Γ, containing N elements. Then there is a subset of Q, $\{m_0\} \cup \{m_{ks}: 1 \leq k \leq r^2, 1 \leq s \leq r\}$, satisfying the following conditions. Let $P_0 = \{m_0\}$, and for $1 \leq k \leq r^2$ let*

(1) $$P_k = P_{k-1} \cup \{p + m_{ks} - m_{kt}: p \in P_{k-1}, 1 \leq s < t \leq r\} \cup \{m_{ks}: 1 \leq s \leq r\}.$$

Then for each k $(1 \leq k \leq r^2)$ the condition

(2, j) $$p + m_{ks} - m_{kt} \notin Q \quad \text{if} \quad p \in P_{k-1} \quad \text{and} \quad 1 \leq s < t \leq j$$

holds for $1 \leq j \leq r$.

Proof. Let m_0 be the maximum element of Q. Let $k \geq 1$. It suffices to describe the choice of the set $\{m_{ks}: 1 \leq s \leq r\}$, supposing that P_{k-1} has been identified. We shall require that

(3) $$m_{ks} > m_{kt} \quad \text{if} \quad s < t.$$

Enumerate Q in order: $n_1 > n_2 > \cdots > n_N$. Let $m_{k1} = n_1$. Suppose now that for $1 \leq s < j$, m_{ks} has been chosen suitably, so that (2, $j - 1$) and (3) hold. Let m_{kj} be the largest element of Q that is smaller than $m_{k,j-1}$ and such that (2, j) holds. To require (2, j) is to impose the following conditions on the choice of m_{kj}: that for each $p \in P_{k-1}$,

$$p + m_{ks} - m_{kj} \notin Q \quad \text{if} \quad 1 \le s \le j - 1.$$

That rules out fewer than $(j-1)N(p)$ choices for m_{kj}, where $N(p) = \#\{n \in Q: n \ge p\}$. Let $m_{ks} = n_{b(k,s)}$ and let $A_j = \sum_{p \in P_j} N(p)$. Then

$$\begin{aligned} b(k, j) &\le b(k, j-1) + (j-1)A_{k-1} \\ &\le 1 + \tfrac{1}{2}j(j-1)A_{k-1}. \end{aligned}$$

Evidently the choice of the elements m_{ks} can be completed, without exhausting Q, if $A_{r^2-1} < r^{-2}N$. Note that $A_0 = 1$. By looking at (1) we find that for $0 < k < r^2$,

$$\begin{aligned} A_k &\le A_{k-1}\left(1 + \frac{r(r-1)}{2}\right) + \sum_{j=1}^{r} b(k, j) \\ &\le A_{k-1}\left(1 + \frac{r^2}{2} + \frac{1}{2}\sum_{j=1}^{r} j^2\right) \\ &\le (r+1)^3 A_{k-1} \le (r+1)^{3r^2} < r^{-2}N, \end{aligned}$$

by the relationship of r to N. The Lemma is proved. □

1.3.5. Lemma. *Let $r \ge 3$. For $1 \le s \le r$, let $w_s \in \mathbb{C}$ and let $|w_s| = 1$. Let $\beta = \sum_{s=1}^{r} w_s$, and write $\sum_{s<t} w_s \bar{w}_t$ as $u + iv$, where u and v are real. Let*

$$\alpha = 1 - \frac{2}{r^2} - \frac{u + iv}{r^3}.$$

Then $|\alpha| + r^{-5/2}|\beta| \le 1$.

Proof. Note that $|\beta|^2 = r + 2u \le r^2$, $-(r/2) \le u \le (r^2/2)$, and $u^2 + v^2 < r^4/4$. Also,

$$\begin{aligned} |\alpha|^2 &= \left(1 - \frac{2}{r^2} - \frac{u}{r^3}\right)^2 + \frac{v^2}{r^6} \\ &< \left(1 - \frac{2}{r^2}\right)^2 - \frac{2u}{r^3}\left(1 - \frac{2}{r^2}\right) + \frac{1}{4r^2} \\ &= 1 - \frac{4}{r^2} + \frac{4}{r^4} + \frac{1}{4r^2} + u\left(\frac{4 - 2r^2}{r^5}\right) \\ &< 1 - \frac{3}{r^2} - \frac{u}{r^3}, \end{aligned}$$

and thus

$$|\alpha| + \frac{|\beta|}{r^{5/2}} < 1 - \frac{3}{2r^2} - \frac{u}{2r^3} + \frac{(r+2u)^{1/2}}{r^{5/2}}$$
$$= 1 - \left[\frac{(r+2u)^{1/2}}{2r^{3/2}} - \frac{1}{r}\right]^2 - \frac{1}{4r^2} < 1.$$

The Lemma is proved. □

1.3.6. Theorem. *Let r and N be positive integers, with*

$$3 \le r \le (\log N/4 \log\log N)^{1/2}.$$

Let $f \in B(\Gamma)$, where Γ is a discrete ordered group. Suppose that the set $Q = \{n \in \Gamma : |f(n)| \ge 1\}$ is finite and contains at least N elements, and that $|f(n)| \le e^{-r}$ for all $n \notin Q$. Then $\|f\|_{B(\Gamma)} \ge r^{1/2}/4$.

Proof. We shall construct a sequence $\{g_k\}$ of trigonometric polynomials on G, let $g = g_{r^2}$, and establish these properties.

(i) $\|g\|_{C(G)} \le 1$.
(ii) $\sum_{n \in Q} f(n)\hat{g}(n) > r^{1/2}(\frac{1}{2} - 1/2e^2)$.
(iii) $\sum \{|\hat{g}(n)| : n \notin Q\} < e^r$.

The last two conditions imply that $|\sum_{n \in \Gamma} f(n)\hat{g}(n)| > r^{1/2}/4$, and the Theorem follows.

Obtain the set $\{m_{ks} : 1 \le k \le r^2, 1 \le s \le r\} \subset Q$ and the sets P_j as Lemma 1.3.1 provides. Let

$$g_0(x) = \frac{|f(n_1)|}{f(n_1)} \langle x, n_1 \rangle.$$

When g_{k-1} has been defined, let

$$w_s(x) = \frac{|f(m_{ks})|}{f(m_{ks})} \langle x, m_{ks} \rangle,$$

$$\beta(x) = \sum_{s=1}^{r} w_s(x),$$

$$u(x) + iv(x) = \sum_{s<t} w_s(x)\overline{w_t(x)},$$

$$g_k = g_{k-1}(1 - r^{-2} - r^{-3}(u + iv)) + r^{-5/2}\beta.$$

Each g_k satisfies (i), by induction; for each x, $|g_k(x)| \le 1$ because $|g_{k-1}(x)| \le 1$, by 1.3.5.

Let $I_k = \sum_{n \in Q} f(n)\hat{g}_k(n)$. Then $I_0 = 1$, and $I_k \geq (1 - 2r^{-2})I_{k-1} + r^{-3/2}$, so that

$$I_{r^2} \geq r^{-3/2}\left(\sum_{j=0}^{r^2-1} (1 - 2r^{-2})^j\right) + (1 - 2r^{-2})^{r^2}$$
$$> r^{1/2}\left(\frac{1}{2} - \frac{1}{2e^2}\right),$$

and (ii) holds. Inequality (iii) also holds, since, summing over *all* $n \in \Gamma$, we find that

$$\sum |\hat{g}_k(n)| \leq \sum |\hat{g}_{k-1}(n)| \left[1 - \frac{1}{r^2} + \frac{r(r-1)}{2r^3}\right] + r^{-3/2}$$
$$\leq \sum |\hat{g}_{k-1}(n)| \left(1 + \frac{1}{r}\right) \leq \left(1 + \frac{1}{r}\right)^{r^2} < e^r.$$

The Theorem is proved. □

The reader who wishes to do so may now proceed to Section 1.4. We should like now to present a recent result of S. K. Pichorides, if only for the sake of the techniques involved. For the circle group, this work gets still closer to the Littlewood Conjecture, and also leads to a version of 1.3.6. Four lemmas are needed.

1.3.7. Lemma. *There exist constants A_1 and A_2 such that if $m_1, \ldots, m_d$ are integers, and if the set $\{\sum_{j=1}^{d} \varepsilon_j m_j \colon \varepsilon_j = 0, 1, \text{ or } -1\}$ contains 3^d distinct elements, then for any measurable function f,*

$$\left(\sum_{j=1}^{d} |\hat{f}(m_j)|^2\right)^{1/2} \leq A_1 \int_T |f|(\log^+|f|)^{1/2} + A_2.$$

For the proof, we refer the reader to the part of Theorem XII.7.6 in Zygmund [1] that has to do with $\{m_j\}$ such that $m_{j+1} \geq 3m_j$ for each j. That condition implies our hypothesis on $\{m_j\}$, which is all that is needed in Zygmund's argument.

Let us make a remark that will prepare the reader for the way we shall use 1.3.7. For an integer $m \neq 0$, let $t(m)$ be the largest integer k such that 2^k divides m. In Case A of the proof of 1.3.11 below, a sequence $\{m_j\} \subseteq Z$ will be obtained such that $t(m_{j+1}) \geq t(m_j) + 2$ for each j. Such a sequence $\{m_j\}$ satisfies the hypothesis of Lemma 1.3.7, as one may easily verify.

1.3.8. Lemma. *Let $P \in C(T)$. Let $p_m(x) = \text{Re}[e^{imx}P(x)]$. Then $\lim_{m \to \infty} \|p_m\| = (2/\pi)\|P\|$ (where the norms are the $L^1(T)$-norms).*

Proof. Since P is continuous,

$$\frac{1}{m}\sum_{j=0}^{m-1}\left|P\left(\frac{2\pi j}{m}\right)\right| \to \|P\|$$

and

$$\|p_m\| - \sum_{j=0}^{m-1}\frac{1}{2\pi}\int_{2\pi j/m}^{2\pi(j+1)/m}\left|\operatorname{Re}\left(P\left(\frac{2\pi j}{m}\right)e^{imx}\right)\right| dx \to 0 \quad \text{as} \quad m \to \infty.$$

Fix m and let $x_j = 2\pi j/m$, $I_j = [2\pi j/m, 2\pi(j+1)/m]$. Then $\operatorname{Re}(P(x_j)e^{imx}) = |P(x_j)|\cos(mx + \theta_j)$ for some θ_j and all x, so

$$\begin{aligned}\frac{1}{2\pi}\int_{I_j}|\operatorname{Re} P(x_j)e^{imx}|\, dx &= |P(x_j)|\frac{1}{2\pi}\int_{I_j}|\cos(mx + \theta_j)|\, dx \\ &= |P(x_j)|\frac{1}{2\pi m}\int_0^{2\pi}|\cos x|\, dx = \frac{2}{\pi m}|P(x_j)|.\end{aligned}$$

The Lemma follows. □

1.3.9. Lemma. *Let $P(x) = \sum_{n\in E} c_n e^{inx}$, where E is a finite set and $|c_n| \geq 1$ for each $n \in E$. Let*

$$P_o(x) = \tfrac{1}{2}(P(x) - P(x+\pi)) = \sum_{\substack{n\in E\\ n \text{ odd}}} c_n e^{inx},$$

and let $P_e = P - P_o$. Then

$$\|P\| \geq \tfrac{1}{2}(\|P_e\| + \|P_o\|) + \frac{\pi}{32\|P\|}. \tag{4}$$

If P is real-valued and denoted by p, then

$$\|p\| \geq \tfrac{1}{2}(\|p_e\| + \|p_o\|) + \frac{1}{4\|p\|}. \tag{5}$$

Proof. For real-valued p, let $Y = \{x: p_e(x)p_o(x) \geq 0\}$, and let Y' be the complement of Y. Then $Y' = Y + \pi$, so $|Y| = \frac{1}{2}$. Note that π is a period for both $|p_e|$ and $|p_o|$. Evidently

$$\int_Y |p| = \int_Y (|p_e| + |p_o|) = \tfrac{1}{2}(\|p_e\| + \|p_o\|). \tag{6}$$

Let $q(e^{it}) = e^{-in^*t}p(t)$, where $n^* = \min E$. Then $q(z)$ is analytic and $|q(0)| = |c_{n^*}| \geq 1$. Therefore, using Jensen's inequality and the convexity of log, we find that

$$0 \leq \log|q(0)| \leq \int_T \log|p| = \frac{1}{2}\int_T \log|p(x)|2\chi_Y(x)\,dx$$
$$+ \frac{1}{2}\int_T \log|p(x)|2\chi_{Y'}(x)\,dx$$
$$\leq \frac{1}{2}\log\int_T |p|2\chi_Y + \frac{1}{2}\log\int_T |p|2\chi_{Y'}.$$

Thus $1 \leq (2\int_Y |p|)^{1/2}(2\int_{Y'}|p|)^{1/2}$ and

(7) $$\int_{Y'} |p| \geq \frac{1}{4\|p\|}.$$

Conditions (6) and (7) imply (5).

Considering arbitrary P, and even m, if $p_m(x) = \mathrm{Re}(e^{imx}2P(x))$, then $p_{me}(x) = \mathrm{Re}(e^{imx}2P_e(x))$ and $p_{mo}(x) = \mathrm{Re}(e^{imx}2P_o(x))$. For m sufficiently large, $\hat{P}(k-m) = 0$ for all $k \leq 0$; hence $|\hat{p}_m(n)| \geq 1$ for $n - m \in E$; and therefore (5) applies to p_m. By Lemma 1.3.8, and since $\|p_m\| \leq 2\|P\|$,

$$\frac{2}{\pi}\|2P\| \geq \frac{1}{2}\left(\frac{2}{\pi}\|2P_e\| + \frac{2}{\pi}\|2P_o\|\right) + \frac{1}{8\|P\|}.$$

Condition (4) follows, and the Lemma is proved. □

1.3.10. Lemma. *Let $P(x) = \sum_{n\in E} c_n e^{inx}$, where E is a finite set and $|c_n| \geq 1$ for each $n \in E$. Write $P = P_o + P_e$ as in 1.3.9. If $\|P_e\|_2^2 - \|P_o\|_2^2 \geq k\|P\|_2^2$, then $\|P\| \geq \|P_o\| + \frac{1}{4}k$.*

Proof. Note that $\int_T |P(t)|\,dt = \int_T |P_e(t) + P_o(t)|\,dt = \int_T |P(t+\pi)|\,dt = \int_T |P_e(t) - P_o(t)|\,dt$. Therefore

$$\|P\| - \|P_o\| = \frac{1}{2}\int_T (|P_e + P_o| + |P_e - P_o| - 2|P_o|)$$
$$\geq \frac{1}{2}\left[\int_T ((|P_e + P_o| + |P_e - P_o|)^2 - 4|P_o|^2)\right]\Big/ 4\sum_{n\in E}|c_n|$$

(because $\|u - v\|_1 \geq \|u^2 - v^2\|_2^2/\|u + v\|_\infty$ when $u - v \geq 0$)

$$= \left[\int_T (|P_e|^2 - |P_o|^2 + |P_e + P_o|\cdot|P_e - P_o|)\right]\Big/ 4\sum_{n\in E}|c_n|$$
$$\geq k\sum_{n\in E}|c_n|^2/4\sum_{n\in E}|c_n| \geq \tfrac{1}{4}k.$$

1.3.11. Theorem. *Let $P(x) = \sum_{n \in E} c_n e^{inx}$, where $\# E \geq N$ and $|c_n| \geq 1$ for all $n \in E$. Then $\|P\|_1 \geq C(\log\|P\|_2)^{1/2}$, where C is a positive constant independent of P, E, and N.*

Proof. We proceed by induction on N. We may suppose that $N > 10$, and that $\|P\| \geq C(\log\|P\|_2)^{1/2}$ whenever $\# E < N$. It suffices to show that the same inequality holds when $\# E = N$. The induction step will require C to be small, but not in a way that depends on P, E, or N.

We may suppose that E contains only positive integers, some odd and some even. Let $E_k = \{n \in E\colon 2^k \text{ divides } n\}$. Then $E = E_0 \supseteq E_1 \supseteq \cdots \supseteq E_j \neq \varnothing$, $E_{j+1} = \varnothing$, for some $j \geq 0$. Let $P_k(x) = \sum_{n \in E_k} c_n e^{inx}$. Then $\|P_k\| \leq \|P_{k-1}\|$, because

$$P_k(x) = \tfrac{1}{2}[P_{k-1}(x) + P_{k-1}(x + 2^{-(k-1)}\pi)].$$

Pick $k_1 > k_2 > \cdots > k_t$ such that $\{k_r\}_{r=1}^t = \{k\colon E_k \neq E_{k+1}\} = \{k\colon E \text{ contains an odd multiple of } 2^k\}$. Then $E = E_{k_1} \supsetneqq E_{k_2} \supsetneqq \cdots \supsetneqq E_{k_t}$, where $t \geq 2$. Let

$$X_r = E_{k_r} \backslash E_{k_{r+1}};\; X_r' = E_{k_{r+1}}.$$

We change notation and write P_r for P_{k_r}:

$$P_r(x) = P_{ro}(x) + P_{re}(x) = \sum_{n \in X_r} c_n e^{inx} + \sum_{n \in X_r'} c_n e^{inx}.$$

For each r such that $1 \leq r < t$, we may suppose that $\|P_{re}\|_2 \geq \|P_{ro}\|_2$, since otherwise we may add 2^{k_r} to each $n \in E$.

Note that $\|P\|_2^2 = \sum_{r=1}^t \|P_{ro}\|_2^2 = \sum_{r=1}^{s-1} \|P_{ro}\|_2^2 + \|P_s\|_2^2$ for each $s \leq t$.

Case A, when $\|P_{ro}\|_2^2 < \|P\|_2^2/4(\log\|P\|_2)^4$ *for all* $r \leq 2(\log\|P\|_2)^4$. In this case, since $t > 2(\log\|P\|_2)^4$, we may select $m_j \in X_{2j-1}$ for $1 \leq j \leq d$, where $d \geq \frac{1}{4}(\log\|P\|_2)^4$. (Since $N > 10$, $(\log\|P\|_2)^4 > 1$; and whenever $x > 1$ there is an integer $d \geq \frac{1}{4}x$ such that $2d - 1 \leq 2x$.) By Lemma 1.3.7,

$$d^{1/2} \leq \left(\sum_{j=1}^d |c_{m_j}|^2\right)^{1/2} \leq A_1 \int |P|(\log^+|P|)^{1/2} + A_2.$$

Since $\log^+|P|$ is bounded by $\log \sum |c_n|$ and hence by $\log\|P\|_2^2$, it follows that

$$\tfrac{1}{2}(\log\|P\|_2)^2 \leq A_2 + \sqrt{2}A_1\|P\|(\log\|P\|_2)^{1/2}$$

and hence $(\log\|P\|_2)^{3/2} \leq A_3\|P\|$ for some constant A_3. So Case A is very easy, so far as the lower bound on $\|P\|$ is concerned; we could, in fact, have used a larger exponent than 4 in describing the condition, and obtained a larger one than 3/2 in the conclusion.

Case B, when $\|P_{ro}\|_2^2 \geq \|P\|_2^2/4(\log\|P\|_2)^4$ *for some* $r \leq 2(\log\|P\|_2)^4$. Let s be the smallest such r, so that $\|P_s\|_2^2 \geq \frac{1}{2}\|P\|_2^2$.

Since $\|P\| \geq \|P_s\|$, it suffices now to find a lower bound for $\|P_s\|$; but for convenience we replace P_s by $Q(x) = P_s(2^{-k_s}x)$, which has the same norm. We know that $\|Q_e\|_2^2 \geq \|Q_o\|_2^2 \geq \|P\|_2^2/4(\log\|P\|_2)^4$, and that $\|Q\|_2^2 \geq \frac{1}{2}\|P\|_2^2$. We distinguish two subcases of Case B.

Case B_1, when $\|Q_o\|_2^2 \geq \frac{1}{4}\|Q\|_2^2$. By the inductive hypothesis,

$$\tfrac{1}{2}(\|Q_e\| + \|Q_o\|) \geq C(\log\|Q_o\|_2)^{1/2} \geq C(\log\|P\|_2 - \log 2\sqrt{2})^{1/2}$$
$$\geq C(\log\|P\|_2)^{1/2}\left(1 - \frac{\log 2\sqrt{2}}{\log\|P\|_2}\right).$$

We may suppose, of course, that $\|Q\| < C(\log\|P\|_2)^{1/2}$. By Lemma 1.3.9, then,

$$\|Q\| \geq \tfrac{1}{2}(\|Q_e\| + \|Q_o\|) + \frac{\pi}{32\|Q\|}$$
$$\geq C(\log\|P\|_2)^{1/2} + \left\{\frac{\pi}{32C(\log\|P\|_2)^{1/2}} - \frac{C\log 2\sqrt{2}}{(\log\|P\|_2)^{1/2}}\right\},$$

and the quantity in brackets is nonnegative provided $C \leq (\pi/32\log 2\sqrt{2})^{1/2}$. (Case B_1 is *the* critical one; if in Lemma 1.3.9, the quantity $\pi/32\|P\|$ could be replaced by some absolute constant, the Littlewood Conjecture could be proved.)

Case B_2, when $\|Q_o\|_2^2 < \|Q\|_2^2/4$. We apply Lemma 1.3.10, with $k = \frac{1}{2}$ and Q in the role of P:

$$\|Q\| \geq \|Q_o\| + \tfrac{1}{8} \geq C\left[\log\frac{\|P\|_2}{(\log\|P\|_2)^4}\right]^{1/2} + \tfrac{1}{8}$$
$$\geq C(\log\|P\|_2)^{1/2}\left[1 - \frac{\log(\log\|P\|_2)^4}{\log\|P\|_2}\right] + \tfrac{1}{8}$$
$$= C(\log\|P\|_2)^{1/2} + \left\{\frac{1}{8} - \frac{C\log(4\log\|P\|_2)}{(\log\|P\|_2)^{1/2}}\right\}.$$

The quantity in brackets is nonnegative if C is small enough. Theorem 1.3.11 is proved. □

A result like 1.3.11, for any discrete Γ, always leads to a version of 1.1.3, as the following technical proposition shows.

1.3.12. Proposition. *Let Γ be a discrete abelian group, and let $\{C_N\}$ be a sequence such that $C_N \to \infty$. Then (A) implies (B), where (A) and (B) are the following statements.*

(*A*) *Let f be a function on Γ such that for every $n \in \Gamma$, either $|f(n)| \geq 1$ or $f(n) = 0$. If $N = \#\{n \in \Gamma : |f(n)| \geq 1\}$, then $\|f\|_{B(\Gamma)} \geq C_N$.*

(*B*) *Let $\varepsilon_N = C_N/8 \cdot 3^N$, and let $a_N = C_N/8N$. Let $f \in B(\Gamma)$ and suppose that*

$$N = \#\{n \in \Gamma : |f(n)| \geq 1\} \quad \textit{and} \quad \#\{n \in \Gamma : \varepsilon_N < |f(n)| < 1\} < a_N N.$$

Then $\|f\|_{B(\Gamma)} \geq C_N/4$.

Proof. We claim that whenever $E \subseteq \Gamma$ and $\#E = N$, there exists a trigonometric polynomial F on G such that

$$(8) \qquad \|F\|_{L^1(G)} = 1, \ \# \operatorname{supp} \hat{F} \leq 3^N,$$
$$\tfrac{1}{2} \leq \hat{F} \leq 1 \text{ on } E, \quad \text{and} \quad 0 \leq \hat{F} \leq 1 \text{ on } \Gamma.$$

Assuming that to be the case, we assume (A) and let f satisfy the hypothesis of (B). Let $E = \{n \in \Gamma : |f(n)| \geq 1\}$ and obtain F as in (8). The function $2\hat{F}f\chi_E$ satisfies the hypothesis of (A), and therefore its $B(\Gamma)$-norm is greater than C_N. Therefore there exists $g \in L^\infty(G)$ such that $\|g\|_\infty \leq 1$ and

$$\left| \sum_{n \in E} 2\hat{F}(n)\hat{g}(n)f(n) \right| > C_N.$$

Let $h = 2F * g$. Then $\|h\|_\infty \leq 2$,

$$\left| \sum_{n \in E} \hat{h}(n)f(n) \right| > C_N, \quad \text{and} \quad \left| \sum_{n \notin E} \hat{h}(n)f(n) \right| < 2a_N N + 2 \cdot 3^N \varepsilon_N.$$

In view of the definition of a_N and ε_N, it follows that

$$\left| \sum_{n \in \Gamma} \hat{h}(n)f(n) \right| > C_N/2,$$

and hence $\|f\|_{B(\Gamma)} > C_N/4$.

It remains to find F satisfying (8). For $x \in G$, let

$$F(x) = c \prod_{n \in E} (1 + \tfrac{1}{2}\langle x, n\rangle + \tfrac{1}{2}\langle x, -n\rangle),$$

where c is chosen so that $\|F\|_1 = \hat{F}(0) = 1$. Evidently $0 < c \leq 1$ and $0 \leq \hat{F} \leq 1$ on Γ. Let n_o be one of the elements of E, and let

$$d(x) = \prod_{\substack{n \in E \\ n \neq n_o}} (1 + \tfrac{1}{2}\langle x, n\rangle + \tfrac{1}{2}\langle x, -n\rangle).$$

Then $F(x) = c(1 + \frac{1}{2}\langle x, n_o\rangle + \frac{1}{2}\langle x, -n_o\rangle)d(x)$. Therefore

$$\hat{F}(0) = c[\hat{d}(0) + \hat{d}(n_o)],$$

since $\hat{d}(n_o) = \hat{d}(-n_o)$. Also, then

$$\begin{aligned}\hat{F}(n_o) &= c[\hat{d}(n_o) + \tfrac{1}{2}\hat{d}(0) + \tfrac{1}{2}\hat{d}(2n_o)] \\ &= \hat{F}(0) - c[\tfrac{1}{2}\hat{d}(0) - \tfrac{1}{2}\hat{d}(2n_o)].\end{aligned}$$

Since $\hat{d} \geq 0$ everywhere and $\hat{d}(0) \geq 1$,

$$\hat{F}(n_o) \geq 1 - \frac{c}{2} \geq \tfrac{1}{2}.$$

Now F clearly satisfies (8), and the Proposition is proved. □

1.3.13. Remark. To prove 1.3.11, if suffices to show that with $N = \#E$,

$$\|P\| \geq C(\log N)^{1/2}. \tag{9}$$

In fact, (9) implies the stronger conclusion that

$$\|P\| \geq C'(\log \textstyle\sum |c_n|)^{1/2}. \tag{10}$$

Proof. If $\log \sum |c_n| \leq 2 \log N$, then (9) implies (10) with $C' = C/\sqrt{2}$. If $\log \sum |c_n| > 2 \log N$, then $\sum |c_n| \leq N \max |c_n| \leq N\|P\|$, so that $\log \sum |c_n| \leq \log N + \log\|P\| \leq \frac{1}{2}(\log \sum |c_n|) + \|P\|$, and it follows that $\|P\| \geq \frac{1}{2}(\log \sum |c_n|)$. The Remark is proved. □

Remarks and Credits. Paul Cohen [1] was the one who proved that if $P(x) = \sum_{n \in E} c_n e^{inx}$, where $E \subseteq Z$, $\#E = N$, and $|c_n| \geq 1$ for $n \in E$, then $\|P\|_1 \geq C_N$, where $C_N \to \infty$. His argument gave $C_N = C(\log N/\log \log N)^{1/8}$. Davenport [1] modified Cohen's proof and got $C_N = C(\log N/\log \log N)^{1/4}$; Theorem 1.3.6 and its proof represent Davenport's version of things, but carried out in the more general setting, as Hewitt and Zuckerman [1] pointed out that it might as well be. Pichorides [1] improved Davenport's proof, obtaining $C_N = C(\log N/\log \log N)^{1/2}$; again, the procedure works in the general setting of 1.3.6. The reason we presented Davenport's proof procedure and not that of Pichorides [1] is that the latter is not suited to the purposes of Section 1.4.

Recently, Fournier [2] discovered an advanced version of Cohen's proof which yields $C_N = C(\log N)^{1/2}$ (and gives other results as well). It appears that his procedure will do everything that Davenport's will do, and do it somewhat better. As Fournier has pointed out to us (Remark 1.3.13), his result implies Theorem 1.3.11. But the proof of 1.3.11 that we have given here is due to Pichorides [4, 5]. We included it because the technique is different and probably more promising than the others in view. (Pichorides [4] preceded Fournier [2], but gave only $\|P\|_1 \geq C(\log N)^{1/2}$ in the case $|c_n| = 1$ for all n; Pichorides [5] came later and provided a refinement that yielded 1.3.11.) Proposition 1.3.12 is from a conversation of Katznelson and McGehee.

For other contributions to the Littlewood Conjecture, see Pichorides [2], Roth [1], and Salem [2]. A comprehensive treatment of the subject of norms of exponential sums as of late 1976 appears in the Orsay lecture notes of Pichorides [3]. A more recent contribution is due to Dixon [1].

The remark that small nonzero coefficients can be tolerated off E, as provided for in our formulations of 1.1.3 and 1.3.6, without changing anything essential, is due to Kahane [10]. As he pointed out, one may prove from 1.1.3 by construction that there exists $f \in C_0(Z)$ such that for every permutation p of Z, $f \circ p \notin B(Z)$.

Theorem 1.3.6 has a significance also for Fourier transforms on the line. If S is a Lebesgue-measurable set, let $|S|$ denote its measure. Let $L_0(R)$ be the class of Lebesgue-measurable functions f such that $|\{x: |f(x)| > \varepsilon\}|$ is finite for every $\varepsilon > 0$. McGehee [6] proved two results which, when improved by the use of Pichorides [1], read as follows.

1.3.14. Theorem. *Let $f \in L_0(R)$, and let r be a positive integer. Let $0 < a < r^{-5r}$ and $0 < \varepsilon < 2^{-r}/50$. Let*

$$E = \{x: |f(x)| \geq 1\}, F = \{x: \varepsilon < |f(n)| < 1\},$$

and suppose that $|F| < (a/4)|E|$. Then there exists a discrete measure μ such that $\|\hat{\mu}\|_\infty \leq 1$ and $|\int f\, d\mu| > C\sqrt{r}$, where C is independent of f and r. Thus if $f \in A(R)$, $\|f\|_{A(R)} > C\sqrt{r}$.

1.3.15. Corollary. *There exists $f \in C_o(R)$ with compact support, such that for every measure-preserving transformation $h: R \to R$, $f \circ h \notin A(R)$.*

It is not known whether some satisfactory version of 1.3.14 holds for T. It is not known whether 1.3.15 remains true with T in the role of R.

1.4. Transforms of Continuous Measures

1.4.1. Theorem. *Let r be an integer greater than 2. Let G be a compact abelian group with dual group Γ. Let μ be a strongly continuous measure. Let*

$$Q = Q(\mu) = \{n \in \Gamma: |\hat{\mu}(n)| \geq 1\}$$

and suppose that $|\hat{\mu}(n)| \leq e^{-r}$ for $n \notin Q$.

(*A*) *If* $\|\mu\| < r^{1/2}/4$, *then* Q *is a finite set.*
(*B*) *If* $\|\mu\| < r^{1/2}/4$, *and* Γ *is torsion-free, and* N *is an integer such that*

$$r \leq (\log N/4 \log \log N)^{1/2},$$

then $\# Q < N$.

Note that 1.4.1 is an explicit version of 1.1.4, restricted to the case when $K = 1$. (The case of arbitrary finite K follows easily.)

Proof that Part (*A*) *implies Part* (*B*). The proof of 1.3.6 had two parts. First, since Γ was torsion-free (being a discrete ordered group), we were able to show (by 1.3.4) that if $N \leq \# Q < \infty$, then:

(1) A set $\{m_0\} \cup \{m_{ks}: 1 \leq k \leq r^2, 1 \leq s \leq r\} \subset Q$ can be selected so that, if $P_0 = \{m_0\}$ and

$$P_k = P_{k-1} \cup \{p + m_{ks} - m_{kt}: p \in P_{k-1}, 1 \leq s < t \leq r\}$$
$$\cup \{m_{ks}: 1 \leq s \leq r\},$$

for $1 \leq k \leq r^2$, then for each k,

$$P_{k-1} + m_{ks} - m_{kt} \cap Q = \varnothing \quad \text{for} \quad 1 \leq s < t \leq r.$$

Second, we showed that if (1) holds, and if $|\hat{\mu}(n)| \leq e^{-r}$ for $n \notin Q = Q(\mu)$, then $\|\mu\| \geq r^{1/2}/4$. Thus (A) $\Rightarrow$ (B). $\square$

We shall offer two proofs of Part (A). The second one is superfluous, especially since it works only for torsion-free Γ, but we include it anyway because of the interesting techniques that it uses. Both proofs use Norbert Wiener's Theorem A.2.2 about the mean-square of a Fourier–Stieltjes transform.

Proof 1 of 1.4.1(*A*). We shall suppose that Q is infinite and deduce that $\|\mu\| \geq r^{1/2}/4$. To do so, it suffices to establish condition (1).

Let us point out an equivalent formulation of the strong continuity hypothesis. When Λ is a subgroup of Γ, let $\varphi = \varphi_\Lambda$ be the quotient map from G onto $G/\Lambda^\perp$, which is the dual group of Λ. Then φ induces a mapping $\Phi = \Phi_\Lambda: M(G) \to M(G/\Lambda^\perp)$ such that for $v \in M(G)$ and $f \in C(G/\Lambda^\perp)$,

$$\int_{G/\Lambda^\perp} f(x + \Lambda^\perp)\, d(\Phi v) = \int_G f(\varphi(x))\, dv,$$

and such that $(\Phi v)\hat{}(\lambda) = \hat{v}(\lambda)$ for $\lambda \in \Lambda$. To say that μ is a strongly continuous measure in $M(G)$ is to say that for every infinite subgroup Λ of Γ, $\Phi_\Lambda \mu$ is a continuous measure.

It is easy (and not really necessary) to reduce to the case of metrizable G or, equivalently, countable Γ. Simply let Λ be the subgroup of Γ generated by some countably infinite subset of Q. Then Λ is countable, its dual $G/\Lambda^\perp$ is metrizable, $Q(\Phi_\Lambda \mu)$ is infinite, $\Phi_\Lambda \mu$ is strongly continuous, $|(\Phi_\Lambda \mu)\hat{}\,(\lambda)| \le e^{-r}$ for $\lambda \notin Q(\Phi_\Lambda \mu)$, and $\|\Phi_\Lambda \mu\| \le \|\mu\|$. The reduction is then clear. Its advantage is that when G is metrizable, the unit ball in $M(G)$, endowed with the weak$*$ topology, is metrizable, so that all nets therein may be replaced by sequences.

The measures in the set $\bar{Q}\mu = \{\bar{\gamma}\mu : \gamma \in Q\}$ all have norm at least one, since $|(\bar{\gamma}\mu)\hat{}\,(0)| = |\hat{\mu}(\gamma)| \ge 1$ for $\gamma \in Q$. Let ν be a weak$*$ cluster point of minimal norm. Then

$$(2) \qquad |\hat{\nu}(\gamma)| \le e^{-r} \quad \text{for} \quad \gamma \notin Q(\nu) = \{\lambda : |\hat{\nu}(\lambda)| \ge 1\}.$$

Choose $\{\gamma_n\} \subset Q = Q(\mu)$ such that $\bar{\gamma}_n \mu \to \nu$ weak$*$ and such that $\{\gamma_n\}$ is an infinite set (we may insist on that because ν is a *cluster* point). For every $\rho \in \Gamma$, $\bar{\gamma}_n \bar{\rho}\mu \to \bar{\rho}\nu$ weak$*$ and thus $\hat{\mu}(\rho + \gamma_n) \to \hat{\nu}(\rho)$. It follows that if $\rho \in Q(\nu)$, then $\rho + \gamma_n \in Q$ eventually, and therefore $\bar{\rho}\nu$ is a weak$*$ cluster point of $\bar{Q}\mu$. Therefore $\|\sigma\| = \|\nu\|$ for every element σ of the weak$*$ closure Y of the set $Y_0 = \{\bar{\rho}\nu : \rho \in Q(\nu)\}$. Evidently $Y = E\nu$, where E is a family of functions of modulus one on the support of ν. The norm and weak$*$ topologies of $M(G)$ coincide on such a set $E\nu$. In proving that, we may suppose that $\nu \ge 0$ and $\|\nu\| = 1$; it suffices to observe that if e_δ, $e \in E$ and $e_\delta \nu \to e\nu$ weak$*$, then $\|e_\delta \nu - e\nu\| = (\int |e_\delta - e|\, d\nu)^2 \le \int |e_\delta - e|^2\, d\nu = 2 - \int (e\bar{e}_\delta + \bar{e}e_\delta)\, d\nu \to 0$.

Therefore in the norm topology, Y is compact and Y_0 is dense in Y. In particular, Y is covered by a finite number of the sets $U_a = \{\omega \in M(G): \|\omega - \bar{a}\nu\| < 1 - e^{-r}\}$ with $a \in Q(\nu)$:

$$(3) \qquad Y \subset \bigcup_{k=1}^{m} U_{a_k}, \quad \{a_k\} \subset Q(\nu).$$

We shall use that fact to prove that $Q(\nu)$ is a finite union of cosets of some subgroup. First we define an equivalence relation among elements of Γ:

$$a \sim b \Leftrightarrow Q(\nu) - a = Q(\nu) - b.$$

Note that if $a \in Q(\nu)$ and $a \sim b$, then $b \in Q(\nu)$, so that $Q(\nu)$ is a union of equivalence classes. If $\|\bar{a}\nu - \bar{b}\nu\| < 1 - e^{-r}$, evidently $\gamma + a \in Q(\nu)$ if and only if $\gamma + b \in Q(\nu)$, in view of (2). In view of (3), it follows that $Q(\nu)$ is the union of a finite number of equivalence classes. Let F be one of them, and let $a \in F$. Then $0 \in F - a$, and to prove that $F - a$ is a group it suffices to show that if b and c belong to $F - a$ then $b - c \in F - a$; that is, if $b + a \sim c + a$ then $b - c + a \sim a$. That is clear, since $Q(\nu) - (b - c + a) = Q(\nu) - (b + a) + (c + a) - a = Q(\nu) - (c + a) + (c + a) - a = Q(\nu) - a$. If $a \in F$, then

$$b \in F - a \Leftrightarrow b + a \sim a \Leftrightarrow b \sim 0.$$

The latter condition is independent of F. It follows that every equivalence class F is a coset of the same subgroup. Call it Λ.

Let $a \in \Gamma$. We claim that the intersection $(\Lambda + a) \cap \{\gamma_n\}$ is finite. Supposing otherwise, we reach a contradiction, as follows. Let $\{\lambda_n\}$ be a subsequence of $\{\gamma_n\}$ contained in $\Lambda + a$. Since $\bar{a}\mu$ is strongly continuous, $\Phi_\Lambda(\bar{a}\mu) \in M_c(G/\Lambda^\perp)$. The measure v is the weak$*$ limit of $\bar{\lambda}_n\mu = \overline{(\lambda_n - a)}(\bar{a}\mu)$. The characters $\lambda_n - a$ belong to Λ and $\overline{(\lambda_n - a)}\Phi_\Lambda(\bar{a}\mu) \to \Phi_\Lambda(v)$ weak$*$. Since

$$\Phi_\Lambda(v)^\wedge = \hat{v}\Big|_\Lambda$$

is the transform of a continuous measure, it follows from A.2.2 that $\inf\{|\hat{v}(\lambda)|: \lambda \in \Lambda\} = 0$, which contradicts the fact that $\Lambda \subset Q(v)$.

We can now use the fact that $(\Lambda + a) \cap \{\gamma_n\}$ is finite for every $a \in \Gamma$ to establish (1). Choose $m_0 \in Q = Q(\mu)$ arbitrarily. Let $1 \leq k \leq r^2$ and suppose that P_{k-1} has been identified. Now we shall choose $\{m_{ks}\}$ in such a way that

$$\text{(4)} \qquad m_{ks} \in \{\gamma_n\} \backslash [P_{k-1} - Q(v)] \quad \text{for} \quad 1 \leq s \leq r,$$

and

$$\text{(5)} \qquad (P_{k-1} + m_{ks} - m_{kt}) \cap Q = \varnothing \quad \text{for} \quad 1 \leq s < t \leq r.$$

The set $[P_{k-1} - Q(v)]$ is a finite union of cosets of Λ, so that there are infinitely many elements of $\{\gamma_n\}$ outside it, and we may indeed insist on (4). That condition implies that

$$|\hat{v}(p - m_{ks})| \leq e^{-r} \quad \text{for} \quad p \in P_{k-1}, 1 \leq s \leq r.$$

First we choose m_{kr} arbitrarily, satisfying (5). Let $1 \leq j < r$ and suppose that the selection of $\{m_{ki}: j < i \leq r\}$, consistent with (4) and (5), has been completed. Then we choose m_{kj}, satisfying (4) and such that

$$|(\overline{m}_{kj}\mu)^\wedge - \hat{v}| < 1 - e^{-r} \quad \text{on} \quad \bigcup_{j<i} (P_{k-1} - m_{ki}).$$

Let $\gamma = p + m_{kj} - m_{ki}$, where $p \in P_{k-1}$ and $j < i \leq r$. Then $|\hat{\mu}(\gamma)| = |(\overline{m}_{kj}\mu)^\wedge(p - m_{ki})| < (1 - e^{-r}) + |\hat{v}(p - m_{ki})| \leq 1$. Therefore $\gamma \notin Q$, and the choice of m_{kj} is consistent with (5).

Thus the choice of $\{m_{ks}\}$ may be carried out suitably, (1) is established, and 1.4.1 is proved. □

We shall now give a second proof of 1.4.1(A), assuming this time that Γ is torsion-free. We require some preliminaries regarding random walks in Z^n. A hyperplane $H \subset R^n$ will be called *rational* if for some $z \in Z^n$, $z + H$ is a subspace of R^n containing $n-1$ linearly independent vectors from Z^n; equivalently, if for some $z \in Z^n$, $(z+H) \cap Z^n$ is isomorphic to Z^{n-1}.

1.4.2. Lemma. *Let $\{q_i\}_{i=1}^{\infty}$ be a sequence in Z^n, where $n \geq 2$, and let S be a finite subset of Z^n such that $q_{i+1} - q_i \in S$ for all i. Then for each positive integer N there is a rational hyperplane $H \subset Z^n$ such that $q_j \in H$ for at least N values of j.*

Proof. Suppose for the moment that $n = 2$.

We may suppose that each q_i is nonzero and that $q_i \to \infty$. Let $a = (a_1, a_2)$ be a cluster point of the sequence of unit vectors $q_i/\|q_i\|$.

Case 1, when a_1, a_2 are rational. Let $H_0 = \{(x, y): a_1 y = a_2 x\}$. The translates of H_0 by elements of Z^2—that is, the translates that are rational hyperplanes—may be enumerated, H_i for $i \in Z$, so that $\operatorname{dist}(H_i, H_0) = |i|d$, where $d = \operatorname{dist}(H_0, H_1)$. If the Lemma is false, then for some N and every line H there are at most $N-1$ values of j such that $q_j \in H$. Fix J, k and consider the set $\{q_j: J < j \leq J + (2k+1)(N-1) + 1\}$. At least one of its elements occurs on an H_i with $|i| > k$. Call it q_j. Then $q_j = q_J + \sum_{i=J+1}^{j} s_i$, where $s_i \in S$. Let $M = 1 + \max\{\|s\| : s \in S\}$. The dot product $\langle q_j | a\rangle$ may be estimated as follows.

$$|\langle q_j | a\rangle| \leq |\langle q_J | a\rangle| + (2k+1)(N-1)M.$$

Let A be the angle $\angle\, q_j, a$. Then

$$\tan A = \frac{\operatorname{dist}(q_j, H_0)}{|\langle q_j | a\rangle|} > \frac{kd}{|\langle q_J | a\rangle| + (2k+1)NM} > \frac{d}{3NM}$$

for $k = k(J)$ sufficiently large. Let H', H'' be lines with rational slopes that pass through 0 and make angles α and $-\alpha$ with H_0, where $0 < |\alpha| < \arctan(d/3MN)$. Since $\{q_i/\|q_i\|\}$ clusters at a, there are infinitely many J such that q_J lies between H' and H''. Every such J has a successor j such that q_j does *not* lie between them. Therefore for infinitely many i, q_i and q_{i+1} are not on the same side of H' (if that is not true for H', it is true for H''). Therefore H' and the rational translates of H' that lie within distance M of H' contain q_i for infinitely many i. Therefore some one of them also does.

Case 2, when a_1, a_2 are not both rational. Suppose that the Lemma fails for N. Let $M = 1 + \max\{\|s\| : s \in S\}$. Note that both a_1, a_2 are nonzero. There exist infinitely many triples of integers U, u_1, u_2 such that $U > 0$ and $|Ua_i - u_i| < U^{-1/2}$ for $i = 1, 2$ (see Cassels [1, p. 14]), so that if $u = (u_1, u_2)$,

$$\left\| a - \frac{u}{U} \right\| < \sqrt{2}\, U^{-3/2}. \tag{6}$$

We fix u and U, requiring that U be large enough so that

$$16\sqrt{2}U^{-3/2} < \frac{1}{8MNU} \tag{7}$$

and

$$\frac{U}{2} < \|u\| < 2U. \tag{8}$$

Let $w = (-u_2, u_1)$. Let x be a rational number that lies between the two numbers in (7). Let H^+, H^- be the lines generated, respectively, by $u + xw$ and $u - xw$. Under our supposition, which is that the Lemma fails for N, we shall prove that

$$\left|\frac{\langle q_i|w\rangle}{\langle q_i|u\rangle}\right| < x \qquad \text{for infinitely many } i \tag{9}$$

and

$$x < \left|\frac{\langle q_j|w\rangle}{\langle q_j|u\rangle}\right| \qquad \text{for infinitely many } j. \tag{10}$$

It follows that the walk $q_i \to q_{i+1}$ crosses the line H^+ (if not the line H^-) infinitely often, so that some rational translate of H^+ contains q_i for infinitely many i. We will have thus reached a contradiction.

Condition (9) holds because

$$\begin{aligned} |\langle a|u\rangle| &= \left|\left\langle \frac{u}{U}\middle| u\right\rangle + \left\langle a - \frac{u}{U}\middle| u\right\rangle\right| \\ &\geq U^{-1}\|u\|^2 - \sqrt{2}U^{-3/2}\|u\| \\ &> (U/4) - 2\sqrt{2}U^{-1/2} > U/8, \end{aligned}$$

and

$$|\langle a|w\rangle| = \left|\left\langle a - \frac{u}{U}\middle| w\right\rangle\right| \leq \sqrt{2}U^{-3/2}\|u\| < 2\sqrt{2}U^{-1/2}.$$

Thus

$$\left|\frac{\langle a|w\rangle}{\langle a|u\rangle}\right| \leq \frac{2\sqrt{2}U^{-1/2}}{U/8} = 16\sqrt{2}U^{-3/2} < x.$$

Since $\{q_i/\|q_i\|\}$ accumulates at a, (9) follows. To prove (10), we shall show that if

$$\left|\frac{\langle q_J|w\rangle}{\langle q_J|u\rangle}\right| < \frac{1}{8MNU},$$

then J has a successor j such that

$$\frac{1}{8MNU} < \left|\frac{\langle q_j|w\rangle}{\langle q_j|u\rangle}\right|.$$

Let H_0 be the line generated by u. Let $\{H_i\}_{i\in Z}$ be an enumeration of the lines parallel to H_0 that intersect Z^2, such that $\mathrm{dist}(H_i, H_0) = |i|d$, where $d = \mathrm{dist}(H_1, H_0)$. Under our supposition that the Lemma fails for N, for each k and J at least one of the points $\{q_j: J < j \leq J + (2k+1)(N-1)+1\}$ occurs on an H_i with $|i| > k$. Call it q_j. Since $|i|d \leq \mathrm{dist}(q_j, H_0) = |\langle q_j|w\rangle|/\|w\|$, it follows that

$$|\langle q_j|w\rangle| > kd\|w\|.$$

Since $q_j = q_J + \sum_{i=J+1}^{j} s_i$, where $s_i \in S$,

$$|\langle q_j|u\rangle| \leq |\langle q_J|u\rangle| + (2k+1)(N-1)M2U.$$

Therefore

$$\left|\frac{\langle q_j|w\rangle}{\langle q_j|u\rangle}\right| \geq \frac{kd\|w\|}{|\langle q_J|u\rangle| + 2(2k+1)(N-1)MU} > \frac{d\|w\|}{8MNU}$$

for sufficiently large $k = k(J)$.

The square of width $\|u\| = \|w\|$, with sides u and w, has plane area $\|w\|^2$. It contains that many points of Z^2, and that many lines H_i intersect it. Therefore $d\|w\|^2 = \|w\|$, $d = 1/\|w\|$, so

$$\left|\frac{\langle q_j|w\rangle}{\langle q_j|u\rangle}\right| > \frac{1}{8MNU}.$$

The proof of the Lemma is complete for when $n = 2$.

Let $n > 2$ and fix N. Let π be the projection from R^n onto some two-dimensional coordinate plane. Apply the two-dimensional case of the Lemma to the sequence $\{\pi q_i\}$ and the finite set πS. Then there is a rational line L such that $\pi q_j \in L$ for at least N values of j. For all those j, the rational hyperplane $\pi^{-1}(L)$ contains q_j. The Lemma is proved. □

1.4.3. Corollary. *Let S be a finite subset of Z^n, $n \geq 2$. For every positive integer N, there is another one, $N' = N'(n, S, N)$, such that for every sequence*

$\{q_i: 1 \le i \le N'\} \subset Z^n$ with $q_{i+1} - q_i \in S$ for each i, there exists a rational hyperplane H such that $q_j \in H$ for at least N values of j.

Proof. Suppose that the statement fails for N. Then for each $N' > N$ there is a sequence $\{q_{N',i}: 1 \le i \le N'\}$, with $q_{N',i+1} - q_{N',i} \in S$ for each i, that meets every H in at most $N - 1$ points. We may require that $q_{N',1} = 0$ for all N'. We set $q_1 = 0$ and proceed by an easy induction argument to construct a sequence $\{q_i\}_{i=1}^{\infty}$ such that for every $k > N$,

$$q_i = q_{N',i} \quad \text{for} \quad 1 \le i \le k$$

for infinitely many N'. But then $\{q_i\}$ would be a counterexample to 1.4.2; it cannot meet any H in more than $N - 1$ points because no initial segment $\{q_i\}_{i=1}^{k}$ can. The Corollary is proved. □

Proof 2 of 1.4.1(A), assuming Γ torsion-free. Again, the plan is to show that if Q is infinite, then (1) holds.

This time, we are assuming that Γ is torsion-free and thus may be ordered. We revise the statement of (1), adding a technical condition that will be of use, as follows:

(I) A set $\{m_0\} \cup \{m_{ks}: 1 \le k \le r^2, 1 \le s \le r\} \subset Q$ can be selected so that for some ordering of Γ,

(11) $$m_{ks} > m_{kt} \quad \text{if} \quad s < t;$$

and

(12) $$p + m_{ks} - m_{kt} \notin Q \quad \text{if} \quad p \in P_{k-1}, 1 \le s \le t \le r, \text{ and } 1 \le k \le r^2.$$

We prove it first for $G = T^n$, by induction on n.

The case when $G = T$. We may suppose that $Q \cap Z^+$ is infinite. Let m_0 be an arbitrary element of Q. Let $k \ge 1$; it suffices to show that a suitable choice of m_{ks} for $1 \le s \le r$ can be made, supposing that P_{k-1} is identified. Begin by letting m_{kr} be any element of Q that is greater than $\max\{|p|: p \in P_{k-1}\}$. Now suppose that $i \ge 1$ and that for $i < j \le r$, m_{kj} has been chosen consistently with (11) and (12). Suppose that

(13) $$\text{there is no suitable choice for } m_{ki},$$

which is to say that for every $q \in Q$ that exceeds $m_{k,i+1}$,

(14) $$p + q - m_{kt} \in Q$$

for some $p \in P_{k-1}$ and some $t \in (i, r]$. The integers $p - m_{kt}$ that can occur here are all negative; let $-M$ be the smallest one. Then condition (13) implies that if $q \in Q$ and $q > m_{k,i+1}$, there exists $q' \in Q \cap [q - M, q)$. It follows that for arbitrarily large L, there are at least L elements of Q between $m_{k,i+1}$ and $m_{k,i+1} + (L + 1)M$. It follows that

$$\limsup_{N\to\infty} \frac{1}{2N+1} \sum_{|n|\leq N} |\hat{\mu}(n)|^2 \geq (1/2M),$$

which cannot be the case, because μ is a continuous measure. We have reached a contradiction for (13), and completed the proof for the case when $G = T$.

The case when $G = T^n, n \geq 2$. Let π be a projection onto a coordinate axis such that πQ is an infinite set. We may suppose that $\pi Q \cap Z^+$ is infinite. Put a lexicographic order on Z^n such that $\pi(m) > \pi(n)$ if $m > n$. Then (11) will follow from

(15) $$\pi(m_{ks}) > \pi(m_{kt}) \quad \text{for} \quad s < t.$$

Let m_0 be an arbitrary element of Q and let $k \geq 1$. If $k > 1$, suppose that for $1 \leq h < k$ and $1 \leq s \leq r$, the elements m_{hs} have been chosen consistently with (15) and (12). Now to select m_{ks} for $1 \leq s \leq r$. Let m_{kr} be any element of Q such that

$$\pi(m_{kr}) > \pi(p) \quad \text{for all} \quad p \in P_{k-1}.$$

Now suppose that $i \geq 1$, and that m_{kj} has been chosen suitably for $i < j \leq r$, and that

(16) there is no suitable choice for m_{ki}.

Condition (16) may be rephrased: for every $q \in Q$ such that $\pi(q) > \pi(m_{k,i+1})$, there exists $q' \in Q$ such that $q' - q \in S$, where S is the finite set $\{p - m_{kj}: p \in P_{k-1}, 1 < j \leq r\}$, so that $\pi(q) - M \leq \pi(q') < \pi(q)$, where $-M = \min\{\pi(s): s \in S\}$. Let N' be the integer provided by 1.4.3. Let $q_1 \in Q$ with $\pi(q_1) > \pi(m_{k,i+1}) + (N' + 1)M$. Applying (16) repeatedly, we find a sequence of distinct points $\{q_i: 1 \leq i \leq N'\} \subset Q$ such that $q_{i+1} - q_i \in S$ for $1 \leq i < N'$. Then there exists a rational hyperplane H such that $q_j \in H$ for at least N values of j.

Let $z \in Z^n$ be such that $L = (z + H) \cap Z^n$ is a subgroup of Z^n, isomorphic to Z^{n-1}. Let ψ be the quotient map from T^n onto $T^n/L^\perp$. Let $z\mu$ be the measure $\langle x, z\rangle d\mu(x)$. Let $v \in M(\psi T^n)$ be such that $v(E) = (z\mu)(\psi^{-1}(E))$ for Borel sets E, so that $\hat{v}(m) = \hat{\mu}(m - z)$ for $m \in L$. Since ψT^n is isomorphic to T^{n-1}, the inductive hypothesis (I) holds with v and ψT^{n-1} in the roles of μ and G. Let the set selected as provided by (I) be denoted with primes: $\{m_0'\} \cap \{m_{ks}': 1 \leq k \leq r^2, 1 \leq s \leq r\}$. Those points are in $Q(v) = L \cap (z + Q)$ where Q is of course $Q(\mu)$. Now we throw out the choices of m_0 and m_{kj} made so far, and define m_0 to be $m_0' - z$, m_{kj} to be $m_{kj}' - z$, for $1 \leq j \leq r, 1 \leq k \leq r^2$. Then

(12) holds; and (11) holds, at least for some ordering of Z^n. Condition (I) and the Theorem have been proved for the cases when $G = T^n$.

If Γ is finitely generated, then Γ is isomorphic to Z^n. Otherwise, let Q_1 be a subset of Q that contains at least N elements, and let Λ be the subgroup of Γ generated by Q_1. Let $\nu = \Phi_\Lambda \mu$. We know that (I) holds for $Q(\nu) = Q(\mu) \cap \Lambda$. The set $\{m_0\} \cup \{m_{ks}\}$ thus provided will work also for $Q(\mu)$. The Theorem is proved. □

Remarks and Credits. The first proof of 1.4.1 is due to Ramsey and Wells [1], but their procedure owes much to Glicksberg [1], who studied the related situation in which 0 is an isolated point of the set $\{0\} \cup \hat{\mu}(\Gamma)$. The second proof, applicable to the case of torsion-free Γ, is due to Ramsey [1], except that de Leeuw and Katznelson [2] proved a version of 1.4.1 for the case when $G = T$. Their work, treated in the next section, was the point of departure for Ramsey's.

Recently, by a further exploitation of the procedures we have presented, Pigno and Smith [1] have proved an informative generalization of 1.1.4, in which the condition of strong continuity is omitted and μ can be treated somewhat as an idempotent measure is treated in the proof of 1.1.1.

Consider the question: When $\mu \in M_c(G)$, on how large a set E can $\hat{\mu}$ be bounded away from zero? Kaufman [8] showed that the cases when E is non-Sidon are plentiful indeed. Katznelson [6] found a case when E is dense in $b\Gamma$ (see Theorem 7.6.1). A related result is Theorem 7.5.7 (ii).

If $\hat{\mu} \in B(Z)$ and $\hat{\mu}$ vanishes on intervals (m_k, n_k) for $k = 1, 2, \ldots$, and if $\inf_k(m_k - n_k)/n_{k-1} > 0$, then the sums $\sum |\hat{\mu}(m_k)|^2$ and $\sum |\hat{\mu}(n_k)|^2$ are finite. For proofs, see either Meyer [3, p. 533] or the later papers, Fournier [1] and Fournier [2, Theorem 6].

1.5. The Two Sides of a Fourier Transform

Our concern here is for measures $\mu \in M(T)$, and the behavior of $\hat{\mu}(n)$ for n near $-\infty$, and for n near $+\infty$. First we present two familiar old results.

1.5.1. Proposition. *Let $\nu, \rho \in M(T)$ and suppose that $\nu \ll \rho$. Then*

$$\lim_{n\to+\infty} \hat{\rho}(n) = 0 \Rightarrow \lim_{n\to+\infty} \hat{\nu}(n) = 0,$$

$$\lim_{n\to-\infty} \hat{\rho}(n) = 0 \Rightarrow \lim_{n\to-\infty} \hat{\nu}(n) = 0,$$

$$\lim_{N\to+\infty} \|\hat{\rho}\|_{B[N,\infty)} = 0 \Rightarrow \lim_{N\to+\infty} \|\hat{\nu}\|_{B[N,\infty)} = 0,$$

and

$$\lim_{N\to-\infty} \|\hat{\rho}\|_{B(-\infty,N]} = 0 \Rightarrow \lim_{N\to-\infty} \|\hat{\nu}\|_{B(-\infty,N]} = 0.$$

Proof. Recall that for $E \subset Z$, $B(E)$ is the algebra of restrictions to E of functions in $B(Z)$, with norm

$$\|f\|_{B(E)} = \inf\{\|g\|_{B(Z)}: g|_E = f\}.$$

Since $\nu \in L^1(\rho)$, for each $\varepsilon > 0$ there is a trigonometric polynomial $p(x) = \sum_{j=1}^k a_j e^{in_jx}$ such that $\|\nu - p\rho\| < \varepsilon$. Then

$$|\hat{\nu}(n)| < \varepsilon + \sum |a_j||\hat{\rho}(n - n_j)| \quad \text{for} \quad n \in Z,$$

$$\|\hat{\nu}\|_{B(E)} < \varepsilon + \sum |a_j|\|\hat{\rho}\|_{B(E-n_j)} \quad \text{for} \quad E \subset Z.$$

The Proposition follows.

1.5.2. Proposition. *Let* $\mu \in M(T)$. *If*

(1) $$\hat{\mu}(n) \to 0 \quad \text{as} \quad n \to -\infty,$$

then

(2) $$\hat{\mu}(n) \to 0 \quad \text{as} \quad n \to +\infty.$$

If

(3) $$\|\mu\|_{B(-\infty, N]} \to 0 \quad \text{as} \quad N \to -\infty$$

then

(4) $$\|\mu\|_{B[N, +\infty)} \to 0 \quad \text{as} \quad N \to +\infty,$$

and in fact

(5) $$\|\mu\|_{B(\{n: |n| \geq N\})} \to 0 \quad \text{as} \quad N \to +\infty.$$

Proof. Recall these elementary facts. Condition (5) holds if and only if $\hat{\mu} \in A(Z)$, $\mu \in L^1(T)$. If $\nu \in M(T)$ and $\nu_1(E) = \overline{\nu(E)}$ for Borel sets E, then $\hat{\nu}_1(n) = \overline{\hat{\nu}(-n)}$. If $\sigma \in M(T)$ and σ is real-valued, then $\hat{\sigma}(n) = \overline{\hat{\sigma}(-n)}$. Therefore $\|\hat{\sigma}\|_{B(E)} = \|\hat{\sigma}\|_{B(-E)}$ for $E \subset Z$.

Let $\mu \in M(T)$ and let $\sigma = |\mu|$. Since σ is absolutely continuous with respect to μ, and vice versa, the implications (1) $\Rightarrow$ (2) and (3) $\Rightarrow$ (4) follow from 1.5.1.

If (3) holds, then for each $\varepsilon > 0$ there exists $\mu_\varepsilon \in M(T)$ such that $\hat{\mu}_\varepsilon = \hat{\mu}$ on $(-\infty, N]$ for some N and $\|\mu_\varepsilon\| < \varepsilon$. By the F. and M. Riesz Theorem (see Hoffman [1, p. 47] or Duren [1, p. 41]), $\mu - \mu_\varepsilon$ is absolutely continuous, so $\|\hat{\mu} - \hat{\mu}_\varepsilon\|_{B(\{n: |n| \geq N\})} \to 0$ as $N \to \infty$. Therefore $\|\hat{\mu}\|_{B(\{n: |n| \geq N\})} < \varepsilon$ for all sufficiently large N. Condition (5) follows. □

The main result of this Section is the following, quantitatively explicit version of 1.1.5.

1.5.3. Theorem. *For* $\varepsilon,\ C > 0$, *let* $\delta = \delta(\varepsilon, C) = \varepsilon/[2(2C/\varepsilon)^{(2C/\varepsilon)^2 \ln 2}]$. *If* $\mu \in M(T)$, $\|\mu\| \leq C$, *and* $\limsup_{n\to-\infty} |\hat{\mu}(n)| < \delta$, *then* $\limsup_{n\to+\infty} |\hat{\mu}(n)| < \varepsilon$.

Proof. Suppose that α is a nonzero accumulation point of the sequence $\{\hat{\mu}(n) = \int e^{-inx}\, d\mu : n > 0\}$. It suffices to show that $|\alpha| < \varepsilon$. Since $\{e^{-inx} : n > 0\}$ is a bounded sequence in $L^2(|\mu|)$, there is a subsequence converging weakly to some $\varphi \in L^2(|\mu|)$. Then $\int \varphi\, d\mu = \alpha$. For each $k > 0$, φ is in the weak closure of $\{e^{-inx} : n > k\}$, and therefore in the norm closure of its convex hull, so that there is a convex combination g_k of these characters such that $\int |\varphi - g_k|^2\, d\mu < 1/k$ (see Dunford and Schwartz [1, V.3.14]). Then $|\varphi| \leq 1$ a.e. since $|g_k| \leq 1$ a.e. for all k. Since all the elements in sight are bounded, products preserve convergence, so that for each $m > 0$, $g_k^{m+1}(\bar{g}_k)^{m-1}\bar{g}_u \to |\varphi|^{2m}\varphi$ in $L^2(|\mu|)$, as $k, u \to \infty$. For each m and each $\eta > 0$, we may choose k large enough so that

$$(6)\qquad \left| \int g_k^{m+1}(\bar{g}_k)^{m-1}\bar{g}_u\, d\mu - \int |\varphi|^{2m}\varphi\, d\mu \right| < \eta \quad \text{for} \quad u \geq k.$$

There exists n_0 such that $|\int e^{inx}\, d\mu| < \delta$ for $n \geq n_0$. With k fixed and u sufficiently large, the integrand on the left in (6) is a convex combination of characters e^{inx} with $n \geq n_0$. It follows that

$$\left| \int |\varphi|^{2m}\varphi\, d\mu \right| \leq \delta \quad \text{for} \quad m \geq 1.$$

Let $p_\varepsilon(x) = 1 - (1 - x)^n = -\sum_{m=1}^{n} \binom{n}{m}(-x)^{n-m}$, where $n = [(2C/\varepsilon)^2 \ln (2C/\varepsilon)]$. If $(\varepsilon/2C)^2 \leq x \leq 1$, then $0 \leq 1 - p_\varepsilon(x) = (1 - x)^n \leq (1 - (\varepsilon/2C)^2)^n < \varepsilon/2C$. Thus for $0 \leq t \leq 1$, $0 \leq (1 - p_\varepsilon(t^2))t < \varepsilon/2C$, and

$$\begin{aligned} |\alpha| = \left| \int \varphi\, d\mu \right| &\leq \left| \int (1 - p_\varepsilon(|\varphi|^2))\varphi\, d\mu \right| \\ &\quad + \left| \int p_\varepsilon(|\varphi|^2)\varphi\, d\mu \right| \\ &\leq C \cdot \max |(1 - p_\varepsilon(|\varphi|^2))\varphi| \\ &\quad + \sum_{j=1}^{n} \binom{n}{j} \left| \int |\varphi|^{2j}\varphi\, d\mu \right| \\ &\leq \frac{\varepsilon}{2} + 2^n\delta < \varepsilon. \end{aligned}$$

The Theorem follows. □

Remarks. The Theorem may be applied to show that for $\|\mu\| \le C$, if $\hat{\mu}(n)$ is close to some finite set $\{a_1, \ldots, a_k\}$ for n near $-\infty$, then the same is true for n near $+\infty$. One simply applies the Theorem to the measure

$$(2C)^{-k} \mathop{\bigstar}_{j=1}^{k} (\mu - a_j\delta_0)$$

We do not know whether the function $\delta(\varepsilon, C)$ is best possible in some sense, but there are easy examples to show that δ depends non-trivially on C. Let

$$\mu_1 = \prod_{j=1}^{\infty} (1 - \sin 3^j x),$$

where the infinite product denotes the weak $*$ limit of the partial products. Let $\mu_2 = \delta_0 - 2i\mu_1$. Then $\hat{\mu}_2(3^j) = 2$ for all j, $\hat{\mu}_2(0) = 1 - 2i$, and $|\hat{\mu}_2(n)| \le 3/2$ for all other n. Let $\mu_3 = \frac{1}{2}(\mu_2 - (1 - 2i)m_T)$. Then $\hat{\mu}_3(3^j) = 1$ for all j and $|\hat{\mu}_3(n)| \le \frac{3}{4}$ for all other n. The convolution powers of μ_3 show that δ depends on $\|\mu\|$:

$$\limsup_{n \to -\infty} |\hat{\mu}_3^m(n)| = (\tfrac{3}{4})^m.$$

The next theorem is a version of 1.1.3. Its proof uses only 1.5.3, the weak $*$ compactness of the unit ball of $M(T)$, and some Riesz products.

1.5.4. Theorem. *For each $C > 0$ there exists $\eta = \eta(C) > 0$ such that if μ is a real continuous measure in $M(T)$, $\|\mu\| \le C$, and $q = \{n: \eta < |\hat{\mu}(n)| < 1\}$ is a finite set, then $Q = \{n: |\hat{\mu}(n)| \ge 1\}$ is also a finite set.*

Proof. Given $C > 0$, let $\eta > 0$ be sufficiently small so that $\eta < \delta(.99, C)$ and so that condition (7), which appears at the end of this proof, is satisfied for some positive integer k. Let $\mu \in M_c(T)$ be real-valued, so that $\hat{\mu}(-n) = \overline{\hat{\mu}(n)}$ for all $n \in Z$. Suppose that $\|\mu\| \le C$ and that q is finite, but that Q is infinite.

Since $\lim_{N \to \infty} (2N + 1)^{-1} \sum_{|n| \le N} |\hat{\mu}(n)|^2 = 0$, for each j there exists $n_j \in Q$ such that $n_j > j$ and $|\hat{\mu}(n)| < \eta$ for $n_j - j \le n < n_j$. Let $d\mu_j(x) = e^{-in_jx}\, d\mu(x)$. Then $|\hat{\mu}_j(0)| \ge 1$ but $|\hat{\mu}_j(n)| < \eta$ for $-j \le n < 0$. We may suppose that $\{\mu_j\}$ converges weak $*$ to some measure λ with $\|\lambda\| \le C$. Since $\hat{\mu}_j(n) \to \hat{\lambda}(n)$ for all $n \in Z$,

$$|\hat{\lambda}(0)| \ge 1 \quad \text{and} \quad |\hat{\lambda}(n)| \le \eta \quad \text{for all } n < 0.$$

Since $n_j \to \infty$, $\{n: \eta < |\hat{\lambda}(n)| < 1\} = \varnothing$. Since $\eta < \delta(.99, C)$,

$$\limsup_{n \to +\infty} |\hat{\lambda}(n)| < 0.99.$$

Therefore the set $\{n: |\hat{\lambda}(n)| > \eta\}$ is finite, and contained in $[0, N]$ for some N. Thus if $M > N$,

$$|\hat{\mu}_j(n)| = |\hat{\mu}(n + n_j)| \le \eta \quad \text{for} \quad N < n \le M$$

when j is sufficiently large. Therefore the sequence $\{n_j\}$ has an infinite subsequence—we denote it also by $\{n_j\}$—such that

$$n_1 > N, n_{j+1} > 3n_j,$$

and

$$|\hat{\mu}(-n)| = |\hat{\mu}(n)| \leq \eta \quad \text{for} \quad n_j + N < n \leq \sum_{i=1}^{j} n_i$$

as well as for

$$n_j - \sum_{i=1}^{j-1} n_i \leq n < n_j.$$

Write $\hat{\mu}(n_j)$ as $r_j e^{i\theta_j}$, where $r_j \geq 1$ and θ_j is real. Then $\hat{\mu}(-n_j) = r_j e^{-i\theta_j}$. Let k be a positive integer satisfying condition (7), found below, and let

$$\varphi(x) = \prod_{j=1}^{k}\left(1 + \frac{i}{\sqrt{k}}\cos(n_j x - \theta_j)\right).$$

Then $\hat{\varphi}(0) = 1$ and

$$\hat{\varphi}(\pm n_j) = \frac{i}{2\sqrt{k}}\exp(\mp i\theta_j)$$

so that

$$\hat{\varphi}(n_j)\hat{\mu}(n_j) + \hat{\varphi}(-n_j)\hat{\mu}(-n_j) = \frac{ir_j}{2\sqrt{k}}.$$

Excepting those in the set $J = \{0\} \cup \{\pm n_j: 1 \leq j \leq k\}$, the integers in the support of $\hat{\varphi}$ are of the form

$$\sum_{j=1}^{k} \varepsilon_j n_j \quad \text{where} \quad \varepsilon_j = 0, 1 \quad \text{or} \quad -1 \quad \text{and} \quad \sum |\varepsilon_j| \geq 2.$$

For every such integer n, $|\hat{\mu}(n)| \leq \eta$. Note that

$$\sum_{n \in Z} |\hat{\varphi}(n)| = \left(1 + \frac{1}{\sqrt{k}}\right)^k < e^{\sqrt{k}}$$

and

$$\|\varphi\|_{C(T)} \leq \left(1 + \frac{1}{k}\right)^k < e.$$

Therefore

$$Ce \geq \left| \int \varphi(-x) d\mu(x) \right| = |\sum \hat{\varphi}(n)\hat{\mu}(n)|$$

$$\geq \frac{1}{\sqrt{k}} \sum r_j - |\mu(0)| - \eta e^{\sqrt{k}} \geq \sqrt{k} - C - \eta e^{\sqrt{k}}.$$

If η and k have been chosen so that

$$C(e + 1) < \sqrt{k} - \eta e^{\sqrt{k}}, \tag{7}$$

then we have reached a contradiction. Thus Q must be a finite set, and the Theorem is proved. Note that the relationship between η and C is similar to the corresponding relationship in 1.3.6. □

Theorem 1.5.4 has a corollary that generalizes the Idempotent Theorem for the case of T. We recommend the proof as an exercise; it uses only 1.5.4 and the elementary properties of the transforms of continuous and of discrete measures.

1.5.5. Corollary. *For each $C > 0$ there exists $\tau = \tau(C) > 0$ such that if $\mu \in M(T)$, $\|\mu\| < C$, and*

$$\limsup_{|n| \to \infty} |\hat{\mu}(n) - \hat{\mu}(n)^2| < \tau,$$

then there is an idempotent measure ν such that $|\hat{\mu}_d(n) - \hat{\nu}(n)| < \frac{1}{10}$ for all n and $\limsup_{|n| \to \infty} |\hat{\mu}_c(n)| < \frac{1}{10}$.

Credits. 1.5.1 and 1.5.2 are due to Rajchman [1]. For some interesting extensions, see Glicksberg [3]. The other results are due to deLeeuw and Katznelson [3]. Their paper proves 1.5.5 and generalizes 1.5.3 somewhat.

Leaving aside the explicit identification of $\delta(\varepsilon)$, 1.5.3 is equivalent to the following result. For the proof, see Holbrook [1].

1.5.6. Theorem. *For each $\varepsilon > 0$ there exists $\delta > 0$ such that if T is a contraction on a Hilbert space H, and if g, h belong to the unit ball of H, then*

$$\lim_{n \to \infty} |\langle T^n h, g \rangle| \leq \delta \Rightarrow \lim_{n \to \infty} |\langle (T^*)^n h, g \rangle| \leq \varepsilon,$$

where T^* is the adjoint of T.

1.6. Transforms of Rudin-Shapiro Type

Our purpose here is to discuss the prevalence of measures $\mu \in M(G)$, supported on finite sets, for which the ratio of $\|\mu\|_M$ to $\|\mu\|_{PM} = \|\hat{\mu}\|_{C(\Gamma)}$ is relatively large; and to place in perspective the technical result 1.6.3, which is fundamental to the construction treated in Section 1.7. An extensive treatment of further, related material appears in Section 11.6.

1.6.1. Proposition. *Let $\mu \in M(E)$, $\mu \neq 0$, where $E = \{x_1, \ldots, x_N\} \subset G$. Then $1 \leq \|\mu\|_M/\|\mu\|_{PM} \leq \sqrt{N}$.*

Proof. The left-hand inequality is clear, since $|\hat{\mu}(\gamma)| \leq \|\mu\|_M$ for every $\gamma \in \Gamma$. If $\mu = \sum_{x \in E} a_x \delta_x$, then $|\hat{\mu}(\gamma)|^2 = |\sum_{x \in E} a_x \langle x, -\gamma\rangle|^2 = \sum |a_x|^2 + \sum_{x \neq y} a_x \bar{a}_y \langle x - y, -\gamma\rangle$ for $\gamma \in \Gamma$. Those quantities are also meaningful for $\gamma \in b\Gamma$, the Bohr compactification. Integration with respect to the Haar measure of $b\Gamma$ (see (3) in Section 1.4) yields the equality

$$\int |\hat{\mu}(\gamma)|^2 \, dm_{b\Gamma}(\gamma) = \sum |a_x|^2.$$

Therefore $\|\mu\|_{PM}^2 \geq \sum |a_x|^2 \geq N^{-1}(\sum |a_x|)^2$, by the Schwarz Inequality. □

In his 1951 master's thesis, Shapiro [1] showed that the inequality of 1.6.1 is about the best possible. His example was essentially as follows. Let $x_1, \ldots, x_n$ be elements of G such that the set $E = \{\sum_{j=1}^n \varepsilon_j x_j : \varepsilon_j = 0 \text{ or } 1\}$ contains $N = 2^n$ distinct elements. For example, we might take $G = T$, $x_j = 2^{-j}$; or $G = Z$, $x_j = 2^{j-1}$, $E = \{0, 1, \ldots, N-1\}$. Define two sequences of measures μ_k, ν_k inductively by these conditions:

$$\hat{\mu}_0 \equiv 1, \; \hat{\nu}_0 \equiv 1,$$

$$\hat{\mu}_k(\gamma) = \hat{\mu}_{k-1}(\gamma) + \langle x_k, \gamma\rangle \hat{\nu}_{k-1}(\gamma),$$

$$\hat{\nu}_k(\gamma) = \hat{\mu}_{k-1}(\gamma) - \langle x_k, \gamma\rangle \hat{\nu}_{k-1}(\gamma).$$

Each of the two measures μ_n and ν_n assigns mass $+1$ or -1 to each point of E, so that $\|\mu_n\|_M = \|\nu_n\|_M = 2^n$. But both $\|\mu_n\|_{PM}$ and $\|\nu_n\|_{PM}$ are bounded by $2^{(n+1)/2}$, because

$$\begin{aligned} |\hat{\mu}_n(\gamma)|^2 + |\hat{\nu}_n(\gamma)|^2 &= 2(|\hat{\mu}_{n-1}(\gamma)|^2 + |\hat{\nu}_{n-1}(\gamma)|^2) \\ &= \cdots = 2^n(|\hat{\mu}_0(\gamma)|^2 + |\hat{\nu}_0(\gamma)|^2) = 2^{n+1}. \end{aligned}$$

Thus for either measure, the ratio of M- to PM-norm is at least $(N/2)^{1/2}$.

Rudin [12] presented the same example in a 1960 paper, and the trigonometric polynomials $\hat{\mu}_n$ and $\hat{\nu}_n$ are called *Rudin–Shapiro* polynomials.

For an arbitrary closed set $F \subseteq G$, consider the norm-decreasing inclusion mapping: $A(F) \subseteq C_0(F)$, and its adjoint: $M(F) \subseteq PM(F)$. The bicontinuity of either is equivalent to the surjectivity of the first. Let $\alpha(F)$ denote the norm of the inverse of either inclusion. What 1.6.1 says is that $1 \leq \alpha(F) \leq \sqrt{\# F}$. Now consider the set E, defined above, in the role of F. By the duality just pointed out, the measures μ_n have a second interpretation. Let $f_n(x) = \mu_n(\{x\})$ for $x \in E$. Then $\| f_n \|_{C(E)} = 1$, but

$$\| f_n \|_{A(E)} = \sup \left\{ \frac{|\int f_n \, d\nu|}{\|\nu\|_{PM}} : \nu \in M(E), \nu \neq 0 \right\}$$

$$\geq \frac{|\int f_n \, d\mu_n|}{\|\mu_n\|_{PM}} = \frac{2^n}{2^{(n+1)/2}} = 2^{(n-1)/2} = (N/2)^{1/2}.$$

If a finite set E is independent—that is, if whenever $u_1, \ldots, u_N$ are integers, $\sum u_j x_j = 0 \Rightarrow u_j x_j = 0$ for each j—then $\alpha(E) \leq 2$ (see Rudin [1, 5.6.7]). Thus it is no surprise that in the example above, where $\alpha(E)$ is large, the points of the set E have many nontrivial arithmetic relations; the set E participates actively in the algebraic structure of G. It may be shown by probabilistic methods that under such conditions, the measures μ for which the ratio of norms is large are plentiful. Davie [1] provided a particularly attractive formulation of this principle. He considered the case when the participation of E in the group is maximal, when E *is* a group. First, a lemma.

1.6.2. Lemma. *For $1 \leq j \leq N$, let a_j be a complex number. Let $\varepsilon = (\varepsilon_1, \ldots, \varepsilon_N)$ be a list of N independent random variables, each with two possible values, and let $f(\varepsilon) = |\sum_{j=1}^N \varepsilon_j a_j|$. Let P(E) denote the probability of the event E. Suppose that either*

(*a*) $$P(\varepsilon_j = 1) = P(\varepsilon_j = -1) = \tfrac{1}{2} \quad \textit{for each } j,$$

or

(*b*) $$P(\varepsilon_j = 2) = \tfrac{1}{3} \quad \textit{and} \quad P(\varepsilon_j = -1) = \tfrac{2}{3} \quad \textit{for each } j.$$

Then

$$P(f(\varepsilon) > A[\textstyle\sum |a_j|^2 \log N]^{1/2}) < AN^{-3},$$

where A is some independent constant.

Proof. We may suppose that the numbers a_j are real and that $\sum a_j^2 = 1$. Let $\mathscr{E}$ denote the expectation (or integral) over the finite probability space consisting of the 2^N points ε. Then for each $\lambda > 0$,

$$\mathscr{E}(e^{\lambda f(\varepsilon)}) < \mathscr{E}(\exp(\lambda \textstyle\sum \varepsilon_j a_j)) + \mathscr{E}(\exp(-\lambda \textstyle\sum \varepsilon_j a_j))$$

$$= \begin{cases} 2\prod(\frac{1}{2}e^{\lambda a_j} + \frac{1}{2}e^{-\lambda a_j}) \leq 2e^{\lambda^2} & \text{in case } (a), \\ \prod(\frac{1}{3}e^{2\lambda a_j} + \frac{2}{3}e^{-\lambda a_j}) + \prod(\frac{1}{3}e^{-2\lambda a_j} + \frac{2}{3}e^{\lambda a_j}) \leq 2e^{2\lambda^2} & \text{in case } (b). \end{cases}$$

Here we have used the inequalities $\frac{1}{2}(e^x + e^{-x}) \le e^{x^2}$, $\frac{1}{3}(e^{2x} + 2e^{-x}) \le e^{2x^2}$ for real x. In both cases,

$$\mathscr{E}(e^{\lambda f(\varepsilon)}) \le 2e^{2\lambda^2},$$

and it follows that

$$P(\lambda f(\varepsilon) > 2\lambda^2 + 3 \log N) < 2/N^3.$$

Putting $\lambda = (3 \log N)^{1/2}$ gives the desired result. □

1.6.3. Proposition. *Let G be an abelian group with N elements: $G = \{x_j: 1 \le j \le N\}$. Let ε be as in 1.6.2. Let $\mu_\varepsilon = \sum_{j=1}^N \varepsilon_j \delta_{x_j}$. In either case (a) or case (b),*

$$P(\|\mu_\varepsilon\|_{PM} > A(N \log N)^{1/2}) < AN^{-2}. \tag{1}$$

In case (a) if N is a multiple of 2, and in case (b) if N is a multiple of 3, there exists ε such that

$$\|\mu_\varepsilon\|_{PM} < 2A(N \log N)^{1/2} \quad \textit{and} \quad \hat{\mu}_\varepsilon(0) = 0. \tag{2}$$

Proof. $\|\mu_\varepsilon\|_{PM} = \sup_{\gamma\in\Gamma} |\hat{\mu}_\varepsilon(\gamma)| = \sup_{\gamma\in\Gamma} |\sum \varepsilon_j \langle x_j, -\gamma\rangle|$. By 1.6.2, applied with $\langle x_j, -\gamma\rangle$ in the role of a_j, $P(|\hat{\mu}_\varepsilon(\gamma)| > A(N \log N)^{1/2}) < AN^{-3}$. Since there are N elements of Γ, (1) follows. Therefore there exists ε such that $\|\mu_\varepsilon\|_{PM} < A(N \log N)^{1/2}$. In case (*a*), if N is a multiple of 2 then so is $\hat{\mu}_\varepsilon(0) = \sum \varepsilon_j$. In case (*b*), if N is a multiple of 3 then $\sum \varepsilon_j$ also is. Evidently by changing ε_j to the other allowed value for a certain number of values of j, we may obtain (2). The Proposition is proved. □

Remark. 1.6.3(a) may be restated as follows: If E is a group with N elements, then

$$\text{meas}\{f \in \{-1 + 1\}^E: \|f\|_{B(E)} \le (N/A^2 \log N)^{1/2}\} < AN^{-2}.$$

Since $\|f\|_{B(E)} \le N^{1/2}$ for *all* $f \in \{-1, +1\}^E$, that result is rather striking, since it says that most of the functions f have $B(E)$-norm near the maximum possible.

The condition of "participation by E in the group structure," the requirement of many arithmetic relations among the points of E, may be weakened and $\alpha(E)$ still be proved large. Let $x_1, \ldots, x_n \in G$ and consider the arithmetic mesh

$$H = \left\{\sum_{j=1}^n u_j x_j: u_j \quad \text{is an integer and} \quad \sum |u_j| \le N\right\}.$$

If $K \ge 1$ and $\#(E \cap H) \ge Kn \log N$, then

$$\text{meas}\{f \in \{-1, 1\}^E: \|f\|_{B(E)} \le c\sqrt{K}\} \le e^{-n}N^{-2},$$

where c is an independent constant. For that result, we refer to Lemma 11.6.6 and its proof.

The results above say, in rough terms, that when a finite set E is arithmetically thick, then there are many functions of modulus one on E whose $B(E)$-norm is large. In particular then, $\alpha(E)$ is large. The following recent result involves no arithmetic conditions on E, but says simply that if $\alpha(E)$ is large, then there are quite a few functions of modulus one on E whose $B(E)$ norm is large.

1.6.4. Theorem. *Let E be a finite subset of G. Consider T as the multiplicative group $\{z: |z| = 1\}$, so that T^E is the multiplicative group of functions of modulus one on E. Then for $m = 1, 2, \ldots,$*

$$\operatorname{meas}\{f \in T^E: \|f\|_{B(E)} < \alpha(E)^{1/m}\} < 1/m.$$

1.6.5. Lemma. *Let A and B be subsets of the group $Z_p = \{0, 1, \ldots, p-1\} =$ Z mod p, where p is a prime. Let $0 \in B$. Then $\#(A + B) \geq \min(p, \#A + \#B - 1)$.*

Proof. Let $r = \#A$, $s = \#B$. We may suppose that $r + s - 1 \leq p$. Note that if the pair A, B is a counterexample to the Lemma, then $r \geq 2$ and $s \geq 2$. Supposing the Lemma to be false, choose a counterexample pair A, B such that s is minimal. We claim that there exists $a^* \in A$ such that $a^* + B \not\subset A$. Let $A = \{a_1, \ldots, a_r\}$, $B = \{0, b_1, \ldots, b_{s-1}\}$. If the claim is false, then for each j there exists k such that $a_j + b_1 = a_k$. Evidently the map: $j \to k$ is one-to-one. Then $\sum_{j=1}^{r}(a_j + b_1) = \sum_{k=1}^{r} a_k$, so $rb_1 = 0$, which can't be since $r < p$. The claim is proved.

Let $B'' = \{b'' \in B: a^* + b'' \notin A\}, A' = A \cup \{a^* + b'': b'' \in B''\}, B' = B \backslash B''$. Then $A' + B' \subset A + B$, $\#A' + \#B' = r + s$, and $\#B' < s$. The minimality of s is contradicted, and the Lemma is proved. □

For real x, let $\tilde{x} = \min\{n \in Z: x \leq n\}$.

1.6.6. Corollary. *If $B \subset Z_p$, p is a prime, and $\#B = s \geq 2$, then*

$$\widetilde{\left(\frac{p-1}{s-1}\right)} B = Z_p.$$

Proof. Since $nB = Z_p$ if and only if $n(B - x) = Z_p$, we may suppose that $0 \in B$, so the Corollary follows easily from the Lemma. □

1.6.7. Proposition. *Let U be an open subset of T^n with Haar measure u. Then $([u^{-1}] + 1)U = T^n$, where mU means $\{\sum_{k=1}^{m} x_k: x_k \in U\}$, and T =* R mod 2π.

Proof for $n = 1$. Let $u^{-1} < v < [u^{-1}] + 1$. For sufficiently large primes p, there is a set $B \subset \{0, 1, \ldots, p-1\}$ such that, if $s = \#B$, then

$$\frac{s-1}{p-1} > \frac{1}{v} \tag{3}$$

and

$$U' \subset U, \quad \text{where} \quad U' = \bigcup_{b \in B} \left[\frac{2\pi b}{p}, \frac{2\pi(b+1)}{p}\right]. \tag{4}$$

By 1.6.6, $(p - 1/s - 1)\check{}B = Z_p$. Therefore $(p - 1/s - 1)\check{}U' = T$. By (3) and (4) and the choice of v, $([u^{-1}] + 1)U = T$. □

Proof for $n > 1$. Viewing T^n as $[0, 2\pi)^n$, endow it with the metric $d(x, y) = \max\{|x_j - y_j|: 1 \le j \le n\}$. Let $\varepsilon > 0$, and let $p, p_1, \ldots, p_n$ be distinct primes such that $p_i \varepsilon > 2\pi$ for $1 \le i \le n$ and $p > \prod_{i=1}^n p_i$. Let λ be the point in T^n whose jth coordinate is $2\pi(\prod_{i=1}^n p_i)/pp_j$. The subgroup $\{k\lambda: 0 \le k \le p-1\}$ has order p and is ε-dense in T^n.

Given an open set $U \subset T^n$ with $m_{T^n}(U) = u$, we may choose $\varepsilon > 0$ sufficiently small so that there is a set $B \subset \{0, 1, \ldots, p-1\}$ such that if $\#B = s$, then (3) holds and

$$U' \subset U, \quad \text{where} \quad U' = \bigcup_{b \in B} \{x \in T^n: d(b\lambda, x) \le \varepsilon\}. \tag{5}$$

It follows that $([u^{-1}] + 1)U = T^n$. □

It is easy to prove 1.6.4 from 1.6.7.

Remarks and Credits. Theorem 1.6.4 is an observation of L. T. Ramsey. Lemma 1.6.5 is a special case of an old result, which one may find in Halberstam and Roth [1, p. 49]; 1.6.7 was discovered by Macbeath [1].

It is well known that the Lipschitz class $\Lambda_r(T)$ is contained in $A(T)$ if and only if $r > \frac{1}{2}$. An application of the Rudin–Shapiro measures, in which they are precisely what is needed, appears in Kahane and Katznelson [3], where an f in $\Lambda_{1/2}(T)$ is constructed such that only the constants belong to the intersection of $A(T)$ with the algebra generated by f in A(T). The measures are used also in Katznelson [8], where a function $f \in C(T)$ is constructed such that the Haar measure of the set $\{x: f(x) = g(x)\}$ is zero for every $g \in A(T)$.

1.7. A Separable Banach Space That Has No Basis

A sequence $\{e_k\}_{k=1}^{\infty}$ in a Banach space X is a *basis* for X if for every $x \in X$ there is a unique sequence of scalars $\{x_k\}$ such that the series $\sum x_k e_k$ converges in norm to x. If X has a basis, then it also has the *approximation property*: the identity operator I can be approximated uniformly on compact subsets of X by bounded linear operators of finite rank (that is, with finite-dimensional range). The problem of whether every X has a basis, or at least enjoys the approximation property, was a famous problem of Banach space structure theory. Enflo [1] solved it in 1972, and Davie [1] later gave a much shorter proof that uses the same idea. We present Davie's work because it is such a striking use of objects familiar in Fourier analysis. The procedure is reminiscent of the construction by Rider [3] of a subalgebra of $A(Z)$ that is not the closed span of the idempotents that it contains.

Incidentally, Grothendieck [2] showed in 1955 that these three assertions are equivalent:

(1) Every Banach space has the approximation property.

(2) If $A = \{a_{ij}: i, j = 1, 2, \ldots\}$ is an infinite matrix such that $\sum_i \sup_j |a_{ij}| < \infty$ and $A^2 = 0$, then trace $(A) = 0$.

(3) If f is continuous on the unit square $[0, 1] \times [0, 1]$ and

$$\int_0^1 f(x, t) f(y, t)\, dt = 0$$

for all $x, y \in [0, 1]$, then $\int_0^1 f(t, t)\, dt = 0$.

The construction presented below may be used to give a specific counterexample to any one of the three; see Davie [2].

1.7.1. Theorem. *Let $2 < p \leq \infty$. There exists a closed subspace X of l^p that does not have the approximation property.*

Proof. Let G be the disjoint union $\bigcup_{k=0}^{\infty} G_k$, where G_k is an abelian group with $3 \cdot 2^k$ elements. We shall consider l^p as $L^p(G)$, where G has the counting measure. By 1.6.3(b), the dual group of G_k may be partitioned into two sets:

$$\Gamma_k = \{\sigma_j^k: 1 \leq j \leq 2^k\} \cup \{\tau_j^k: 1 \leq j \leq 2^{k+1}\},$$

such that if p_k is the polynomial on G_k,

$$p_k(x) = 2 \sum \langle x, \sigma_j^k \rangle - \sum \langle x, \tau_j^k \rangle,$$

then $\|p_k\|_\infty \leq c(k2^k)^{1/2}$, where c is an independent constant. For $k \geq 1$ and $1 \leq j \leq 2^k$, define the function e_j^k on G as follows.

$$e_j^k(x) = \begin{cases} 0 \quad \text{for} \quad x \notin G_{k-1} \cup G_k, \\ \langle x, \tau_j^{k-1} \rangle \quad \text{for} \quad x \in G_{k-1}, \\ \varepsilon_j^k \langle x, \sigma_j^k \rangle \quad \text{for} \quad x \in G_k, \end{cases}$$

where ε_j^k is either $+1$ or -1 and will be specified later. Let X be the closed linear span in $L^p(G)$ of the functions e_j^k. Let $\{d_j^k\}$ be the orthogonal system dual to $\{e_j^k\}$, defined by the condition that $d_j^k(e_i^m) = \delta_{km}\delta_{ji}$. Then every $f \in X$ has a pointwise convergent series: $f(x) = \sum_{j,k} d_j^k(f) e_j^k(x)$. Explicitly for $x \in G_k$,

$$f(x) = \sum_{j=1}^{2^k} d_j^k(f) e_j^k(x) + \sum_{j=1}^{2^{k+1}} d_j^{k+1}(f) e_j^{k+1}(x).$$

That sum is essentially the Fourier series for the restriction f_k of f to G_k, because of the definition of the functions e_j^k. Thus

$$d_j^{k+1}(f) = \hat{f}_k(\tau_j^k) = 3^{-1}2^{-k} \sum_{x \in G_k} f(x)\langle x, -\tau_j^k\rangle$$

and

$$d_j^k(f) = \varepsilon_j^k \hat{f}_k(\sigma_j^k) = 3^{-1}2^{-k} \sum_{x \in G_k} f(x)\varepsilon_j^k\langle x, -\sigma_j^k\rangle.$$

Let $B(X)$ be the space of bounded linear operators on X. For $k \geq 0$, let $X_k = \operatorname{span}\{e_j^k : 1 \leq j \leq 2^k\}$. For $T \in B(X)$, let $\beta^k(T)$ be the trace of T relative to X_k:

$$\beta^k(T) = 2^{-k} \sum_{j=1}^{2^k} d_j^k(Te_j^k).$$

We claim that $\{\beta^k\}$ converges in norm in $B(X)^*$ to an element β which is continuous on $B(X)$ with respect to the topology of uniform convergence on compact sets. The Theorem will follow from that claim, since it is clear that β annihilates all finite rank operators, whereas $\beta(I) = 1$.

We shall prove first that $\{\beta^k\}$ is a Cauchy sequence in $B(X)^*$. For $T \in B(X)$,

$$\begin{aligned}\beta^{k+1}(T) - \beta^k(T) &= 2^{-k-1} \sum_{j=1}^{2^{k+1}} 3^{-1}2^{-k} \sum_{x \in G_k} (Te_j^{k+1})(x)\langle x, -\tau_j^k\rangle \\ &\quad - 2^{-k} \sum_{j=1}^{2^k} 3^{-1}2^{-k} \sum_{x \in G_k} (Te_j^k)(x)\varepsilon_j^k\langle x, -\sigma_j^k\rangle \\ &= 3^{-1}2^{-2k-1} \sum_{x \in G_k} (T\varphi_x^k)(x),\end{aligned}$$

where φ_x^k is a function on G:

$$\varphi_x^k = \sum_{j=1}^{2^{k+1}} \langle x, -\tau_j^k\rangle e_j^{k+1} - 2\sum_{j=1}^{2^k} \varepsilon_j^k\langle x, -\sigma_j^k\rangle e_j^k.$$

Thus

$$\varphi_x^k(y) = \begin{cases} -2\sum_{j=1}^{2^k} \varepsilon_j^k\langle x, -\sigma_j^k\rangle\langle y, \tau_j^{k-1}\rangle & \text{for} \quad y \in G_{k-1}, \\ \sum_{j=1}^{2^{k+1}} \langle y - x, \tau_j^k\rangle - 2\sum_{j=1}^{2^k} \langle y - x, \sigma_j^k\rangle & \text{for} \quad y \in G_k, \\ \sum_{j=1}^{2^{k+1}} \varepsilon_j^{k+1}\langle x, -\tau_j^k\rangle\langle y, \sigma_j^{k+1}\rangle & \text{for} \quad y \in G_{k+1}, \\ 0 \quad \text{elsewhere.} \end{cases}$$

Since $\varphi_x^k(y) = -p_k(y - x)$ for $y \in G_k$, φ_x^k is bounded by $c(k2^k)^{1/2}$ on G_k. It follows from 1.6.2(a) that the signs $\{\varepsilon_j^k : k \geq 0, 1 \leq j \leq 2^k\}$ may be chosen so that for every k, φ_x^k is similarly bounded on $G_{k-1} \cup G_{k+1}$. Then $\|\varphi_x^k\|_\infty \leq c(k2^k)^{1/2}$. Since $\|\varphi_x^k\|_2$ is proportional to $2^{k/2}$, Hölder's inequality yields that for $2 < p < \infty$,

$$\|\varphi_x^k\|_p \leq \|\varphi_x^k\|_\infty^{(p-2)/p} \|\varphi_x^k\|_2^{2/p} \leq Ck^{(p-2)/2p} 2^{k/2}.$$

Therefore $|(T\varphi_x^k)(x)| \leq \|T\| C(k2^k)^{1/2}$ for $2 < p \leq \infty$, so that regardless of p,

$$|\beta^{k+1}(T) - \beta^k(T)| \leq Ck^{1/2} 2^{-k/2} \|T\|$$

for some constant C. Let $\beta = \lim_{k \to \infty} \beta^k = \beta^0 + \sum_{k=0}^{\infty} (\beta^{k+1} - \beta^k)$.

To prove that β is continuous as claimed, it suffices to show that for every $\varepsilon > 0$ there is a neighborhood of 0 in $B(X)$ of the form $\{T : \|Tf\| < \varepsilon$ for all $f \in K\}$, where K is a compact set in X. Given $\varepsilon > 0$, let

$$K = \{2^{-3k/4} \varphi_x^k : x \in G_k, k = 0, 1, \ldots\} \cup \{0\}.$$

If $\|Tf\| < \varepsilon$ for all $f \in K$, so that in particular $\|T\varphi_x^k\| < 2^{3k/4}\varepsilon$, then

$$|\beta(T)| < |\beta^0(T)| + \sum |(\beta^{k+1} - \beta^k)T| \leq \|T\varphi_0^0\| + \sum_k 3^{-1} 2^{-2k-1} \sum_{x \in G_k} \|T\varphi_x^k\|$$

$$< \varepsilon + \sum_k C\varepsilon 2^{-k/4} < C'\varepsilon.$$

(The norm is the p-norm, where $2 < p \leq \infty$). The Theorem is proved. □

1.8. Restrictions of Fourier–Stieltjes Transforms to Sets of Positive Haar Measure

When f is a function defined on all or part of Γ, what conditions on f suffice to assure that f agrees on its domain with the transform of some measure $\nu \in M(G)$? This Section discusses answers of a certain kind to that question. The main technique is to consider the relation between $M(G)$ and $M(bG)$. Although G may not always be measurable in bG, there is a natural inclusion of $M(G)$ in $M(bG)$. To wit, one may identify a measure $\tau \in M(G)$ with $\tau^{(b)}$, defined as follows. Let A be a σ-compact subset of G (and hence of bG) such that $|\tau|(G \backslash A) = 0$. Then for every Borel set $E \subseteq bG$, let $\tau^{(b)}(E) = \tau(E \cap A)$. For more details, if desired, see Hewitt and Ross [1, Theorem 33.19]. There is also a natural decomposition of each measure $\sigma \in M(bG)$, as follows. Let $\{K_j\}$ be a sequence of compact subsets of G such that $\sup_j |\sigma|(K_j) = \sup\{|\sigma|(K) : K$ is a compact subset of $G\}$, and let $A = \bigcup_j K_j$. Then A is a Borel subset of bG. Let ν be the restriction of σ to A; equivalently, let $\nu \in M(G)$

be determined by the requirement that $\nu(K) = \sigma(K)$ for every compact set $K \subseteq G$. Let $\rho = \sigma - \nu$. Then $\sigma = \rho + \nu$, where ρ annihilates every compact subset of G, and $\nu \in M(G)$. The last theorem of the Section provides another approach to that decomposition.

We shall use Bochner's Theorem, that every continuous positive-definite function on Γ equals the transform of some positive measure on G. For a standard proof, see Rudin [1, 1.4.3] or Hewitt and Ross [1, Section 33]. An elegant new proof appears in Lumer [1].

We shall first prove directly the following well-known theorem, and then discuss a newer and stronger result.

1.8.1. Theorem. *Let f be a measurable complex-valued function defined on Γ. If*

$$\left|\int f\, d\mu\right| \leq c\|\hat{\mu}\|_\infty \quad \textit{for every} \quad \mu \in M_1(\Gamma), \tag{1}$$

then there exists $\nu \in M(G)$ such that $\|\nu\| \leq c$ and $f = \hat{\nu}$ a.e.

To prove the Theorem, it suffices to deal with continuous f. For if that case has been dealt with, we may proceed as follows. First, note that if $p \in L^1(\Gamma)$, $\|p\|_1 \leq 1$, and f satisfies the hypotheses, then $p * f$ satisfies (1), since for $\mu \in M_1(\Gamma)$,

$$\left|\int (p * f)(x)\, d\mu(x)\right| = \left|\iint f(x - y)p(y)\, dy\, d\mu(x)\right|$$
$$\leq \int\left|\int f(x) d\mu(x + y)\right| |p(y)| dy \leq c\|\hat{\mu}\|_\infty \|p\|_1.$$

Let $\{p_\alpha\}$ be an approximate identity in $L^1(\Gamma)$ such that $\|p_\alpha\|_1 \leq 1$ and $\hat{p}_\alpha \to 1$ uniformly on each compact subset of G. Since the continuous function $p_\alpha * f$ satisfies (1), there exists $\nu_\alpha \in M(G)$ such that $\|\nu_\alpha\| \leq c$ and $\hat{\nu}_\alpha = p_\alpha * f$ on Γ. Then $\{\nu_\alpha\}$ has a weak$*$ accumulation point $\nu \in M(G)$ such that $\|\nu\| \leq c$ and $\hat{\nu} = f$ a.e. on Γ.

Having noted that f may be assumed continuous, we present two proofs of the Theorem. The first is a simple one that works for the circle group.

Proof 1, for the case $\Gamma = T$. It suffices to show that $\sum_{n=-K}^{K} |\hat{f}(n)| \leq c$ for every positive integer K. Choose a_n such that $a_n \hat{f}(-n) = |\hat{f}(-n)|$, and let $p_K(x) = \sum_{n=-K}^{K} a_n e^{inx}$. Then

$$\sum_{n=-K}^{K} |\hat{f}(n)| = \sum \hat{f}(-n)a_n = \int_T f p_K = (f p_K)\hat{}\,(0). \tag{2}$$

Let μ_N be the probability measure that assigns mass $1/N$ to each of the points $2\pi j/N$ $(1 \leq j \leq N)$. Then $\hat{\mu}_N(n) = 1$ if N divides n, and 0 otherwise. For $N > 2K$, $\|(p_K \cdot \mu_N)^\wedge\|_\infty = 1$, so that $|\int_T f p_K \, d\mu_N| \leq c$. But $\int_T f p_K = \lim_{N \to \infty} \int_T f p_K \, d\mu_N$. Therefore the quantity (2) is bounded by c, and the proof is complete. □

Proof 2. Note that $\{\hat{\mu}: \mu \in M_1(\Gamma)\}$ is a subspace of $AP(G)$, which is identifiable with $C(bG)$. Condition (1) implies that the mapping $\hat{\mu} \to \int f \, d\mu$ is a linear functional with norm bounded by c on that subspace. By the Riesz Representation Theorem there is a measure $\sigma \in M(bG)$ with $\|\sigma\| \leq c$ such that $\int f \, d\mu = \int_{bG} \hat{\mu} \, d\sigma$ for $\mu \in M_1(\Gamma)$, and hence $\hat{\sigma} = f$ on Γ. Of course, what we want is a measure in $M(G)$ with these properties. Let g be the Radon–Nikodym derivative of $|\sigma|$ with respect to σ, so that $d|\sigma| = g \, d\sigma$. We may suppose that $|g(x)| = 1$ for all $x \in bG$. There is a sequence of trigonometric polynomials g_n on bG that converge to g in $L^1(|\sigma|)$. For each n, $f_n(\gamma) = \int_{bG} \langle \gamma, y \rangle g_n(y) d\sigma(y)$ is a finite linear combination of translates of $\hat{\sigma}$ and hence is continuous on Γ. Evidently f_n converges uniformly to F, where

$$F(\gamma) = \int_{bG} \langle \gamma, -y \rangle g(y) d\sigma(y) = |\sigma|^\wedge(\gamma) \quad \text{for} \quad \gamma \in \Gamma.$$

Since F is a continuous positive-definite function on Γ, it is the transform of some positive measure $\tau \in M(G) \subseteq M(bG)$. Since $\hat{\tau} = |\sigma|^\wedge$, it follows that $\tau = |\sigma|$ and that $f = (g^{-1}\tau)^\wedge$ on Γ. Proof 2 is complete. □

To prove the more general Theorem 1.8.4, we need to lay some groundwork.

Definition. A function $g: \Gamma \to C$ *averages to zero* if for every $\varepsilon > 0$, and every neighborhood N of 0 in Γ, there is a discrete probability measure λ with support in N such that $\|\lambda * g\|_\infty < \varepsilon$.

The next result is generalized by Theorem 1.8.6. We postpone the more general version because we do not know a short proof of it.

1.8.2. Theorem. *Let G be a discrete abelian group. If $\rho \in M(bG)$ and $|\rho|(K) = 0$ for every finite set $K \subseteq G$, then $\hat{\rho}$ averages to zero.*

Proof. We may suppose that $\|\rho\|_M = 1$. Let $0 < \varepsilon < \frac{1}{2}$ and let N be a neighborhood of 0 in Γ. Let Y be a compact symmetric neighborhood of 0 in Γ such that $Y + Y \subset N$. Let $\Delta = m_\Gamma(Y)^{-2} \chi_Y * \chi_Y$, so that $\Delta \geq 0$, $\hat{\Delta} \geq 0$, and $\|\Delta\|_{L^1(\Gamma)} = \hat{\Delta}(0) = 1$. There is a finite symmetric neighborhood U of 0 in G such that $\sum\{\hat{\Delta}(x): x \notin U\} < \varepsilon$. Let J be a compact symmetric neighborhood of 0 in bG such that $(J + J) \cap U = \{0\}$ and $|\rho|(U + J + J) < \varepsilon$. Let μ be the measure on Γ_d such that $\hat{\mu} = m_{bG}(J)^{-1}\chi_J * \chi_J$, where the convolution occurs on bG. Then $\hat{\mu}(0) = 1 = \|\mu\|_M$, $\Delta\mu \geq 0$, and

$$\|\Delta\mu\|_M = \hat{\Delta} * \hat{\mu}(0) = \hat{\Delta}(0)\hat{\mu}(0) + \sum\{\hat{\Delta}(x)\hat{\mu}(x) : x \in G\backslash U\}$$

since $\hat{\mu}$ vanishes on U except at 0. Therefore $1 - \varepsilon < \|\Delta\mu\|_M < 1 + \varepsilon$. For $x \in G\backslash(U + J + J)$,

$$\begin{aligned}|\Delta * \hat{\mu}(x)| &= |\sum\{\hat{\Delta}(x - y)\hat{\mu}(y) : y \in (J + J) \cap G\}| \\ &\leq \sup_y |\hat{\mu}(y)| \cdot \sum\{\hat{\Delta}(u) : u \notin U\} < \varepsilon.\end{aligned}$$

Let $\lambda = \Delta\mu/\|\Delta\mu\|_M$. Then $\hat{\lambda}$ is bounded by $\varepsilon/(1 - \varepsilon) < 2\varepsilon$ on $G\backslash(U + J + J)$ and bounded by 1 everywhere, whereas $|\rho|(U + J + J) < \varepsilon$. Therefore $\|\lambda * \hat{\rho}\|_\infty \leq \|\hat{\lambda}\rho\|_M < 3\varepsilon$. The Theorem is proved. □

1.8.3. Theorem. *Let Γ be a locally compact abelian group. Let g be a bounded complex-valued function on Γ that averages to zero. If g is measurable on a measurable set $E \subseteq \Gamma$, then $g = 0$ a.e. on E.*

Proof. Suppose it is not the case that $g = 0$ a.e. on E. We may assume that g is real, and that g is positive on some compact set $E_1 \ni 0$. By Lusin's Theorem there is a set $F \subseteq E_1$ such that $|F| > 0$ and g is continuous on F. We may assume that $0 \in F$, and that for some compact neighborhood I of 0 and some $r > 0$, $|I \cap F| > 0$ and $g(\gamma) > r$ for all $\gamma \in I \cap F$. Let $m = \inf\{g(\gamma) : \gamma \in \Gamma\}$ and let

$$g_1(\gamma) = \begin{cases} g(\gamma) & \text{if } \gamma \in F, \\ m & \text{if } \gamma \in (I - I)\backslash F, \\ 0 & \text{otherwise.}\end{cases}$$

Then g_1 is measurable on Γ, and $g \geq g_1$ on $I - I$. Let

$$H(t) = \int_{I \cap F} g_1(\gamma - t)\,dm_\Gamma(\gamma).$$

Then $H(0) \geq r|I \cap F| > 0$ and H is continuous, so that there exists a neighborhood N of 0 such that $N \subseteq I$ and $H(t) \geq (r/2)|I \cap F|$ for $t \in N$. Let λ be a discrete probability measure with support in N such that $\|g * \lambda\|_\infty < r/4$. If $\gamma \in I \cap F$, then

$$\int g_1(\gamma - t)\,d\lambda(t) \leq \int g(\gamma - t)\,d\lambda(t) < \|g * \lambda\|_\infty < \frac{r}{4}.$$

Integrating over $I \cap F$ with respect to Haar measure, one obtains the inequality

$$\frac{r}{2}|I \cap F| \le \int_{I \cap F} \int g_1(\gamma - t)d\lambda(t)d\gamma < \frac{r}{4}|I \cap F|.$$

We have reached a contradiction, and Theorem 1.8.3 is proved. □

1.8.4. Theorem. *Let f be a measurable complex-valued function defined on a measurable set $E \subseteq \Gamma$. If*

(3) $$\left|\int f\, d\mu\right| \le c\|\hat{\mu}\|_\infty \quad \textit{for every} \quad \mu \in M_1(E),$$

then there exists $\nu \in M(G)$ such that $\|\nu\| \le c$ and $f = \hat{\nu}$ a.e. on E.

Proof. It suffices to prove the Theorem in the case when E is compact. For suppose that we know the Theorem in that case, and consider an arbitrary measurable E. Let $\mathscr{C}$ be the family of compact subsets of E, ordered by inclusion. For each $K \in \mathscr{C}$, there exists $\nu_K \in M(G)$ such that $\hat{\nu}_K = f$ a.e. on K and $\|\nu_K\| \le c$. Then $\{\nu_K\}$ has a subnet that converges weak$*$ to some $\nu \in M(G)$ with $\|\nu\| \le c$. Let $\{\nu_K\}$ denote the subnet. Then for every $g \in L^\infty(\Gamma)$ with compact support contained in E, $\hat{g} \in C_0(G)$ and hence $\int_\Gamma fg = \lim_K \int_\Gamma \hat{\nu}_K g = \lim_K \int_G \hat{g}\, d\nu_K = \int_G \hat{g}\, d\nu = \int_\Gamma \hat{\nu} g$. Therefore $\hat{\nu} = f$ a.e. on E, and the general case is proved.

Now suppose that 1.8.4 has been proved for all compact Γ. Let f, Γ, E satisfy the hypothesis of the theorem, supposing in addition only that E is compact. Let F be a compact neighborhood of 0 that contains E, and let Λ be the closed subgroup of Γ generated by F. Then Λ is isomorphic to a group $C \times Z^m \times R^n$, where C is compact, $m \ge 0$, and $n \ge 0$ (see Hewitt and Ross [1, Theorem 9.8]). Let $H = \hat{\Lambda} = G/\Lambda^\perp \cong \hat{C} \times T^m \times R^n$. Let $q: G \to G/\Lambda^\perp$ be the natural quotient mapping. For each integer $N > 0$, let Λ_N be the discrete subgroup $\{0\} \times (NZ)^m \times (NZ)^n$ of Λ, and let $j_N: \Lambda \to \Lambda/\Lambda_N$ be the natural quotient map. Note that $(\Lambda/\Lambda_N)^\wedge = \Lambda_N^\perp = \hat{C} \times (T_N)^m \times (N^{-1}Z)^n$, where $T_N = \{e^{it}: e^{iNt} = 1\}$. For N sufficiently large, j_N is a homeomorphism on E; and for each $\varepsilon > 0$, we may take N sufficiently large so that in addition,

$$\sup\{|\hat{\mu}(x)|: x \in H\} < (1 + \varepsilon)\sup\{|\hat{\mu}(x)|: x \in \Lambda_N^\perp\} \quad \text{for all} \quad \mu \in M_1(E).$$

By our supposition, since Λ/Λ_N is compact, there is a measure $\alpha_\varepsilon \in M(\Lambda_N^\perp) \subset M(H)$ such that $\|\alpha_\varepsilon\| \le c(1 + \varepsilon)$ and $\hat{\alpha}_\varepsilon = fj_N^{-1}$ a.e. on E. Now α_ε may be considered an element of $M_d(H)$ such that $\|\alpha_\varepsilon\| \le c(1 + \varepsilon)$ and $\hat{\alpha}_\varepsilon = f$ a.e. on E. Let α be a weak$*$ accumulation point in $M(H)$ of $\{\alpha_{1/k}\}_{k=1}^\infty$. Then $\|\alpha\| \le c$ and $\hat{\alpha} = f$ a.e. on E. Finally, let ν be any element of $M(G)$ such that

$\alpha = \nu \circ q^{-1}$ and $|\alpha| = |\nu| \circ q^{-1}$. Then $\hat{\nu} = f$ a.e. on E and $\|\nu\| = \|\alpha\| \leq c$. Thus 1.8.4 follows when it has been proved in the case of compact Γ (discrete G).

We are now prepared to complete the proof. Let f, Γ, and E be as in the hypothesis. Note that $\{\hat{\mu}: \mu \in M_1(E)\}$ is a subspace of $AP(G)$, which is identifiable with $C(bG)$. Condition (3) states that the mapping $\hat{\mu} \to \int f\, d\mu$ is a linear functional, with norm bounded by c, on that subspace. Therefore there is a measure $\sigma \in M(bG)$, with $\|\sigma\| \leq c$, such that $\int f\, d\mu = \int_{bG} \hat{\mu}\, d\sigma$ for every $\mu \in M_1(E)$; in particular, $f = \hat{\sigma}$ on E. Since we may assume that G is discrete, the decomposition of σ is especially easy to describe. Let ν be the measure on G such that $\nu(\{x\}) = \sigma(\{x\})$ for each $x \in G$. Let $\rho = \sigma - \nu$. Then 1.8.2 implies that $\hat{\rho}$ averages to 0 on Γ. Since $\hat{\rho} = f - \hat{\nu}$ on E, it is measurable on E, and by 1.8.3 $\hat{\rho}$ must vanish a.e. on E. Thus $f = \hat{\nu}$ a.e. on E, and Theorem 1.8.4 is proved. □

1.8.5. Theorem. *Let G be a locally compact abelian group. Let X denote $M(bG)$. Let*

$$X_c = \{\sigma \in X : \hat{\sigma} \text{ is continuous on } \Gamma\},$$

$$X_0 = \{\sigma \in X : \hat{\sigma} \text{ averages to zero}\}.$$

For every measure $\sigma \in X$, there exists a unique measure $\nu \in X_c$ such that $\sigma - \nu \in X_0$. The mapping $P: \sigma \to \nu$ is a linear projection from X to X_c, with norm one.

Remarks. Evidently X_c and X_0 are closed subspaces of X. Theorem 1.8.3 implies that $X_c \cap X_0 = \{0\}$.

Assuming 1.8.5 to be true, we point out that the decomposition of σ that 1.8.5 provides is the same one that we described at the beginning of the Section. For let $\rho = \sigma - \nu$, and let K be a compact subset of G. It is easy to see that $L^1(\rho) \subseteq X_0$; in particular, $\rho|_K$ is in X_0. But $\rho|_K$ is also in $M(G)$, hence in X_c. Therefore $\rho|_K = 0$ and $|\rho|(K) = 0$. It follows that ν is indeed the measure in $M(G)$ that agrees with σ on the compact subsets of G.

1.8.6. Theorem. *Let G be a locally compact abelian group. If $\sigma \in M(bG)$ and $|\sigma|(K) = 0$ for every compact set $K \subseteq G$, then $\hat{\sigma}$ averages to zero.*

How 1.8.5 implies 1.8.6. Obviously $P\sigma = 0$, and therefore $\sigma \in X_0$.

Proof of 1.8.5. Let $B(X)$ denote the space of bounded linear operators on X, with the weak operator topology. Hereinafter, let X have the weak topology. Then $B(X)$ may be considered as a subset, with the relative topology, of the product space $\prod_{\sigma \in X} X_\sigma$, where each X_σ is X. For $\gamma \in \Gamma$, let T_γ be the element of $B(X)$ such that $T_\gamma \sigma = \gamma\sigma$. For $V \subset \Gamma$, let $T_V = \{T_\gamma: \gamma \in V\}$, $T_V \sigma = \{T_\gamma \sigma: \gamma \in V\}$. Note that T_γ preserves norm, $T_0 = I$, $\gamma \to T_\gamma$ is continuous, and $(\gamma_1, \gamma_2) \to T_{\gamma_1 + \gamma_2} = T_{\gamma_1} \circ T_{\gamma_2}$ is separately continuous.

If $V \subset \Gamma$, then the norm closure of $T_V \mu$ is weakly compact (see Edwards [1, 4.22]; Grothendieck [1]; or Dunford and Schwartz [1, IV.13.22]). It follows that for each $\sigma \in X$, the set $K_\sigma = \cap\{\overline{co}(T_V \sigma): V$ is a neighborhood of 0 in $\Gamma\}$ is weakly compact (Dunford and Schwartz [1, V.6.4]). (Note that if $0 \in K_\sigma$, then $\sigma \in X_0$).

Let $S = \cap_V (T_V)^-$. Then S is a commuting semigroup of continuous linear maps of K_σ onto itself (because if $\lambda \in K_\sigma$, then $S\lambda \subset \cap_V \overline{co} T_{V+V} \sigma = \cap_V \overline{co} T_V \sigma = K_\sigma$), and S contains the identity. By the Markov–Kakutani Fixed Point Theorem (Dunford and Schwartz [1, V.10.6]), there exists $v \in K_\sigma$ such that $Av = v$ for every $A \in S$. Let $P\sigma = v$.

We shall show that $v \in X_c$, which holds if and only if the map $\gamma \to T_\gamma v$ is continuous. Let W be an open neighborhood of v. There is a neighborhood V of 0 in Γ such that $T_V v \subset W$; if there were not, then for every V there would exist an $A_V \in T_V$ such that $A_V v \notin W$. Let A be a cluster point of $\{A_V\}$. Then $A \in S$, but $Av \notin W$—impossible, since $Av = v$ for every A in S. So P maps X into X_c.

We claim that v is the only element of $K_\sigma \cap X_c$. Suppose that τ is another one. Let W be a convex neighborhood of 0 in X. Since v and τ are in X_c, there is a neighborhood V of 0 in Γ such that $T_V \subset \tau + W$ and $T_V \tau \subset \tau + W$. Since v and τ are in K_σ, for every $\varepsilon > 0$ there exist $a_i > 0$, $b_j > 0$, $u_i \in V$ and $v_j \in V$ such that $\sum a_i = \sum b_j = 1$, $\|v - \sum_i a_i T_{u_i} \sigma\| < \varepsilon$, and $\|\tau - \sum_j b_j T_{v_j} \sigma\| < \varepsilon$. Let $v_1 = \sum_j b_j T_{v_j} v$, $\tau_1 = \sum_i a_i T_{u_i} \tau$. Then $v_1 \in co T_V v$ and $\tau_1 \in co T_V \tau$, so that $v_1 = v + v_2$ and $\tau_1 = \tau + \tau_2$, where $v_2, \tau_2 \in W$. Now

$$\left\| v_1 - \sum_j \sum_i b_j a_i T_{v_j} T_{u_i} \sigma \right\| < \varepsilon, \quad \text{and} \quad \left\| \tau_1 - \sum_i \sum_j a_i b_j T_{u_i} T_{v_i} \sigma \right\| < \varepsilon.$$

Therefore $\|v_1 - \tau_1\| < 2\varepsilon$, so that $v - \tau = v_1 - \tau_1 + \tau_2 - v_2 \in 2\varepsilon B_1 + W - W$, where B_1 is the unit ball. Since W and ε may be arbitrarily small, $v = \tau$.

Thus for $\sigma \in X$, $P\sigma$ is the unique element of $K_\sigma \cap X_c$. When $\sigma \in X_c$, then $\sigma \in X_c \cap K_\sigma$, so P is a projection. Since $K_{c\sigma} = cK_\sigma$, it is evident that $P(c\sigma) = cP(\sigma)$.

To show that P is additive, it suffices to show that if $\sigma, \tau \in X$, then $P\sigma + P\tau \in K_{\sigma+\tau}$. Let W be an open convex neighborhood of 0 in X. Since $P\sigma$, $P\tau$ are in X_c, there exists a neighborhood V of 0 in Γ such that $T_V P\sigma \subset P\sigma + W$ and $T_V P\tau \subset P\tau + W$. Let $\varepsilon > 0$. Let $a_i > 0$, $b_j > 0$, u_i, $v_j \in V$ such that $\sum a_i = \sum b_j = 1$, $\|P\sigma - \sum_i a_i T_{u_i} \sigma\| < \varepsilon$ and $\|P\tau - \sum_j b_j T_{v_j} \tau\| < \varepsilon$. Then the measures $\sigma_1 = \sum_j b_j T_{v_j} P\sigma$ and $\tau_1 = \sum_i a_i T_{u_i} P\tau$ are elements of $co T_V P\sigma$ and $co T_V P\tau$, respectively, and thus $\sigma_1 = P\sigma + \sigma_2$ and $\tau_1 = P\tau + \tau_2$ where σ_2, $\tau_2 \in W$. Evidently $\|\sigma_1 + \tau_1 - \sum_{i,j} a_i b_j T_{u_i + v_j}(\sigma + \tau)\| < 2\varepsilon$. That sum belongs to $coT_{V+V}(\sigma + \tau)$, so $P\sigma + P\tau = \sigma_1 + \tau_1 - \sigma_2 - \tau_2 \in 2\varepsilon B_1 - W - W + coT_{V+V}(\sigma + \tau)$. It follows that $P\sigma + P\tau$ is in $K_{\sigma+\tau}$ and hence equals $P(\sigma + \tau)$. The proof that P is linear and the proof of 1.8.5 are complete. □

Credits and Remarks. Other results related to the subject of this Section appear in Chapter 12.

We recommend the survey of positive-definite functions by Stewart [1]. To his long bibliography we would add two papers of R. A. Horn [1, 2].

Bochner [2] (1934) proved Theorem 1.8.1 in the case when $\Gamma = R$ and f is continuous (as we have seen, it is easy to reduce the theorem to the case of continuous f). Phillips [1] (1950) gave a different proof of the theorem, also for $\Gamma = R$, and he dealt with the more general case of Banach space-valued functions. The theorem as stated, with a proof not using bG but depending on approximate identities, was published first by Takeda [1] (1953), following some correspondence between Takeda and Phillips, and later independently by Eberlein [1] (1955). The authors who followed Bochner all used an idea of Schoenberg [1] (1934). Our Proof 2 is essentially Bochner's and Proof 1 is folklore.

Theorem 1.8.4 was proved by Kreĭn [1] (1943) in the case when $\Gamma = R$ and E is an interval. For an English-language account of his proof, see Ahiezer [1, p. 154–159]. Kreĭn's method works for a larger class of sets E, and this flexibility was exploited by Rosenthal [5] (1967) to obtain a proof for the case when $\Gamma = R^n$ or T^n and E is arbitrary. DeLeeuw and Glicksberg learned of Rosenthal's work and pointed out to him that some methods of theirs [1] (1965) would provide a proof for the general case. We learned of their approach from seminar talks that Rosenthal gave in Berkeley in 1969. Theorems 1.8.3 and 1.8.5 and their proofs come from those talks.

Independently of deLeeuw and Glicksberg, Domar [1] (1970) gave a proof of Rosenthal's theorem that is related to theirs, but is more direct. The proof of 1.8.2 is an adaptation of Domar's; we prove it for arbitrary compact Γ, whereas Domar did it for $\Gamma = R$.

The first to publish a proof for the general 1.8.4 was Doss [1] (1971). His methods share the spirit of Kreĭn's and Rosenthal's but amount to more than a routine generalization. The idea is to show that f determines a linear functional, bounded by c, on a subspace of $C_0(G)$ (namely, the space of transforms of functions $g \in L^\infty(\Gamma)$ that have compact support within E), and then to apply the Riesz Representation Theorem to obtain ν. That proof becomes considerably easier if one first carries out the reduction of 1.8.4 to the case of compact Γ.

Chapter 2

A Proof That the Union of Two Helson Sets Is a Helson Set

2.1. Introduction

The question, whether the union of two Helson sets is a Helson set, resisted answering for some time. S. W. Drury and N. Th. Varopoulos solved the problem in 1970, and we now know that if $H = H_1 \cup H_2$ where H_1 and H_2 are Helson subsets of G, then

$$\alpha(H) \leq \frac{3^{3/2}}{2}(\alpha(H_1)^3 + \alpha(H_2)^3).$$

One may still hope for simpler proofs and better inequalities.

The union question is easy to answer when H_1 and H_2 are disjoint and there exists a function $g \in B(G)$ that equals one on H_1 and zero on H_2. Given $f \in C_o(H)$, we know of course that there exist $f_j \in A(G)$ such that $f_j = f$ on H_j. The function $gf_1 + (1 - g)f_2$ belongs to $A(G)$ and equals f on H. It follows that H is a Helson set. Or one may prefer to argue with reference to the dual spaces, proceeding as follows. Let $\mu \in M(H)$. Then $\mu = \mu_1 + \mu_2$, where $\mu_j \in M(H_j)$. For each $\varepsilon > 0$, there exist $\gamma_j \in \Gamma$ and $\theta_j \in R$ such that

$$e^{i\theta}\hat{\mu}_j(\gamma_j) \geq \|\mu_j\|_{PM}(1 - \varepsilon) \geq \alpha(H_j)^{-1}\|\mu_j\|_M(1 - \varepsilon).$$

Let $h = (\bar{\gamma}_1 e^{i\theta_1} g) + (1 - g)\bar{\gamma}_2 e^{i\theta_2}$. Then

$$\|h\|_{B(H)}\|\mu\|_{PM} \geq \int h\, d\mu \geq \sum_{j=1}^{2} \alpha(H_j)^{-1}\|\mu_j\|_M(1 - \varepsilon)$$
$$\geq \|\mu\|_M(1 - \varepsilon)(\max_j \alpha(H_j))^{-1}.$$

It follows that H is a Helson set, and in fact

$$\alpha(H) \leq \|h\|_{B(H)} \max(\alpha(H_1), \alpha(H_2)).$$

In either argument, estimating $\alpha(H)$ depends on the norm of g. If H_1 and H_2 are compact and disjoint, then the function g is always available, but as far as one can tell by elementary considerations, the norm of g may get large as the distance from H_1 to H_2 gets smaller. The reader can no doubt see the difficulty in dealing with the case of non-disjoint H_1 and H_2 in a non-discrete G.

It suffices to have the function g equal to one on H_1 and merely small on H_2. When H_1 is a lacunary subset of Z, a Riesz product provides such a g; see A.1.4.

The following statement deals with a quite special situation. We denote by $Gp(H)$ the group generated by a set H.

2.1.1. Proposition. *Let H_1 and H_2 be countable compact Helson subsets of the circle group T. Suppose that there exists $x \in T$ such that*

(1) πx *is irrational,*

(2) $Gp(x) \cap Gp(H_1) = \{0\}$,

(3) $Gp(H_1 + x) \cap Gp(H_2 + x) = \{0\}$.

Then $\alpha(H) = \max(\alpha(H_1), \alpha(H_2))$, *where* $H = H_1 \cup H_2$.

Proof 1. The argument depends on the following theorem from elementary Diophantine approximation theory (see Cassels [1, Section III.5]). For $1 \leq j \leq J$, let x_j and θ_j be real numbers. Suppose that whenever $u_1, \ldots, u_J$ are integers and $\sum u_j x_j \equiv 0 \bmod 2\pi$, then also $\sum u_j \theta_j \equiv 0 \bmod 2\pi$. Then for each $\delta > 0$ there exists an integer a such that $|e^{iax_j} - e^{i\theta_j}| < \delta$ for $1 \leq j \leq J$.

The Proposition will be proved when we show that if $\mu_j \in M(H_j)$ and $\mu = \mu_1 + \mu_2$, then $\|\mu\|_{PM} = \|\mu_1\|_{PM} + \|\mu_2\|_{PM}$. Since the measures with finite support are dense in $M(H)$, we may suppose that the support of μ_j is a finite set, and we may write:

$$\hat{\mu}_1(m) = \sum_{j=1}^{J} c_j e^{ims_j}, \qquad \hat{\mu}_2(n) = \sum_{j=1}^{J} d_j e^{int_j}.$$

It suffices to prove that for every pair $m, n \in Z$ and every $\varepsilon > 0$, there exists $r \in Z$ such that

$$|\hat{\mu}(r)| > (1 - \varepsilon)(|\hat{\mu}_1(m)| + |\hat{\mu}_2(n)|).$$

Let $m \in Z, \theta \in R$. By (1) and (2), if $u_1, \ldots, u_J$, and u are integers and $\sum u_j s_j + ux \equiv 0 \bmod 2\pi$, then $u = 0$ and hence in particular $\sum u_j m s_j + u\theta \equiv 0 \bmod 2\pi$. Therefore if $\delta > 0$, there exists $m' \in Z$ such that $|e^{im's_j} - e^{ims_j}| < \delta$ for $1 \leq j \leq J$ and $|e^{im'x} - e^{i\theta}| < \delta$. If δ is small enough, the first J inequalities make $\hat{\mu}_1(m')$ close to $\hat{\mu}_1(m)$.

Given $m, n \in Z$ and $\varepsilon > 0$, choose θ so that $|e^{i\theta}\hat{\mu}_1(m) + e^{inx}\hat{\mu}_2(n)| = (|\hat{\mu}_1(m)| + |\hat{\mu}_2(n)|)$. It follows from the discussion above that there exists $m' \in Z$ such that

$$|e^{im'x}\hat{\mu}_1(m') + e^{inx}\,\hat{\mu}_2(n)| \geq (1 - \varepsilon)(|\hat{\mu}_1(m)| + |\hat{\mu}_2(n)|).$$

Note that

$$e^{im'x}\hat{\mu}_1(m') = \sum c_j e^{im'(s_j+x)},$$
$$e^{inx}\hat{\mu}_2(n) = \sum d_j e^{in(t_j+x)}.$$

By (3), if $u_1, \ldots, u_J, v_1, \ldots, v_J$ are integers and $\sum u_j(s_j + x) + \sum v_j(t_j + x) \equiv 0 \bmod 2\pi$, then each of the two sums equals $0 \bmod 2\pi$, and hence

$$\sum u_j m'(s_j + x) + \sum v_j n(t_j + x) \equiv 0 \bmod 2\pi.$$

Therefore for each $\delta > 0$ there exists $r \in Z$ such that $|e^{ir(s_j+x)} - e^{im'(s_j+x)}| < \delta$ and $|e^{ir(t_j+x)} - e^{in(t_j+x)}| < \delta$ for $1 \le j \le J$, which, if δ is small enough, makes

$$\begin{aligned} |\hat{\mu}(r)| &= |e^{irx}\hat{\mu}_1(r) + e^{irx}\hat{\mu}_2(r)| \\ &\ge (1 - \varepsilon)|e^{im'x}\hat{\mu}_1(m') + e^{inx}\hat{\mu}_2(n)| \\ &\ge (1 - \varepsilon)^2(|\hat{\mu}_1(m)| + |\hat{\mu}_2(n)|). \quad \square \end{aligned}$$

Proof 2. This argument is close to the previous one but more abstract. Let G be the discrete subgroup of T_d that is generated by $H_1 \cup H_2 \cup \{x\}$, and let Γ be the dual group of G. For $\mu \in M(G)$,

$$\|\mu\|_{PM(T)} = \|\mu\|_{PM(G)} = \max\{|\hat{\mu}(\gamma)| : \gamma \in \Gamma\}.$$

Let $\mu_j \in M(H_j)$, and choose $\gamma_j \in \Gamma$ such that $|\hat{\mu}_j(\gamma_j)| = \|\mu_j\|_{PM}$. Let θ be a real number such that

$$|e^{i\theta}\langle x, -\gamma_1\rangle\hat{\mu}_1(\gamma_1) + \langle x, -\gamma_2\rangle\hat{\mu}_2(\gamma_2)| = \|\mu_1\|_{PM} + \|\mu_2\|_{PM}.$$

It follows from (1) and (2) that there exists $\gamma_3 \in \Gamma$ such that

$$\langle t, \gamma_3\rangle = 1 \text{ for } t \in Gp(H_1), \quad \text{and} \quad \langle x, -\gamma_3\rangle = e^{i\theta}.$$

Then

$$|(\delta_x * \mu_1)\hat{}\,(\gamma_1 + \gamma_3) + (\delta_x * \mu_2)\hat{}\,(\gamma_2)| = \|\mu_1\|_{PM} + \|\mu_2\|_{PM}.$$

It follows from (3) that there exists $\gamma \in \Gamma$ such that $\langle t, \gamma\rangle = \langle t, \gamma_1 + \gamma_3\rangle$ for all $t \in Gp(H_1 + x)$ and $\langle t, \gamma\rangle = \langle t, \gamma_2\rangle$ for all $t \in Gp(H_2 + x)$. Then

$$\begin{aligned} |\hat{\mu}(\gamma)| &= |\hat{\mu}_1(\gamma) + \hat{\mu}_2(\gamma)| \\ &= |(\delta_x * \mu_1)\hat{}\,(\gamma) + (\delta_x * \mu_2)\hat{}\,(\gamma)| = \|\mu_1\|_{PM} + \|\mu_2\|_{PM}. \quad \square \end{aligned}$$

Remark. Let h_1, h_2 be distinct nonzero elements of T such that πh_j is irrational, and let $H_j = \{-h_j, h_j\}$. Then $\alpha(H_j) = 1$. The hypothesis of 2.1.1 fails. The conclusion also fails, and in fact $\alpha(H) \geq \sqrt{2}$. For let μ be the measure such that

$$\hat{\mu}(n) = \tfrac{1}{4}(e^{inh_1} + e^{-inh_1} + e^{inh_2} - e^{-inh_2}).$$

Then $\|\mu\|_M = 1$, but $|\hat{\mu}(n)| = \frac{1}{2}|\cos nh_1 + i \sin nh_2| \leq 1/\sqrt{2}$ for all n.

With those few special cases to suggest the nature of the problem, we now present the general results and their proof. For $E \subseteq G$, let

$$E^* = \left\{\sum_{k=1}^{n} u_k x_k : n \text{ finite}, x_k \in E, u_k \in Z, \sum_{k=1}^{n} u_k = 1\right\}.$$

Then E^* is a subset of the group generated by E, and $E^* = \bigcup_{N=1}^{\infty} E^N$, where

$$E^N = \left\{\sum_{k=1}^{n} u_k x_k \in E^* : \sum_{k=1}^{n} |u_k|^2 \leq 2N\right\}.$$

Note that $E^1 = E$, and that if E is a singleton, then $E^* = E$.

2.1.2. Theorem. *If $H_1, \ldots, H_p$ are Helson sets, then their union H is also a Helson set. Let $\alpha_j = \alpha(H_j)$ and suppose that $\alpha_1 = \min \alpha_j$.*

A. *In every case*

$$\alpha(H) \leq \frac{3^{3/2}}{2}\left(\left(\sum_{j=1}^{p} \alpha_j^2\right)\left(\sum_{j=2}^{p} \alpha_j^2\right)\right)^{1/2}.$$

B. *If $H_j^* \cap H_i \subseteq H_j$ for every pair i, j, then*

$$\alpha(H) \leq \sum \alpha(H_j)^2.$$

C. *Let $N \geq 1$. If $H_j^N \cap H_i \subseteq H_j$ for every pair i, j, then*

$$\alpha(H) \leq (2N + 1)^{1 + (1/2N)}(2N)^{-1}\left(\left(\sum_{j=1}^{p} \alpha_j^2\right)\left(\sum_{j=2}^{p} \alpha_j^2\right)\right)^{1/2N}.$$

Note that Part A is just the case $N = 1$ of Part C; and that Part B follows easily from Part C. The theorem has to do with the relationship of a Helson set to other Helson sets in the group. We shall now state what appears to be a stronger theorem, one that has to do with the relationship of a Helson set to the whole group. It makes reference to functions ω_N ($N = 1, 2, \ldots$) which will be defined later. For the moment, let it suffice to say that ω_N is a decreasing function defined on $(0, 1]$ such that $\omega_N(1) = 1$ and $\omega_N(\varepsilon) \leq \varepsilon^{-(1/2N)}$.

2.1.3. Theorem. *Let E be a Helson set in the locally compact abelian group G. Let $\beta > \alpha(E)$ and $0 < \varepsilon \le 1$. Let $\varphi \in C_o(E)$ with $\|\varphi\|_{C_o(E)} = 1$. Then for every closed set F disjoint from E^N, there exists $f \in A(G)$ such that*

(i) $f = \varphi$ *on* E,
(ii) $|f| \le \beta^2\varepsilon$ *on* F,
(iii) $\|f\|_A \le \beta^2\omega_N(\varepsilon)$.

If F is disjoint from E^*, there exists $f \in A(G)$ satisfying (i) and (ii), such that

(iii)′ $\|f\|_A \le \beta^2$.

Note that the last sentence of the theorem follows from the rest.

How 2.1.3 implies 2.1.2. It suffices, assuming 2.1.3, to prove Part C of 2.1.2. Fix N, let $\beta_j > \alpha_j$, and let $\beta^2 = \sum_{j=1}^{p} \beta_j^2$. Let $\mu \in M(H)$, $\mu \neq 0$. It suffices to show that

$$\frac{\|\mu\|_M}{\|\mu\|_{PM}} \le \frac{(2N+1)^{1+(1/2N)}}{2N} \beta^2(\beta^2 - \alpha_1^2)^{1/2N}. \tag{4}$$

For each j, let μ_j be the restriction $\mu|_{H_j}$ and let $\nu_j = \mu - \mu_j$. If $f \in A(G)$, then

$$\|f\|_A \|\mu\|_{PM} \ge \left|\int_{H_j} f\, d\mu_j\right| - \left|\int_{H\backslash H_j} f\, d\nu_j\right|. \tag{5}$$

Let $\eta > 0$. There exists $\varphi \in C_o(H_j)$ with norm one such that $|\int \varphi\, d\mu_j| \ge \|\mu_j\|_M(1-\eta)$. Let $0 < \varepsilon \le 1$. There exists a compact set F disjoint from H_j^N such that $|\nu_j|(H\backslash(H_j \cup F)) < \eta\beta_j^{-2}\varepsilon^{1/2N}$. Apply 2.1.3 to obtain f. According to (5),

$$\begin{aligned}\beta_j^2\varepsilon^{-1/2N}\|\mu\|_{PM} &\ge \|\mu_j\|_M(1-\eta) - \eta - \beta_j^2\varepsilon(\|\mu\|_M - \|\mu_j\|_M)\\ &\ge \|\mu_j\|_M(1 - \eta + \alpha_1^2\varepsilon) - \eta - \beta_j^2\varepsilon\|\mu\|_M.\end{aligned}$$

This inequality holds for every $\eta > 0$. Therefore

$$\beta_j^2\varepsilon^{-1/2N}\|\mu\|_{PM} \ge \|\mu_j\|_M(1 + \alpha_1^2\varepsilon) - \beta_j^2\varepsilon\|\mu\|_M.$$

We sum over the index j, and obtain the inequality

$$\beta^2\varepsilon^{-1/2N}\|\mu\|_{PM} \ge \|\mu\|_M(1 - (\beta^2 - \alpha_1^2)\varepsilon),$$

which in another form is:

$$\frac{\|\mu\|_M}{\|\mu\|_{PM}} \le \beta^2(\varepsilon^{1/2N} - \varepsilon^{1+(1/2N)}(\beta^2 - \alpha_1^2))^{-1}. \tag{6}$$

It is a calculus exercise to show that the choice of ε that minimizes the right-hand side is $\varepsilon = ((2N + 1)(\beta^2 - \alpha_1^2))^{-1}$. Then (6) gives (4). The implication 2.1.3 $\Rightarrow$ 2.1.2 is established.

As we shall show, the following statement is equivalent to Theorem 2.1.3.

2.1.3*. Theorem. *Let E be a Helson set in the locally compact abelian group G. Let $0 < \varepsilon \leq 1$, $\alpha = \alpha(E)$, $\omega = \omega_N(\varepsilon)$. Then*

$$\|\mu_E\|_M \leq \frac{\alpha^2\varepsilon}{1 + \alpha^2\varepsilon}[\varepsilon^{-1}\omega\|\mu\|_{PM} + \|\mu\|_M] \tag{7}$$

for every measure $\mu \in M(G)$ such that $|\mu - \mu_E|(E^N) = 0$; or, equivalently, (7) holds for every $\mu \in M(E \cup F)$ whenever F is a closed set disjoint from E^N.

Proof that 2.1.3 and 2.1.3 are equivalent.* Standard Banach space techniques suffice, but it takes a bit of thought to see 2.1.3 in the right way. Fix E, G, N, F, $\alpha = \alpha(E)$, and $0 < \varepsilon \leq 1$ as in the hypothesis of 2.1.3. Let $\omega = \omega_N(\varepsilon)$. For each $t \in [0, 1)$, consider the following condition.

(P_t) For every $\varphi \in C_o(E)$ with norm one, and every $\beta > \alpha$, there exists $f \in A(G)$ such that

$$(Q_t)\begin{cases} \text{(i) } |f - \varphi| \leq t \text{ on } E, \\ \text{(ii) } |f| \leq (1 - t)\beta^2\varepsilon \text{ on } F, \\ \text{(iii) } \|f\|_A \leq (1 - t)\beta^2\omega. \end{cases}$$

Clearly (P_0) is precisely the conclusion of 2.1.3. As t increases, inequalities (ii) and (iii) become stronger, and inequality (i) becomes weaker, but the changes cancel out: the truth value of (P_t) is independent of t. The proof should be obvious to most readers; it goes as follows. Fix $t \in (0, 1)$. If f satisfies (Q_0), then $(1 - t)f$ satisfies (Q_t), so that $(P_0) \Rightarrow (P_t)$. If (P_t) holds, fix φ and then let f approximate φ as in (Q_t); let $f_0 = f$. Apply (P_t) with $(\varphi - f_0)/\|\varphi - f_0\|_{C_o(E)}$ in the role of φ, obtaining f_1 such that $|\varphi - f_0 - f_1| \leq t^2$ on E, $|f_0 + f_1| \leq (1 + t)(1 - t)\beta^2\varepsilon$ on F, and $\|f_0 + f_1\|_A \leq (1 + t)(1 - t)\beta^2\omega$. Proceed inductively, obtaining $\{f_k\} \subseteq A(G)$ such that $|\sum_{k=0}^n f_k - \varphi| \leq t^{n+1}$ on E, $|\sum_{k=0}^n f_k| \leq (\sum_{k=0}^n t^k)(1 - t)\beta^2\varepsilon$ on F, and $\|\sum_{k=0}^n f_k\|_A \leq (\sum_{k=0}^n t^k)(1 - t)\beta^2\omega$. If $g = \sum_{k=0}^\infty f_k$, then $g = \varphi$ on E, $|g| \leq \beta^2\varepsilon$ on F, and $\|g\|_A \leq \beta^2\omega$, so that g satisfies (Q_0). Therefore $(P_t) \Rightarrow (P_0)$.

When using or proving Theorem 2.1.3, then, we are free to use (P_t) for whatever choice of t is most convenient. It turns out that for each choice of α and E, there is a value of t such that (P_t) is easy to deal with. The equivalence of 2.1.3 and 2.1.3* is the following result.

2.1.4. Proposition. *Let E, G, N, F, $\alpha = \alpha(E)$, $0 < \varepsilon \leq 1$ be as in the hypothesis of Theorem 2.1.3. Let $\omega = \omega_N(\varepsilon)$. Let $t = \alpha^2\varepsilon/(1 + \alpha^2\varepsilon)$. Then the condition ($P_t$) holds if and only if*

$$\|\mu_E\|_M \leq t(\varepsilon^{-1}\omega\|\mu\|_{PM} + \|\mu\|_M) \quad \textit{for all } \mu \in M(E \cup F). \tag{P_t^*}$$

Proof. When t has that value, then $(1 - t)\alpha^2\varepsilon = t$, and (P_t) states that for every $\varphi \in C_o(E)$ with norm one and every $\beta > \alpha$, there exists $f \in A(G)$ such that

(i) $|f - \varphi| \le t$ on E,

(ii) $|f| \le \dfrac{\beta^2}{\alpha^2} t$ on F,

(iii) $\varepsilon\omega^{-1}\|f\|_A \le \dfrac{\beta^2}{\alpha^2} t.$

Consider the Banach space $X = A(G) \times C_o(E \cup F)$ with norm

$$\|(g, \psi)\| = \max(\varepsilon\omega^{-1}\|g\|_A, \|\psi\|_{C_o(E \cup F)}).$$

Let V be the subspace of X, $V = \{(f, f): f \in A(G)\}$. For $\varphi \in C_o(E)$ with norm one, let $\tilde{\varphi}$ be the element $(0, \varphi_1)$ of X such that $\varphi_1 = \varphi$ on E and $\varphi_1 = 0$ on F. Evidently (P_t) holds if and only if

(P_t') $\quad$ $\operatorname{dist}(\tilde{\varphi}, V) \le t$ $\quad$ for every $\varphi \in C_o(E)$ with norm one,

because

$$\operatorname{dist}(\varphi, V) = \inf\{\max[\varepsilon\omega^{-1}\|f\|_A, \|f - \varphi\|_{C_o(E)}, \|f\|_{C_o(F)}]: f \in A(G)\}.$$

The dual space of X is $X^* = PM(G) \times M(E \cup F)$, with norm

$$\|(S, \mu)\| = \varepsilon^{-1}\omega\|S\|_{PM} + \|\mu\|_M.$$

The annihilator of V in X^* is $V^\perp = \{(-\mu, \mu): \mu \in M(E \cup F)\}$. Therefore

$$\begin{aligned}\operatorname{dist}(\tilde{\varphi}, V) &= \sup\{|L(\tilde{\varphi})|: L \in V^\perp, \|L\| \le 1\}\\ &= \sup\left\{\left|\int_E \varphi\, d\mu\right|: \mu \in M(E \cup F), \varepsilon^{-1}\omega\|\mu\|_{PM} + \|\mu\|_M \le 1\right\}.\end{aligned}$$

It is now obvious that $(P_t^*) \Leftrightarrow (P_t')$. The proposition is proved and the equivalence of 2.1.3 and 2.1.3* is established. ☐

2.1.5. Proposition. (*a*) *To prove* 2.1.3, *it suffices to prove it in the case when E is compact.*

(*b*) *To prove* 2.1.3 *in the case when G is discrete, it suffices to prove it in the case when E is finite.*

(*c*) *To prove* 2.1.3 *in the case when E is metrizable, it suffices to prove it in the case when E is totally disconnected.*

(*d*) *To prove* 2.1.3, *it suffices to prove it in the case when G is compact.*

Proof. Our argument uses the equivalence of 2.1.3 and 2.1.3*.

(a) To prove condition (7), it suffices to prove it for a set of measures that is dense in $\{\mu \in M(G): |\mu - \mu_E|(E^N) = 0\}$. The measures μ in that set such that μ_E has compact support are dense.

(b) is a special case of (a).

(c) When E is metrizable, the measures μ such that μ_E has totally disconnected support are dense.

(d) Let G be arbitrary. By (a), we may suppose that E is compact. Let θ be the inclusion map of G in its Bohr compactification bG. The set θE is compact because E is compact. The inclusion $M(G) \subseteq M(bG)$ is isometric for both norms. Therefore $\alpha(\theta E) = \alpha(E)$, and Theorem 2.1.3* applied to the case of θE in bG clearly implies Theorem 2.1.3* in the case of E in G. The Proposition is proved. □

2.2. Definition of the Functions ω_N

For $n \geq 1, N \geq 1$, let E_n be the set consisting of the n canonical generators of the group Z^n, and let $\omega_{N,n}(\varepsilon) = \inf\{\|f\|_A: f \in A(Z^n),\ f = 1 \text{ on } E_n, \text{ and } |f| \leq \varepsilon \text{ off } E_n^N\}$. We shall show that $\omega_{N,n}(\varepsilon) \leq \varepsilon^{-1/2N}$. Let H be the hyperplane, $H = \{u = (u_1, \ldots, u_n) \in Z^n: \sum_{k=1}^n u_k = 1\}$. There exists $h \in B(Z^n)$ such that $h = 1$ on H, $h = 0$ off H, and $\|h\|_B = 1$. Note that $H \cap E_n^1 = E_n$. Let $p(u) = e^{-(t/2)|u|^2}$, where $t > 0$ and $|u|^2 = \sum u_k^2$. Then $\|p\|_A = 1$. Let $f = e^{(t/2)}ph$. Then $\|f\|_A \leq e^{t/2}$, $f = 1$ on E_n, $f = 0$ off H, and $|f| \leq e^{-(t/2)(|u|^2 - 1)}$ for all u. If we choose $t = -N^{-1} \log \varepsilon$, then $\|f\|_A \leq \varepsilon^{-1/2N}$, and $u \notin E_n^N \Rightarrow u \notin H$ or $(|u|^2 - 1) \geq 2N \Rightarrow |f(u)| \leq \varepsilon$. Let

$$\omega_N(\varepsilon) = \sup_{n \geq 1} \omega_{N,n}(\varepsilon).$$

Let E be the set of the canonical generators of Z^∞. It follows from weak$*$ compactness of balls in $B(Z^\infty)$ that

$$\omega_N(\varepsilon) = \inf\{\|f\|_B: f = 1 \text{ on } E, |f| \leq \varepsilon \text{ off } E^N\};$$

$$\omega_N(\varepsilon) \leq \varepsilon^{-1/2N}.$$

It is easy to see that ω_N is convex, and hence continuous, and also log-subadditive:

$$\omega_N\left(\frac{\delta + \varepsilon}{2}\right) \leq \tfrac{1}{2}[\omega(\delta) + \omega(\varepsilon)],$$

$$\omega_N(\delta\varepsilon) \leq \omega_N(\delta)\omega_N(\varepsilon).$$

In view of the definition of ω_N, it is clear that Theorem 2.1.3 holds in the very special case when $G = Z^n$, $E = E_n$, and $\varphi = 1$ on E. Theorem 2.1.3 will be proved by means of a step-by-step reduction to this case.

2.3. Transferring the Problem from One Group to Another

The purpose of this section is to prove Theorem 2.3.2, stated below, which makes possible the key step in the proof of 2.1.3. Let us mention first a couple of elementary prototypes of 2.3.2. One is a well-known fact (see Rudin [1, Section 2.7]). Let $E \subset H$, where H is a closed subgroup of G. Since "$A(E)$" would be ambiguous notation, let us write $A(E, X)$ for the quotient algebra of $A(X)$ consisting of the restrictions to E of functions in $A(X)$. Then $A(E, H) = A(E, G)$, that is, the two algebras are isometrically isomorphic by the natural map. That result is a special case of the following one.

2.3.1. Proposition. *Let X and Y be locally compact abelian groups. Let π be a continuous homomorphism from X into Y. If E is a compact set in X, then the mapping $f \to f \circ \pi$ gives a norm-decreasing homomorphism from $A(\pi E, Y)$ into $A(E, X)$. If π is one-to-one, it induces an isometric isomorphism of the two algebras.*

The proof is not difficult; see deLeeuw and Herz [1].

2.3.2. Theorem. *Let G and H be locally compact abelian groups. Let π be a continuous homomorphism from H into G. Let E be a compact Helson set in G, and let θ be a continuous map from E into H such that $\pi \circ \theta$ is the identity on E. Then if $h \in A(H)$, $\beta > \alpha(E)$, $\delta > 0$, and V is a neighborhood of the identity in G, there exists $g \in A(G)$ such that*

(i) $|h \circ \theta - g| < \delta$ *on E,*

(ii) $|g(x)| \le \beta^2 \sup_{\pi y \in x+V} |h(y)|$ *if* $\pi^{-1}(x + V) \neq \varnothing$, *and* $g(x) = 0$ if $\pi^{-1}(x + V) = \varnothing$.

(iii) $\|g\|_A \le \beta^2 \|h\|_A$.

2.3.3. Lemma. *Let G, H be locally compact abelian groups. Let E be a compact Helson set in G, and let θ be a continuous map from E into H. Then if $f \in A(G \times H)$, $\beta > \alpha(E)$, and $\varepsilon > 0$, there exists $g \in A(G)$ such that*

(i) $|f(x, \theta x) - g(x)| < \varepsilon$ *for $x \in E$,*

(ii) $|g(x)| \le \beta^2 \sup_{y \in H} |f(x, y)|$ *for all $x \in G$,*

(iii) $\|g\|_A \le \beta^2 \|f\|_A$.

How Theorem 2.3.2 follows from the Lemma. Let h be the function given in the hypothesis of the theorem. Let $k \in A(G)$ be such that $\|k\|_{A(G)} = k(0) = 1$ and $k = 0$ off $-V$. Then $k(x)h(y)$ is an element of $A(G \times H)$, and since $(x, y) \to (x - \pi y, y)$ is an automorphism of the group $G \times H$, the function $f(x, y) = k(x - \pi y)h(y)$ is also in $A(G \times H)$, with norm bounded by $\|h\|_A$. Now $f(x, y) = 0$ except when $x - \pi y \in -V$, that is, when $\pi y \in x + V$; and $f(x, \theta x) = h(\theta x)$ for $x \in E$. The function g provided by the Lemma satisfies the conclusion of the Theorem.

Given a Banach space B, $1 \le p < \infty$, let $L^p(G, B)$ denote the space of strongly measurable B-valued functions on G that are p-integrable, in the sense of the Bochner integral with respect to Haar measure dx, with norm $\|u\|_p = (\int_G \|u(x)\|_B^p\, dx)^{1/p}$ (see Hille and Phillips [1, p. 58–92] for a treatment of the Bochner integral). Let $C_o(G, B)$ denote the sup-normed algebra of continuous B-valued functions on G that vanish at infinity.

Proof of Lemma 2.3.3. We may suppose that $\|f\|_{A(G \times H)} = 1$, so that there exists $p \in L^1(\hat{G} \times \hat{H})$ with norm one such that

$$f(x, y) = \int_{\hat{H}} \int_{\hat{G}} \langle x, \zeta \rangle \langle y, \eta \rangle p(\zeta, \eta) d\zeta\, d\eta \quad \text{for} \quad x \in G, y \in H.$$

Let

$$\varphi(x, \eta) = \int_{\hat{G}} \langle x, \zeta \rangle p(\zeta, \eta) d\zeta.$$

Then

(1) $$\varphi \in L^1(\hat{H}, A(G)), \qquad \int_{\hat{H}} \|\varphi(\cdot, \eta)\|_{A(G)}\, d\eta = 1,$$

and

(2) $$f(x, y) = \int_{\hat{H}} \langle y, \eta \rangle \varphi(x, \eta) d\eta.$$

The desired function $g \in A(G)$ will be produced as an average of the elements in the range of the $A(G)$-valued function $\eta \to \varphi(\cdot, \eta)$, in the following sense:

(3) $$g(x) = \int_{\hat{H}} \psi(x, \eta) \varphi(x, \eta) d\eta \quad \text{for} \quad x \in G.$$

In order to obtain condition (i), ψ will be chosen so that for $x \in E$, $\langle \theta x, \eta \rangle - \psi(x, \eta)$ is small in a sense that will make the quantity

$$|f(x, \theta x) - g(x)| = \left| \int_{\hat{H}} [\langle \theta x, \eta \rangle - \psi(x, \eta)] \varphi(x, \eta) d\eta \right|$$

less than ε for $x \in E$. The function ψ will be chosen so that, furthermore,

(4) $$\psi \in C_o(\hat{H}, A(G)) \quad \text{and} \quad \sup_\eta \|\psi(\cdot, \eta)\|_{A(G)} \le \beta^2,$$

(5) $$\psi \in C_o(G, A(\hat{H})) \quad \text{and} \quad \sup_x \|\psi(x, \cdot)\|_{A(\hat{H})} \le \beta^2.$$

Condition (iii) of the Lemma follows from (1), (3) and (4). By (5), there exists h such that

(6) $$h \in C_o(G, L^1(H)), \qquad \sup_x \|h(x, \cdot)\|_{L^1(H)} \le \beta^2,$$

and

$$\psi(x, \eta) = \int_H \langle y, \eta \rangle h(x, y) dy$$

for $x \in G$. Then

$$g(x) = \int_H f(x, y) h(x, y) dy \quad \text{for} \quad x \in G,$$

and condition (ii) of the Lemma follows.

Now to define ψ. There exists a continuous function $k: \hat{H} \to \mathbb{C}$ with compact support K such that $\|k\|_{L^2(\hat{H})} \le 1$ and

(7) $$\int_{\hat{H}} |1 - (k * k^*)(\eta)| \, |\varphi(x, \eta)| d\eta < \varepsilon/3 \quad \text{for all } x,$$

where $k^*(\eta) = \overline{k(-\eta)}$. Since θE is compact, the set

$$U = \{\eta \in \hat{H}: |1 - \langle \theta x, \eta \rangle| < \varepsilon/3 \quad \text{for} \quad x \in E\}$$

is a neighborhood of the identity in $\hat{H}$. Since K is compact, it is covered by the union of a finite collection of translates of U: $K \subseteq \bigcup_{i=1}^n (\eta_i + U)$. Let $K_1 = K \cap (\eta_1 + U)$, $K_{i+1} = (K \cap (\eta_{i+1} + U)) \setminus \bigcup_{j=1}^i K_j$ for $1 \le i < n$. Then K is the union of n disjoint Borel sets $K_1, \ldots, K_n$ such that on each of the K_i's, each of the characters θx (for $x \in E$) is virtually constant:

(8) $$|\langle \theta x, \eta_i \rangle - \langle \theta x, \eta \rangle| < \varepsilon/3 \quad \text{for} \quad x \in E, \eta \in K_i.$$

For each i, there is a function $q_i \in A(G)$ such that $\|q_i\|_A \le \beta$ and $q_i(x) = \langle \theta x, \eta_i \rangle$ for $x \in E$. Let q be the Borel measurable, $A(G)$-valued function defined on $\hat{H}$ as follows:

$$q(x, \eta) = \begin{cases} q_i(x) & \text{for} \quad \eta \in K_i, \\ 0 & \text{for} \quad \eta \notin K. \end{cases}$$

Then $k(\eta)q(x, \eta)$ belongs to $L^2(\hat{H}, A(G))$ with norm bounded by β. Let $\psi = (kq) * (kq)^*$, where the convolution is carried out on $\hat{H}$. Condition (4) is now clear, because if $F, G \in L^2(\hat{H}, B)$, where B is any Banach algebra, then the convolution

$$F * G(\eta) = \int_{\hat{H}} F(\eta - \zeta)G(\zeta)d\zeta \tag{9}$$

(where the multiplication is that of B), is clearly an element of $C_o(\hat{H}, B)$, with $\|F * G\|_{C_o(\hat{H}, B)} \leq \|F\|_{L^2(\hat{H}, B)}\|G\|_{L^2(\hat{H}, B)}$. To obtain (5), consider the two norm-decreasing inclusions:

$$L^2(\hat{H}, A(G)) \subseteq L^2(\hat{H}, C_o(G)) \subseteq C_o(G, L^2(\hat{H})).$$

The first is obvious, and so is the second, since if $w(x, \eta) \in L^2(\hat{H}, C_o(G))$, then

$$\sup_x \int_{\hat{H}} |w(x, \eta)|^2\, d\eta \leq \int_{\hat{H}} (\sup_x |w(x, \eta)|^2) d\eta.$$

Now if $F, G \in L^2(\hat{H}, A(G))$, then (9) becomes

$$F * G(x, \eta) = \int_{\hat{H}} F(x, \eta - \zeta)G(x, \zeta)d\zeta.$$

Viewing F and G as elements of $C_o(G, L^2(\hat{H}))$ we see that for each $x \in G$, $F * G(x, \cdot)$ is an element of $A(\hat{H})$ with norm bounded by

$$\|F(x, \cdot)\|_{L^2(\hat{H})}\|G(x, \cdot)\|_{L^2(\hat{H})}.$$

Condition (5) is now clear.

It remains to obtain condition (i) of the Lemma. Let

$$t(x, \eta) = \langle \theta x, \eta \rangle.$$

Then

$$\begin{aligned}(tk) * (tk)^*(x, \eta) &= \int_{\hat{H}} \langle \theta x, \eta - \zeta \rangle k(\eta - \zeta)\overline{\langle \theta x, -\zeta \rangle k(-\zeta)} d\zeta \\ &= \langle \theta x, \eta \rangle (k * k^*)(\eta).\end{aligned}$$

For $(x, \eta) \in E \times \hat{H}$,

$$\begin{aligned}|(tk) * (tk)^* - (qk) * (qk)^*| &\leq |((t - q)k) * (tk)^*| + |(qk) * ((q - t)k)^*| \\ &\leq \frac{\varepsilon}{3} \|k\|_{L^2(\hat{H})}\|tk\|_{L^2(\hat{H})} + \frac{\varepsilon}{3} \|qk\|_{L^2(\hat{H})}\|k\|_{L^2(\hat{H})} \\ &\leq 2\varepsilon/3.\end{aligned}$$

Thus for all $x \in E$,

$$\begin{aligned}|f(x, \theta x) - g(x)| &= \left| \int_{\hat{H}} (\langle \theta x, \eta \rangle - \psi(x, \eta))\varphi(x, \eta) d\eta \right| \\ &= \left| \int_{\hat{H}} [\langle \theta x, \eta \rangle (1 - k * k^*(\eta)) \right. \\ &\qquad \left. + (tk) * (tk)^*(\eta) - (qk) * (qk)^*(\eta)] \varphi(x, \eta) d\eta \right| \\ &\leq \varepsilon/3 + 2\varepsilon/3 = \varepsilon;\end{aligned}$$

(i) is established, and the Lemma is proved. □

2.4. Proof of Theorem 2.1.3

Let E be a Helson set in G. We shall say that *the pair E, G has property one* if whenever $\gamma > 1$ and $0 < \varepsilon \leq 1$, then for every closed set F disjoint from E^N, there exists $q \in A(G)$ such that $q = 1$ on E, $|q| \leq \varepsilon$ on F, and $\|q\|_A < \gamma\omega_N(\varepsilon)$. The statement that E, G has property one is equivalent to what Theorem 2.1.3 says when $\alpha(E) = 1$ and $\varphi = 1$.

Theorem 2.3.2 makes possible the following reduction. It is the key step in the proof of 2.1.3.

2.4.1. Lemma. *Let E be a compact Helson set in a locally compact abelian group G. Suppose that there exist another locally compact abelian group H, a continuous homomorphism π from H into G, and a continuous map θ from E into H such that $\pi \circ \theta$ is the identity on E, $\alpha(\theta E) = 1$, and the pair θE, H has property one. Then Theorem 2.1.3 holds in the case of E and G.*

Proof. The argument is straightforward. Let φ, β, and ε be as in the hypothesis of 2.1.3. Choose $\gamma > 1$ and $\delta > 0$ so that the conditions (1), found below, are satisfied. Let F be a closed set in G disjoint from E^N, and let V be a neighborhood of the identity in G such that $E^N \cap (F - V) = \varnothing$. Since $\alpha(\theta E) = 1$, there exists $p \in A(H)$ such that $p = \varphi \circ \pi$ on θE and $\|p\|_A < \gamma$. Since π is a continuous homomorphism, one knows that $(\theta E)^N \cap \pi^{-1}(F - V) = \varnothing$. Since θE, H has property one, there exists $q \in A(H)$ such that $q = 1$ on θE, $|q| \leq \varepsilon$ on $\pi^{-1}(F - V)$, and $\|q\|_A < \gamma\omega_N(\varepsilon)$. Let $h = pq$. Then $h = \varphi \circ \pi$ on θE (so that $h \circ \theta = \varphi$ on E), $|h| \leq \varepsilon\|p\|_A < \varepsilon\gamma$ on $\pi^{-1}(F - V)$, and $\|h\|_A < \gamma^2\omega_N(\varepsilon)$. Now apply 2.3.2 with $\gamma\alpha = \gamma\alpha(E)$ in place of β, obtaining $g \in A(G)$ such that

$$\begin{aligned}|\varphi - g| = |h \circ \theta - g| < \delta \quad &\text{on } E, \\ |g| \leq (\gamma\alpha)^2\varepsilon\gamma = \gamma^3\alpha^2\varepsilon \quad &\text{on } F, \\ \|g\|_A \leq (\gamma\alpha)^2\|h\|_A \leq \gamma^4\alpha^2\omega_N(\varepsilon).&\end{aligned}$$

Since $\|\varphi - g\|_{C(E)} < \delta$, there exists $g_1 \in A(G)$ such that $g_1 = \varphi - g$ on E and $\|g_1\|_A < \alpha\delta$. Let $f = g + g_1$. Then f satisfies conditions (i), (ii), and (iii) of 2.1.3, provided that γ and δ were chosen so that

$$(1) \qquad \gamma^3\alpha^2\varepsilon + \alpha\delta < \beta^2\varepsilon \quad \text{and} \quad \gamma^4\alpha^2\omega_N(\varepsilon) + \alpha\delta < \beta^2\omega_N(\varepsilon).$$

The Lemma is proved. □

Proof of 2.1.3 in the case when E is finite and G is compact. If E has n elements, let θ be a one-to-one map from E onto the set E_n in Z^n. Let π be the homomorphism from Z^n into G determined by the requirement that $\pi \circ \theta$ be the identity on E. The result follows from 2.4.1 and the definition of ω_N in Section 2.2.

We shall now define the *free compact abelian group* $\Gamma(E)$ *generated by* a compact $E \subset G$. Let $\hat{\Gamma} = \hat{\Gamma}(E)$ denote the group of all continuous functions from E into $T = \{z : |z| = 1\}$, with pointwise multiplication, and the discrete topology. Let $\Gamma = \Gamma(E)$ denote the (compact) dual group of $\hat{\Gamma}$. Consider the following natural maps:

$$\theta : E \to \Gamma; \langle f, \theta x\rangle = f(x) \quad \text{for} \quad x \in E, f \in \hat{\Gamma}.$$

$$\hat{\pi} : \hat{G} \to \hat{\Gamma}; (\hat{\pi}\eta)(x) = \langle x, \eta\rangle \quad \text{for} \quad \eta \in \hat{G}, x \in E.$$

$$\pi : \Gamma \to G; \langle \eta, \pi\gamma\rangle = \langle \hat{\pi}\eta, \gamma\rangle \quad \text{for} \quad \gamma \in \Gamma, \eta \in \hat{G}.$$

Then π is a continuous homomorphism, θ is a homeomorphism, $\pi \circ \theta$ is the identity on E, and $\alpha(\theta E) = 1$. If we knew that θE, $\Gamma(E)$ had property one, we would have all the hypotheses of 2.4.1 (with $H = \Gamma(E)$). The following statement is now obvious from 2.4.1 and 2.1.5(a): To prove 2.1.3, it suffices to show that if E is a compact Helson set in a locally compact abelian group G, then the pair θE, $\Gamma(E)$ has property one. The next lemma takes care of a special case.

2.4.2. Lemma. *Let E be a compact, totally disconnected Helson set in a compact abelian group G. Then the pair* θE, $\Gamma(E)$ *has property one.*

Proof. Let j denote a finite collection of disjoint clopen sets K whose union is E. Call two points of E equivalent if they belong to the same K. Let E_j be the finite set of equivalence classes. Then $\hat{\Gamma}(E_j)$ is the group of all functions from E_j into T. Its elements may be thought of as the functions in $\hat{\Gamma}(E)$ that are constant on each set in j.

Since E is totally disconnected, we may find a sufficiently large family J of partitions j (we may take *all* partitions, for example) so that the union $\hat{\Gamma}_0(E) = \bigcup_{j \in J} \hat{\Gamma}(E_j)$ forms a subgroup of $\hat{\Gamma}(E)$ which is dense in $\hat{\Gamma}(E)$ with respect to the topology of uniform convergence on E. Finally, let $\Gamma_0(E)$ be the dual group of the discrete group $\hat{\Gamma}_0(E)$. Note that if J is given the refinement ordering, we have E and $\Gamma_0(E)$ as inverse limits:

$$E = \varprojlim E_j, \Gamma_0(E) = \varprojlim \Gamma(E_j).$$

For each j, there is a homomorphism

$$\pi_j: \Gamma_0(E) \to \Gamma(E_j); \qquad \pi_j \gamma = \gamma \Big|_{\hat{\Gamma}(E_j)} \quad \text{for} \quad \gamma \in \Gamma_0(E).$$

Consider the following natural maps:

$$\theta_0: E \to \Gamma_0(E); \qquad \langle f, \theta_0 x \rangle = f(x) \quad \text{for} \quad x \in E, f \in \hat{\Gamma}_0(E).$$

$$\psi: \Gamma(E) \to \Gamma_0(E); \qquad \langle f, \psi\gamma \rangle = \langle f, \gamma \rangle \quad \text{for} \quad f \in \hat{\Gamma}_0(E).$$

Note that $\theta_0 x$ is the restriction of θx to $\hat{\Gamma}_0(E)$; that $\alpha(\theta_0 E) = 1$; and that ψ has a nontrivial kernel if E is infinite. Let Δ denote this kernel; it is the annihilator of $\hat{\Gamma}_0(E)$ in $\Gamma(E)$.

We continue to use additive notation for all the groups in sight except the groups $\hat{\Gamma}$ and $\hat{\Gamma}_0$.

Claim. Let F be a closed subset of $\theta E + \Delta$ that is disjoint from θE. Then for each $\delta > 0$, there exists $k \in A(\Gamma(E))$ such that

$$(2) \qquad |1 - k| < \delta \quad \text{on } \theta E, \quad |k| < \delta \quad \text{on } F, \quad \text{and} \quad \|k\|_A = 1.$$

Proof of claim. The map $(x, y) \to \theta x + y$ is one-to-one and hence is a homeomorphism of $\theta E \times \Delta$ and $\theta E + \Delta$, because if $y_1, y_2 \in \Delta$ and $x_1, x_2 \in E$, and if $\theta x_1 - \theta x_2 + y_1 - y_2 = 0$, then $\theta(x_1 - x_2) \perp \hat{\Gamma}_0$, so that $x_1 = x_2$ and hence $y_1 = y_2$. Therefore there is a closed set $K \subset \Delta$, not containing 1, such that $F \subset \theta E + K$. Choose $p \in A(\Gamma(E))$ such that

$$(3) \qquad p(0) = 1, p = 0 \quad \text{on } K, \quad \text{and} \quad \|p\|_A = 1.$$

Then $p = \sum_{f \in \hat{\Gamma}(E)} \hat{p}(f) f$, $\hat{p} \geq 0$, $\sum \hat{p}(f) = 1$. For each $f \in \hat{\Gamma}(E)$, there exists $\varphi_f \in \hat{\Gamma}_0(E)$ such that $|f - \varphi_f| < \delta$ on E. Note that $|\bar{\varphi}_f f - 1| < \delta$ on E, and that $\varphi_f(\theta x + y) = \varphi_f(\theta x)$ if $y \in \Delta$. Let $k = \sum \hat{p}(f)\bar{\varphi}_f f$. Then

$$(4) \qquad |k(\theta x + y) - p(y)| = |\textstyle\sum \hat{p}(f)[\bar{\varphi}_f(x) f(x) - 1] f(y)| < \delta \quad \text{for} \quad x \in E \quad \text{and} \quad y \in \Delta.$$

Then (2) follows from (3) and (4), and the claim is proved.

Let F be a closed subset of $\Gamma(E)$ that is disjoint from $(\theta E)^N$. Given $\gamma > 1$ and $0 < \varepsilon \leq 1$, by the continuity of ω_N we may choose ε_0 so that

$$\text{(5)} \qquad \varepsilon_0 < \varepsilon, \qquad \omega_N(\varepsilon_0) < \gamma\omega_N(\varepsilon).$$

Then choose $\delta > 0$ so that

$$\text{(6)} \qquad \begin{gathered} \varepsilon_0(1 + \delta) \leq \varepsilon, (1 + \delta)\omega_N(\varepsilon_0)2\delta \leq \varepsilon, \\ (1 + \delta)^2\omega_N(\varepsilon_0) < \gamma\omega_N(\varepsilon). \end{gathered}$$

By the claim, and the fact that $\alpha(\theta E) = 1$, there exists $f_1 \in A(\Gamma(E))$ such that $f_1 = 1$ on θE, $|f_1| < 2\delta$ on $F \cap (\theta E + \Delta)$, and $\|f_1\|_A < 1 + \delta$. Let $F_0 = F \cap \{x: |f_1(x)| \geq 2\delta\}$. Then $\psi\theta E = \theta_0 E$ and ψF_0 are disjoint compact sets in $\Gamma_0(E)$. There exists $j \in J$ such that $\pi_j(\theta_0 E)^N = (\pi_j\theta_0 E)^N = (\theta_0 E_j)^N$ and $\pi_j\psi F_0$ are disjoint compact sets in $\Gamma(E_j)$. But $\pi_j\theta_0 E = \theta_0 E_j$ is a finite set, and Theorem 2.1.3 is already known to hold in this case; so since $\alpha(\theta_0 E_j) = 1$, there exists $f_0 \in A(\Gamma(E_j))$ such that $f_0 = 1$ on $\theta_0 E_j$, $|f_0| \leq \varepsilon_0$ on $\pi_j\psi F_0$, and $\|f_0\|_A \leq (1 + \delta)\omega_N(\varepsilon_0)$. Let $q = (f_0 \circ \pi_j \circ \psi) \cdot f_1$. Then $q = 1$ on θE, $|q| \leq \max(\varepsilon_0(1 + \delta), (1 + \delta)\omega_N(\varepsilon_0)2\delta) \leq \varepsilon$ on F, and $\|q\|_A \leq (1 + \delta)^2\omega_N(\varepsilon_0) < \gamma\omega_N(\varepsilon)$. Lemma 2.4.2 is proved. □

It is clear from 2.4.1 and the discussion of free compact abelian groups that to prove 2.1.3, it suffices to prove the following result.

2.4.3. Lemma. *Let E be a Helson set in a compact abelian group G, with $\alpha(E) = 1$. Then the pair E, G has property one.*

Proof. Proposition 2.1.5(c) and Lemmas 2.4.1 and 2.4.2 show that Theorem 2.1.3 holds in the case when E is metrizable. In particular, Lemma 2.4.3 holds in this case. We shall prove the Lemma in the general case by a reduction.

Let $\gamma > 1, 0 < \varepsilon \leq 1$. Let F be a closed set disjoint from E^N. We must show that there exists $q \in A(G)$ such that $q = 1$ on E, $|q| \leq \varepsilon$ on F, and $\|q\|_A < \gamma\omega_N(\varepsilon)$.

Since E^N and F are compact, there is a finite set $P \subset \hat{G}$ that separates E^N and F, in the sense that if $x \in E^N$ and $y \in F$, there exists $\chi \in P$ such that $\langle x, \chi\rangle \neq \langle y, \chi\rangle$. Let Q be a subset of $\hat{G}$ such that $P \cup Q$ generates $\hat{G}$. Let i, j be the natural homomorphisms from G into T^P and T^Q, respectively:

$$\langle f, ix\rangle = \prod_{p \in P} \langle p, x\rangle^{f(p)} \quad \text{for} \quad x \in G, f \in Z^P;$$
$$\langle g, jx\rangle = \prod_{q \in Q} \langle q, x\rangle^{g(q)} \quad \text{for} \quad x \in G, g \in Z^Q.$$

Then $\pi_0(x) = (ix, jx)$ gives a one-to-one continuous homomorphism from G into $T^P \times T^Q$. By Proposition 2.3.1, the map $g \to g \circ \pi_0$ gives an isometric isomorphism of $A(\pi_0 G, T^P \times T^Q)$ and $A(G)$. We do not know that $\alpha(iE) = 1$

with respect to the group T^P, but since T^P is a finite-dimensional torus, we do know that iE is metrizable. Also, $(iE)^N = iE^N$ and iF are disjoint compact sets in T^P. Let V be a neighborhood of the identity in T^P such that $(iE)^N \cap (iF - V) = \varnothing$. Define $\Gamma(iE), \pi_1 : \Gamma(iE) \to T^P, \theta_1 : iE \to \Gamma(iE)$ as before. Now $\theta_1 iE$ is metrizable and $\alpha(\theta_1 iE) = 1$ so that if $\gamma_1 > 1$ and $0 < \varepsilon_1 \leq 1$, there exists $f \in A(\Gamma(iE))$ such that

$$\begin{gathered} f = 1 \quad \text{on } \theta iE, \\ |f| \leq \varepsilon_1 \quad \text{on } \pi_1^{-1}(iF - V), \\ \|f\|_A \leq \gamma_1 \omega_N(\varepsilon_1). \end{gathered}$$

Define $h \in A(\Gamma(iE) \times T^Q)$ by letting $h(x, y) = f(x)$. Then

$$\begin{gathered} h = 1 \quad \text{on } \theta iE \times T^Q \\ |h| \leq \varepsilon_1 \quad \text{on } \pi_1^{-1}(iF - V) \times T^Q, \\ \|h\|_A = \|f\|_A \leq \gamma_1 \omega_N(\varepsilon_1). \end{gathered}$$

Let π be the homomorphism from $\Gamma(iE) \times T^Q$ into $T^P \times T^Q$ defined by letting $\pi(x, y) = (\pi_1 x, y)$. Let θ map $\pi_0 E$ into $\Gamma(iE) \times T^Q$ by: $\theta(u, v) = (\theta_1 u, v)$. Then by Theorem 2.3.2 (using γ_1 in the role of β) there exists $g \in A(T^P \times T^Q)$ such that

$$\begin{gathered} |1 - g| < \delta \quad \text{on } \pi_0 E, \\ |g| \leq \gamma_1^2 \varepsilon_1 \quad \text{on } iF \times T^Q, \\ \|g\|_A \leq \gamma_1^3 \omega_N(\varepsilon_1). \end{gathered}$$

Assuming that γ_1, ε_1, and δ are chosen appropriately in the above procedure, a slight adjustment of $g \circ \pi_0$ gives the desired q.

Lemma 2.4.3 is proved, and thus Theorem 2.1.3 is proved. □

2.5. Remarks and Credits

Until the results of this chapter were discovered, the best partial answers to the union question were those of D. Rider [1] and of P. Malliavin and M.-P. Malliavin-Brameret [1]. The setting for Rider's work was an arbitrary discrete group. The class of sets for which he settled the union question, the Stechkin sets, may or may not include all Sidon sets. For an account of his work, see López and Ross [1, Chapter 2]. The Malliavins settled the question for countable discrete groups in which every nonzero element has the same order p, a prime. They used a combinatorial result of A. Horn [1] to prove that the Sidon sets in such groups are precisely the finite unions of independent sets. For an exposition of this work and further interesting discussion of the pre-Drury state of affairs, see Lindahl and Poulsen [1, Chapters V and VI].

S. W. Drury [4] answered the union question completely for the case of discrete G, by means of Riesz products and a convolution device. N. Th. Varopoulos [13, Addendum; 14 and 16], using Drury's method together with machinery of his own, settled the union question for all metrizable G. A short note by Lust [1] extended the result to all G. Stegeman [3] improved the quantitative aspects of the work. The reader who wishes to know the Varopoulos approach should read the polished and completely general version by Saeki [7].

The proof given in this Chapter is that of Herz [9]. Lemma 2.3.3 is a generalization of Drury's convolution device, and might be called the Drury-Herz Lemma. We prefer the Herz proof, basically because the superstructure he erects over E (Section 2.4) remains within the category of locally compact abelian groups; Varopoulos uses $\hat{\Gamma}(E)$ also, but gives it the topology of uniform convergence on E.

The quantitative aspects of the results obtained by Herz and in Saeki's treatment are equally good. Incidentally, the role played by the Gaussian kernel in Section 2.2 can be filled equally well by Riesz products.

Herz [9, p. 216] has shown that for ε near zero, the estimate $\omega_N(\varepsilon) \leq \varepsilon^{-1/2N}$ may be improved. Therefore there are smaller bounds for $\alpha(H)$ than Theorem 2.1.2 provides, in the case when the numbers $\alpha(H_j)$ are very large.

Proposition 2.1.1 and the remark following it are due to Galanis [4]. Similar results are found in Galanis [1, 2]. In the same vein, and more to be recommended as mental preparation for the complicated procedures of this Chapter, is the short paper of Bernard and Varopoulos [1]. One of their results is the description of a compact continuum in T^∞ which is a Kronecker set, and whose union with a singleton is an independent set but *not* a Kronecker set.

If E is a Helson set, it is natural to ask whether every $\varphi \in C_o(E)$ is the restriction to E of a positive-definite element of $A(G)$. The answer is affirmative if we impose the obviously necessary conditions that $0 \notin E$, and that $\varphi(x) = \overline{\varphi(-x)}$ if both x and $-x$ belong to E.

2.5.1. Theorem. *Let E be a Helson subset of a locally compact abelian group G such that $E = -E$ and $0 \notin E$. Let F be a closed set disjoint from $E \cup \{0\}$. Let $\varphi \in C_o(E)$ such that $\varphi(x) = \overline{\varphi(-x)}$. Then for every $\varepsilon > 0$ and $\beta > 1$, there exists a positive-definite function $f \in A(G)$ such that*

(i) $f = \varphi$ *on* E,
(ii) $|f| \leq \beta^2\varepsilon$ *on* F,
(iii) $\|f\|_A \leq 16\beta^2\alpha(E)^4\varepsilon^{-1}$.

That theorem is due to Smith [1]. His work derives from that of Varopoulos and Herz, and that of Drury [6], who obtained the result in the case when G is discrete. That work of Drury is exposed in López and Ross [1, Chapter 3].

The quantitative aspects of Theorems 2.1.2 and 2.1.3 are not known to be the best possible. For some relevant constructions, see Körner [1] and [3], particularly Section 7 of each paper.

Rider [7] gives a probabilistic variation on Drury's union proof which minimizes the use of Riesz products.

Drury [5] presents his convolution device in a general setting and discusses its applications, discovered by Varopoulos [17, 19], to peak sets for uniform algebras.

The union problem and its solution are not the only interesting work on Sidon sets. Déchamps-Gondim [1] proved the following result, among others. Let G be connected, Γ metrizable, E a Sidon set in Γ_d. Then for every non-void open set $U \subset G$, there is a constant C such that $\sum_{\gamma \in E} |\hat{p}(\gamma)| \leq C \sup_{x \in U} |p(x)|$ for every trigonometric polynomial p with spectrum contained in E. For some worthwhile footnotes on Déchamps-Gondim's work, see Blei [2], Ross [1], and Déchamps-Gondim [2]. An exposition of her work appears in López and Ross [1, Chapter 8].

Let E be a subset of discrete Γ. Rudin ([10]; or see [1, 5.7.7]) proved that if E is a Sidon set, then there exists a constant K such that for $2 < p < \infty$ and for every E-polynomial f, $\|f\|_p \leq K\sqrt{p}\|f\|_2$. Using Gaussian processes, Pisier [1] proved the converse. Pisier [2] adapts the same treatment to the setting studied by Déchamps-Gondim, described above.

Now we should like to mention a number of results that have to do with Helson sets, though not particularly with the union problem.

McGehee and Woodward [1] have studied manifolds in R^n that are Helson sets, or that are Sidon sets in R^n_d—how smooth they can be, and how small the constants can be. That work, like McGehee [5] which it supersedes, uses constructions to prove the existence results. Nevertheless it is based on the ideas that are present in the earlier Baire category arguments of Kahane [11] and Varopoulos [15]. The work of the latter paper was refined by Stegeman [2]; a full treatment appears in Stegeman [4].

Let K be a compact set of reals, $\Lambda = \{\lambda_n\}$ a sequence of positive reals. Helson and Kahane [2] call K *appropriate to* Λ if for some k and every bounded sequence $\{d_k, d_{k+1}, \ldots\}$, there exists $\mu \in M(K)$ such that $\hat{\mu}(\lambda_n) = d_n$ for all $n \geq k$. They considered the questions of what sets K are appropriate to a given Λ, and to what sequences Λ a given K is appropriate. (This work is evidently related to that of Déchamps-Gondim, but here the concern is with rather thin sets K.) Further results appear in Kahane [2, p. 136–142] and Kaufman [11].

The following results are of interest from the point of view of Banach space theory. Rosenthal [2] showed that a compact subset E of metrizable G is a Helson set if and only if for every closed set $F \subseteq E$ there is a projection from $A(E)$ onto $\{f \in A(E): f^{-1}(0) \supseteq F\}$. Varopoulos [20] proved, among other results, that if $A(E)$ is isomorphic, as a Banach space, to $C(X)$ for some compact topological space X, then E is a Helson set.

Finally we quote a result that has to do with interpolation, though not particularly with interpolation sets. Let Λ be a closed and either discrete or compact subset of a second-countable G. Let E be such a subset of Γ. Generalizing work of Beurling [1], G. S. Shapiro [1] proved the following result. Suppose that for each $\gamma \in \Gamma$ there exists $v_\gamma \in M(\Lambda)$ such that

$$\hat{v}_\gamma(x) = \langle x, \gamma \rangle \quad \text{for all} \quad x \in E,$$

and $\|v_\gamma\|_M \leq C$, where C is independent of γ. Then there exists a bounded linear projection $B_\Lambda \colon M(G) \to M(\Lambda)$ such that

$$(B_\Lambda \mu)\hat{} = \hat{\mu} \text{ on } E \text{ for each } \mu \in M(G).$$

Chapter 3

Harmonic Synthesis

3.1. Introduction

For any of the spaces natural to our subject, the two opposite processes of "analysis" and "synthesis" of elements may be described. For example, the "analysis" of an element $f \in L^p(T)$ occurs when we take its Fourier transform and thereby decompose it into its basic components, to wit the functions $t \mapsto \hat{f}(n)e^{int}$ where $n \in Z$ and $\hat{f}(n) \neq 0$. "Synthesis" is performed when we reconstitute f from its parts—an easy matter, since the Cesáro sums, which are combinations of the parts, converge to f in norm. However, the words of our title have a more limited use in this book. When we write of the analysis and synthesis of elements, the space that we usually have in mind is the space of pseudomeasures, $PM(G) \triangleq L^\infty(\Gamma)$, equipped with the weak$*$ topology. For the most part, we restrict the Chapter to the case of $G = T, \Gamma = Z$.

The interest of the subject lies in the fact that it is not always easy, or even possible, to synthesize a pseudomeasure $S \in PM = PM(T)$ from the class of objects obtained when S is analyzed. That class may be defined simply as the set of point-masses δ_x for which x belongs to the support of S. We shall now present another, fancier description of the analysis process, and then prove that the two descriptions are equivalent.

For $S \in PM$, let V_S be the weak$*$ closed subspace of PM generated by $\{g_n S: n \in Z\}$, where $g_n(t) = e^{int}$. In other words, V_S is the weak$*$ closure in PM of $\{gS: g \in A\}$. In still other words, $\hat{V}_S = \{\hat{U}: U \in V_S\}$ is the weak$*$ closed, translation-invariant subspace of l^∞ generated by $\hat{S}$. It will become clear that V_S is one-dimensional if and only is S is a multiple of a point-mass, δ_x for some $x \in T$. The point masses are thus entitled to be called the basic elements of PM. The analysis of an element $S \in PM$ consists in finding the masses δ_x that are "part" of S in the sense that they belong to V_S. It is easy to see that the following four conditions are equivalent (note that for f and g in A, $\langle f, gS \rangle = \langle \bar{g}, \bar{f}S \rangle$).

(a) $\delta_x \notin V_S$.
(b) There exists $f \in A$ such that $f(x) \neq 0$ and $\langle f, gS \rangle = 0$ for all $g \in A$.
(c) There exists $f \in A$ such that $f(x) \neq 0$ and $\langle g, fS \rangle = 0$ for all $g \in A$.
(d) There exists $f \in A$ such that $f(x) \neq 0$ and $fS = 0$.

We may restate the definition of the support of S, denoted by supp S, by asserting the equivalence of these two conditions:

(e) $x \notin \operatorname{supp} S$.
(f) There is a neighborhood W of x such that $hS = 0$ for every $h \in A$ with support in W.

We claim that

$$\operatorname{supp} S = \{x: \delta_x \in V_S\}. \tag{1}$$

To prove (1) it suffices to show that $(f) \Rightarrow (d)$ and $(c) \Rightarrow (f)$. The first implication is clear; consider the second. Given f as in (c), let W be a neighborhood of x on which f is bounded away from 0. Let $F \in A$ such that $F = f$ on W and F has no zeros on T. Then $G \equiv 1/F \in A$ by 9.1.1. Let $h \in A$ with support in W. Then for every $g \in A$, $\langle g, hS\rangle = \langle g\bar{h}, S\rangle = \langle g\bar{h}\bar{f}\bar{G}, S\rangle = \langle g\bar{h}\bar{G}, fS\rangle$, which equals zero according to (c). It follows that $hS = 0$. Thus (c) implies (f).

The analysis of S, then, is a matter of identifying supp S, and is not much of a problem. The synthesis of S occurs, and we say that S *admits synthesis*, if S is the weak $*$ limit of some net in the linear span of $\{\delta_x: x \in \operatorname{supp} S\}$. Evidently S fails to admit synthesis precisely when there exists $f \in A$ such that $\langle f, S\rangle$ is nonzero even though $\langle f, \delta_x\rangle = 0$ (that is, $f(x) = 0$) for every $x \in \operatorname{supp} S$.

In the study of synthesis of pseudomeasures, one may consider two other kinds of objects: functions $f \in A$, and closed sets $E \subseteq T$. One wishes to know under what additional hypothesis—on S, on f, or on E—the conditions

$$\operatorname{supp} S \subseteq E \quad \text{and} \quad f = 0 \text{ on } E \tag{2}$$

will imply that

$$\langle f, S\rangle = 0. \tag{3}$$

Evidently a pseudomeasure S admits synthesis precisely when $(2) \Rightarrow (3)$ for every f and E. A function $f \in A$ *obeys synthesis* if $(2) \Rightarrow (3)$ for every S and E. A set E *obeys synthesis*, or is a *set of synthesis*, if $(2) \Rightarrow (3)$ for every S and f.

We fill this paragraph with useful tautologies. The space $N(E)$ is the weak $*$ closed subspace of PM generated by $\{\delta_x: x \in E\}$. Evidently $PM(E) \supseteq N(E)$ for every E; a pseudomeasure S obeys synthesis if and only if $S \in N(\operatorname{supp} S)$; and a set E obeys synthesis if and only if $PM(E) = N(E)$. The *hull of* an ideal I in A is the set $h(I) = \cap\{f^{-1}(0): f \in I\}$. The *kernel of* a set F is the ideal $I(F) = \{f \in A: f^{-1}(0) \supseteq F\}$; evidently $h(I(F))$ is the closure of F. For a (closed) set E, $I_0(E)$ and $I(E)$ are respectively the smallest and largest closed ideals whose hull is E; they are the annihilators in A of $PM(E)$ and of $N(E)$, respectively. Evidently a function $f \in A$ obeys synthesis if and only if $f \in I_0(f^{-1}(0))$; a set E obeys synthesis if and only if $I(E) = I_0(E)$, which holds if and only if $\|f\|_{A(E_\varepsilon)} \to 0$ as $\varepsilon > 0$ for every $f \in I(E)$, where $E_\varepsilon = \{t \in T: \operatorname{dist}(t, E) \leq \varepsilon\}$.

It is well known that a one-point set $\{x_o\}$ obeys synthesis (see A.3). Thus if $f \in I(\{x_o\})$, then $\|f\|_{A([x_o - t, x_o + t])} \to 0$ as $t \to 0$. We claim that a singleton is therefore a strong Ditkin set, that is, there is a sequence $\{e_n\} \subseteq J(\{x_o\})$ that serves as an approximate identity for $I(\{x_o\})$. Let $\{V_\lambda\}$ denote the de la Vallée Poussin kernel (see A.5), let $\tau_n(x) = V_{1/n}(x - x_0)$, and let $e_n = 1 - \tau_n$. Then for $f \in I(\{x_o\})$, $\|f - fe_n\|_A = \|f\tau_n\|_A \leq 3\|f\|_{A(x_o + [-1/n, +1/n])} \to 0$ as $n \to \infty$; the claim is proved. A consequence is that whenever (2) holds and (3) fails, then supp fS must be a perfect set. For if x_o is an isolated point of supp fS, then $\tau_n fS$ is independent of n for sufficiently large n, and $\|\tau_n fS\|_{PM} \to 0$, so that $\tau_n fS = 0$ for all sufficiently large n. Fix such an n. Presumably there exists $g \in A$, with support in $x_o + [-1/n, +1/n]$, such that $\langle g, fS\rangle \neq 0$; but $\langle g, fS\rangle = \langle g\tau_n, fS\rangle = \langle g, \tau_n fS\rangle = 0$, a contradiction. It follows that if a set E disobeys synthesis, its boundary must contain a perfect set.

All of the foregoing discussion generalizes readily to the case when an arbitrary locally compact abelian group G takes the place of T. In fact, except for the last paragraph, it generalizes to the case when an arbitrary regular commutative semisimple Banach algebra B takes the place of $A(T)$. We may consider B as an algebra of functions on its maximal ideal space Δ. The elements of the Banach space dual B^* may be considered as distributions on Δ, with supports and synthesis defined essentially as above. We shall say that the algebra B is an *algebra of synthesis* if every element of B^* admits synthesis; or, equivalently, if every closed subset of Δ is the hull of only one closed ideal.

It is easy to find algebras that are not of synthesis. Consider, for example, the algebra $C^1(T)$ of continuously differentiable functions, with norm $\|f\| = \|f\|_{C(T)} + \|f'\|_{C(T)}$. Then $I_o = \{f\colon f(0) = f'(0) = 0\}$ and $I = \{f\colon f(0) = 0\}$ are two distinct closed ideals whose hull is the singleton $\{0\}$, which therefore is not a set of synthesis. The distribution that disobeys synthesis in this case is δ_0', the formal derivative of the point mass δ_0, defined by

$$\langle f, \delta_0'\rangle = f'(0);$$

evidently it does not annihilate every element of I. This distribution will not directly help us prove that synthesis fails in $A(T)$; it is not a pseudomeasure, since

$$(\delta_0')\hat{}(n) = in;$$

the only pseudomeasures supported by a singleton are the multiples of the point-mass, and all measures obey synthesis; and anyway, singletons are strong Ditkin sets!

Until the 1948 paper of Laurent Schwartz [1], it was not known whether the algebras $L^1(G)$, for non-compact locally compact abelian groups G, were algebras of synthesis or not. He proved that for $n \geq 3$, the sphere S^{n-1} in R^n is not a set of synthesis. The example above in $C^1(T)$ is close in spirit to Schwartz's, since on the sphere the critical distribution is one which, restricted to a subspace of $A(R^n)$ that consists of smooth functions, is the functional that takes the average on the sphere of the normal derivative! It was over a decade later that Malliavin [1, 2, 3] proved that none of the algebras $L^1(G)$ are algebras of synthesis. Some years later, Varopoulos introduced methods for "transferring" proofs of the failure of synthesis from one group to another, so that in particular one can obtain Malliavin's theorem from Schwartz's work, as we shall explain in Section 11.2.

We started from the point of view that the problem of synthesis has to do with when and how a pseudomeasure can be synthesized. Alternatively, one might write a chapter under the same title whose theme would be the ideal structure of $A(\Gamma)$. One question to ask is whether every closed ideal I is *principal*—that is, whether there is some element f such that I is the smallest closed ideal containing f. Aharon Atzmon [4] answered that question negatively in 1972, using a product of two 4-spheres in R^{10}. By using more spheres in higher dimensions, together with an extension of Varopoulos's transfer method, Atzmon showed that, furthermore, for every non-discrete Γ, $A(\Gamma)$ contains a closed ideal that is not even finitely generated. That work, which was preceded by the partial results of Atzmon [1, 2], is presented in Section 11.2.

As for the synthesis of pseudomeasures, the strongest condition is that of bounded synthesis by discrete measures. A set E obeys *bounded synthesis with constant* c if for every $S \in PM(E)$ there is a sequence $\{\mu_k\} \subseteq M(E)$ such that $\|\mu_k\|_{PM} \leq c\|S\|_{PM}$ and $\mu_k \to S$ weak$*$. The set E obeys *bounded synthesis by discrete measures with constant* c if the sequence $\{\mu_k\}$ can always be found in $M_d(E)$.

In Section 3.2 we present the most basic results about when and how synthesis is obeyed, including the modified Herz criterion for bounded synthesis by discrete measures, metric conditions on sets $E \subseteq T$, and Lipschitz conditions on functions $f \in A$. We also construct a set E that obeys bounded synthesis with constant 1, but does not obey bounded synthesis by discrete measures.

In Section 3.3 we survey briefly the three methods by which the failure of synthesis is proved in other chapters of the book, and we then present yet another proof, the one that gives the most nearly explicit description of a subset of the circle group that disobeys synthesis.

An outstanding problem in the subject is to determine whether the union of two sets of synthesis is also a set of synthesis. We know of no promising approach to the question. It would be answered affirmatively if we knew that every set E of synthesis has the property that, for every $f \in I(E)$ and every $\varepsilon > 0$, there exists $g \in J(E)$ such that $\|f - fg\|_A < \varepsilon$. A set with this property is called a *Ditkin* (*or Calderon*, or *Wiener-Ditkin*) *set*.

A closed set $E \subseteq T$ is a *strong Ditkin* set if there is a sequence $\{u_n\} \subseteq I_o(E)$ such that $u_n f \to f$ for every $f \in I(E)$. Relatively few sets enjoy this rather strong property; note that it implies factorization in $I(E)$. We discuss the literature on this topic at the end of this section.

Remarks and Credits. A decade after the discovery by L. Schwartz [1] in 1948 of the failure of synthesis for spheres in R^n, for $n \geq 3$, Herz [2] proved that a circle in R^2 *obeys* synthesis. The kinds of ideals pointed out by Schwartz, whose hulls are spheres, were described in detail by Varopoulos [8] in 1966.

In 1958, Dixmier [2] pointed out that various manifolds other than spheres have the same synthesis properties, and for the same reasons. In 1962 Herz [4] combined the methods of Fourier analysis and differential geometry to study the Fourier transforms of volume and surface-area forms on smooth manifolds in R^n, $n \geq 3$, and proved a result that he had stated in his 1960 essay on synthesis (Herz [3]), to wit that if the surface of a smooth convex body has everywhere-positive Gaussian curvature, then it disobeys synthesis. That result follows from the main theorem of the paper, Theorem 3, and as Herz learned later, essentially the same theorem is found in a 1950 paper by Hlawka [1]. Some further advances in the area are due to Littman [1]. Applications to the problem of counting lattice points in a convex set appear in Herz [5].

The results of Herz [2] and Varopoulos [8] were generalized by Domar [2] in 1971, using a different method. He studied certain subsets E of an arbitrary C^∞ $(n-1)$-dimensional manifold in R^n, without multiple points and with non-vanishing Gaussian curvature. He considered the algebras

$$A_\alpha(R^n) = \left\{ f\colon \int_{R^n} |f(x)|(1 + |x|)^\alpha \, dx < \infty \right\}$$

for $n \geq 2$ and α real. As he points out, there are numerous interesting open questions.

Domar [7] has constructed a C^∞ curve in the plane that disobeys synthesis.

Both Malliavin's and Schwartz's proofs that synthesis fails make use of the fact that $A(\Gamma)$ contains a closed subalgebra isomorphic to an algebra of differentiable functions on an interval of the real line. But that observation is worth only so much; Atzmon [3] shows that every nondiscrete locally compact abelian group Γ has a compact subset E such that $A(E)$ is an algebra of synthesis even though it contains a regular, self-adjoint subalgebra that is isomorphic to an algebra of C^∞ functions (and which therefore is not of synthesis). The same paper goes on to discuss the connection between the failure of synthesis in the algebra, and the existence of a closed subalgebra with bounded point derivations.

The question Atzmon [4] considered for $A(\Gamma) = L^1(G)$ has its analogue for $L^p(G)$, with $1 < p < 2$. Atzmon [5, 6] showed that whenever G is noncompact, $L^p(G)$ has a closed translation-invariant subspace that is not the closed span of the translates of any single function. The proof uses not the spheres of R^n, but an adaptation of Malliavin's methods. Those methods had worked also for Lohoué [3], who showed that synthesis fails in the algebras $A_p(G)$, defined in Section 10.2; and for Bloom [1], who considered the algebras $L^1 \cap L^p(G)$.

Strong Ditkin sets. Wik [1] studied strong Ditkin sets on the circle group. At the time, the only sets known to be strong Ditkin sets were singletons, closed intervals, and finite unions of strong Ditkin sets. He proved that (1) every strong Ditkin set with zero Lebesque measure is finite; (2) every strong Ditkin set with positive Lebesque measure contains at least one interval; and (3) every strong Ditkin set consists of (i) a union of closed intervals, (ii) the cluster points of (i), if any, and (iii) a countable set whose cluster points, if any, are included among the endpoints of the intervals of (i) and the points of (ii).

Rosenthal [4] proved that a closed nowhere dense subset E of R is a strong Ditkin set if and only if E differs by a finite set from some finite union of sets of the form $\{an + b: n \in Z\}$.

Wik used the proof of 1.3.12, the Cohen-Davenport procedure, to obtain (1), whereas Rosenthal [4] derived it directly from the Idempotent Theorem 1.1.1. In fact, Rosenthal showed that if G is R^n, T^n, or any compact metrizable group in which the union of the finite subgroups is dense, then a closed set $E \subseteq G$ with empty interior is a strong Ditkin set only if it belongs to the coset ring of G_d. Schreiber [1] and Gilbert [1, 2] independently proved the converse of this result. All this work is extended nicely in Saeki [10], where strong Ditkin sets are defined, and the ones without interior characterized, in the setting of an arbitrary locally compact abelian group.

Meyer and Rosenthal [1] show that a polygon in R^n (for example, $\{(x, y): |x| + |y| \leq 1\}$ in R^2) is never a strong Ditkin set, and that the closure of a polygon's exterior is one if and only if the polygon is convex.

Meyer (see [2] or [3, p. 570]) showed that sets like $[-1, 0] \cup \{2^{-n}: n = 1, 2, \ldots\}$ are strong Ditkin sets in R.

3.2. When Synthesis Succeeds

We shall show that there is a natural way to approximate in norm each function $f \in A = A(T)$ by piecewise linear functions; and a natural way to approximate weak* each pseudomeasure $S \in PM = PM(T)$ by measures with finite support. We shall then establish the Herz Criterion for a set $E \subseteq T$ to be a set of synthesis.

Let f be a function in A. For each positive integer q, let m_q denote the Haar measure on the q-element subgroup T_q of T:

$$m_q = \frac{1}{q} \sum_{j=0}^{q-1} \delta_{2\pi j/q}.$$

Let $\Delta_q(x) = \max(0, 1 - (q|x|/2\pi))$ for $|x| \leq \pi$. Let f_q be the piecewise linear function that agrees with f on T_q:

$$f_q = q \cdot (f m_q) * \Delta_q.$$

We shall show that

$$f_q \in A, \tag{1}$$

$$\|f_q\|_A \leq \|f\|_A, \tag{2}$$

and

$$\|f_q - f\|_A \to 0 \quad \text{as} \quad q \to \infty. \tag{3}$$

Condition (1) is immediate from the fact that A contains all Lipschitz functions (see 11.1.1 and 11.5.1, or Kahane [2, Section II.6]), but we shall obtain (1) and (2) simultaneously from Theorem 1.8.1 (the short "Proof 1" covers the case at hand). Note that $\hat{m}_q = \chi_{qZ}$, and that if $j \in Z$ and $dv(x) = e^{-ijx}\, dm_q(x)$, then $\hat{v} = \chi_{j+qZ}$. Thus

$$\begin{aligned} \sum_{k \in Z} \hat{\Delta}_q(j + kq) &= \|\Delta_q * (e^{-ijx} m_q)\|_A \\ &= (\Delta_q * e^{-ijx} m_q)\hat{\ }(0) \qquad (4)\\ &= \frac{1}{q} \end{aligned}$$

for each $j \in Z$. Note also that

$$\langle f_q, \mu \rangle = \langle q f m_q * \Delta_q, \mu \rangle = \langle f, q(\Delta_q * \mu) m_q \rangle \tag{5}$$

for every $\mu \in M$; and

$$(q(\Delta_q * \mu) m_q)\hat{\ }(j) = q \sum_{k \in Z} \hat{\Delta}_q(j - kq)\hat{\mu}(j - kq) \tag{6}$$

for every $j \in Z$ and $\mu \in M$. Therefore

$$|\langle f_q, \mu \rangle| \leq \|f\|_A \|\mu\|_{PM} \tag{7}$$

for every $\mu \in M$. Now Theorem 1.8.1 implies (1) and (2). Since the mappings $f \mapsto f_q$ are linear and norm-decreasing, (3) will follow for all $f \in A$ if we prove it for f in some dense subspace. We recall an elementary estimate that holds when $g \in A$ and $g' \in L^2$:

$$\begin{aligned} \|g\|_A &= |\hat{g}(0)| + \sum_{n \neq 0} |n\hat{g}(n)| \cdot |n|^{-1} \\ &\leq |\hat{g}(0)| + \left(\sum_{n \neq 0} n^{-2}\right)^{1/2} \left(\sum_{n \neq 0} |n\hat{g}(n)|^2\right)^{1/2} \\ &\leq \|g\|_1 + \frac{\pi}{\sqrt{3}} \|g'\|_2. \end{aligned}$$

It follows that (3) holds for all $f \in C^1$, and hence for all $f \in A$.

Having proved (1)–(3), we now know that (5)–(7) hold for all $\mu \in PM$, not just for $\mu \in M$. For an arbitrary $\mu \in PM$, and each integer $q > 0$, let μ_q be the measure which appears in (6),

$$\mu_q = q(\Delta_q * \mu)m_q.$$

Then $\mu_q \in M(T_q)$, and by (6) and (4),

$$\|\mu_q\|_{PM} \leq \|\mu\|_{PM}. \tag{8}$$

The fact that $\langle f, \mu_q \rangle = \langle f_q, \mu \rangle$ tends to $\langle f, \mu \rangle$ whenever $f \in A$ may be restated:

$$\mu_q \to \mu \quad \text{weak}* \text{ in} \quad PM = A^*. \tag{9}$$

Note that condition (9) is equivalent to the statement:

$$\sup_q \|\mu_q\|_{PM} < \infty, \quad \text{and} \quad \hat{\mu}_q(j) \to \hat{\mu}(j) \quad \text{for each} \quad j \in Z.$$

Thus one may obtain (9), as well as (8), without any mention of f or $\{f_q\}$, simply by using (4) and the fact that

$$q\hat{\Delta}_q(j) \to 1 \quad \text{as} \quad q \to \infty, \quad \text{for every} \quad j \in Z. \tag{10}$$

A variant on these procedures is obtained if T_q is replaced by some translate $T_q + t_q$ and $dm_q(x)$ by $dm_q(x - t_q)$. Obviously we may suppose that $t_q \to 0$. The effect is to multiply $\hat{f}_q(k)$ and $\hat{\mu}_q(k)$ both by $e^{ikt}q$. Evidently (1)–(3), (8), and (9) remain true. It is that modified procedure that will prove the following Theorem.

3.2.1. The Herz Criterion. *Let E be a closed subset of T. Let $\{q_n\}$ be a sequence of positive integers tending to infinity, and let $\{t_{q_n}\}$ be a null sequence. Suppose that for every n, each point of the set $\{x + t_{q_n}: x \in T_{q_n}\}$ either belongs to E or has distance at least $2\pi/q_n$ away from E. Then E is a set of bounded synthesis by discrete measures, with constant* 1.

Proof. If $\mu \in PM(E)$ and dist $(x, E) \geq 2\pi/q$, then $\Delta_q * \mu(x) = 0$. Thus the hypothesis implies that for every n, $\mu_{q_n} \in M(E \cap (T_{q_n} + t_{q_n})) \subseteq M_d(E)$. Since $\mu_{q_n} \to \mu$ weak$*$, the result follows. $\square$

If we want merely to show that E is a set of synthesis, we may argue as follows. For $f \in I(E)$, the sets $E_n = f_{q_n}^{-1}(0)$ have finite boundary and hence are sets of synthesis. Therefore there exist functions g_n such that $g_n^{-1}(0)$ is a neighborhood of E_n (and hence of E) and such that $\| f_{q_n} - g_n\|_A < n^{-1}$, say. Then $\{g_n\} \subseteq J(E)$ and $\|g_n - f\|_A \to 0$.

We shall now describe a modification in the approximation procedures and show that the hypothesis of 3.2.1 can be weakened. Let $0 \leq \varepsilon < 1$, $\varepsilon_q \to \varepsilon < \frac{1}{2}$. Let

$$\tau_q(x) = \max(0, 1 - (q|x|/2\pi(1 - \varepsilon_q)) \quad \text{for} \quad |x| \leq \pi.$$

Then

(11) $$\|\tau_q\|_A = \hat{\tau}_q(0) = 1,$$

(12) $$\sum_{k \in Z} \hat{\tau}_q(j + kq) = (\tau_q * e^{-ijx}\, dm_q)^\wedge(0) = \frac{1}{q} \quad \text{for each } j \in Z,$$

and

(13) $$q\hat{\tau}_q(j) \to 1 - \varepsilon \quad \text{as} \quad q \to \infty, \quad \text{for each} \quad j \in Z.$$

We now define a variant of μ_q: let

(14) $$\nu_q = q(\tau_q * \mu)m_q.$$

Then

(15) $$\hat{\nu}_q(j) = q \sum_{k \in Z} \hat{\tau}_q(j - kq)\hat{\mu}(j - kq) \quad \text{for each} \quad j \in Z.$$

In particular, $\|\nu_q\|_{PM} \leq \|\mu\|_{PM}$. If $\varepsilon = 0$, then the result of the earlier procedure again holds: $\nu_q \to \mu$ weak∗. If $\varepsilon > 0$, we can make no such claim, but from (11)–(13) and (15) it is clear that some subsequence of $\{\nu_q\}$ converges weak∗ to some $\mu_1 \in PM$ such that

$$\|\mu_1\|_{PM} \leq \|\mu\|_{PM} \quad \text{and} \quad \|\mu - \mu_1\|_{PM} \leq 2\varepsilon\|\mu\|_{PM}.$$

Repeating the process with $\mu - \mu_1$ in the role of μ, we obtain $\mu_2 \in PM$, as a limit of discrete measures, such that

$$\|\mu_2\|_{PM} \leq 2\varepsilon\|\mu\|_{PM} \quad \text{and} \quad \|\mu - \mu_1 - \mu_2\|_{PM} \leq (2\varepsilon)^2\|\mu\|_{PM};$$

and so forth. Thus we may write $\mu = \sum_1^\infty \mu_k$, where

$$\sum \|\mu_k\|_{PM} \leq (1 - 2\varepsilon)^{-1}\|\mu\|_{PM} \quad \text{and} \quad \mu_k = \text{weak}*\lim_{m \to \infty} \nu_{k,q(m)},$$

where $\nu_{k,q(m)} \in M(T_{q(m)})$ with PM-norm bounded by $(2\varepsilon)^{k-1}\|\mu\|_{PM}$. By diagonalizing we may conclude that there exist measures $\mu_n \in M(T_{q(n)})$ such that

$$\|\mu_n\|_{PM} \leq (1 - 2\varepsilon)^{-1}\|\mu\|_{PM}, \quad \text{and} \quad \mu_n \to \mu \text{ weak}*. \quad \square$$

We leave to the reader the proof of the following result.

3.2.2. The Modified Herz Criterion. *Let E be a closed subset of T. Let $\{q_n\}$ be a sequence of positive integers tending to infinity. Let $\{\varepsilon_n\}$ be a sequence such that $0 \leq \varepsilon_n < 1$ and $\varepsilon_n \to \varepsilon < \frac{1}{2}$. Let $\{t_n\}$ be a null sequence. Suppose that for each n, every point of the set $\{x + t_n: x \in T_{q_n}\}$ either is in E or has distance at least $2\pi(1 - \varepsilon_n)/q_n$ away from E. Then E is a set of bounded synthesis by discrete measures, with constant $(1 - 2\varepsilon)^{-1}$.*

Next we shall prove it possible for a set to obey bounded synthesis and yet not obey bounded synthesis by discrete measures. To do so it suffices to exhibit an M_o-set E of bounded synthesis that is independent. For if E is independent, so that $\|\mu\|_M = \|\mu\|_{PM}$ for every $\mu \in M_d(E)$, then if E obeyed bounded synthesis by discrete measures it would follow that $PM(E) = M(E)$; in particular, E would be a Helson set.

3.2.3. Theorem. *There exists an independent M_o-set $E \subseteq T$ such that for every $\mu \in PM(E)$, there exists a sequence $\{\mu_n\} \subseteq M_c(E)$ such that $\|\mu_n\|_{PM} \leq \|\mu\|_{PM}$ and $\mu_n \to \mu$ weak* in PM. In particular, E obeys bounded synthesis with constant 1.*

To understand the proof, it is helpful to consider first the following simpler result, which follows from either of 3.2.1 and 3.2.2.

3.2.4. Proposition. *For every closed set E, there exists a countable set H such that $E \cup H$ is a set of bounded synthesis by discrete measures, and such that the accumulation points of H are contained in E.*

Proof. For a closed set $F \subseteq T$, let

$$\mathscr{H}(F, q) = \left\{x = \frac{2\pi j}{q}: j = 0, 1, \ldots, q - 1; \operatorname{dist}(x, F) < \frac{2\pi}{q}\right\}.$$

Choose a sequence $\{q_k\}$ such that $0 < q_k < q_{k+1}$ and q_k divides q_{k+1}. Let $E_1 = E$ and proceed inductively to define

$$H_k = \mathscr{H}(E_k, q_k), \quad E_{k+1} = E_k \cup H_k.$$

Let $H = \bigcup_{k=1}^{\infty} H_k$. It is easy to show that for each k, $\mathscr{H}(E \cup H, q_k)$ equals H_k and hence is contained in $E \cup H$. Thus the hypothesis of 3.2.1 is satisfied and the Proposition is proved. ☐

Given a set E, we may undertake to get the same result with a slightly smaller H by using 3.2.2, since if $\varepsilon > 0$, the set

$$(16) \quad \mathscr{H}(F, \varepsilon, q) = \left\{x = \frac{2\pi j}{q}: j = 0, 1, \ldots, q - 1; \operatorname{dist}(x, F) \leq \frac{2\pi(1 - \varepsilon)}{q}\right\}$$

is perhaps smaller than $\mathscr{H}(F, q)$. Suppose that we select $\{q_k\}$ as before, and let $0 \leq \varepsilon_k \to \varepsilon < \frac{1}{2}$, and proceed as before, but letting $H_k = \mathscr{H}(E_k, \varepsilon_k, q_k)$. The difficulty is that $\mathscr{H}(E \cup H, \varepsilon_k, q_k)$ may be larger than H_k. That can be prevented by choosing q_{k+1} judiciously after H_k is determined, taking advantage of the fact that we wrote "$\leq$" in (16). A similar technical matter arises in the following argument.

Proof of 3.2.3. The idea is to begin with an independent set and attach to it by an inductive procedure a sequence of sets, similar to the sets of (16) for the purpose of achieving synthesis, and yet different from them so that independence is maintained.

Let U, V, and W be disjoint perfect sets whose union is independent; and suppose that every portion of V is an M_o-set (for an existence proof, see Section 4.7). We shall design inductively a sequence of sets E_k, such that $E_k \subseteq E_{k+1}$, and then let $E = \bigcup_{k=1}^{\infty} E_k$. With each k will be associated a set $V_k \subseteq V$ and an integer q_k, such that $q_{k+1} > q_k$.

Let $E_1 = U$, $q_1 = 100$, $V_0 = \varnothing$. When E_j, q_j, and V_{j-1} have been defined for $1 \leq j \leq k$, let $H_k = \mathscr{H}(E_k, 2^{-k}, q_k)$ and choose a portion V_k of $V \backslash \bigcup_{j=1}^{k-1} V_j$ such that

$$(17) \qquad d_k \equiv \operatorname{diam} V_k < \frac{1}{4}\left(\frac{2\pi 2^{-k}}{q_k}\right).$$

Pick $w_k \in Gp(W)$ (meaning, the group generated by W) such that

$$(18) \qquad V_k - w_k \subseteq (-d_k, d_k).$$

Let $\rho_k \in PF \cap M^+(V_k - w_k)$ such that $\|\rho_k\|_M = \hat{\rho}_k(0) = 1$. Choose J_k sufficiently large so that

$$(19) \qquad q_k^{1/2}|\hat{\rho}_k(j)| \leq k^{-1} \quad \text{for} \quad |j| \geq J_k.$$

Choose $\delta > 0$ sufficiently small so that

$$(20) \qquad \delta < \frac{1}{4}\left(\frac{2\pi 2^{-k}}{q_k}\right)$$

and

$$(21) \qquad 2\pi\delta q_k^{1/2} J_k < k^{-1}.$$

For each $x \in H_k$, choose $t_x \in Gp(W)$ such that

$$(22) \qquad |x - t_x| < \delta.$$

Let

$$E_{k+1} = E_k \cup \bigcup_{x \in H_k} (t_x + V_k - w_k).$$

It follows from (17), (18), (20), and (22) that

$$\operatorname{dist}(y, H_k) < \frac{1}{2}\left(\frac{2\pi 2^{-k}}{q_k}\right) \quad \text{for each} \quad y \in E_{k+1} \backslash E_k.$$

Therefore $\mathscr{H}(E_{k+1}, 2^{-k}, q_k) = H_k$. Furthermore, there exists $\varepsilon_k > 0$ such that $\varepsilon_k < \varepsilon_{k-1}/2$ (if $k > 1$) and such that if

$$(23) \qquad E \supseteq E_{k+1} \quad \text{and} \quad \operatorname{dist}(y, E_{k+1}) < \varepsilon_k \quad \text{for each} \quad y \in E,$$

then $\mathscr{H}(E, 2^{-k}, q_k) = H_k$. Now choose q_{k+1} large enough so that

$$\frac{2\pi}{q_{k+1}} < \frac{\varepsilon_k}{2}.$$

The inductive procedure is now fully described. We are assured that for each k,

$$2\pi \sum_{j=k+1}^{\infty} q_j^{-1} < \varepsilon_k,$$

so that the set $E = \bigcup_{k=1}^{\infty} E_k$ satisfies (23) for every k. Evidently E is an M_o-set. To show that E is independent, it suffices to show that each E_{k+1} is independent, and that follows from Lemma 3.2.5 below.

Let $\mu \in PM(E)$. For each k, let

$$\tau_k(x) = \max\left(0, 1 - \frac{q_k|x|}{(1 - 2^{-k})2\pi}\right) \quad \text{for} \quad |x| \leq \pi.$$

Let

$$\mu_k = q_k(\tau_k * \mu)m_k,$$

where m_k is Haar measure on T_{q_k}. Then $\mu_k \in M(H_k)$, $\|\mu_k\|_{PM} \leq \|\mu\|_{PM}$, and $\mu_k \to \mu$ weak$*$ as $k \to \infty$. (But of course $T_{q_k} \cap E = \varnothing$, and μ_k is not in $M(E)$ unless $\mu_k = 0$).

We may write μ_k in the form

$$\mu_k = \sum\{a_x \delta_x : x \in H_k\}.$$

Since $H_k \subseteq T_{q_k}$, it follows from 1.6.1 that

$$(24) \qquad \|\mu\|_M = \sum |a_x| \leq q_k^{1/2} \|\mu_k\|_{PM} \leq q_k^{1/2} \|\mu\|_{PM}.$$

Now we perturb μ_k just a bit, obtaining

$$\mu_k' = \sum\{a_x \delta_{t_x} : x \in H_k\}.$$

Then

$$(25) \qquad |\hat{\mu}_k'(j)| \le \sum |a_x| \le q_k^{1/2} \|\mu\|_{PM} \quad \text{for all} \quad j \in Z.$$

Also for all j, $|\hat{\mu}_k'(j) - \hat{\mu}_k(j)| \le \sum |a_x(e^{ijt_x} - e^{ijx})| \le 2\pi \sum |a_x j(t_x - x)|$. It follows from (21), (22), and (24) that

$$(26) \qquad |\hat{\mu}_k'(j) - \hat{\mu}_k(j)| < k^{-1} \|\mu\|_{PM} \quad \text{for} \quad |j| \le J_k.$$

Let $\nu_k = \mu_k' * \rho_k$. Then $\nu_k \in M(H_k + V_k - w_k) \subseteq M(E)$. By (19) and (25),

$$(27) \qquad |\hat{\nu}_k(j)| \le k^{-1} \|\mu\|_{PM} \quad \text{for} \quad |j| \ge J_k.$$

For all j, $|\hat{\nu}_k(j) - (\mu_k * \rho_k)\hat{}(j)| = |(\hat{\mu}_k'(j) - \hat{\mu}_k'(j))\hat{\rho}_k(j)|$. It follows from (26) and (27) that

$$\|\nu_k\|_{PM} \le \|\mu_k\|_{PM}(1 + k^{-1}).$$

Since $d_k \to 0$ and $\rho_k \in M(-d_k, d_k)$, $\hat{\rho}_k(j) \to 1$ as $k \to \infty$ for each j. It follows, then, from (26), the definition of ν_k, and the fact that $\mu_k \to \mu$ weak$*$, that

$$\hat{\nu}_k(j) \to \hat{\mu}(j) \quad \text{as} \quad k \to \infty, \quad \text{for each} \quad j \in Z.$$

Thus $\nu_k \to \mu$ weak$*$ and $\{\nu_k\} \subseteq M_c(E)$. Letting $\mu_k = (1 + k^{-1})^{-1}\nu_k$, we find that the Theorem is proved. □

We have used the following lemma, part (B). Its proof is an easy exercise, using the definition of independence.

3.2.5. Lemma. *(A) If $A_1, \ldots, A_k$ are disjoint sets whose union is independent, and if $x_j \in GpA_j$ and $\sum_{j=1}^k x_j = 0$, then each x_j is zero.*

(B) Let U, V, and W be disjoint sets whose union is independent. Let $\{V_j\}$ be a sequence of disjoint portions of V. Let $\{t_{jm} : 1 \le j \le k, 1 \le m \le m(j)\} \subseteq GpW$. Then $U \cup \bigcup_{j=1}^k \bigcup_{m=1}^{m(j)} (V_j + t_{jm})$ is an independent set.

A set E is a set of synthesis when $M_d(E)$ is weak$*$ dense in $PM(E)$. In the example of 3.2.4, $M_d(E)$ is not weak$*$ *sequentially* dense, but the set of all weak$*$ limits of sequences in $M_d(E)$ includes $M(E)$ (as is the case of arbitrary E), and $M(E)$ is weak$*$ sequentially dense in $PM(E)$. The next result, for which we refer the reader to Katznelson and McGehee [4], makes clear that the process of taking sequential limits may sometimes have to be repeated more than twice to obtain the weak$*$ closure. See Section 4.3 for the definition of *order*.

3.2.6. Theorem. *For every countable ordinal number a, there exists a set of synthesis $E \subseteq T$ such that the order of $M(E)$ in $PM(E)$ is a.*

Metric conditions that imply synthesis. For $\varepsilon > 0$, let

$$E_\varepsilon = \{x: \operatorname{dist}(x, E) \leq \varepsilon\},$$

and

$$g_\varepsilon(x) = \max\left(0, \frac{2\pi}{\varepsilon}\left(1 - \frac{|x|}{\varepsilon}\right)\right) \quad \text{for} \quad |x| \leq \pi. \tag{28}$$

Then

$$\hat{g}_\varepsilon(0) = \|g_\varepsilon\|_1 = 1; \qquad g_\varepsilon(0) = \|\hat{g}_\varepsilon\|_1 = \frac{2\pi}{\varepsilon};$$

and

$$\|\hat{g}_\varepsilon\|_p \leq \|\hat{g}_\varepsilon\|_1^{1/p} \leq C_p \varepsilon^{-1/p} \quad \text{for} \quad p \geq 1,$$

where C_p is a constant depending on p. For $S \in PM(E)$, $g_\varepsilon * S$ is a function in A with support contained in E_ε, and $\langle f, S\rangle = \lim_{\varepsilon \to 0}\langle f, g_\varepsilon * S\rangle$ for $f \in A$. If $S \in PM$ and $f \in I(E)$, then

$$\langle f, g_\varepsilon * S\rangle = \int_{E_\varepsilon \setminus E} f(x)(g_\varepsilon * S)(x)dx. \tag{29}$$

A variety of conditions, imposed on f, on E, on S, or on some combination of the three, will suffice to imply that this integral tends to zero with ε and hence that $\langle f, S\rangle = 0$. We shall point out a few of them.

One may estimate (29) as follows:

$$\begin{aligned} |\langle f, g_\varepsilon * S\rangle| &\leq \left(\int_{E_\varepsilon \setminus E} |f(x)|^2\, dx\right)^{1/2}\left(\int |g_\varepsilon * S(x)|^2\, dx\right)^{1/2} \\ &\leq \max_{x \in E_\varepsilon} |f(x)| \cdot (\text{meas } E_\varepsilon \setminus E)^{1/2} \left(\sum |\hat{g}_\varepsilon(n)\hat{S}(n)|^2\right)^{1/2}. \end{aligned} \tag{30}$$

Let $2 \leq r \leq \infty$ and $2/r + 2/s = 1$. Then $2 \leq s \leq \infty$ and

$$\begin{aligned} \left(\sum |\hat{g}_\varepsilon(n)\hat{S}(n)|^2\right)^{1/2} &\leq \|\hat{g}_\varepsilon\|_s \|\hat{S}\|_r = \|\hat{g}_\varepsilon\|_{2r/(r-2)} \|\hat{S}\|_r \\ &= O(\varepsilon^{(2-r)/2r}) \end{aligned} \tag{31}$$

as $\varepsilon \to 0$, provided $S \in l^r$. In every case, $\hat{S} \in l^\infty$, and therefore $\langle f, S \rangle = 0$ whenever

$$\text{(32)} \qquad \liminf_{\varepsilon \to 0} \left[\max_{x \in E_\varepsilon} |f(x)| (\text{meas}(E_\varepsilon \backslash E))^{1/2} \varepsilon^{-1/2} \right] = 0.$$

3.2.7. Theorem.

(a) *If* $\hat{S} \in l^2$, *then* S *admits synthesis.*
(b) *If* $f \in A \cap \Lambda_{1/2}$, *then* f *obeys synthesis.*
(c) *If* $f \in A \cap \Lambda_\alpha \cap I(E)$, $S \in PM(E)$, $\hat{S} \in l^r$, *and* $\text{meas}(E_\varepsilon \backslash E) = o(\varepsilon^t)$, *and if*

$$\gamma = \alpha + \frac{t}{2} + \frac{1}{r} - \frac{1}{2} \geq 0,$$

then $\langle f, S \rangle = 0$.

Proof.

(a) If $\hat{S} \in l^2$, then (30) and (31) imply that

$$\langle f, g_\varepsilon * S \rangle = O(\max_{x \in E_\varepsilon} |f(x)| (\text{meas } E_\varepsilon \backslash E)^{1/2})$$

as $\varepsilon \to 0$.
(b) If $f \in \Lambda_{1/2}$, then $\max_{x \in E_\varepsilon} |f(x)| = O(\varepsilon^{1/2})$ as $\varepsilon \to 0$ and (31) follows.
(c) In this case, (30) and (31) imply that $\langle f, g_\varepsilon * S \rangle = O(\varepsilon^\gamma)$ as $\varepsilon \to 0$. □

3.2.8. Theorem. *Let* $f \in A \cap \Lambda_\alpha(T)$.

(a) *If* $\alpha > \frac{1}{3}$ *and* f *is real-valued, then for almost all* y *in the range of* f *(in the sense of Lebesgue measure on the line),* $f - y$ *obeys synthesis.*
(b) *If* $\alpha > \frac{1}{4}$, *then for almost all* z *in the range of* f *(in the sense of Lebesgue measure on the plane),* $f - z$ *obeys synthesis.*

Remark. Part (*b*) is meaningful only when the range of f has positive measure. That there exist such $f \in A \cap \Lambda_\alpha$ (for any $\alpha < 1$) was proved by Kahane, Weiss, and Weiss [1, Theorem II]. Their examples are space-filling lacunary series, which may be taken to be in Λ_α, as one may learn from Katznelson [1, bottom of p. 110] or Zygmund [1, II.4.9].

Proof. (a) For each $N = 1, 2, \ldots$, the graph of f is contained in the union of N rectangles of length $2\pi N^{-1}$ and height bounded by $cN^{-\alpha}$, where c depends only on f. Let R_N denote this union. Its area is bounded by $2\pi c N^{-\alpha}$, which is to say

$$\int \text{meas}\{x \colon (x, y) \in R_N\} dy \leq 2\pi c N^{-\alpha}.$$

Therefore

$$\text{meas}\{y \colon \text{meas}\{x \colon (x, y) \in R_N\} \neq o(N^{-\alpha} \log N) \quad \text{as} \quad N \to \infty\} = 0.$$

Thus for each y outside a null set, the set $E^y = f^{-1}(y)$ is covered by the union of $o(N^{1-\alpha} \log N)$ intervals of length $2\pi N^{-1}$, and $E^y_{2\pi/N}$ by at most three times as many. For almost all y, then,

$$\text{meas } E^y_\varepsilon = o(\varepsilon^\alpha |\log \varepsilon|) \quad \text{as } \varepsilon \to 0,$$

and since $\alpha > \frac{1}{3}$, condition (31) follows, with $f - y$ and E^y in the roles of f and E.

(b) The proof is similar to that of (a). The graph of f is contained in the union R_N of N cylinders of thickness $2\pi N^{-1}$ and radius bounded by $cN^{-\alpha}$. The volume of R_N is therefore bounded by $2\pi^2 cN^{-2\alpha}$. Consequently

$$\text{meas}\{z\colon \text{meas}\{x\colon (x, z) \in R_N\} \neq o(N^{-2\alpha} \log N) \quad \text{as} \quad N \to \infty\} = 0.$$

Finally we find that for almost all z in the range of f,

$$\text{meas } E^z_\varepsilon = o(\varepsilon^{2\alpha} |\log \varepsilon|) \quad \text{as } \varepsilon \to 0,$$

and since $\alpha > \frac{1}{4}$, condition (32) follows, with $f - z$ and E^z in the roles of f and E. □

3.2.9. Theorem. *Every function in A of bounded variation obeys synthesis.*

Proof. Let $f \in A \cap BV$ and $S \in PM(E)$, where $E \subseteq f^{-1}(0)$. It suffices to show that $\langle f, S\rangle = 0$. Since $\langle f, S\rangle = \langle f, S - \hat{S}(0)\delta_x\rangle$ if $x \in E$, we may suppose that $\hat{S}(0) = 0$. The measure df is continuous. Since f is continuous and vanishes on E, $|df|(E) = 0$. For continuous functions h,

$$\int g\, df = \sum_I \int_I g\, df,$$

where one sums over the set of open intervals I whose union is the complement of E. Note that

$$\int_I df = 0 \quad \text{for each } I,$$

since f vanishes at both endpoints of I. Let F be the formal primitive of S: $\hat{F}(n) = \hat{S}(n)/in$ for $n \neq 0$, $\hat{F}(0) = 0$. If F is continuous, then $\langle F, df\rangle = \sum_I \int_I F\, df$, which vanishes because F is constant on each I, and the proof is finished: $\langle S, f\rangle = \langle F, df\rangle = 0$. However, in general the best one can say about F is that $F \in L^2$, and we must argue as follows. Note that $F_\varepsilon = g_\varepsilon * F$ is the primitive of $S_\varepsilon = g_\varepsilon * S$; where g_ε is as defined by (28). Since F_ε is continuous and $f \in A$,

$$\sum_I \int_I F_\varepsilon\, df = \langle F_\varepsilon, df\rangle = \langle S_\varepsilon, f\rangle \underset{\varepsilon \to 0}{\to} \langle S, f\rangle$$

It suffices now to show that the sum tends to 0. Let $m(I, \varepsilon)$ be the value of F_ε at the midpoint of I. Let $F_\varepsilon^* = F_\varepsilon - m(I, \varepsilon)$ on I. Then

$$\int_I F_\varepsilon \, df = \int_I F_\varepsilon^* \, df.$$

Now $S_\varepsilon(x) \le \sum |\hat{S}(n)\hat{g}_\varepsilon(n)| \le \|\hat{S}\|_\infty \varepsilon^{-1}$. Therefore

$$|F_\varepsilon(u) - F_\varepsilon(v)| \le \|\hat{S}\|_\infty \varepsilon^{-1} |u - v| \quad \text{for all } u, v.$$

If $I = [a, b]$ and $b - a \ge 2\varepsilon$, then $F_\varepsilon^* = 0$ on $[a + \varepsilon, b - \varepsilon]$ and $|F_\varepsilon^*| \le \|\hat{S}\|_\infty$ elsewhere on $[a, b]$. If $b - a \le 2\varepsilon$, then $|F_\varepsilon^*| \le \|\hat{S}\|_\infty$ everywhere on $[a, b]$. For each I, then,

$$\left| \int_I F_\varepsilon^* \, df \right| \le \|S\|_{PM} |df|([a, a + \varepsilon] \cup [b - \varepsilon, b]),$$

which tends to zero as $\varepsilon \to 0$ since df is a continuous measure. The Theorem follows. □

Remarks and Credits. The discussion of the sequence $\{f_q\}$ is from Kahane [2, Section V.3]. The Criterion 3.2.1 is from Herz [1]; generalizations to more groups were proved by Herz [3] and appear in Rudin [1, Section 7.4]. The sets in T that satisfy the Criterion with all $t_{q_n} = 0$ were characterized by Rosenthal [6] as the sets of the form $\bigcap_{n=1}^\infty F_n$, where each F_n is a finite union of intervals whose endpoints belong to T_{q_n}, where

$$F_n \subseteq F_{n+1}, \quad q_n < q_{n+1}, \quad q_n | q_{n+1},$$

and

$$F_n \cap T_{q_{n+1}} \subseteq F_{n+1}.$$

The Modified Criterion was known to many in the fifties but did not appear in print until the 1974 paper of Katznelson and Körner [1], whose application of it appears in Section 12.5.

The Herz Criterion shows that for every integer $r \ge 3$, the set $E_r = \{\sum_{j=1}^\infty \varepsilon_j r^{-j} : \varepsilon_j = 0 \text{ or } 1\}$ is a set of bounded synthesis. Meyer [1, Chapter VII] shows that the same is true whenever r is a Pisot number.

Theorem 3.2.3 is a sample of the great array of constructions to be found in Körner [2, Section 9]. In that work one finds also a union of two sets of bounded synthesis that is not a set of bounded synthesis; the two sets intersect in a singleton. This construction of course bears a relation to the open problem of whether the union of two sets of synthesis is also one. Theorem 3.2.6 is a byproduct of an effort to solve this problem.

Some of the ideas giving sufficient conditions for synthesis trace back to Agmon and Mandelbrojt [1]. In treating Theorems 3.2.7 through 3.2.9 we have followed Kahane and Salem [1, Section IX.6] and Kahane [2, Section V.5]. The use of the integral (29) is called the method of Beurling and Pollard (see Pollard [1]). Theorem 3.2.9 is due to Katznelson.

The result 3.2.7(a) is sharp; for $r > 2$, there exists $S \in PM$ with $\hat{S} \in l^r$ that does not admit synthesis (see Kahane and Salem [1, p. 121]). The result 3.2.7(b) is sharp; for $\alpha < \frac{1}{2}$, there exist $E \subseteq T$ and $f \in I(E)$, disobeying synthesis, such that $\max_{x \in E_\varepsilon} |f(x)| = 0(\varepsilon^\alpha)$. The result 3.2.8(a) may be improvable, but for $\alpha < \frac{1}{4}$ there exists $f \in \Lambda_\alpha$ such that the set $\{y: f - y$ disobeys synthesis$\}$ has positive measure. For a treatment of such results, see Kahane [2, Sections V.6–V.8].

We conclude the Section with the statement of five more theorems, giving only references for proofs.

3.2.10. Theorem. *Kronecker sets obey synthesis.*

That Theorem is the same as 4.5.4.

3.2.11. Theorem. *If* $\sum_{j=1}^{\infty}(t_{j+1}/t_j)^2 < \infty$, *then the set* $\{\sum_{j=1}^{\infty} \varepsilon_j t_j : \varepsilon_j = 0$ *or* $1\}$ *is a set of synthesis.*

For a proof, see Meyer [1, p. 242] or Kahane [2, Section IX.3].

3.2.12. Theorem. *Consider the set*

$$J = \{x = (x_n) \in [0, 1]^\infty : \sum |x_n| \leq 1\},$$

with the product topology, as a metric space. Then for all x outside a set of the first category in J, the set

$$K_x = \left\{\sum_{n=1}^{\infty} \varepsilon_n x_n : \varepsilon_n = 0 \text{ or } 1\right\}$$

is a set of synthesis.

That result is a sample of the category-type results to be found in Saeki [17].

3.2.13. Theorem. *A simple* C^3 *curve in* R^3 *with non-vanishing torsion is a set of bounded synthesis.*

See Domar [5] for the proof.

3.2.14. Theorem. *If* $S \in PM(R)$, *and* $\hat{S}$ *is of bounded variation on* $(-\infty, -a)$ *and on* (a, ∞) *for some* $a > 0$, *then S admits synthesis.*

See Atzmon [7] for the proof, and Atzmon [8] for related work.

3.3. When Synthesis Fails

Three ways to produce sets disobeying synthesis are discussed elsewhere in the book. (1) The combination of Theorems 4.5.3 and 4.6.3 implies that every M-set has a subset E that disobeys synthesis (and, furthermore, such that $\alpha(E) = 1$). (2) Section 11.2 derives the existence of a set E disobeying synthesis in T (and in any nondiscrete group) from the fact that the spheres disobey synthesis in R^n if $n \geq 3$. (3) Theorem 12.3.4 shows that the existence of a countable closed set $E \subseteq T$ with a certain property leads to a set P that disobeys synthesis in the Bohr compactification of T_d. It may be that such an E can be produced by a finitistic construction, but unfortunately we know how to obtain it only by using the existence of a set disobeying synthesis in T!

Each of those three approaches has its charms and advantages. We now present a fourth, which gives the most nearly explicit description of a set disobeying synthesis in T.

Let B be a semisimple regular commutative self-adjoint Banach algebra with identity. Consider B as an algebra of continuous functions on its maximal ideal space. Let B^* be the Banach space dual of B. Consider B^* as a module over B; thus for $h \in B$ and $\mu \in B^*$, $h\mu$ is defined by:

$$\langle f, h\mu \rangle = \langle fh, \mu \rangle \quad \text{for } f \in B.$$

3.3.1. Theorem. *Let N be a positive integer. Let f be a real-valued function in B, and let μ be a nonzero element of B^*. Suppose that*

(1) $$\int_{-\infty}^{\infty} \|e^{iuf}\mu\|_{B^*} |u|^N \, du < \infty.$$

Then there exists $y \in R$ such that the closed ideals in B generated by $(f - y)^n$, for $n = 1, 2, \ldots, N + 1$, are all distinct; and the sets $f^{-1}(y)$ and $(f^{-1}(y) \cap$ supp μ*) are not sets of synthesis.*

Proof. We may suppose that

(2) $$\langle 1, \mu \rangle \neq 0,$$

since we know that $\langle h, \mu \rangle \neq 0$ for some $h \in B$ and may replace μ by $h\mu$ without affecting the hypothesis. Note that $\varphi(u) = \langle 1, e^{iuf}\mu \rangle$ is a continuous function of u, that $\varphi(0) \neq 0$, and $\varphi \in L^1(R)$. Choose $y \in R$ so that $\hat{\varphi}(y) \neq 0$, that is,

(3) $$\int_{-\infty}^{\infty} \langle 1, e^{iu(f-y)}\mu \rangle du \neq 0,$$

By (1), the B^*-valued integral

$$S_n = \int_{-\infty}^{\infty} (iu)^n e^{iu(f-y)} \mu \, du \tag{4}$$

converges for $0 \leq n \leq N$. For $m \geq 1$, let I_m be the closed ideal generated by $(f - y)^m$; then $I_m \supseteq I_{m+1}$, and to show that $I_1, \ldots, I_N$, and I_{N+1} are all distinct, it suffices to show that for $1 \leq n \leq N$,

$$S_n \not\perp I_n, \qquad \text{but } S_n \perp I_m \text{ if } n < m. \tag{5}$$

In other words, $\langle g(f - y)^n, S_n \rangle$ is nonzero for some $g \in B$, but

$$\langle g(f - y)^m, S_n \rangle = 0$$

for all $g \in B$ if $n < m$. The support of S_n is contained in supp μ, by (4), and in $f^{-1}(y)$, by (5). Consequently neither $f^{-1}(y)$ nor $(f^{-1}(y) \cap \operatorname{supp} \mu)$ is a set of synthesis.

Let $g \in B$. If $1 \leq n \leq N$ and $n \leq m$, then

$$\begin{aligned}
\langle g(f - y)^m, S_n \rangle &= \int_{-\infty}^{\infty} \langle g(f - y)^m, e^{iu(f-y)}\mu \rangle (iu)^n \, du \\
&= -i \int_{-\infty}^{\infty} (iu)^n \, d\langle g(f - y)^{m-1}, e^{iu(f-y)}\mu \rangle \\
&= -n \int_{-\infty}^{\infty} \langle g(f - y)^{m-1}, e^{iu(f-y)}\mu \rangle (iu)^{n-1} \, du \qquad (6) \\
&= \cdots = (-1)^n n! \int_{-\infty}^{\infty} \langle g(f - y)^{m-n}, e^{iu(f-y)}\mu \rangle du \\
&= (-1)^n n! \langle g(f - y)^{m-n}, S_0 \rangle
\end{aligned}$$

If $n = m$, this is $(-1)^n n! \langle g, S_0 \rangle$, which does not vanish for all $g \in B$, since $\langle 1, S_0 \rangle \neq 0$, by (3). If $n < m$, then (6) equals

$$(-1)^n n! \int_{-\infty}^{\infty} d\langle g(f - y)^{m-n-1}, e^{iuf}\mu \rangle,$$

which vanishes for every $g \in B$. Condition (5) and the Theorem are proved. □

We shall now apply the Theorem with a quotient algebra of $A(T)$ in the role of B.

A *symmetric set* is one of the form

$$E = \left\{ \sum_{j=1}^{\infty} \varepsilon_j t(j) \colon \varepsilon_j = 0 \text{ or } 1 \right\},$$

where for the sake of neatness we require that $t(j) > 0$ and $\sum_{k>j} t(k) < t(j)$ for all j; and $\sum_{j=1}^{\infty} t(j) < 2\pi$.

3.3.2. Theorem. *Every symmetric set E has a subset that is not a set of synthesis for the algebra $A = A(T)$.*

Proof. The argument uses a Bernoulli convolution, such as we discuss in Section 6.6. Let μ be the probability measure with support E defined as the weak* limit of the convolution products

$$\mathop{\bigstar}_{j=1}^{k} \frac{1}{2}(\delta_0 + \delta_{t(j)}).$$

(If $\psi(\sum \varepsilon_j t(j)) = \sum \varepsilon_j 2^{-j}$, then $\mu \circ \psi^{-1}$ is Lebesgue measure on the interval $[0, 1]$; therefore μ is sometimes called the Lebesgue measure on E.) We shall find a real-valued $f \in A$ such that for some $c_o > 0$,

$$\|e^{iuf}\mu\|_{PM} = O(\exp(-c_o u^{1/2})) \quad \text{as } u \to \infty. \tag{7}$$

By 3.3.1, that more than suffices.

We may suppose without loss of generality that

$$\sum_{i=1}^{\infty} \left(\frac{t(j+1)}{t(j)} \right)^2 < \infty. \tag{8}$$

Then by Theorem 11.4.2, $A(E)$ is isomorphic as a Banach algebra to the tensor algebra $\hat{\bigotimes}_{j=1}^{\infty} C(\{0, 1\})$ by the obvious mapping. In particular, the sequence of functions on E,

$$\tilde{\varphi}_m\left(\sum \varepsilon_j t(j)\right) = \begin{cases} 0 & \text{if } (\varepsilon_{2m-1}, \varepsilon_{2m}) = (0, 0), (0, 1), \text{ or } (1, 0); \\ 1 & \text{if } (\varepsilon_{2m-1}, \varepsilon_{2m}) = (1, 1) \end{cases}$$

for $m = 1, 2, \ldots$, is bounded in $A(E)$. For each m we may let φ_m denote an extension of $\tilde{\varphi}_m$ to T such that $\varphi_m \in A$ and

$$\sup_m \|\varphi_m\|_A < \infty.$$

The measure μ may be represented as follows:

$$\mu = \text{weak}* \lim_{n\to\infty}\left(\underset{m=1}{\overset{n}{\bigstar}} \mu_m\right),$$

where

$$\mu_m = (\tfrac{1}{4})(\delta_0 + \delta_{t(2m-1)} + \delta_{t(2m)} + \delta_{(t(2m-1)+t(2m))})$$

Let $f = \sum_{m=1}^{\infty} m^{-2}\varphi_m$. It is an easy exercise to show that

$$e^{iuf}\mu = \text{weak}* \lim_{n\to\infty}\left(\underset{m=1}{\overset{n}{\bigstar}} \nu_m^u\right), \tag{9}$$

where

$$\nu_m^u = e^{iu\varphi_m/m^2}\mu_m = \tfrac{1}{4}(\delta_0 + \delta_{t(2m-1)} + \delta_{t(2m)} + e^{iu/m^2}\delta_{(t(2m-1)+t(2m))}).$$

For each $k \in Z$, $16|\hat{\nu}_m^u(k)|^2$ has the form $|1 + e^{ia} + e^{ib} + e^{i(a+b+c)}|^2$, where a and b depend on k but c does not. Expanding that expression and gathering terms together in an obvious way, we find that it is bounded by $8 + 4|1 + e^{ic}|$, which in turn is bounded by 12 provided $\cos c \leq -\tfrac{1}{2}$. Thus

$$\|\nu_m^u\|_{PM} \leq r = \frac{\sqrt{3}}{2} \quad \text{provided} \quad \cos\left(\frac{u}{m^2}\right) \leq \frac{1}{2}.$$

For each u sufficiently large, the quantity u/m^2 lies in the interval $[2\pi/3, 4\pi/3]$, and in particular $\cos(u/m^2) \leq -\tfrac{1}{2}$, for at least $c_o u^{1/2}$ values of m (for some fixed $c_o > 0$). Therefore for each u sufficiently large,

$$\|\nu_m^u\|_{PM} \leq r$$

for at least $c_o u^{1/2}$ values of m. By (9), and since $\|\nu_m^u\|_{PM} \leq 1$ for every m,

$$\|e^{iuf}\mu\|_{PM} \leq r^{c_o u^{1/2}} \quad \text{for all sufficiently large } u.$$

Condition (7) follows, and the Theorem is proved.

Explicitly, for each y satisfying (3), the set

$$E \cap f^{-1}(y) = \left\{\sum_{j=1}^{\infty} \varepsilon_j t(j) \colon \sum_{m=1}^{\infty} \frac{\varepsilon_{2m-1}\varepsilon_{2m}}{m^2} = y\right\}$$

is not a set of synthesis. It is easy now to argue that $E \cap f^{-1}(y)$ disobeys synthesis for every $y \in (0, \pi^2/6)$. $\square$

Remarks and Credits. The probabilistic proof of the failure of synthesis found in Kahane [2, Section V.6] gives the best results as far as Lipschitz conditions satisfied by an f in $I(E)\backslash I_0(E)$ are concerned, and shows in particular the sharpness of Theorem 3.2.11.

Saeki [15] gives an elegant proof that the sum of two Kronecker sets may disobey synthesis.

Saeki [16] shows that whenever $\mu, \nu \in M_c(G)$ and $\mu * \nu(E) \neq 0$, then E contains a set that disobeys synthesis.

The three references just cited, by Kahane and Saeki, are our primary recommendations for further reading.

Theorem 3.3.1 is due to Malliavin [1]. Theorem 3.3.2 is due to Kahane and Katznelson [2, Theorem 7], and the proof given here was introduced by Katznelson [1, Section VIII.7] to prove the failure of synthesis for the algebra $A(D)$, where D is the Cantor group (which, after all, looks like a symmetric set)! In spirit, however, the proof derives from Richards [1]. Nicely done variants on the procedure appear in Saeki [9,11].

The reader may wonder whether the algebras $A(D)$ and $A(E)$ (where E is a symmetric set satisfying (8)) are isomorphic as Banach algebras. The answer is that they are not. In fact, no group algebra is isomorphic to a restriction algebra of another, except in trivial cases (see Rosenthal [1, Corollary 1.10, p. 28]).

In Section 11.2, we show how the failure of synthesis for $R^n (n \geq 3)$ implies it for every non-discrete group, via the transfer techniques of Varopoulos. As Katznelson [1, Section VIII.10] explains, the implication remains true, and the "transfer" is easier, if one begins with D instead of R^n. The procedure is treated also in Kahane [2, Section VIII.7]. However, for obtaining the other ideal-structure phenomena of Section 11.2, there is as yet no substitute for the music of the spheres.

Malliavin [5] was first to prove that every M-set contains a set that disobeys synthesis. For a variant on his argument, see Filippi [1]. Kahane and Katznelson [2, Theorem 6] proved the same theorem with a somewhat weaker hypothesis. See Kahane [2, Section V.8] for a full discussion.

Kaufman [7] gives an alternative proof of Malliavin's theorem (that synthesis fails in $A(T)$), using gap series.

Kahane [1, Chapter XIII] gives another probabilistic treatment of Malliavin's Theorem; the results are not quite so sharp as those of the later Kahane [2], but the context is rich in probabilistic technique.

Chapter 4

Sets of Uniqueness, Sets of Multiplicity

4.1. Introduction

Since the action in this Chapter takes place on the circle group T, we write A for $A(T)$, PF for $PF(T)$, and so forth.

A closed set $E \subseteq T$ is:

a *U-set* if $PF(E) = \{0\}$;
an *M-set* if $PF(E) \neq \{0\}$;
a *U_1-set* if $PF \cap N(E) = \{0\}$;
an *M_1-set* if $PF \cap N(E) \neq \{0\}$;
a *U_0-set* if $PF \cap M(E) = \{0\}$;
an *M_0-set* if $PF \cap M(E) \neq \{0\}$.

A U-set is called also a set of uniqueness; an M-set, a set of multiplicity. By its definition, the space PF of pseudo-functions S is in one-to-one correspondence with the space of sequences $\hat{S} \in C_0(Z)$. A *trigonometric series* is one of the form $\sum_{n=-\infty}^{\infty} \hat{S}(n)e^{inx}$ where $S \in PF$. It is only a formal series, its actual pointwise convergence being in question. But it turns out (Theorem 4.2.1) that an open interval $I \subseteq T$ is disjoint from the support of a pseudo-function if and only if the corresponding series converges to zero for every $x \in I$. It follows that a set E is a set of uniqueness if and only if the trivial trigonometric series is the only one that converges to zero on the complement of E. It was in consideration of the latter condition that the term "set of uniqueness" was first used.

If E is finite or countably infinite, it is easy to show that E is a U-set; on the other hand, if E has positive measure, then it is not a U-set, since its characteristic function (times Haar measure) is a nonzero pseudofunction. Therefore the interest lies in studying the perfect sets of measure zero to determine which of them are U-sets, and which of them support nonzero pseudofunctions of various kinds.

In Section 4.2 we shall discuss properties of the support of a pseudo-measure and show that on T it has several interesting equivalent formulations.

Note that A is the dual space of PF and that $N(E) \cap PF$ is the annihilator of $I(E)$ in PF. Then it is easy to see that E is a U_1-set if and only if $I(E)$ is weak$*$ dense in A. A set E is a U_1'*-set* if $I(E)$ is weak$*$ sequentially dense in A. In Section 4.3 we shall discuss sets of these types, and show that to obtain the weak$*$ closure of $I(E)$ by repeatedly taking the weak$*$ limits of sequences is sometimes a slow process—in a sense to be specified.

We point out in passing that if every portion of E is an M_1-set, then $I(E)$ is the annihilator in A of $PF \cap N(E)$, and therefore $A(E)$ is the Banach space dual of $PF \cap N(E)$.

The inclusion $M(E) \subseteq N(E)$ is proper precisely when E is non-Helson, and the inclusion $N(E) \subseteq PM(E)$ is proper precisely when E disobeys synthesis (see Section 3.1). The fact that these inclusions are sometimes proper makes it plausible that the three classes of M_o-sets, M_1-sets, and M-sets are distinct, and that is indeed the case. In Section 4.4 we present Pyateckiĭ-Šapiro's remarkable construction of an M_1-set that is not an M_o-set. In Section 4.6 we produce an M-set that is not an M_1-set.

In Section 4.5 we prove among other results that every Helson set is a U_1'-set and hence a U_1-set; and that every Kronecker set is a set of synthesis and hence a U-set.

For years, the question of whether every Helson set is a set of synthesis, or at least a U-set, seemed intractable. Finally in 1971 Thomas Körner constructed an M-set E with $\alpha(E) = 1$. In Section 4.6 we present Robert Kaufman's later, short proof that for every M-set F there is an M-set $E \subseteq F$ such that $\alpha(E) = 1$. Such a set E is evidently not an M_1-set, since a Helson set is a U_1-set. Also, such an E is evidently not a set of synthesis. (The subject of harmonic synthesis is treated at length in Chapter 3.)

In Section 4.7, we construct an independent M_o-set.

Remarks and credits. The early papers on the subject include Cantor [1] (1872), Young [1] (1909), and Menshov [1] (1916). For more historical information, consult Bary [1, vol. 2, Chapter XIV] and Zygmund [1, Notes to Section IX.6].

This Chapter will neglect a number of interesting results in the subject, but we will mention some of them now and give references.

Within certain classes of sets (constructed in a certain regular fashion), there are metric "thinness" conditions which suffice to imply uniqueness, and which are known to be best possible. But there are no necessary metric conditions, because among the "thicker" sets the question of which ones are U-sets becomes a delicate matter, involving arithmetic properties, as the following sample result shows. For $2 < r < \infty$, let $E_r = \{\sum_{j=1}^{\infty} \varepsilon_j r^{-j} : \varepsilon_j = 0 \text{ or } 1\}$. Then E_r is a U-set if and only if r is a Pisot number.

For results that show relationships between metric conditions satisfied by sets E and the presence of nonzero pseudofunctions of various kinds in $PM(E)$, see Kahane and Salem [1, Chapters V and VIII], Kahane [12], Kaufman [9, 10, 13, 14], McGehee [1], and Körner [6]. Körner [8] is an improvement of McGehee [1] and corrects the erroneous Corollary A-2 therein. For the result about Pisot numbers, see Meyer [1, Section III.3] or the older and longer proof in the elegant monograph by Salem [1].

Note that if $S \in PM(T)$ and $\hat{S} \in l^2$, then the support of S must have positive Lebesgue measure. Ivašev-Musatov [3; see also 1, 2] showed that on the other hand, $\hat{S}$ may be arbitrarily close to being in l^2 and yet $m_T(\text{supp } S) = 0$. A nice exposition of that result appears in Körner [4]. Körner's formulation is as follows.

4.1.1. Theorem. *Let $\{\varphi(n)\}_{n=1}^{\infty}$ be a sequence of positive numbers such that these conditions hold.*

(A) $\sum \varphi(n)^2 = \infty$,
(B) *There exists $K > 1$ such that for all n, $K^{-1}\varphi(n) \leq \varphi(r) \leq K\varphi(n)$ whenever $n \leq r \leq 2n$.*
(C) $n\varphi(n) \to \infty$ *as* $n \to \infty$.

Then there exists a positive nonzero measure μ whose support has Lebesgue measure zero such that $\hat{\mu}(n) = O(\varphi(|n|))$ as $|n| \to \infty$.

In Körner [5] it is shown that conditions (B) and (C) can be relaxed somewhat but not dispensed with. In fact, there exists a decreasing positive convex sequence $\{\varphi(n)\}$ such that if $\mu \in M(T)$, $\mu \neq 0$, and $\hat{\mu}(n) = O(\varphi(|n|))$ as $|n| \to \infty$, then the support of μ must be the whole circle T! Körner [1] presents theorems like 4.1.1 in which thinness conditions on supp μ are obtained—in exchange for placing more restrictions on φ.

For $2 < q < \infty$, Gregory [1] studied the condition that $M(E)$ should contain a nonzero measure μ such that $\hat{\mu} \in l^q$. Katznelson [4] showed that for each $\varepsilon > 0$, there is a set E, whose complement has measure less than ε, such that $PM(E)$ contains no nonzero element μ such that $\hat{\mu} \in \bigcup_{p<2} l^p$. His work was generalized by de Michele and Soardi [4]. Kahane and Katznelson [4] proved that if $\varepsilon_n > 0$ and $\varepsilon_n \to 0$, then there exists a set V, whose complement has measure zero, such that if $\hat{S}(n) \leq \varepsilon_{|n|}$ for all n and $\sum \hat{S}(n)e^{int}$ converges to zero for all $t \notin V$, then $\hat{S}(n) = 0$ for all n. B. Connes [1] improves their work by making the complement have zero Hausdorff measure with respect to any given Hausdorff determining function.

4.2. The Support of a Pseudomeasure

A pseudomeasure S is *null on* an open set I if $\langle f, S\rangle = 0$ for every f in A with support contained in I. It is easy to prove that S is null on the union of the open sets on which S is null; the *support of* S is the complement of that union.

When $f \in A$ and $S \in PM$, the product fS is the pseudomeasure such that $(fS)\hat{}(m) = (\hat{f} * \hat{S})(m) = \langle \overline{\overline{f(t)}e^{imt}}, S\rangle$. Evidently S is null on I if and only if $fS = 0$ for every f in A with support contained in I.

Those statements about nullity and supports of pseudomeasures would remain true if A were replaced by a smaller, but dense, class of test functions, such as C^∞.

With each $S \in PM$ we associate a function $\mathscr{S}$, defined as follows. Let

$$\mathscr{S}^+(z) = \sum_{n=0}^{\infty} \overline{\hat{S}(n)} z^n \quad \text{for} \quad |z| < 1,$$

$$\mathscr{S}^-(z) = \sum_{n=-\infty}^{-1} \overline{\hat{S}(n)} z^n \quad \text{for} \quad |z| > 1.$$

The function $\mathscr{S}(z)$ is defined to equal $\mathscr{S}^+(z)$ for $|z| < 1$ and $\mathscr{S}^-(z)$ for $|z| > 1$. It is analytic on the set $D = \{z: |z| \neq 1\}$. If $x \in T$ and there is a planar neighborhood V of e^{ix} such that $\mathscr{S}$ extends analytically to $D \cup V$, then x is a *regular point for S*. If every point of an open set $I \subset T$ is a regular point for S, then S is *regular on I*. In the theorem below, we establish that "null" and "regular" are equivalent.

For $S \in PM$, let $F = F_S$ be the second formal integral of S, to wit,

$$F_S(t) = \tfrac{1}{2}\hat{S}(0)t^2 - \sum_{n \neq 0} n^{-2}\hat{S}(n)e^{int}.$$

4.2.1. Theorem about Support. *Consider an open interval $I \subset T$. If $S \in PM$, then the following three conditions are equivalent.*

(i) *S is null on I.*
(ii) *S is regular on I.*
(iii) *F_S is linear on I.*

Also, the following two conditions are equivalent.

(iv) $\sum_{n=-\infty}^{\infty} \hat{S}(n)e^{int} = 0$ *for all $t \in I$.*
(v) *$S \in PF$ and S is null on I.*

Proof. (i) ⇒ (iii). For $a, t \in T$ and $h > 0$, let $V_{a,h}(t) = \Delta_h(t - a)$, where Δ_h is as defined by A.5(3). In other words, let $V_{a,h}(t) = h^{-1}\max\{0, 1 - |(t - a)h^{-1}|\}$. Then

$$\hat{V}_{a,h}(n) = e^{-ina}\left[\frac{\sin(nh/2)}{nh/2}\right]^2,$$

$$V'_{a,h}(t) = \begin{cases} h^{-2} & \text{on } (a - h, a), \\ -h^{-2} & \text{on } (a, a + h), \end{cases}$$

$$dV'_{a,h} = h^{-2}\,\delta_{a-h} - 2h^{-2}\,\delta_a + h^{-2}\,\delta_{a+h}.$$

Note that the second derivative of $V_{a,h}$ is a discrete measure. The coefficient at n of this measure is $(-n^2)\hat{V}_{a,h}(n)$. Let $[a - h, a + h]$ be an arbitrary closed interval contained in I. It suffices to show that then $F(a + h) + F(a - h) - 2F(a) = 0$, where $F = F_S$. Now

$$\begin{aligned}\int F\,dV'_{a,h} &= h^{-2}[F(a+h) + F(a-h) - 2F(a)]\\ &= h^{-2}\left[\hat{S}(0)h^2 - \sum_{n\neq 0} n^{-2}\hat{S}(n)e^{ina}(2\cos nh - 1)\right]\\ &= \hat{S}(0) + \sum_{n\neq 0}(-n^{-2})\hat{S}(n)(-n^2)\hat{V}_{a,h}(n)\\ &= \langle V_{a,h}, S\rangle,\end{aligned}$$

which vanishes, because S is null on I.

(*iii*) ⇒ (*ii*). Let $F = F_S$, and let

$$F^+(w) = \tfrac{1}{2}\hat{S}(0)w^2 - \sum_{n=1}^{\infty} n^{-2}\hat{S}(n)e^{inw} \quad \text{for} \quad |e^{inw}| \le 1,$$

$$F^-(w) = \sum_{n=-1}^{-\infty} n^{-2}\hat{S}(n)e^{inw} \quad \text{for} \quad |e^{iw}| \ge 1.$$

Then for $t \in T$, $F(t) = F^+(t) - F^-(t)$, and by hypothesis, $F(t) = at + b$ for some a, b and all $t \in I$. Consider the analytic function $\mathscr{F}(w)$, defined for $e^{iw} \in D \cup e^{iI}$ as follows.

$$\mathscr{F}(w) = F^+(w) \quad \text{for} \quad |e^{iw}| \le 1,$$

$$\mathscr{F}(w) = F^-(w) + aw + b \quad \text{for} \quad |e^{iw}| \ge 1.$$

The second derivative of $\mathscr{F}$ on this region is the desired extension $\mathscr{S}(e^{iw})$.

(*ii*) ⇒ (*i*). For $u > 0$, let $\mathscr{S}_u(t) = \mathscr{S}^+(e^{-u+it}) - \mathscr{S}^-(e^{u+it})$. Then $\mathscr{S}_u$ is associated with the pseudomeasure S_u, where $\hat{S}_u(n) = \hat{S}(n)e^{-u|n|}$. Then for $f \in A$,

$$\langle f, S\rangle = \lim_{u\to 0}\langle f, S_u\rangle = \lim_{u\to 0}\int f(t)S_u(t)dt.$$

Since S is regular on I, S_u converges to zero uniformly on each compact subset of I. Therefore $\langle f, S\rangle = 0$ if the support of f is such a set.

(*iv*) ⇒ (*v*). Since $\sum \hat{S}(n)e^{int}$ means $\lim_{N\to\infty}\sum_{n=-N}^{N}\hat{S}(n)e^{int}$, and since $\hat{S}(n)e^{int} + \hat{S}(-n)e^{-int}$ may be written in the form

$$r_n\cos(nt + a_n) + is_n\sin(nt + b_n),$$

where $r_n \geq 0$ and $s_n \geq 0$, in order to show that condition d implies that $S \in PF$ it suffices to show that if

(1) $$\sum_{n=0}^{\infty} r_n \cos(nt + a_n) = 0 \quad \text{for} \quad t \in I,$$

then $r_n \to 0$. Otherwise there would be a subsequence bounded away from 0, say $r_{n_k} \geq r > 0$ for $k \geq 1$. We may suppose that $n_{k+1} > 10n_k$. Then there is a sequence of closed intervals I_k, such that $I \supset I_1 \supset \cdots \supset I_k \supset I_{k+1} \supset \cdots$ and $\cos(n_k t + a_k) > \frac{1}{2}$ for $t \in I_k$. At any point t in $\bigcap_{k=1}^{\infty} I_k$, the series (1) fails to converge, contrary to hypothesis.

To show that S is null on I, we shall use the following result.

4.2.2. Lemma. *Let F be continuous on an open interval I, and let $\Delta_2 F(t, h) = F(t + h) + F(t - h) - 2F(t)$. If $\lim_{h \to 0} h^{-2}\Delta_2 F(t, h) \geq 0$ for every $t \in I$, then F is convex on I.*

Proof. For $\varepsilon > 0$, let $F_\varepsilon(t) = F(t) + \varepsilon t^2$. Then $\overline{\lim}_{h \to 0} h^{-2}\Delta_2 F_\varepsilon(t, h) \geq 2\varepsilon$. If F_ε were not convex, there would be a number b such that the function $g(t) = F_\varepsilon(t) - bt$ has a local maximum at some $t_0 \in I$. It would follow that

$$0 \geq \overline{\lim}_{h \to 0} h^{-2}\Delta_2 g(t_0, h) = \overline{\lim}_{h \to 0} h^{-2}\Delta_2 F_\varepsilon(t_0, h) \geq 2\varepsilon,$$

which cannot be true. Therefore F_ε is convex for every $\varepsilon > 0$, so that F must also be convex. The Lemma is proved. □

To finish the proof that (iv) ⇒ (v), we shall apply the lemma to F and to $-F$, where $F = F_S$. In fact we shall show that

(2) $$\lim_{n \to 0} h^{-2}\Delta_2 F(t, h) = 0 \quad \text{for every} \quad t \in I.$$

It follows that both F and $-F$ are convex, so that F must be linear and S must be null on I.

Let $w(x) = -x^{-2} \sin x$, $w(0) = 1$, and fix $t \in I$. Let $Q_n(t) = \sum_{k=-n}^{n} \hat{S}(k)e^{ikt}$. Then

$$\begin{aligned}\frac{\Delta_2 F(t, 2h)}{4h^2} &= \sum_{n \neq 0} \hat{S}(n)w(nh)e^{int} + \hat{S}(0)\\ &= Q_0(t) + \sum_{n=1}^{\infty} [Q_n(t) - Q_{n-1}(t)]w(nh)\\ &= \sum_{n=0}^{\infty} [w(nh) - w((n+1)h)]Q_n(t).\end{aligned}$$

Note that

$$\sum_{n=0}^{\infty} |w(nh) - w((n+1)h)| = \sum_{n=0}^{\infty} \left| \int_{nh}^{(n+1)h} w'(x)dx \right| \leq \int_0^{\infty} |w'(x)|dx.$$

The latter integral has a finite value; call it C. Then

$$\left| \sum_{n \geq N} [w(nh) - w((n+1)h)]Q_n(t) \right| \leq C \sup_{n \geq N} |Q_n(t)|$$

which tends to zero as N becomes large; also, each summand tends to zero as h tends to zero. Condition (2) follows. The proof that (iv) $\Rightarrow$ (v) is complete.

(v) $\Rightarrow$ *(iv)*. It suffices to suppose that $0 \in I$ and show that $\sum_n \hat{S}(n) = 0$. We use the Dirichlet kernel:

$$D_n(t) = 1 + 2\sum_{j=1}^{n} \cos jt = (\sin nt)\left(\cot \frac{t}{2}\right) + \cos nt.$$

Then $\sum_{k=-n}^{n} \hat{S}(k) = \langle S, D_n \rangle$. If $h \in A$ and $h(t) = \cot(t/2)$ except for t in a neighborhood of 0 contained in I, then

$$\langle S, D_n \rangle = \langle S, h(t)\sin nt \rangle + \langle S, \cot nt \rangle = \langle hS, \sin nt \rangle + \langle S, \cos nt \rangle.$$

Since $S \in PF$, and therefore also $hS \in PF$, it follows that $\langle S, D_n \rangle$ tends to zero as $n \to \infty$.

The Theorem about Support is proved. □

Remark. The material of this section has long been well known, and may be found along with other results as well as historical and bibliographical data in Bary [1, vol. 2, Chapter XIV] and in Zygmund [1, vol. 1, Chapter IX]. Other treatments appear in Kahane and Salem [1, Appendices I–IV] and Benedetto [1, Chapter 3].

4.3. The Weak * Closure of $I(E)$

Let us recall some facts from the theory of Banach spaces and weak topologies thereon. Let Y be a linear subspace of the dual space B^* of a Banach space B. Let $R = \sup\{\|x\|_B : x$ is in the Y closure of the unit ball of $B\}$, $r = \sup\{a$: the B closure of the unit ball in Y contains the ball of radius a in $B^*\}$,

$$s = \inf\left\{\frac{\sup\{|f(x)|/\|f\|_{B^*} : f \in Y, f \neq 0\}}{\|x\|_B} : x \in B, x \neq 0\right\},$$

and

$$t = \inf\left\{\frac{\|x + z\|_{B^{**}}}{\|x\|_{B^{**}}} : x \in iB, x \neq 0, z \in Y^{\perp}\right\},$$

where i is the natural inclusion map, $i: B \to B^{**}$. Note that $r, s, t \in [0, 1]$ and $1 \leq R \leq \infty$.

4.3.1. Theorem. *Let B, Y, R, r, s, and t be as above. Then $R^{-1} = r = s = t$. If B is separable, then $R < \infty$ if and only if Y is weak∗ sequentially dense in B^*.*

For a proof, see Dixmier [1]. (As it affects separable B, the Theorem is implicit in Banach [1, Annexe].) We call the common value of the four quantities R^{-1}, r, s, and t the *weak∗ density number of Y in B*.

If Y is weak∗ dense in B^*, then $iB \cap Y^{\perp} = \{0\}$ (in B^{**}); therefore there are well-defined projections

(1) $$p: iB + Y^{\perp} \to iB \quad \text{and} \quad q: iB + Y^{\perp} \to Y^{\perp}.$$

If p is unbounded, then $t = 0$; otherwise, $t = \|p\|^{-1}$.

When Y is a linear subspace of B^*, let Y_1 denote the set of all weak∗ limits in B^* of sequences in Y. Define Y_a, for every ordinal number a, by the inductive rule

$$Y_a = \left(\bigcup_{b<a} Y_b\right)_1.$$

The *order* of Y is the first ordinal number a such that Y_a is weak∗ closed. It is not difficult to show that the order of Y never exceeds the first uncountable ordinal; and that if B is separable and $Y \subset B^*$, then the order of Y is a countable ordinal.

Now let us consider those principles in the case $B = PF$, $Y = I(E)$. Evidently E is a U_1-set if and only if $I(E)$ is weak∗ dense in A, and by definition, E is a U'_1-set precisely when $I(E)$ is weak∗ sequentially dense in A. In this section, we shall prove the following theorems.

4.3.2. Theorem. *Let $s(E)$ be the weak∗ density number of $I(E)$ in A. Let*

$$\eta(E) = \inf\left\{\frac{\limsup_{|n|\to\infty} |\hat{S}(n)|}{\|S\|_{PM}} : S \in N(E),\ S \neq 0\right\}.$$

Then $s(E) = \eta(E)/(1 + \eta(E))$.

4.3.3. Theorem. *There exists a U_1-set that is not a U'_1-set.*

4.3.4. Theorem. *Every U_1-set is a countable union of U'_1-sets.*

4.3.5. Theorem. *For every countable ordinal a, there is a set $E \subset T$ such that the order of $I(E)$ in A is a.*

Proof of 4.3.2. If E is an M_1-set, then evidently both $\eta(E)$ and $s(E)$ are zero. Otherwise, we know that $I(E)$ is weak∗ dense in A, so that the projections p

and q, as in (1), are well-defined—where $B = PF$ and $Y = I(E)$. Recall that $N(E) = I(E)^{\perp}$. Let $\eta = \eta(E)$. It suffices to show that

$$\|p\| = \frac{1 + \eta}{\eta}. \tag{2}$$

Let $\varepsilon > 0$. Then there exists $U \in I(E)^{\perp}$ and an integer m_0 such that

$$\sup_{|n| > m_0} |\hat{U}(n)| < (\eta + \varepsilon)\sup_n |\hat{U}(n)|.$$

For $m > m_0$, let $S_m \in PF$ be defined by

$$\hat{S}_m(n) = \begin{cases} -(1 + \eta + \varepsilon)\hat{U}(n) & \text{for} \quad |n| \le m \\ 0 & \text{for} \quad |n| > m. \end{cases}$$

Then we see that

$$\|p\| \ge \frac{\|S_m\|_{PM}}{\|S_m + U\|_{PM}} \ge \frac{(1 + \eta + \varepsilon)\sup_{|n| \le m} |\hat{U}(n)|}{(\eta + \varepsilon)\sup_n |\hat{U}(n)|}.$$

Since m may be taken arbitrarily large, and ε arbitrarily small, it follows that $\|p\|$ is at least $(1 + \eta)/\eta$. By a similar argument, $\|q\|$ is at least η^{-1}. But whenever $S \in PF$ and $U \in I(E)^{\perp}$,

$$\frac{\|U\|_{PM}}{\|S + U\|_{PM}} \le \frac{\|U\|_{PM}}{\limsup_{|n| \to \infty} |\hat{U}(n)|} \le \frac{1}{\eta},$$

so that $\|q\| = \eta^{-1}$ and hence $\|p\| \le 1 + \eta^{-1}$; (2) follows, and Theorem 4.3.2 is proved. □

We shall use the following lemma in the construction of a U_1-set that is not a U_1'-set.

4.3.6. Lemma. *For $N = 3, 4, \ldots$, let E_N be the closed, perfect set consisting of all the points $2\pi x$ such that $x \in [0, 1]$ and the N-ary expansion of x requires no 1's:*

$$E_N = \left\{2\pi \sum_{j=1}^{\infty} \varepsilon_j N^{-j} \colon \varepsilon_j = 0, 2, 3, \ldots, N - 2, \text{or } N - 1\right\}.$$

Then $\eta(E_N) > 0$, but $\lim_{N \to \infty} \eta(E_N) = 0$.

Proof. Let g be a function in A, vanishing outside the interval $(2\pi N^{-1}, 4\pi N^{-1})$, such that $g \ge 0$ and $\hat{g}(0) = 1$. Let $g_k(x) = g(N^k x)$ for $k \ge 0$. Then

$\|g_k\|_A = \|g\|_A$, $g_k \in I_0(E_N)$, and $\hat{g}_k(0) = 1$ for every k. Furthermore, $\hat{g}_k$ vanishes except on multiples of N^k, and $\hat{g}_k(pN^k) = \hat{g}(p)$. Since $g_k \to 1$ weak$*$, E_N is a U'_1-set, and hence by Theorems 4.3.1 and 4.3.2, $\eta(E_N) > 0$.

(We may argue more directly. For $S \in PM(E_N)$, and every $k \geq 0$,

$$0 = \langle g_k, S \rangle = \hat{S}(0) + \sum_{p \neq 0} \hat{g}(p)\hat{S}(pN^k).$$

It follows that

$$\eta(E_N) \geq \left(\sum_{p \neq 0} |\hat{g}(p)| \right)^{-1},$$

a positive number which depends on the choice of g, which in turn is affected by the size of N. That argument is a standard use of the fact that E_N is an "H-set"; see Kahane and Salem [1, Proof of Theorem III, p. 58] and Salem [1, pp. 49–52].)

It remains to show that $\eta(E_N) \to 0$. It suffices to show that

(3) $$\frac{\limsup_{|t| \to \infty} |\hat{\mu}_N(t)|}{\|\mu_N\|_{PM}} = 0(N^{-1}) \quad \text{as} \quad N \to \infty,$$

where μ_N is the *Lebesgue measure on* E_N, defined as follows. Fix N. For $k \geq 1$, let λ_k be the measure that assigns mass $1/(N-1)$ to each of the $N-1$ points $\{2\pi j N^{-k}: j = 0, 2, 3, \ldots, N-1\}$. Then $\|\lambda_k\|_{PM} = \hat{\lambda}_k(0) = \|\lambda_k\|_M = 1$, and $\hat{\lambda}_k(t) = \hat{\lambda}_1(tN^{-k})$. For $n \geq 1$, the support of the measure $\mu_{N,n} = \lambda_1 * \lambda_2 * \cdots * \lambda_n$ is the set

$$\left\{ \sum_{j=1}^{n} \varepsilon_j N^{-j} \colon \varepsilon_j = 0, 2, 3, \ldots, \text{or } N-1 \right\}.$$

The sequence $\{\mu_{N,n}\}_{n=1}^{\infty}$ converges weak$*$ to a measure μ, which is μ_N. It is a positive measure in $M(E_N)$, with $\|\mu\|_M = \|\mu\|_{PM} = \hat{\mu}(0) = 1$. Furthermore,

$$\hat{\mu}(t) = \prod_{k=1}^{\infty} \hat{\lambda}_k(t) = \prod_{k=1}^{n} \hat{\lambda}_1(tN^{-k}) = \hat{\lambda}_1(t)\hat{\mu}(tN^{-1}) = \hat{\lambda}_1(t)\hat{\lambda}_2(t)\hat{\mu}(tN^{-2}),$$

and

$$\hat{\mu}(-t) = \overline{\hat{\mu}(t)}.$$

Also, therefore,

$$|\hat{\mu}(t)| \leq |\hat{\lambda}_1(t)\hat{\lambda}_2(t)|$$

and

$$|\hat{\mu}(t)| \leq |\hat{\mu}(tN^{-1})|.$$

Therefore we can prove (3) by showing that there exist c and $a > 0$ such that $|\hat{\lambda}_1(t)\hat{\lambda}_2(t)| \le cN^{-1}$ for $a \le t \le Na$, where c is independent of N. Now

$$\hat{\lambda}_1(t)\hat{\lambda}_2(t) = \frac{1}{(N-1)^2} \prod_{k=1}^{2} \left(\left(\sum_{j=0}^{N-1} e^{-ijt/N^k} \right) - e^{-t/N^k} \right)$$

$$= \frac{1}{(N-1)^2} \prod_{k=1}^{2} \left(\frac{1 - e^{-it/N^{k-1}}}{1 - e^{-it/N^k}} - e^{-it/N^k} \right).$$

The product of the two factors may be written as the sum of four summands, one of which is

$$\frac{1 - e^{-it}}{1 - e^{-it/N^2}},$$

which on the interval $[\pi N, \pi N^2]$ is bounded by N. The sum of the other three is bounded by $2N + 1$. Therefore we have what we need, with $a = \pi$ and $c < 12$. The Lemma is proved. □

Proof of 4.3.3. Let $F_N = \{rx + s : x \in E_N\}$, where E_N is the set we just constructed, and where r and s are positive real numbers chosen so that $F_N \subset (1/(N+1), 1/N)$. Then $\eta(F_N) = \eta(E_N)$. Let $F = \{0\} \cup \bigcup_{N=3}^{\infty} F_N$. Then $\eta(F) = 0$, so that F is not a U_1'-set. And F is a U_1-set, because otherwise there would be a nonzero element $S \in PF \cap N(F)$. Let $f_N \in A$ such that f_N agrees with χ_{F_N} on F. Then $f_N S \in PF \cap N(F_N)$, and it must be nonzero for some N—but this contradicts the fact that every F_N is a U_1-set. Theorem 4.3.3 is proved. □

There is a subtlety in the argument just given: How does one know that $f_N S$ belongs to $N(F_N)$, and not merely that it is in the larger class $PM(F)$? One may prove easily that $f_N S$ vanishes when restricted to $F \setminus F_N$ because f_N vanishes thereon and S obeys synthesis; and that $f_N S \in N(F)$ because $f_N \in A$ and $S \in N(F)$. However, one may handle the question better by noting that F obeys synthesis. For by the Herz criterion (Theorem 3.2.1), E_N obeys synthesis, and therefore so does F_N. The remaining step is easy, but we shall now give the details of it anyway.

4.3.7. Lemma. *If F_N is a set of synthesis contained in the interval $(1/(N+1), 1/N)$, then the set $F = \{0\} \cup \bigcup_{N=1}^{\infty} F_N$ is also a set of synthesis.*

Proof. We must show that if $f \in I(F)$ and $\varepsilon > 0$, then there exists $g \in J(F)$ such that $\|f - g\|_A < \varepsilon$. For each $h > 0$, we define V_h on $[-\pi, \pi]$ to equal $\hat{K}_h$ as defined by A.5(1):

$$V_h(x) = \max(0, 1 - h^{-1}|x|).$$

Since $f(0) = 0$, we may choose $h > 0$ small enough so that $\|fV_h\|_A < \varepsilon/2$ (because a singleton obeys synthesis—see A.3). Let $f_0 = f(1 - V_h)$, and pick M so that $f_0(x) = 0$ for $|x| \le 1/(M+1)$. For $1 \le N \le M$, choose $h_N \in J(F \backslash F_N)$ so that $\sum_{N=1}^{M} h_N = 1$ for $1/(M+1) \le |x| \le \pi$. Thus $f_0 = \sum_{N=1}^{M} h_N f_0$. Since each F_N is a set of synthesis, there exist $f_N \in J(F_N)$ such that

$$\|f_0 - f_N\|_A \le \frac{\varepsilon}{2M\|h_N\|_A}.$$

Let $g = \sum_{N=1}^{M} f_N h_N$. Then $g \in J(F)$ and

$$\|f - g\|_A \le \|f - f_0\|_A + \left\| \sum_{N=1}^{M} h_N(f_0 - f_N) \right\|_A < \varepsilon.$$

Lemma 4.3.7 is proved. □

Proof of 4.3.4. If a is a limit ordinal, and J is an ideal, then the symbol J_{a-1} has no meaning; let us define it, in this case, and only for the duration of this proof, to mean the ideal $\bigcup_{b<a} J_b$. Note that if $J_{a-1} = A$, then there exists $d < a$ such that $1 \in J_d$ and hence $J_d = A$.

Let E be a U_1-set. We shall prove that E is a countable union of U_1'-sets by induction on the order of $I(E)$ in A. If the order is 1, then E itself is a U_1'-set. Assume now that the order of $I(E)$ is $a > 1$, and that the desired conclusion holds for all sets F such that the order of $I(F)$ is less than a. Let $I = I(E)$. Since I_{a-1} is weak$*$ sequentially dense in A, $h(I_{a-1})$ is a U_1'-set, contained in E. It suffices now to show that if F is a closed subset of $E \backslash h(I_{a-1})$, then the order d of $I(F)$ is less than a. Let K denote $I(F)$.

$$I \subset K;$$

so

$$h(K_{a-1}) \subset h(I_{a-1});$$

but

$$h(K_{a-1}) \subset F \subset E \backslash h(I_{a-1});$$

so

$$h(K_{a-1}) = \varnothing \quad \text{and} \quad K_{a-1} = A.$$

Therefore $d < a$. Theorem 4.3.4 is proved. □

4.3.8. Lemma. *If $x \in E$, $\varepsilon > 0$, and a is an ordinal number, then $x \in h(I(E)_a)$ if and only if $x \in h[I(E \cap (x - \varepsilon, x + \varepsilon))_a]$.*

Proof. Let g be a function in A that equals 1 at x and vanishes off $(x - \varepsilon, x + \varepsilon)$. Evidently $fg \in I(E)$ if $f \in I(E \cap (x - \varepsilon, x + \varepsilon))$. It is easy to prove by induction that

$$f \in I(E \cap (x - \varepsilon, x + \varepsilon))_a \Rightarrow fg \in I(E)_a,$$

and hence that if $x \notin h[I(E \cap (x - \varepsilon, x + \varepsilon))_a]$, then $x \notin h(I(E)_a)$. The converse is obvious. The Lemma is proved. □

4.3.9. Lemma. *If K is an ideal whose hull is a set of synthesis F, then $K_1 = I(F)_1$.*

Proof. If $f \in I(F)_1$, then there exist $f_n \in I(F)$ such that $f_n \to f$ weak∗. Since F obeys synthesis, there exist $g_n \in J(F) \subset K$ such that $\|f_n - g_n\|_A \to 0$. Therefore $g_n \to f$ weak∗, so that $f \in K_1$. Therefore $K_1 = I(F)_1$, and the Lemma is proved. □

Proof of 4.3.5. Since a singleton $\{x\}$ is a set of synthesis, and $\eta(\{x\}) = 1$, every ideal whose hull is a singleton has order 1.

We proceed by induction, considering first the case when a is a limit ordinal. Let b be a one-to-one map of the positive integers onto $\{d: d < a\}$. Let $E = \{0\} \cup \bigcup_{N=1}^{\infty} H_n$, where $I(H_n)$ has order $b(n)$ and $H_n \subset (1/(n + 1), 1/n)$. Using 4.3.8, we find that $h(\bigcup_{d<a} I(E)_d) = \{0\}$. Therefore the order of $I(E)$ is a.

Now consider the case $a = 2$. Let F be defined as in the proof of Theorem 4.3.3. Then $h(I(F)_1) = \{0\}$, $I(F)_2 = A$, and $I(F)$ has order 2.

To deal with the case of an ordinal number $a > 2$ that has a predecessor, we shall make further use of the set F. It is a set of synthesis, by Lemma 4.3.7. It obviously contains a countable dense subset F_0 such that for every $x \in F_0$ there is a nonempty interval $(x, y_x]$ that does not intersect F. For each $x \in F_0$, let G_x be a set such that $G_x \subset [x, y_x]$, the order of $I(G_x)$ is $a - 1$, and

$$h\left(\bigcup_{b<a-1} I(G_x)_b\right) = \{x\}.$$

Let $E = F \cup \bigcup_{x \in F_0} G_x$. Clearly $h(\bigcup_{b<a-1} I(E)_b)$ contains F, and by 4.3.8 it equals F, and therefore since F is a set of synthesis, $I(E)_{a-1} = I(F)_1$ by 4.3.9. Therefore $h(I(E)_{a-1}) = \{0\}$. It follows that the order of $I(E)$ is a. Theorem 4.3.5 is proved. □

Credits. Theorems 4.3.3 and 4.3.4 are due to Pyateckiĭ-Šapiro [1]. Theorems 4.3.2 and 4.3.5 are due to McGehee [3]. Theorem 4.3.5 and its proof provide what seems to be the only known way to exhibit subspaces of l^1 of orders beyond the first infinite ordinal number. Such subspaces of l^∞ were constructed by Sarason [1, 2].

4.4. An M_1-Set That Is Not an M_0-Set

4.4.1. Lemma. *If $\rho \in M \cap PF$ and σ is a measure that is absolutely continuous with respect to ρ, then $\sigma \in PF$. In particular, if $\rho \in M \cap PF$ then $|\rho| \in M \cap PF$.*

Proof. For every $\varepsilon > 0$ there is a trigonometric $p(x) = \sum c_j e^{in_j x}$ such that $\|\sigma - p\rho\|_M < \varepsilon$. Then $(p\rho)\hat{}(n) = \sum c_j \hat{\rho}(n - n_j) \to 0$ as $n \to \infty$, so

$$\limsup_{n \to \infty} |\hat{\sigma}(n)| \leq \varepsilon.$$

It follows that $\hat{\sigma}(n) \to 0$ as $|n| \to \infty$. □

4.4.2. Theorem (Pyateckiĭ-Šapiro [1]). *Let $0.\varepsilon_1\varepsilon_2\ldots$ denote the binary representation of a real number in $[0, 1)$ and let $0 < r < \frac{1}{2}$.*

The closed, perfect set

$$\Upsilon = \Upsilon_r = \left\{0.\varepsilon_1\varepsilon_2\ldots : \sum_{k=1}^{n} \varepsilon_k \leq rn \text{ for } n = 1, 2, \ldots\right\}$$

is an M_1-set, but is not an M_o-set.

Proof. In this section, we shall consider the circle group T as the real numbers modulo one, and modify other conventions accordingly.

If Υ were an M_o-set, then by 4.4.1 it would support a nonzero measure μ that is also a pseudofunction. To see that that cannot be the case, consider a function $f \in A$ such that

$$f(x) = 1 \quad \text{for} \quad 0 \leq x \leq \tfrac{1}{2};$$

$$-c \leq f(x) \leq 1 \quad \text{for all } x, \text{ where } \quad c < \frac{1 - r}{r};$$

and

$$\hat{f}(0) = 0.$$

Let $f_n(x) = n^{-1} \sum_{k=0}^{n-1} f(2^k x)$ for $n = 1, 2, \ldots$. Then

$$f_n(x) \geq 1 - r - cr > 0 \quad \text{for } x \in \Upsilon, \text{ for all } n.$$

Since $\|f_n\|_A = \|f\|_A$, and $\hat{f}_n(k) \to 0$ as $n \to \infty$ for every k, the sequence $\{f_n\}$ converges to 0 in the PF topology of A; in particular, $\int f_n \, d\mu \to 0$. But μ is a positive measure, so that $\int f_n \, d\mu \geq (1 - r - cr)\mu(\Upsilon) > 0$ for all n—a contradiction. So Υ is not an M_o-set.

It remains to show that Υ is not a U_1-set. Here is an outline of the proof. First we shall show that

(i) If Υ is a U_1-set, then Υ is a U_1'-set.

It follows that there is a sequence $\{f_m\} \subset I(\Upsilon)$ such that $f_m \to 1$ in the PF-topology of A. To reach a contradiction, we consider certain measures, as follows.

For $N = 0, 1, 2, \ldots$ and $-1 < t < 1$, with

$$0 < p = \frac{1+t}{2} < 1 \quad \text{and} \quad q = 1 - p,$$

let $\mu_{N,t}$ be the measure supported by the interval $[0, 2^{-N}]$ such that for $n = 1, 2, \ldots$ and $k = 0, 1, \ldots, 2^n - 1$,

$$\mu_{N,t}\left[\frac{k}{2^{N+n}}, \frac{k+1}{2^{N+n}}\right] = p^{n-s}q^s,$$

where

(1)
$$\frac{k}{2^n} = 0.\varepsilon_1 \ldots \varepsilon_n \quad \text{and} \quad s = \sum_{k=1}^{n} \varepsilon_k.$$

The total variation of $\mu_{N,t}$ is always 1, since

$$1 = (p+q)^n = \sum_{s=0}^{n} \binom{n}{s} p^{n-s}q^s.$$

We shall show that

(ii) for every $r' < r$, $\lim_{N\to\infty} \mu_{N,t}(\Upsilon_r) = 1$ uniformly for $t \in [1 - 2r', 1]$ (that is, for $p \in [1 - r', 1]$).

The measures $\mu_{N,0}$ are absolutely continuous with respect to Lebesgue measure and hence belong to PF, so that

(2)
$$\lim_{m\to\infty} \int f_m \, d\mu_{N,0} = \hat{\mu}_{N,0}(0) = 1 \quad \text{for each } N.$$

Note that $\mu_{N,t}$ may be equivalently defined as the weak $*$ limit, as $n \to \infty$, of the discrete measure

$$\mu_{N,t,n} = \sum_{\varepsilon} p^{n-s}q^s\delta_{k2^{-N-n}},$$

where $\varepsilon = (\varepsilon_1, \ldots, \varepsilon_n)$ ranges over 2^n values, with each $\varepsilon_k = 0$ or 1; and where s and k are related to ε as in (1).

Let $e(x) = e^{2\pi i x}$. For all real y,

$$\begin{aligned}\hat{\mu}_{N,t}(-y) &= \int e(yx)d\mu_{N,t}(x) = \lim_{n\to\infty} \hat{\mu}_{N,t,n}(-y)\\ &= \lim_{n\to\infty} \sum_{\varepsilon} e\left(y2^{-N}\sum_{k=1}^{n}\varepsilon_k 2^{-k}\right)p^{n-s}q^s\\ &= \prod_{k=1}^{\infty}(p + qe(y2^{-k-N}))\\ &= \prod_{k=1}^{\infty}\left(1 + \frac{t-1}{2}(1 - e(y2^{-k-N}))\right).\end{aligned}$$

Let $C_y(t) = \hat{\mu}_{N,t}(-y)$. We shall prove that

(iii) there is a planar region D containing the segment $(-1, 1)$, and a uniform bound Q' for the functions C_y on D:

$$|C_y(t)| \le Q' \quad \text{for } y \in R, t \in D.$$

Let $Q = \sup_m \|f_m\|_A$. Consider the entire functions

$$\begin{aligned}F_{m,N}(t) &= \int f_m(x)d\mu_{N,t}(x)\\ &= \sum_j \hat{f}_m(j)\hat{\mu}_{N,t}(j).\end{aligned}$$

Those functions are all bounded by QQ' on D, so that their restrictions to D make up a normal family. For $j = 1, 2, \ldots$, first pick N_j so that

$$|F_{m,N_j}(t)| < \frac{1}{j} \quad \text{for } t \in [1 - 2r', 1), \text{ for all } m.$$

That choice is possible because of (ii), and because $f_m \in I(\Upsilon)$. Then pick m_j such that

$$|F_{m_j,N_j}(0)| > \tfrac{1}{2}.$$

That choice is possible by (2). The sequence $\{F_{m_j,N_j}\}_j$ must have a subsequence that converges, uniformly on each compact subset of D, to some analytic function. But that is impossible, because no analytic function can be zero on $[1 - 2r', 0)$ and nonzero at 0.

The outline of the Theorem's proof is complete. It remains to prove (i), (ii), and (iii).

Proof of (i). If Υ is a U_1-set, then by Theorem 4.3.4 and the Baire category theorem, some portion of Υ must be a U_1'-set. That is, there is an interval J such that $J \cap \Upsilon$ is a non-empty U_1'-set. Let $x = 0.\varepsilon_1\varepsilon_2\ldots$ be a point in $J \cap \Upsilon$. For sufficiently large N, both $x_N = 0.\varepsilon_1\varepsilon_2\ldots\varepsilon_N$ and $x_N + 2^{-N}$ are in $J \cap \Upsilon$. Let

$$\Upsilon' = \{0.\varepsilon_1\ldots\varepsilon_N\delta_1\delta_2\ldots: 0.\delta_1\delta_2\ldots\varepsilon_N \in \Upsilon\}.$$

Then $\Upsilon' \subset J \cap \Upsilon$, so Υ' is a U_1'-set. Since dilations and translations preserve U_1'-sets, and since the mapping $x \to 2^N(x - x_N)$ takes Υ' onto Υ, Υ is a U_1'-set. Statement (i) is proved.

Proof of (ii). The statement is an elementary fact from probability theory; see Feller [1, Section VI.4]. But here is a direct proof.

Let $B_N = [0, 2^{-N}]\setminus\Upsilon$; then $B_N = \bigcup_{n=1}^{\infty} B_{N,n}$, where

$$B_{N,n} = \left\{2^{-N}\cdot(0.\varepsilon_1\varepsilon_2\ldots): \sum_{k=1}^{n} \varepsilon_k > r(n+N)\right\}.$$

Then $\mu_{N,t}(B_{N,n}) = \sum\{\binom{n}{s}p^{n-s}q^s: s > r(n+N)\}$. Calculus techniques show that the quantity $p^{1-r}q^r$, where $p > 0$ and $p + q = 1$, attains its maximum value when $p = 1 - r$, and only then. Let $p_1 = 1 - r$, $q_1 = r$,

$$\beta = \frac{p^{1-r}q^r}{p_1^{1-r}q_1^r}.$$

If $p > p_1$ (that is, $t > 1 - 2r$), then $\beta < 1$ and we can make the following estimates:

$$\sum_{s>r(n+N)} \binom{n}{s} p^{n-s}q^s = \sum_{s>r(n+N)} \binom{n}{s} (p^{1-r}q^r)^{N+n} p^{r(n+N)-s-N} q^{s-r(n+N)}$$

(now observe that $r(n + N) - s - N < 0$ and $s - r(n + N) > 0$)

$$\leq \left[\sum_{s>r(n+N)} \binom{n}{s} (p_1^{1-r}q_1^r)^{N+n} p_1^{r(n+N)-s-N} q_1^{s-r(n+N)}\right]\beta^{N+n}$$

$$= \left[\sum_{s>r(n+N)} \binom{n}{s} p_1^{n-s}q_1^s\right]\beta^{N+n} \leq \beta^{N+n}.$$

Therefore $\mu_{N,t}(T\setminus\Upsilon) \leq \sum_{n=0}^{\infty} \mu_{N,t}(B_{N,n}) \leq \beta^N/(1-\beta) \to 0$ as $N \to \infty$ for every $t \in (1 - 2r, 1)$, and the limit is uniform for $t \in [1 - 2r', 1)$ if $r' < r$. Statement (ii) is proved.

Proof of (iii). We need to show that the entire functions

$$\varphi_\lambda(t) = C_{\lambda 2^{-N}}(t) = \prod_{k=1}^{\infty}\left[1 + \frac{t-1}{2}(1 - e(\lambda 2^{-k}))\right]$$

are uniformly bounded on the region

$$D = \{t = x + iy : 1 - |t|^2 > 4\pi|y|\}.$$

We have

$$|\varphi_\lambda(t)| = \prod_{k=1}^{\infty} |\cos\theta_k - it\sin\theta_k|,$$

where $\theta_k = \pi\lambda 2^{-k}$.

$$\begin{aligned}|\cos\theta - it\sin\theta|^2 &= |\cos\theta + y\sin\theta - ix\sin\theta|^2\\ &= 1 - (1 - |t|^2)\sin^2\theta + y\sin 2\theta\\ &\le \exp(y\sin 2\theta - (1 - |t|^2)\sin^2\theta).\end{aligned}$$

Therefore

$$|\varphi_\lambda(t)|^2 \le \exp\left[y\sum_{k=1}^{\infty}\sin 2\pi\lambda 2^{-k} - (1 - |t|^2)\sum_{k=1}^{\infty}\sin^2\pi\lambda 2^{-k}\right].$$

It suffices to show that

$$(3)\quad \sum_{k=1}^{\infty}\sin 2\pi\lambda 2^{-k} \le 4\pi\sum_{k=1}^{\infty}\sin^2\pi\lambda 2^{-k} + \frac{\pi}{4} \quad \text{for all } \lambda \text{ and for all } t \in D,$$

for then it follows that

$$\begin{aligned}|\varphi_\lambda(t)|^2 &\le \exp\left[(4\pi y - (1 - |t|^2))\sum_{k=1}^{\infty}\sin^2\pi\lambda 2^{-k} + \frac{y\pi}{4}\right]\\ &\le \exp(\pi y/4) < e^{1/16}.\end{aligned}$$

Fix λ. For each k, $\lambda 2^{-k}$ equals an integer plus a fractional part $\{\lambda 2^{-k}\}$, where $0 \le \{\lambda 2^{-k}\} < 1$. Let $[k_1, n_1], [k_2, n_2], \ldots$ be the integer-intervals such that

$$\{\lambda 2^{-k}\} < \tfrac{1}{2} \quad \text{for} \quad k_s \le k \le n_s.$$

For each of the k_s except perhaps k_1 (which may be 1), we have

$$\alpha_s = \{\lambda 2^{-k_s}\} \ge \tfrac{1}{4}.$$

since $\{\lambda 2^{-k_s+1}\} > \frac{1}{2}$, and hence

$$4\pi\alpha_s \leq 4\pi \cdot 4\alpha_s^2.$$

In the case of k_1, even if $\alpha_1 < \frac{1}{4}$,

$$4\pi\alpha_1 \leq 4\pi \cdot 4\alpha_1^2 + \pi/4.$$

$$\sum_{k=k_s}^{n_s} \sin^2 \pi\lambda 2^{-k} \geq \sin^2 \pi\alpha_s \geq 4\alpha_s^2;$$

and

$$\sum_{k=k_s}^{n_s} \sin 2\pi\lambda 2^{-k} = \sum_{m=0}^{n_s-k_s} \sin 2\pi\alpha_s 2^{-m}$$
$$\leq 2\pi\alpha_s \sum_{m=0}^{\infty} 2^{-m} = 4\pi\alpha_s.$$

Therefore

$$\sum_s \sum_{k_s}^{n_s} \sin 2\pi\lambda 2^{-k} \leq \sum_s 4\pi\alpha_s \leq \sum_s 4\pi \cdot 4\alpha_s^2 + \frac{\pi}{4}$$
$$\leq 4\pi \sum_s \sum_{k_s}^{n_s} \sin^2 \pi\lambda 2^{-k} + \frac{\pi}{4}.$$

Then (3) follows, because if k is not in one of the intervals $[k_s, n_s]$, then $\{\lambda 2^{-k}\} > \frac{1}{2}$, so that $\sin 2\pi\lambda 2^{-k} \leq 0 \leq \sin^2 \pi\lambda 2^{-k}$.

Statement (iii) is proved. The proof of the theorem is complete. □

Remark. The theorem is true and the proof works in a variety of settings. For example, the subset

$$\left\{\sum_{j=1}^{\infty} \varepsilon_j 3^{-j} \colon \varepsilon_j = 0 \text{ or } 1 \quad \text{and} \quad \sum_{j=1}^{n} \varepsilon_j \leq rn \text{ for all } n\right\}$$

of the Cantor set has the same property. In the compact Cantor group $\prod_{k=1}^{\infty} Z_2$, if a point is represented as $\varepsilon = (\varepsilon_1, \varepsilon_2, \ldots)$ with each $\varepsilon_k = 0$ or 1, then the set $\{\varepsilon \colon \sum_{k=1}^{n} \varepsilon_j \leq rn \text{ for all } n\}$ also is a U_0-set but not a U_1-set. The nice thing about this setting is that the proof of the analogue to (iii) is very easy.

Lemma 4.4.1 is due to Milicer-Gruzewska [1].

4.5. Results About Helson Sets and Kronecker Sets

4.5.1. Lemma. *If*

$$\varepsilon > \eta(E) = \inf\left\{\frac{\limsup_{|n|\to\infty} |\hat{\mu}(n)|}{\|\mu\|_{PM}} : \mu \in M(E), \mu \neq 0\right\},$$

then there exists a continuous measure $\nu \in M(E)$ such that

(1) $$\limsup_{|n|\to\infty} |\hat{\nu}(n)| < 3\varepsilon\|\nu\|_{PM}.$$

Proof. It follows from A.2.1 that for a continuous measure μ_c and every $p > 0$,

$$\lim_{m\to\infty} \inf_{|n|\geq m} \max\{|\hat{\mu}_c(k)| : n \leq k \leq n + p\} = 0.$$

For an arbitrary $\mu = \mu_c + \mu_d$, since $\hat{\mu}_d$ is almost periodic, it follows that $\|\mu\|_{PM} \geq \|\mu_d\|_{PM}$ and in fact $\limsup_{|n|\to\infty} |\hat{\mu}(n)| \geq \limsup |\hat{\mu}_d(n)| = \|\mu_d\|_{PM}$. We may suppose that $\varepsilon < 1/3$. Choose $\mu \in M(E)$ such that $\|\mu\|_{PM} = 1$ and $\limsup |\hat{\mu}(n)| < \varepsilon$. Then $\|\mu_c\|_{PM} \geq 1 - \varepsilon > 2\varepsilon/3$ and $\limsup |\hat{\mu}_c(n)| < 2\varepsilon$, so that the ratio is bounded by $2\varepsilon/(1 - \varepsilon)$, which is never more than 3ε. The Lemma is proved. □

4.5.2. Theorem. *If E is a Helson set, then $I(E)$ is weak∗ sequentially dense in A. (More briefly: every Helson set is a U'_1-set.)*

We offer first a proof of the simpler result that a Helson set E is a U_o-set (and hence of course a U_1-set). Suppose otherwise, and let ν be a nonzero element of $M(E) \cap PF$. Since $\nu \in PF$, by A.2.1 ν is a continuous measure. Therefore the support of ν is a closed perfect set, whose complement is a countable union of disjoint open intervals. Let x be a point of the support that is not an endpoint of any of the intervals, so that for every $h > 0$, $|\nu|(x, x + h) \neq 0$ and $|\nu|(x - h, x) \neq 0$. Let

$$f(t) = \begin{cases} 1 & \text{for} \quad x < t < x + 1, \\ 0 & \text{for} \quad x - 1 < t \leq x, \end{cases}$$

and define f on the rest of T so that it is continuous except at x. The mapping $\mu \to \int f\, d\mu$ is a bounded linear functional on M, and in particular on $M(E) \cap PF$, on which the M and PF norms are equivalent since E is a Helson set. Since $PF^* = A$, there exists $g \in A$ such that

$$\int f\, d\mu = \int g\, d\mu \quad \text{for} \quad \mu \in M(E) \cap PF.$$

Recall that $\sigma \in PF \cap M(E)$ for every $\sigma \in L^1(\nu)$. By that fact, and by the choice of x, it follows that g equals 0 at some points to the left of x, and equals 1 at some points to the right of x, in every neighborhood of x, and hence cannot be continuous at x. We have reached a contradiction and proved the assertion.

Proof 1 of Theorem 4.5.2. By Theorem 4.3.2, it suffices to show that $\eta(E) > 0$. Let $\beta > \alpha(E)$, $\varepsilon > \eta(E)$, $k \in Z$, $k > 0$, and $\delta > 0$. Let ν be a continuous measure in $M(E)$ satisfying (1). Let $E_1, \ldots, E_k$ be disjoint portions of E whose union is E, and such that if ν_j is the restriction of ν to E_j, then $\|\nu_j\|_M < (1 + \delta)\|\nu\|_M/k$. (We are using the fact that E must be totally disconnected.) There exist $f_j \in A$ such that $f_j = \chi_{E_j}$ on E and $\|f_j\|_A < \beta$. Then $\nu_j = f_j \nu$, $\hat{\nu}_j = \hat{f}_j * \hat{\nu}$, and consequently

$$\limsup_{|n| \to \infty} |\hat{\nu}_j(n)| < 3\varepsilon\beta\|\nu\|_{PM}.$$

We may select integers $n_1, n_2, \ldots, n_k$ such that if $\sigma = \sum_{j=1}^k e^{in_j x}\nu_j$, then

$$\begin{aligned} \|\sigma\|_{PM} &< \max_j \|\nu_j\|_{PM} + 3(k-1)\varepsilon\beta\|\nu\|_{PM} \\ &< \|\nu\|_M(\beta(1+\delta)k^{-1} + 3(k-1)\varepsilon\beta), \end{aligned}$$

whereas $\|\sigma\|_M = \|\nu\|_M$. Therefore

$$\alpha(E) \geq \frac{\|\sigma\|_M}{\|\sigma\|_{PM}} > \frac{1}{\beta((1+\delta)k^{-1} + 3(k-1)\varepsilon)}.$$

It follows that

$$\alpha(E) \geq \frac{k}{\alpha(E)(1 + 3k(k-1)\eta(E))},$$

or

$$\eta(E) \geq \frac{k - \alpha(E)^2}{3k(k-1)\alpha(E)^2} \quad \text{for} \quad k = 1, 2, \ldots.$$

The right-hand side is positive for $k > \alpha(E)^2$, and thus the Theorem is proved. By optimizing the choice of k, one finds a lower bound for $\eta(E)$ that is a function of $\alpha(E)$, on the order of $\alpha(E)^{-4}$. As we shall see, there is another approach that yields a finer result, on the order of $\alpha(E)^{-2}$. □

Proof 2 of Theorem 4.5.2. Let $\varepsilon > \eta(E)$. Then there exists $\mu \in M(E)$ such that

$$\limsup_{|n| \to \infty} |\hat{\mu}(n)| < \varepsilon\|\mu\|_{PM}.$$

For $k \geq 2$, we may choose $n_1, \ldots, n_k$ such that if $|a_j| = 1$ and $\sigma_a = \sum_{j=1}^{k} a_j e^{in_jx}\mu$, then

$$\|\sigma_a\|_{PM} \leq \|\mu\|_{PM}(1 + (k-1)\varepsilon) \leq \|\mu\|_M(1 + (k-1)\varepsilon).$$

Allowing the values $+1$ and -1 for each a_j, there are 2^k values for the vector $a = (a_1, \ldots, a_k)$. We claim that the average of the 2^k values $\|\sigma_a\|_M$ is at least $(k/3)^{1/2}\|\mu\|_M$, so that for at least one value of a,

$$\frac{\|\sigma_a\|_M}{\|\sigma_a\|_{PM}} \geq \frac{(k/3)^{1/2}}{1 + (k-1)\varepsilon}.$$

Therefore

$$\alpha(E) \geq \frac{(k/3)^{1/2}}{1 + (k-1)\eta(E)}$$

or

$$\eta(E) \geq \frac{(k/3)^{1/2} - \alpha(E)}{(k-1)\alpha(E)} \text{ for } k = 1, 2, \ldots.$$

It remains to prove the claim. We regard the a_j as k independent random variables, each equal to $+1$ or -1 and taking each value with probability 1/2. The expectation (average value) of $\|\sigma_a\|_M$ is

$$\mathscr{E}(\|\sigma_a\|_M) = \mathscr{E}\left(\int \left|\sum a_j e^{in_jx}\right| |d\mu|(x)\right) = \int \mathscr{E}\left(\left|\sum a_j e^{in_jx}\right|\right) |d\mu|(x).$$

For a fixed x, write c_j for e^{in_jx} and Y for $|\sum a_j c_j|$, and apply Hölder's inequality. Then

(2) $$\mathscr{E}(Y^2) = \mathscr{E}(Y^{2/3}Y^{4/3}) \leq \mathscr{E}(Y)^{2/3}\mathscr{E}(Y^4)^{1/3}.$$

$$Y^2 = \sum |c_i|^2 + \sum_{i \neq j} c_i \bar{c}_j a_i a_j,$$

and

$$Y^4 = \sum |c_i|^4 + \sum_{i \neq j} (2|c_i|^2|c_j|^2 + c_i\bar{c}_j) a_i^2 a_j^2 + \sum,$$

where $\mathscr{E}(\sum) = 0$.

Since $\mathscr{E}(a_i) = 0$ and $\mathscr{E}(a_i^2) = 1$,

$$\mathscr{E}(Y^2) = \sum |c_i|^2 = k$$

and

$$\mathscr{E}(Y^4) \le \sum |c_i|^4 + 3 \sum_{i \ne j} |c_i|^2 |c_j|^2$$
$$\le 3(\sum |c_i|^2)^2 = 3k^2.$$

Therefore by (2),

$$k \le (\mathscr{E}(Y)^{2/3})(3^{1/3})(k^{2/3}),$$

or

$$\mathscr{E}(Y) \ge (k/3)^{1/2}.$$

This proof of Theorem 4.5.2 is complete. □

4.5.3. Theorem. *Every Kronecker set is a set of synthesis.*

Proof. Let $S \in PM(E)$, where E is a Kronecker set. It suffices to show that S is a measure. Let W be the class of functions $f \in C(T)$ such that $f^{-1}(\{+1, -1\})$ is a neighborhood of E. If it is shown that there is a constant k such that

$$\text{(3)} \qquad |\langle f, S\rangle| \le k \quad \text{for every} \quad f \in W,$$

it will follow that the linear functional $f \to \langle f, S\rangle$ is bounded, with respect to the sup-norm, on the class of all functions $f \in C(T)$ such that for some finite set F, $f^{-1}(F)$ is a neighborhood of E. That class is dense in $C(T)$, and hence it will follow that S is a measure.

Let $f \in W$. Let S_j be the restriction of S to $f^{-1}(j)$, for $j = -1$ and $+1$. Let $h > 0$. Then there exists $n \in Z$ such that $|f(x) - e^{inx}| < h$ for $x \in E$, and hence for all x in some neighborhood of E, which we may take to be of the form $E_{-1} \cup E_1$, where f equals j on E_j. By A.3.1(a), $\|j - e^{inx}\|_{A(E_j)} < h + 0(h^2)$ for each j, so

$$|\langle f, S\rangle| = \left| \hat{S}(n) + \sum_j \langle j - e^{inx}, S_j\rangle \right| \le \|S\|_{PM} + 2(h + 0(h^2))\max_j \|S_j\|_{PM}.$$

Since h is arbitrary, (3) is established with $k = \|S\|_{PM}$, and Theorem 4.5.3 is proved. □

Remarks. The fact that Helson sets are U_0-sets, and the proof we gave, are due to Helson [3]. That early paper, presenting the basic facts about the sets of interpolation for A, accounts for their being called "Helson sets". The method used in Proof 1 of Theorem 4.5.2 derives from Doss [2]. Proof 2 is from Kahane and Salem [1, p. 143], and an exposition appears also in Lindahl and Poulsen [1, Section 1.4]. Theorem 4.5.2, easily generalized, may be used to give an easy proof of the fact that the Fourier transform maps $L^1(G)$ *onto* $C_0(\Gamma)$ only if G is finite; see Friedberg [2].

Theorem 4.5.3 is due to Varopoulos [4]; one may prefer the exposition found in Lindahl and Poulsen [1, Chapter 3]. The proof we have given works for any locally compact abelian group provided E is totally disconnected, as a Kronecker set must be if the group is finite-dimensional (see Rudin [1, p. 104]). A proof of 4.5.3 for the general case is presented by Saeki [4].

4.6. M-Sets Whose Helson Constant Is One

The following preliminary result is related to the argument of Pyateckiĭ-Šapiro in Section 4.4.

4.6.1. Theorem. *For each positive integer N and $0 < \varepsilon < 1$, consider T^N as $[-\pi, \pi)^N$ and let $V_{N,\varepsilon}$ be the set of all $x = (x_1, \ldots, x_N) \in T^N$ such that $|x_k| \le \varepsilon$ for at least $(1 - \varepsilon)N$ values of k. For each $\varepsilon > 0$, there exists N_ε such that if $N \ge N_\varepsilon$, then there exists $F = F_{N,\varepsilon} \in A(T^N)$ such that F vanishes outside $V_{N,\varepsilon}$ and $|\hat{F}(m)| < \varepsilon|\hat{F}(0)|$ for every nonzero $m \in Z^N$.*

Proof. For $0 \le t \le 1$ define $\sigma_t \in M(T)$ by

$$\sigma_t = t\frac{dx}{2\pi} + (1 - t)\delta_0.$$

Fix N. Let λ_t be the product measure $\sigma_t \times \sigma_t \times \cdots \times \sigma_t$ on T^N. Since σ_t gives measure $p = (1 - t) + t\varepsilon/\pi$ to the set $\{x \in T : |x| \le \varepsilon\}$, and measure $q = 1 - p$ to its complement, the λ_t-measure of the complement of $V_{N,\varepsilon}$ may be estimated as follows (see the proof of (ii) in Section 4.4).

$$\lambda_t(T^N \backslash V_{N,\varepsilon}) = \sum_{s > \varepsilon N} \binom{N}{s} p^{N-s} q^s < \left(\frac{p^{1-\varepsilon} q^\varepsilon}{(1 - \varepsilon)^{1-\varepsilon} \varepsilon^\varepsilon}\right)^N.$$

In particular,

(1) $\lambda_t(T^N \backslash V_{N,\varepsilon}) \to 0$ as $N \to \infty$, uniformly for $0 \le t \le \varepsilon$.

Now suppose that the theorem is false, and thus that whenever $F \in A(T^N)$ and F vanishes outside $V_{N,\varepsilon}$, it follows that

$$|\hat{F}(0)| \leq \varepsilon^{-1} \sup\{|\hat{F}(m)| : m \in Z^N, m \neq 0\}.$$

Then the mapping $\hat{F} \to \hat{F}(0)$ is a bounded linear functional on a subspace of $C_o(Z^N \backslash \{0\})$, and hence must be given by an element $\hat{g} = \hat{g}_N \in L^1(Z_N \backslash \{0\})$:

$$\sum_{m \neq 0} |\hat{g}(m)| \leq \varepsilon^{-1},$$

and

$$\hat{F}(0) = \sum_{m \neq 0} \hat{g}(m)\hat{F}(m)$$

for every F in that subspace. It follows that

$$\sum_{m \neq 0} \hat{g}(m)\langle m, x\rangle = 1 \quad \text{for} \quad x \in V_{N,\varepsilon}.$$

Thus,

$$\text{(2)} \qquad \lambda_t(V_{N,\varepsilon}) - \varepsilon^{-1}\lambda_t(T^N \backslash V_{N,\varepsilon}) \leq \int g \, d\lambda_t$$

$$= \sum_{m \neq 0} \hat{g}(m)\hat{\lambda}_t(m)$$

$$= \sum_{m \neq 0} \hat{g}(m)(1 - t)^{k(m)}$$

(where $k(m)$ is the number of nonzero coordinates m_i of $m = (m_1, \ldots, m_N)$)

$$= \sum_{k=1}^{N} C_{N,k}(1 - t)^k$$

where $\sum_{k=1}^N |C_{N,k}| \leq \varepsilon^{-1}$. Let $f_N(z) = \sum_k C_{N,k} z^k$. Then $\{f_N\}$ is a sequence of entire functions which, for $|z| \leq 1$, form a normal family. A subsequence converges to some function that is analytic for $|z| < 1$. But for $0 \leq t \leq \varepsilon$, $f_N(1 - t) \to 1$ uniformly, by (1) and (2), whereas $f_N(0) = 0$. We have reached an impossible conclusion, so Theorem 4.6.1 must be true. □

4.6.2. Theorem *Let g be a real-valued continuous function defined on a neighborhood U of a compact set $E \subset T$, such that $g(U)$ is a finite set. Let Y be an infinite subset of Z, S a pseudofunction with support E, and $\varepsilon > 0$. Then there exist $y_1, \ldots, y_N \in Y$, and $S_1 \in PF$ with support E_1, such that $E_1 \subset E$, $\|S - S_1\|_{PM} < \varepsilon$, and $|e^{ig(x)} - N^{-1} \sum_{k=1}^N e^{iy_k x}| < \varepsilon$ for $x \in E_1$.*

Proof. We claim that the quantity $C = \sup_{n \in Z} \|e^{ing}\|_{A(U)}$ is finite. For if g is constant on each of k disjoint open sets $U_1, \ldots, U_k$ whose union is U, and if $\varphi_j \in A(T)$ such that $\varphi_j = \chi_{U_j}$ on U, then $e^{ing} = \sum_{j=1}^k e^{ing(U_j)}\varphi_j$ on U, and the A-norm of this sum is bounded by $\sum \|\varphi_j\|_A$, regardless of n.

Let $\varepsilon > 0$. For $N > N_\varepsilon$, let $F = F_{N,\varepsilon}$ be the function provided by Theorem 4.6.1, normalized so that $\hat{F}(0) = 1$ and thus $|\hat{F}(m)| < \varepsilon$ for $m \neq 0$. Let $y_k \in Y$ for $1 \leq k \leq N$, and let

$$S_1(dx) = F(g(x) - y_1 x, g(x) - y_2 x, \ldots, g(x) - y_N x)S(dx).$$

Let m denote an element $(m_1, \ldots, m_N)$ of Z^N. Then

$$(3) \qquad (S_1 - S)(dx) = \sum_{m \neq 0} \hat{F}(m)e^{i\sum m_k g(x)}e^{-i(\sum y_k m_k)x}S(dx).$$

Since of course

$$(4) \qquad (e^{-iqx}S(dx))\hat{}\,(n) = \hat{S}(n + q) \quad \text{whenever } n, q \in Z \text{ and } S \in PM,$$
$$\|e^{-i(\sum y_k m_k)x}S(dx)\|_{PM} = \|S\|_{PM} \quad \text{for all } m, y \in Z^N.$$

Therefore the series (3) converges, and $\|S_1 - S\|_{PM} \leq \|F\|_A C\|S\|_{PM}$. Clearly S_1 is a pseudofunction whose support E_1 is a subset of E; and for each $x \in E_1$, $|e^{ig(x)} - e^{iy_k x}| < \varepsilon$ for at least $(1 - \varepsilon)N$ of the N values of k, so that

$$(5) \qquad \left| e^{ig(x)} - N^{-1}\sum_{k=1}^N e^{iy_k x} \right| < 3\varepsilon \quad \text{for every } x \in E_1.$$

We have not yet restricted the choice of the integers y_k, but we shall now do so. Let L be a finite subset of Z^N such that

$$(6) \qquad C\|S\|_{PM} \sum_{m \notin L} |\hat{F}(m)| < \varepsilon.$$

Let $S_m = e^{i\sum m_k g}S$. Then $\sup_n |\hat{S}_m(n)| \leq C\|S\|_{PM}$ and $\hat{S}_m(n) \to 0$ as $|n| \to \infty$ for every m. Therefore we may choose $y_1, \ldots, y_N$ from Y so that the sets

$$Z_m = \left\{ n \in Z : \left| \hat{S}_m\left(n + \sum_{k=1}^N y_k m_k\right) \right| > 1/\text{Card } L \right\},$$

for $m \in L$, are disjoint. In light of the principle (4), it follows that

$$(7) \qquad \left\| \sum_{0 \neq m \in L} \hat{F}(m)e^{i\sum m_k g(x)}e^{-i\sum y_k m_k x}S(dx) \right\|_{PM} \leq \varepsilon(C\|S\|_{PM} + 1).$$

By (3), (6), and (7), then

$$\|S - S_1\|_{PM} < \varepsilon(C\|S\|_{PM} + 2).$$

Theorem 4.6.2 is proved. □

4.6.3. Theorem. *Let Y be an infinite subset of Z, S a pseudofunction with support E, and $\varepsilon > 0$. Then there exists a pseudofunction S^* with support $E^* \subset E$, such that $\|S - S^*\|_{PM} < \varepsilon$ and $\alpha(E^*) = 1$; and in fact, such that*

$$\|\mu\|_M = \sup_{y \in Y} |\hat{\mu}(y)| \quad \text{for } \mu \in M(E^*).$$

Proof. Case 1, when E is totally disconnected. We may choose a sequence $\{g_j\}_{j=0}^{\infty}$ of continuous real-valued functions such that for each j, there is a neighborhood U_j of E such that $g_j(U_j)$ is a finite set; and such that $\{e^{ig_j}|_E\}$ is dense in $\{f \in C(E): |f| \equiv 1\}$. Apply 4.6.2 with g_0, S, and $\varepsilon/2$ in the roles of g, S, and ε, respectively, obtaining S_1 on E_1. Proceed inductively; when S_j and E_j have been chosen for $1 \le j \le k$, apply 4.6.2 with g_j, S_k, and $2^{-k-1}\varepsilon$ in the roles of g, S, and ε, respectively, obtaining a pseudofunction S_{k+1} with support $E_{k+1} \subset E_k \subset \cdots \subset E_1$, such that $\|S_{k+1} - S_k\|_{PM} < 2^{-k-1}\varepsilon$. Finally, let $S^* = \lim_{k \to \infty} S_k$. Then the support of S^* is contained in $\bigcap_{k=1}^{\infty} E_k$, and $\|S - S^*\|_{PM} < \varepsilon$. For $\mu \in M(E^*)$, and $\eta > 0$, there exists j such that $\varepsilon 2^{-j-1} < \eta$ and $|\int e^{ig_j}\, d\mu| \ge \|\mu\|_M(1 - \eta)$. By the jth step of the procedure we know that there is a trigonometric polynomial $p(x) = N^{-1} \sum_{k=1}^{N} e^{iy_k x}$ such that $y_k \in Y$ and $|p(x) - e^{ig_j(x)}| < \eta$ for $x \in E_j$. Therefore

$$\left| N^{-1} \sum_{k=1}^{N} \hat{\mu}(y_k) \right| = \left| \int p\, d\mu \right| \ge \left| \int e^{ig_j} d\mu \right| - \eta\|\mu\|_M,$$

so that for at least one value of k,

$$|\hat{\mu}(y_k)| \ge \|\mu\|_M(1 - 2\eta).$$

Theorem 4.6.3 is proved in Case 1.

The general case. It suffices to prove that if $S \in PF$ and E is the support of S, then for each $\eta > 0$ there exists $S_\eta \in PF$ whose support is a totally disconnected subset of E such that $\|S - S_\eta\|_{PM} < \eta$. Consider $x \in E$ and for each $\varepsilon > 0$, let f_ε be the function that equals one on $[x - \varepsilon, x + \varepsilon]$, vanishes off $[x - 2\varepsilon, x + 2\varepsilon]$, and is linear on each of the two remaining intervals. Then $\|f_\varepsilon\|_A \le 3$, and $\hat{f}_\varepsilon(n) \to 0$ as $\varepsilon \to 0$ for each $n \in Z$. It follows that $\|f_\varepsilon S\|_{PM} = \sup_n |\sum_k \hat{S}(n - k)\hat{f}_\varepsilon(k)| \to 0$ as $\varepsilon \to 0$. Therefore as $\varepsilon \to 0$, S is approached in PM-norm by $(1 - f_\varepsilon)S$, whose support is contained in E and is disjoint from $(x - \varepsilon, x + \varepsilon)$. In view of this remark, it is an easy exercise to show the existence of S_η, as called for above. □

Remarks. This section comes essentially from Kaufman [12], but uses a simplifying idea from Saeki [13], who has generalized Kaufman's procedure to a large class of groups. The role of the set Y is stressed here because of its importance in Katznelson and Körner [1] (see Section 12.5). Körner [1] was first to construct Helson sets of multiplicity. His approach is opposite to Kaufman's, in that he builds the pseudofunction as a weak* limit of measures with finite support. His original proof, which was extremely long, his later short proof, and Kaufman's work all appear in the same issue of *Astérisque.*

4.7. Independent M_o-Sets

4.7.1. Theorem. *Every M_o-set $E \subset T$ has a subset F that is an independent M_o-set.*

Proof. The procedure that we shall use involves the following two simple observations.

(i) If ρ and σ are two positive measures such that $\rho(I) = \sigma(I)$, where $I = [t - d, t + d]$, then

$$\left| \int_I e^{-inx}\, d(\rho - \sigma)(x) \right| = \left| \int_I (e^{-inx} - e^{-int}) d(\rho - \sigma)(x) \right|$$

$$\leq 2\rho(I) \max_{x \in I} |e^{-inx} - e^{-int}|$$

$$\leq 2\rho(I)(|n|d + 0(n^2 d^2)) \quad \text{as } d \to 0.$$

It follows that for every K and $\varepsilon > 0$, there exists $d > 0$ such that if ρ and σ are positive measures, both supported within the union $\bigcup_{j=1}^k I_j$, where for each j, I_j is an interval of diameter no greater than $2d$ and $\rho(I_j) = \sigma(I_j)$, then

$$|\hat{\rho}(n) - \hat{\sigma}(n)| < \varepsilon \|\rho\|_M \quad \text{for } |n| \leq K.$$

(ii) Let U be a finite subset of Z^k. If $(t_1, \ldots, t_k) \in T^k$ and $\sum_{j=1}^k u_j t_j$ is nonzero for each $u = (u_1, \ldots, u_k) \in U$, then there exists $\delta > 0$ such that if $|x_j - t_j| < \delta$ for $1 \leq j \leq k$ and $u \in U$, then $\sum_{j=1}^k u_j x_j \neq 0$.

By 4.4.1, since E is an M_o-set, there exists a probability measure $\mu \in M(E) \cap PF$. The plan is to modify μ, in a procedure requiring an infinite sequence of steps, so as to obtain another probability measure $\nu \in M(E) \cap PF$, with support F such that if $x_j \in F$ and $u_j \in Z$ for $1 \leq j \leq k$, where k is finite, and $\sum u_j^2 \neq 0$, then $\sum u_j x_j \neq 0$. We shall show that given $\varepsilon > 0$ we can pick the measure ν so that, furthermore, $\|\nu - \mu\|_{PM} < \varepsilon$.

Let $\{L_m\}$ and $\{N_m\}$ be sequences of positive integers such that $L_m \to \infty$

and $\sum N_m^{-1} < \varepsilon/3$. Let $\mu_0 = \mu$, $E_0 = E$. In step number m we begin with the probability measure μ_{m-1} and end with another one, $\mu_m \in PF \cap M(E_m)$, where $E_m \subset E_{m-1}$ and

(1) $$\|\mu_m - \mu_{m-1}\|_{PM} < 3N_m^{-1}.$$

The set E_m will be covered by intervals H_{mi}, $1 \le i \le N_m^2$, and will enjoy the property that

(2) $$\begin{cases} \text{if } X = \{x_1, \ldots, x_N\} \subset E_m, \text{ and if the set } X \cap H_{mi} \text{ contains at} \\ \text{most one element for each } i, \text{ and if } |\mu_j| \le L_m \text{ for } 1 \le j \le N_m \text{ and} \\ \sum u_j^2 \ne 0, \text{ then } \sum u_j x_j \ne 0. \end{cases}$$

It follows that $\|\mu_m - \nu\|_{PM} \to 0$ as $m \to \infty$ for some probability measure $\nu \in M(F) \cap PF$, where $F \subset \bigcap E_m \subset E$, such that $\|\nu - \mu\|_{PM} < \varepsilon$.

We will choose the intervals H_{mi} so that

(3) $$\max_i \operatorname{diam}(H_{mi}) \to 0 \quad \text{as} \quad m \to \infty.$$

If k is finite and $X = \{x_1, \ldots, x_k\} \subset F$, then for all sufficiently large m, $k \le N_m$ and $X \cap H_{mi}$ is at most a singleton for each i. It follows that F is independent. Thus if we can carry out step number m as claimed, the Theorem will be proved.

Choose N_m^2 intervals H_{mi} such that

$$\operatorname{supp}(\mu_{m-1}) \subset \bigcup\{H_{mi}: 1 \le i \le N_m^2\} \quad \text{and} \quad \mu_{m-1}(E \cap H_{mi}) = N_m^{-2}$$

for each i.

That selection may be made consistently with (3). Let $\{h_a: 1 \le a \le b\}$ be an enumeration of the subsets of $\{i: 1 \le i \le N_m^2\}$ that contain exactly N_m distinct integers. Then b is the binomial coefficient $\binom{N_m^2}{N_m}$.

The process of modifying μ_{m-1} to obtain μ_m has b substeps. Let $\mu_{m,0} = \mu_{m-1}$. At substep a we modify $\mu_{m,a-1}$ to obtain $\mu_{m,a}$. Finally we will obtain $\mu_{m,b}$ and call it μ_m, its support E_m; and properties (1) and (2) will be satisfied.

Set K_0 equal to 10. Now we shall describe substep a, for $1 \le a \le b$.

Let ρ be the restriction of $\mu_{m,a-1}$ to $H^a = \bigcup\{H_{mi}: i \in h_a\}$. Then $\|\rho\| = N_m^{-1}$, and $\hat{\rho} \in c_0$ by 4.4.1. Enlarge K_{a-1}, if necessary, so that

(4) $$|\hat{\rho}(n)| < N_m^{-1} b^{-1} \quad \text{for } |n| > K_{a-1}.$$

Let $I_1, \ldots, I_k$ be disjoint intervals each contained in H_{mi} for some $i \in h_a$, such that $\rho(H^a) = \sum \rho(I_j)$ and $\rho(I_j) \neq 0$ for each j, and such that if σ is a positive measure supported in $\bigcup I_j$ with $\sigma(I_j) = \rho(I_j)$ for each j, then

$$(5) \qquad |\hat{\rho}(n) - \hat{\sigma}(n)| < N_m^{-1}b^{-1} \quad \text{for } |n| \leq K_{a-1}.$$

Note that $k \geq N_m$.

Choose $t_j \in \operatorname{supp}(\rho) \cap I_j^0$ so that the set $\{t_j: 1 \leq j \leq k\}$ is independent. Then in particular, $\sum_{j=1}^k u_j t_j \neq 0$ for $u \in U$, where

$$U = \{u = (u_1, \ldots, u_k) \in Z^k: |u_j| \leq L_m, \sum u_j^2 \neq 0\}.$$

There exists $\delta > 0$ such that $[t_j - \delta, t_j + \delta] \subset I_j$ for each j and such that if $|x_j - t_j| < \delta$ and $u \in U$, then $\sum u_j x_j \neq 0$. Let

$$\sigma = \frac{1}{kN_m} \sum_{j=1}^k \frac{\chi_{[t_j - \delta, t_j + \delta]}\rho}{\rho([t_j - \delta, t_j + \delta])}.$$

Then (5) holds, $\|\sigma\|_M = N_m^{-1}$, and of course

$$(6) \qquad |\hat{\sigma}(n)| \leq N_m^{-1} \quad \text{for all } n.$$

Thus

$$(7) \qquad |\hat{\sigma}(n) - \hat{\rho}(n)| \leq N_m^{-1}(1 + b^{-1}) \quad \text{for } |n| > K_{a-1}.$$

Pick $K_a > K_{a-1}$ such that

$$(8) \qquad |\hat{\sigma}(n)| < N_m^{-1}b^{-1} \quad \text{for } |n| > K_a.$$

Recall that ρ is the restriction of $\mu_{m,a-1}$ to H^a. Let

$$\mu_{m,a} = \mu_{m,a-1} - \rho + \sigma.$$

Then by (4)–(8),

$$(9) \quad |\hat{\mu}_{m,a}(n) - \hat{\mu}_{m,a-1}(n)| < \begin{cases} N_m^{-1}b^{-1} & \text{for } |n| \leq K_{a-1}, \\ N_m^{-1}(1 + b^{-1}) & \text{for } K_{a-1} < |n| \leq K_a, \\ 2N_m^{-1}b^{-1} & \text{for } |n| > K_a. \end{cases}$$

Note that (9) remains true if K_a is enlarged.

Properties (1) and (2) follow, with $\mu_m = \mu_{m,b}$ and $E_m = \operatorname{supp}(\mu_m)$. The Theorem is proved. □

Credits and Remarks. The existence of independent M_0-sets is due to Rudin [11]. An exposition of Rudin's procedure and the result of Salem that underlies it appears in Benedetto [1, Section 6.3]. Such sets are produced as Brownian images in Kahane [1, Chapter XV]. The proof given here is due to Körner; see Lindahl and Poulsen [1, Section XIII.3] for this and other related constructions. Varopoulos [9] has obtained a result like Rudin's for a large class of groups.

T. W. Körner [1, 2, 3, 4, 5, 6, 7, 8] is a master of exceptional-set construction; in this Chapter and others in the book we provide only a sampling of his work.

Chapter 5

A Brief Introduction to Convolution Measure Algebras

5.1. Elementary Properties

We have two objectives in this chapter: to introduce the general theory of convolution measure algebras and to give examples and applications pertinent to measure algebras on groups. We shall thus make clear the setting in which the action of Chapters 6 through 8 takes place. We shall state without proof some results (the most important ones are Theorems 5.1.1 and 5.3.6). It is not necessary to read their proofs in order to appreciate that action. We begin with some useful terminology.

Let (X, Σ) denote a measurable space, and let $M(X, \Sigma)$ denote the space of all complex-valued bounded measures on Σ. A closed subspace Y of $M(X, \Sigma)$ is an *L-subspace* of $M(X, \Sigma)$ if $\mu \in Y$ and $\nu \ll \mu$ imply that $\nu \in Y$.

Let X be a locally compact Hausdorff space that is also a semigroup. We shall generally write the semigroup operations multiplicatively. If the semigroup operation from $X \times X$ to X is continuous in each factor separately, then X is a *semitopological semigroup.* If the semigroup operation is jointly continuous, then X is a *topological semigroup.*

When X is a topological semigroup, $\hat{X}$ denotes the set of non-zero continuous maps $f: X \to \{z: |z| \leq 1\}$ such that $f(xy) = f(x)f(y)$. Those functions are the continuous *semicharacters* on X.

Let X be a semitopological semigroup and Σ the σ-ring of Borel subsets of X. We denote by $M(X)$ the L-subspace of $M(X, \Sigma)$ that consists of all the regular bounded Borel complex-valued measures. By the Riesz representation theorem (Rudin [2, p. 139]), $M(X) = C_o(X)^*$. (That duality is given by $f \to \int f\, d\bar{\mu}$.) A theorem of B. E. Johnson [4] says that $(x, y) \to f(xy)$ is $\mu \times \nu$ measurable for each $f \in C_o(X)$ and $\mu, \nu \in M(X)$. Therefore the following formula defines a continuous linear functional on $C_o(X)$.

$$L(f) = \iint f(xy)d\bar{\mu}(x)d\bar{\nu}(y).$$

By the Riesz representation theorem, there exists a unique measure $\mu * \nu$, the *convolution* of μ and ν, such that

$$(1) \qquad \int f(s)d(\overline{\mu * \nu})(s) = \iint f(xy)d\bar{\mu}(x)d\bar{\nu}(y), \quad \text{for all} \quad f \in C_o(X).$$

It is obvious from (1) that $\|\mu * \nu\| \leq \|\mu\| \|\nu\|$. With the operation of convolution, $M(X)$ becomes a Banach algebra that is commutative if and only if X is commutative.

An *L-subalgebra* of $M(X)$ is an L-subspace B of $M(X)$ that is closed under convolution. By *commutative convolution measure algebra* (*CCMA*) we shall mean an L-subalgebra of $M(X)$, where X is a commutative semitopological semigroup. An *L-homomorphism* from a CCMA B_1 to another CCMA B_2 is a Banach algebra homomorphism $\varphi: B_1 \to B_2$ such that

(i) if $\mu \in B_1$ and $\mu \geq 0$, then $\varphi(\mu) \geq 0$, and $\|\mu\| = \|\varphi(\mu)\|$; and
(ii) if $\mu \in B_1$, $\mu \geq 0$, and $0 \leq \nu \leq \varphi(\mu)$, then there exists $\omega \in B_1$ such that $\varphi(\omega) = \nu$.

Let B be a CCMA. Then B can be thought of as an L^1-space, where the measure ω having $B = L^1(\omega)$ is usually far from being σ-finite. To see that, let $\{Y_\alpha\}$ be the set of all subsets of non-negative measures in B such that for all α, and all $\mu, \nu \in Y_\alpha$, $\mu \neq \nu$ implies that $\mu \perp \nu$. We say $Y_\alpha \geq Y_\beta$ if for each $\nu \in Y_\beta$ there exists a sequence $\{\mu_j\} \subseteq Y_\alpha$ with $\nu \ll \sum \mu_j/2^j\|\mu_j\|$. By Zorn's Lemma, there exists a maximal $Y_{\alpha(0)}$. We set $\omega = \sum\{\mu: \mu \in Y_{\alpha(0)}\}$. It is easy to see that $B = L^1(\omega)$.

From the preceding we can obtain a representation of elements of B^*. We assume that B is an L-subalgebra of $M(X)$, where X is a semitopological semigroup. For each $\psi \in B^*$ and $\mu \in B$, there exists a μ-measurable function ψ_μ on X such that $\langle \psi, \nu \rangle = \int \overline{\psi_\mu(x)} d\nu$ for all $\nu \ll \mu$. The map $(\mu, x) \to \psi_\mu(x)$ has the following properties:

(2) $$x \to \psi_\mu(x) \text{ is } \mu\text{-measurable for all } \mu \in B;$$

(3) $$\nu \ll \mu \text{ implies } \psi_\nu = \psi_\mu \text{ a.e. } d\nu \text{ for all } \mu \in B;$$

and

(4) $$\sup_{\mu \in B} \operatorname{ess\,sup}_{x \in X} |\psi_\mu(x)| < \infty.$$

Clearly, the left hand side of (4) equals $\|\psi\|$. A mapping $(\mu, x) \to \psi_\mu(x)$ satisfying (2)–(4) is called a *generalized function for* B. It is easy to see that if $\psi_\mu(x)$ is a generalized function for B, then $\mu \to \int \overline{\psi_\mu(x)} d\mu(x)$ defines a continuous linear functional on B.

When $\psi \in \Delta B$, then ψ_μ has two additional properties,

(5) $$0 < \sup_{\mu \in B} \operatorname{ess\,sup}_{x \in X} |\psi_\mu(x)|,$$

and if $\omega \geq 0$ and $\omega^2 \ll \omega$, then

(6) $$\psi_\omega(x)\psi_\omega(y) = \psi_\omega(x + y) \quad \text{a.e.} \quad d(\omega \times \omega).$$

Note that (2)–(6) imply that

$$(5)' \qquad 0 < \sup_{\mu \in B} \operatorname{ess\,sup}_{x \in X} |\psi_\mu(x)| \leq 1.$$

Of course, (5)′ follows at once from (2)–(4) and the fact that $\|\psi\| \leq 1$. Property (5) is obvious. For (6) we argue as follows. Let $\omega \in B$ be such that $\omega \geq 0$ and $\omega^2 \ll \omega$. Let $\sigma, \tau \ll \omega$. We may assume that $0 \leq \sigma, \tau$. Then, by the multiplicative property of ψ, the definition of ψ_ω, $\omega \gg \omega^2$, (3) and (5),

$$\iint \overline{\psi}_\omega(xy) d\sigma(x) d\tau(y) = \int \overline{\psi}_\omega \, d\sigma * \tau = (\sigma * \tau)\hat{}(\psi) = \hat{\sigma}(\psi)\hat{\tau}(\psi)$$

$$= \int \overline{\psi}_\sigma \, d\sigma \int \overline{\psi}_\tau \, d\tau = \int \overline{\psi}_\omega \, d\sigma \int \overline{\psi}_\omega \, d\tau.$$

Now (6) follows by an application of Fubini's theorem.

A *generalized character for B* is a function $\psi: B \times X \to \mathbb{C}$ satisfying (2)–(6).

5.1.1. Theorem. *Let B be a CCMA.*

(i) *Let $\psi \in \Delta B$. Then there exists a unique generalized character ψ' for B such that*

$$(7) \qquad \hat{\mu}(\psi) = \int \overline{\psi'_\mu}(x) d\mu(x), \quad \textit{for all } \mu \in B.$$

(ii) *If ψ' is a generalized character for B, then the map sending μ to the right hand side of (7) defines a multiplicative linear functional on B.*

Proof. (*i*) follows from the discussion above, and (*ii*) is left to the reader. □

Examples. (i) Let $B = L^1(G)$, where G is a locally compact abelian group with dual group Γ. Then the elements of Γ are the generalized characters of B and there is no dependence on μ.

(ii) Let $B = M([0, 1])$, where $[0, 1]$ has maximum multiplication: $st = \max(s, t)$. Then the generalized characters for B are the indicator functions of intervals of the form $[0, x)$ and $[0, x]$. If $[0, 1]$ is identified with the set of (indicator functions for) closed intervals $[0, x]$, then the elements of B are mapped by the Gel'fand transform to the functions of bounded variation. The continuous measures in $M([0, 1])$ do not distinguish between $[0, x)$ and $[0, x]$, so $\Delta M_c([0, 1]) = (0, 1]$, and the Gel'fand transforms of $M_c([0, 1])$ are the continuous functions of bounded variation. Similarly, $L^1([0, 1])\hat{}$ is the set of absolutely continuous functions of bounded variation.

(iii) Let $B = M(G)$ where G is a non-discrete locally compact abelian group. Let $\psi_\mu(x) = 1$ a.e. $d\mu$ if μ is discrete and $\psi_\mu(x) = 0$ a.e. $d\mu$ if μ is continuous. The continuous measures on G are a closed ideal, and the discrete measures are a closed subalgebra. From those facts, it is easy to show that ψ is indeed a generalized character.

(iv) Let $B = M(R^2)$ and let $H = R \times \{0\}$. Let $\psi_\mu = 1$ a.e. $d\mu$ if μ is concentrated on a countable union of cosets of H and $\psi_\mu = 0$ a.e. $d\mu$ if $|\mu|(x + H) = 0$ for all $x \in R^2$. As in (iii), it is not hard to show that ψ is a generalized character.

Examples (iii)–(iv) are special cases of generalized characters arising from Raĭkov systems. Those are discussed in Section 5.2.

In some cases, (6) can be established without having $\mu^2 \ll \mu$. Here is an example. We claim that if $\mu \in M(G)$ and $x \in G$, then

$$(8) \qquad \psi_{\delta(x)*\mu}(z + x) = \psi_{\delta(x)}(x)\psi_\mu(z) \quad \text{a.e. } d\mu(z).$$

To establish (8), let $\nu \ll \mu$. By 5.2.6 below,

$$\begin{aligned}\bar{\psi}_{\delta(x)}(x) \int \bar{\psi}_\mu(z) d\nu &= \int \bar{\psi}_{\delta(x)*\mu}\, d(\delta(x) * \nu) \\ &= \iint \bar{\psi}_{\delta(x)*\mu}(w + z) d\delta(x)(w) d\nu(z) \\ &= \int \bar{\psi}_{\delta(x)*\mu}(x + z) d\nu(z)\end{aligned}$$

Now (8) follows.

The dependence of $\psi_\mu(x)$ on μ as well as on x, as in examples (iii)–(iv), is sometimes distracting. There is a way to "spread out" X so that B is mapped to a CCMA on a new semigroup such that the generalized characters for B become continuous functions on the new semigroup. That result, our next theorem, applies to the more abstract CCMA of Taylor [3], and is very useful, particularly in conceptualizing methods of proof. For its proof see Taylor [1, 3.23] or [3].

5.1.2. Theorem. *Let B be a CCMA with maximal ideal space ΔB. Then there exists a compact commutative topological semigroup S and an L-homomorphism $\check{\varphi}$ of B onto an L-subalgebra of $M(S)$ with the following properties.*

(i) *$\hat{S}$ separates points of S.*
(ii) *For each $\psi \in \Delta B$, there exists a unique $f \in \hat{S}$ such that $\hat{\mu}(\psi) = \int \bar{f}\, d(\check{\varphi}(\mu))$ for all $\mu \in B$.*
(iii) *$\check{\varphi}(B)$ is weak* dense in $M(S)$.*

Properties (i)–(iii) characterize S uniquely up to isomorphism, the map ϕ is an isomorphism if and only if B is semisimple, and the rule $B \to S$ is a functor from the category of CCMA's to that of compact abelian topological semigroups.

The semigroup S is called the *structure semigroup* of B. We now give some examples.

EXAMPLES. (i) Let G be a locally compact abelian group and $B = L^1(G)$. Then the structure semigroup for B is bG, the Bohr compactification of G.

(ii) Let G be as in (i), and let $B = \text{Rad } L^1(G)$, the set of measures $\mu \in M(G)$ such that $\hat{\mu}(\psi) = 0$ for all $\psi \in \Delta M(G) \backslash \Gamma$, where Γ is the dual group of G. Again, bG is the structure semigroup. See 7.2.4 for the existence of singular measures in Rad $L^1(G)$.

(iii) Let X be any discrete abelian semigroup such that $\hat{X}$ separates points of X. Then the structure semigroup of $M(X) = l^1(X)$ is the almost periodic compactification of X. See, for example, Burckel [1, pp. 4, 6]. Note that this example generalizes the discrete case of (i).

(iv) Let $X = [0, 1]$ with maximum multiplication: $st = \max(s, t)$. The structure semigroup of $M(X)$ is the set $\{(x, i): 0 \leq x \leq 1, i = 0, 1\}$, where $(x, i)(y, j) = (\max(x, y), \max(i, j))$, and the topology is the "interval topology" induced by the lexicographical order. That seemingly peculiar form of S is due to the identification of the multiplicative linear functionals on $M(X)$ with integration against indicator functions of intervals $[0, a]$ and $[0, a)$. The reader may work out the details for himself or consult Kuhlmann [1]. For more about $M(X)$, and other algebras on intervals and products of intervals, see Hewitt and Zuckermann [4], Ross [2], Baartz [1], Akst [1], and Kuhlmann [1].

Theorems 5.1.1 and 5.1.2 provide two representations of the maximal ideal space of a CCMA B. We shall usually identify the elements of ΔB with generalized characters f for B and identify them also with the continuous semicharacters on S. That identification of $\hat{S}$ with ΔB gives $\hat{S}$ the Gel'fand topology, or weak topology. It is usually the case that weak convergence in $\hat{S}$ does not imply strong convergence ($\psi_\alpha \to \psi$ strongly if $\|\bar{\psi}_\alpha \mu - \bar{\psi}\mu\| \to 0$ for all $\mu \in B$). See 5.1.4 for cases in which weak convergence does imply strong convergence. The CCMA $B = L^1(R)$, with $S = bR$, gives an example in which the topologies of weak convergence and strong convergence coincide, but for which neither is equivalent to the topology of uniform convergence on compact subsets of S.

Generalized characters ψ such that $|\psi_\mu(x)| = 1$ a.e. $d\mu$ for all $\mu \in B$ are special, as the following result of Brown and Moran [1] shows.

5.1.3. Proposition. *Let B be a CCMA and ψ a generalized character on B. If*

$$|\psi_\mu(x)| = 1 \quad \text{a.e. } d\mu \quad \textit{for all } \mu \in B, \tag{9}$$

then ψ represents an element of the Šilov boundary of B.

Proof. It will suffice to show that if U is a neighborhood of ψ then there exists $\mu \in B$ such that $\hat{\mu}(\psi) = 1$ and $|\hat{\mu}| < 1$ outside U.

Let U be any neighborhood of ψ. Then there exist $n \geq 1$, $\varepsilon > 0$ and $\mu_1, \ldots, \mu_n \in B$ such that the open set $V = \{\rho: |\hat{\mu}_j(\rho) - \hat{\mu}_j(\psi)| < \varepsilon$ for $1 \leq j \leq n\}$ is contained in U. Let

$$\mu = a \mathop{\bigstar}_{j=1}^{n} [\psi_{\mu_j}|\mu_j| + (\psi_{\mu_j}|\mu_j|) * (\psi_{\mu_j}|\mu_j|)],$$

where $a > 0$ is chosen to be such that $\|\mu\| = 1$. Since $\hat{\mu}(\psi) = \sum(\|\mu_j\| + \|\mu_j\|^2) = \|\mu\| \neq 0$, such a choice of a is possible.

If $\rho \in \Delta B$, and $\rho_{\mu_j} \neq \psi_{\mu_j}$ on a set of positive μ_j-measure, for some j, then $|\hat{\mu}(\rho)| < 1$. Thus, $|\hat{\mu}| < 1$ outside of V. Of course, $\hat{\mu}(\psi) = 1 = \|\mu\| = \sup|\hat{\mu}|$. Since $V \subseteq U$, $|\hat{\mu}| < 1$ outside U. The Proposition is proved. □

Remarks. (i) Neither (9) nor the weaker condition

$$(10) \qquad |\psi_\mu(x)|^2 = |\psi_\mu(x)| \text{ a.e. } d\mu \quad \text{for all } \mu \in B$$

is necessary for ψ to be in ∂B. See 6.2.9, from which a proof can easily be extracted.

Formula (10) is not sufficient for $\psi \in \partial B$: consider $B = l^1(N)$, the algebra of absolutely convergent Taylor series. The generalized characters for B are the functions $\varphi_z(n) = z^n$, where $|z| \leq 1$. ($0^0 = 1$ here.) Then $\varphi_0^2 = \varphi_0$, but φ_0 represents evaluation of the absolutely convergent Taylor series $\sum a_n z^n$ at zero, and the resulting multiplicative linear functional is not in the Šilov boundary.

(ii) It is easy to show that each $\varphi \in \Delta B$ has the form $\varphi = \lambda|\varphi|$ where $\lambda \in \Delta B$ and $|\lambda|^2 = |\lambda|$. That is the *polar decomposition* of φ.

(iii) There is an obvious multiplication in ΔB: $(\psi\varphi)_\mu(x) = \psi_\mu(x)\varphi_\mu(x)$ for all $\mu \in B, x \in X$. There is also an exponentiation $z \to \lambda|\varphi|^z$, for Re $z > 0$, where $\varphi = \lambda|\varphi|$ is the polar decomposition of φ. It is easy to see that $z \to \hat{\mu}(\lambda|\varphi|^z)$ is an analytic function of z for all $\mu \in B$. Those functions are discussed further in Chapter 8: see 8.2.1 and Sections 8.5 and 8.6.

5.1.4. Proposition. *Let B be a CCMA with maximal ideal space ΔB. Then:*

(i) *Multiplication is separately continuous in ΔB.*
(ii) *If Λ is an open subgroup of ΔB, then Λ is a locally compact abelian group.*
(iii) *If $\{\psi_\alpha\}$ is a net in ΔB, $\lim \psi_\alpha = \psi$, and $|\psi_\alpha| \leq |\psi|$ for all α, then ψ_α converges strongly to ψ.*
(iv) *If $\psi \geq 0$ and $\psi_\alpha \geq \psi$ with $\psi_\alpha \to \psi$, then ψ_α converges strongly to ψ.*

Proof. (i) is immediate from the definition of the Gel'fand (weak$*$) topology and the multiplication on ΔB.

(ii) Suppose that $\gamma_\alpha \to \gamma$ and $\lambda_\beta \to \lambda$ in Λ. Then for all $\mu \in B$ and all indices α, β

$$|\hat{\mu}(\gamma\lambda) - \hat{\mu}(\gamma_\alpha\lambda_\beta)| \le |\hat{\mu}(\gamma\lambda) - \hat{\mu}(\gamma_\alpha\lambda)| + |\hat{\mu}(\gamma_\alpha\lambda) - \hat{\mu}(\gamma_\alpha\lambda_\beta)|. \tag{11}$$

But $|\hat{\mu}(\gamma_\alpha\lambda) - \hat{\mu}(\gamma_\alpha\lambda_\beta)| \le \|(\overline{\gamma_\alpha\lambda})\mu - (\overline{\gamma_\alpha\lambda_\beta})\mu\| \le \|\bar{\lambda}\mu - \bar{\lambda}_\beta\mu\|$. Since $|\lambda_\beta| = |\lambda|$ for all β,

$$\begin{aligned} \|\bar{\lambda}\mu - \bar{\lambda}_\beta\mu\|^2 &= \left(\int |\bar{\lambda} - \bar{\lambda}_\beta| d|\mu|\right)^2 \le \|\mu\| \int |\bar{\lambda} - \bar{\lambda}_\beta|^2 \, d|\mu| \\ &\le \|\mu\| \int (2|\lambda|^2 - \lambda\bar{\lambda}_\beta - \bar{\lambda}\lambda_\beta) d|\mu| \\ &= 2\|\mu\| \operatorname{Re}[(\lambda|\mu|)\hat{}(\lambda) - (\lambda|\mu|)\hat{}(\lambda_\beta)]. \end{aligned} \tag{12}$$

Since the first summand on the right-hand side of (11) and the last line of (12) tend to zero, $\gamma_\alpha\lambda_\beta$ converges strongly to $\gamma\lambda$, and (ii) is proved.

(iii) We must show that $\|\bar{\psi}_\alpha\mu - \bar{\psi}\mu\| \to 0$ for all $\mu \in B$. We may assume that $\mu \ge 0$. Then

$$\begin{aligned} \|\bar{\psi}_\alpha\mu - \bar{\psi}\mu\|^2 &\le \|\mu\| \int |\psi_\alpha - \psi|^2 \, d\mu \\ &= \|\mu\| \int (|\psi|^2 - \bar{\psi}\psi_\alpha - \psi\bar{\psi}_\alpha + |\psi_\alpha|^2) d\mu \\ &\le \|\mu\| \int (2|\psi|^2 - \bar{\psi}\psi_\alpha - \psi\bar{\psi}_\alpha) d\mu \\ &= \|\mu\| 2 \operatorname{Re}[(\psi\mu)\hat{}(\psi) - (\psi\mu)\hat{}(\psi_\alpha)] \to 0. \end{aligned}$$

(iv) As in (iii) we may assume that $\mu \ge 0$. Then $\|\bar{\psi}_\alpha\mu - \bar{\psi}\mu\| = \int (\bar{\psi}_\alpha - \bar{\psi}) d\mu = \hat{\mu}(\psi_\alpha) - \hat{\mu}(\psi) \to 0$. The Proposition is proved. □

Suggestions for further reading. The best place to begin is with the first three chapters of Taylor's lectures [1]. For the abstract definition of CCMA see Taylor [3]. Brown [1] provides another approach to the structure semigroup and Rennison [1, 2] a third. For results related to the latter two authors' work, see Chow and White [1], McKilligan and White [1], White [1], and Ylinen [1]. For more on Johnson's [4] separate continuity and measurability result, see Wong [1] and Moran [1].

5.2. *L*-Subalgebras and *L*-Ideals

An *L-ideal* of a CCMA is an L-subalgebra that is an ideal. An L-subalgebra B_1 of B is a *prime L-subalgebra* if $B_1^\perp = \{\mu \in B : \mu \perp \nu \text{ for every } \nu \in B_1\}$ is an ideal. An L-ideal B_1 is a *prime L-ideal* if $B_1^\perp$ is an algebra. For example, if G is a nondiscrete group, then $M_d(G)$, the set of discrete measures in $M(G)$, is a prime L-subalgebra, and the set of continuous measures $M_c(G)$ is a prime L-ideal. That is implicit in Examples (iii) and (iv) following 5.1.1. More general versions of those examples are as follows.

Let G and H be locally compact abelian groups, and let $\varphi: H \to G$ be a continuous, one-to-one, group homomorphism. We define a new topology on G by declaring the image of H to be open in G. On each translate of φH, we place the topology of φH, and let τ denote the resulting new topology on G.

Let G_τ denote G equipped with the topology τ. Then G_τ is a (possibly new) locally compact abelian group called a *refinement* of G. The identity map $G_\tau \to G$ is a continuous, one-to-one and onto homomorphism of locally compact abelian groups, and $M(G_\tau)$ may therefore be identified with its image $\check{\varphi}M(G_\tau)$ in $M(G)$. That image is obviously an L-subalgebra. We shall show in (5.2.1) below that the image is prime, but we first generalize the preceding class of examples. A *Raĭkov system* on the locally compact abelian group G is a collection $\mathscr{R}$ of Borel subsets of G such that

(i) if $E \in \mathscr{R}$ and $F \subseteq E$ is an F_σ subset of G, then $F \in \mathscr{R}$;
(ii) if $E_1, E_2, \ldots, \in \mathscr{R}$, then $\bigcup_{j=1}^{\infty} E_j \in \mathscr{R}$;
(iii) if $E_1, E_2 \in \mathscr{R}$, then $E_1 + E_2 \in \mathscr{R}$; and
(iv) $\{x\} \in \mathscr{R}$ for all $x \in G$.

If $\mathscr{R}$ is a Raĭkov system on G, we define

$$A_{\mathscr{R}} = \{\mu \in M(G): |\mu|(E) = \|\mu\| \text{ for some } E \in \mathscr{R}\}; \quad \text{and}$$

$$I_{\mathscr{R}} = \{\mu \in M(G): |\mu|(E) = 0 \text{ for all } E \in \mathscr{R}\}.$$

If $\mathscr{R}$ is the collection of countable subsets of G, then $A_{\mathscr{R}} = M_d(G)$ and $I_{\mathscr{R}} = M_c(G)$. If G_τ is a refinement of G, we let $\mathscr{R}$ be the Raĭkov system generated by any compact τ-neighborhood of 0. It is easy to see that $\check{\varphi}M(G_\tau) = A_{\mathscr{R}}$ in that case.

5.2.1. Theorem. *Let $\mathscr{R}$ be a Raĭkov system on the locally compact abelian group G. Then $A_{\mathscr{R}}$ is a prime L-subalgebra of $M(G)$.*

Proof. That $A_{\mathscr{R}}$ is an L-subalgebra follows easily from the definition of a Raĭkov system. It will suffice to show that $I_{\mathscr{R}}$ is an ideal.

Let $\nu \in I_{\mathscr{R}}$ and $\mu \in M(G)$. Let $E \in \mathscr{R}$. Then

$$|\mu * \nu|(E) \leq |\mu| * |\nu|(E) = \int |\nu|(E - x)d|\nu|(x) = 0$$

since $E + x \in \mathscr{R}$ for all x, and $|\nu|(F) = 0$ for all $F \in \mathscr{R}$. The Theorem is proved. □

Some L-ideals of $M(G)$ that are *not* prime are $L^1(G)$, Rad $L^1(G)$, and $M_o(G)$. Those assertions are easily established using Riesz products and A.7.1.

5.2.2. Theorem. *Let A be an L-subalgebra of the CCMA B. Then A is prime if and only if there exists $\psi \in \Delta A$ such that*

(i) $\psi_\mu = 1$ a.e. $d\mu$ *for all* $\mu \in A$; *and*
(ii) $\psi_\nu = 0$ a.e. $d\nu$ *for all* $\nu \perp A$.

Furthermore, if A is prime, then $\Delta A = \psi \Delta B$ where ψ is as in (i)–(ii).

The proofs of 5.2.2 and of the next result are left to the reader.

5.2.3. Corollary. *Let K be a closed subgroup of the locally compact abelian group G. Then every element of $\Delta M(K)$ has an extension to an element of $\Delta M(G)$.*

Let G be a locally compact abelian group and G_τ a refinement of G. Let $A = M(G_\tau)$. The element $\psi^{(\tau)}$ whose existence is guaranteed by 5.2.2 is the *critical point* associated with the refinement G_τ. See Section 5.3 for the properties of critical points and some of their applications.

Raĭkov systems are not the only way to product prime L-subalgebras of $M(G)$, as the next result shows. For related results, see 6.2.12, 6.2.14, 6.3.10 and 6.3.11.

5.2.4. Theorem. *Let G be a locally compact abelian group and let Λ be a set of characters on G. Let $S(\Lambda)$ be the set of all measures $\mu \in M(G)$ such that every $\gamma \in \Lambda$ is μ-measurable. Then $S(\Lambda)$ is a prime L-subalgebra of $M(G)$.*

We begin the proof of 5.2.4 with two lemmas.

5.2.5. Lemma. *Let G be a locally compact abelian group, let $\mu \in M(G)$ be a non-negative measure, and let f be any complex-valued function on G. For each Borel set $E \subseteq G$, let $\mu_f(E) = \sup\{\mu(F)\colon F \subseteq E,\ F$ compact, f is continuous on $F\}$.*

Then

(i) *μ_f is a regular Borel measure on G;*
(ii) $\mu_f \ll \mu$;
(iii) $(\mu - \mu_f) \perp \mu_f$;
(iv) *whenever $0 \neq \nu \ll \mu - \mu_f$, f is not ν-measurable; and*
(v) *$d\mu_f/d\mu = 0$ or 1 a.e. $d\mu$.*

Proof. (i), (ii), (iii), and (v) are routine and are left to the reader.

(iv) Suppose that $0 \neq \nu \ll \mu - \mu_f$ are that f were ν-measurable. Then Lusin's Theorem (Rudin [2, p. 56]) would imply that there exists a closed subset E on which f is continuous and for which $\nu(E) \neq 0$. We may assume that E is compact.

But then $\mu_f(E) \neq 0$ since f is continuous on E. Hence $\nu \not\perp \mu_f$, which contradicts (iii). That ends the proof of 5.2.5. □

5.2.6. Lemma. *Let $f: G \to \mathbb{C}$, and $\sigma, \tau \in M(G)$. If either side of the following equation exists, then both sides exist and they are equal.*

$$\int \overline{f(z)} d(\sigma * \tau)(z) = \int \overline{f(x + y)} d\sigma(x) d\tau(y).$$

Proof. We may assume that σ and τ are non-negative. For each subset E of G, let $E^* = \{(x, y) \in G \times G : x + y \in E\}$. Since addition is a continuous mapping from $G \times G$ to G, the set E^* is Borel whenever E is Borel. Suppose that E is $\sigma * \tau$-measurable. We claim that E^* is $\sigma \times \tau$-measurable. To prove that, it suffices to show that

(1) $$\inf \sigma \times \tau(F_1) = \sup \sigma \times \tau(F_2)$$

where the infimum is taken over all Borel sets $F_1 \supseteq E^*$ and the supremum over all Borel sets $F_2 \subseteq E^*$. But E is $\sigma * \tau$-measurable, so

$$\inf \sigma * \tau(F_1) = \sigma * \tau(E) = \sup \sigma * \tau(F_2),$$

where F_1 ranges over Borel sets $F_1 \supseteq E$ and F_2 over Borel sets $F_2 \subseteq E$. Since $\sigma * \tau(F) = \sigma \times \tau(F^*)$ for all Borel sets F in G, (1) now follows. Similarly, if E^* is $\sigma \times \tau$-measurable, then E is $\sigma * \tau$-measurable. It will suffice to prove the Lemma for real-valued functions f. Suppose that $\int f(z) d(\sigma * \tau)(z)$ exists. Let $\alpha, \beta \in R$, $\alpha < \beta$, and let $E_{\alpha,\beta} = \{x; \alpha < f(x) \leq \beta\}$. Then $E_{\alpha,\beta}$ must be $\sigma * \tau$-measurable. By the preceding, $E^*_{\alpha,\beta}$ is $\sigma \times \tau$-measurable. Since α and β are arbitrary, the mapping $(x; y) \to f(x + y)$ is $\sigma \times \tau$-measurable, so $\int f(x + y) d\sigma(x) d\tau(y)$ exists. A similar argument completes the proof of the Lemma. □

Proof of 5.2.4. That $S(\Lambda)$ is a closed L-subspace is easily shown. Let $\gamma \in \Lambda$, and $\sigma, \tau \in S(\Lambda)$. We may assume that $\sigma, \tau \geq 0$. Then $\int \gamma(s) d\sigma(s)$ and $\int \gamma(t) d\tau(t)$ are defined, so

$$\int \overline{\gamma(s)} d\sigma(s) \int \overline{\gamma(t)} d\tau(t) = \iint \overline{\gamma(s + t)} d\sigma(s) d\tau(t)$$

is defined by 5.2.6, and therefore γ is $\sigma * \tau$-measurable. It follows that $S(\Lambda)$ is an algebra. We must show that $S(\Lambda)^\perp = \{\mu \in M(G) : \mu \perp \nu \text{ for all } \nu \in S(\Lambda)\}$ is an ideal.

Let $\mu \in S(\Lambda)^{\perp}$, $\mu \neq 0$. For each non-zero $\nu \ll \mu$, there exists $\lambda \in \Lambda$ such that λ is not ν-measurable. However, there may exist a non-zero $\omega \ll \mu$ such that the same λ *is* ω-measurable. But we shall show that μ may be written as a sum $\mu = \sum \sigma_j$, where for each j there exists $\lambda(j) \in \Lambda$ such that $\nu \ll \sigma_j$ and $\nu \neq 0$ imply that $\lambda(j)$ is not ν-measurable. We may assume that $\mu \geq 0$. We define inductively sequences $\{\mu_j\}$, $\{\sigma_j\}$ and $\{\lambda(j)\}$ as follows, letting $\mu_1 = \mu$. When μ_j is defined, if $\mu_j = 0$, let $\sigma_j = 0$, $\lambda(j) = 1$ and terminate the induction. Otherwise choose $\lambda(j) \in \Lambda$ so that $\|\mu_j - (\mu_j)_{\lambda(j)}\| > \frac{1}{2} \sup_{\lambda \in \Lambda} \|\mu_j - (\mu_j)_\lambda\|$. Then let $\mu_{j+1} = (\mu_j)_{\lambda(j)}$ and $\sigma_j = \mu_j - \mu_{j+1}$. That ends the inductive step.

It follows at once that the σ_j are pairwise mutually singular, and that for every j, either $\sigma_j = 0$ or $\lambda(j)$ is non-ν-measurable for every non-zero $0 < \nu \ll \sigma_j$. By the definition of the σ_j,

$$\sum_1^\infty \sigma_j = \sum_1^\infty (\mu_j - \mu_{j+1}) = \mu_1 = \mu. \tag{2}$$

Now let $\tau \in M(G)$. To show that $\tau * \mu \in S(\Lambda)^{\perp}$, it will suffice to show that $\tau * \sigma_j \in S(\Lambda)^{\perp}$ for $j = 1, 2, \ldots$ where the σ_j are as above. There is no loss of generality in assuming that $\tau \geq 0$. Let j be fixed.

By Lemma 5.2.6, $\lambda(j)$ is $\tau * \sigma_j$-measurable if and only if

$$\lambda(j)(x + y) = \lambda(j)(x)\lambda(j)(y)$$

is $\tau \times \sigma_j$-measurable. Since $\lambda(j)$ is non-ν-measurable for all nonzero $\nu \ll \sigma_j$, we conclude that $\lambda(j)$ is not $\tau * \sigma_j$-measurable. That holds also for all non-zero $\tau' \ll \tau$, since products of the form $\tau' * \nu$, where $0 < \nu \ll \sigma_j$, span the set $\{\omega: 0 < \omega \ll \tau * \sigma_j\}$. Therefore $\tau * \sigma_j \in S(\Lambda)^{\perp}$. The Theorem is proved. □

We now give a result analogous to 5.1.4.

5.2.7. Proposition. *Let A be a prime L-subalgebra of the CCMA B. Suppose that $\psi \in \Delta B$ has $\psi_\nu = 0$ a.e. $d\nu$ for all $\nu \in A^{\perp}$. If ψ is a strong boundary point for the uniform closure of $B^\wedge$ in $C(\Delta B)$, then $\psi|_A \in \partial A$.*

Proof. We may identify ΔA with $\{\rho \in \Delta B: \rho_\nu \equiv 0$ a.e. $d\nu$ for all $\nu \in A^{\perp}\}$ by 5.2.2. Let the neighborhood $U \subseteq \Delta A$ of ψ be fixed. Then $V = \{\rho \in \Delta B: \rho|_A \in U\}$ is a neighborhood of ψ in ΔB. Since ψ is a strong boundary point for $\hat{B}^-$, there exists $\omega \in B$ such that $\hat{\lambda}(\psi) = 1$ and $\sup\{|\lambda^\wedge(\sigma)|: \rho \in \Delta B \setminus V\} < \frac{1}{2}$. Let $\omega = \mu + \nu$ where $\mu \in A$ and $\nu \perp A$. Since $\psi_\nu \equiv 0$ a.e. $d\nu$, $|\hat{\mu}(\psi)| = 1$. But then $\hat{\mu}(\rho) = \hat{\omega}(\rho)$ for all $\rho \in \Delta A$. Therefore $\sup\{|\hat{\mu}(\rho)|: \rho \in \Delta A\} \geq 1$ and $\sup\{|\hat{\mu}(\rho)|: \rho \in \Delta A \setminus U\} < \frac{1}{2}$. The Proposition is proved. □

Remarks and Credits. Raĭkov systems were introduced at least as early as Gel'fand, Raĭkov and Šilov [1]. For a guide to the extensive literature on the set of locally compact abelian group topologies that a group may have, consult Rickert [1], Sully [1], and Miller and Rajagapolan [1] and the bibliographies of those three papers. Theorem 5.2.2 is due to Taylor [3]. Theorem 5.2.4 is due to Šreĭder [1] as are Lemmas 5.2.5 and 5.2.6. Whether $S(\Lambda) = M(G_\tau)$ when $\Lambda = (G_\tau)^\wedge$ appears to be an open question, even in the case that τ is the discrete topology.

Saka [1] proved that when $\mathscr{R}$ is a Raĭkov system, $A_{\mathscr{R}} = M(G_\tau)$ if and only if there exists a non-zero $\mu \in A_{\mathscr{R}}$ such that the mapping $x \to \mu(E + x)$ is τ continuous for all compact $E \in \mathscr{R}$. (The measure μ is absolutely continuous with respect to Haar measure on G_τ.) For related matters, see Section 8.2.

Proposition 5.2.7 is abstracted from Saeki and Sato [1, Proof of Theorem 1(a)].

5.3. Critical Point Theory and a Proof of the Idempotent Theorem

Let B be a semisimple CCMA. Then by 5.1.2, B can be identified with an L-subalgebra of $M(S)$ where S is the structure semigroup of B.

We set $\hat{S}^+ = \{\psi \in \hat{S} : \psi \geq 0\}$. An element ψ of $\hat{S}^+$ is a *critical point* if it is isolated in the set $\{\rho \in \hat{S}^+ : \rho \leq \psi\}$. (By 5.1.4(v), ψ is isolated in the strong topology if and only if it is isolated in the Gel'fand topology.)

For a proof of the next result, see Taylor [1, 5.2.4].

5.3.1. Lemma. *Let $\psi \in \hat{S}^+$. Then the following are equivalent.*

(i) *ψ is a critical point.*
(ii) *$\{\rho \in \hat{S} : |\rho| = \psi\}$ is open in $\psi\hat{S} = \{\rho \in \hat{S} : |\rho| \leq \psi\}$.*
(iii) *ψ is a minimal element of some strongly open and closed subset of $\hat{S}^+$.*

Let $\psi \in \hat{S}$ be a critical point. Then $S(\psi) = \{s \in S : \psi(s) = 1\}$ is an open and closed subsemigroup in S. Every compact, commutative topological semigroup is known (Hoffman and Mostert [1, pp. 14–15] or Paalman-de Miranda [1, p. 32]) to have a minimal closed ideal, the *kernel*. The kernel is always a group. We let $K(\psi)$ denote the kernel of $S(\psi)$ and $\Gamma(\psi) = \{\rho \in \hat{S} : |\rho| = \psi\}$. A *group algebra* in B is an L-subalgebra of B that is isomorphic as a CCMA to $L^1(G)$ for some locally compact abelian group G.

5.3.2. Proposition.

(i) *Let G be a maximal group in S. Then G is closed. If $v(G) \neq 0$ for some $v \in B$, then $G = K(\psi)$ for some critical point $\psi \in \hat{S}^+$.*
(ii) *Suppose that N is a group algebra in B. Then there exists a critical point $\psi \in \hat{S}^+$ such that every measure in N is supported on $K(\psi)$.*

Proof. (i) Suppose that G is a subgroup of S. Let $\rho \in \hat{S}$. Then either $|\rho(x)| = 1$ for all $x \in G$ or $|\rho(x)| = 0$ for all $x \in G$. If $\{x_\alpha\} \subseteq G$, and $\lim x_\alpha = x$, then $\rho(x_\alpha) \to \rho(x)$ for all $\rho \in \hat{S}$. Therefore $\rho(x_\alpha^{-1}) = \rho(x)^-$, so each x in the closure of G has an inverse. Therefore maximal groups are closed.

If $\nu(G) \neq 0$ for some $\nu \in B$, then there is a probability measure $\mu \in B$ such that $\mu(G) = 1$. The restriction to G of any $\rho \in \hat{S}$ is either identically zero or equal to a continuous character on G. If $\rho \in \hat{S}^+$, then either $\rho \equiv 1$ on G or $\rho \equiv 0$ on G. Hence for each $\rho \in \hat{S}^+$ either $\hat{\mu}(\rho) = 1$ or $\hat{\mu}(\rho) = 0$. Therefore the set

$$I = \{\rho \in \hat{S}^+ : \hat{\mu}(\rho) = 1\}$$

is strongly open and closed in $\hat{S}^+$. Let ψ be a minimal element of I. Now 5.3.1 implies that ψ is a critical point. Obviously, $\psi \equiv 1$ on G and $S(\psi) \supseteq G$. Furthermore, if $x \in \operatorname{supp} \mu$, then $\psi(x) = 1$.

We now use the minimality of ψ to show that $K(\psi) = G$. Let p be the identity of $K(\psi)$ and q the identity of G. If $p \neq q$, there exists $f \in \hat{S}$ such that $f(p) \neq f(q)$. Since $p^2 = p$, $q^2 = q$ and $pq = p$ (by the minimality of $K(\psi)$), we see that $f(p) = 0$ and $f(q) = 1$. Therefore $f < \psi$ and $f\psi$ is an element of I that is strictly smaller than ψ. That contradiction proves that $K(\psi) = G$.

(ii) Suppose that there is an L-algebra isomorphism from $L^1(G)$ to an L-subalgebra N of B. The structure semigroup of $L^1(G)$ is the Bohr compactification bG of G. It is not difficult to show (Taylor [1, p. 26]) that an L-algebra homomorphism $\varphi\colon B_1 \to B_2$ induces a topological semigroup homomorphism $\tilde{\varphi}$ of the structure semigroup S_1 of B_1 into the structure semigroup S_2 of B_2 and that the elements of φB_1 are supported in $\tilde{\varphi} S_1$. The map $\tilde{\varphi}$ is defined by the property that $\varphi\mu = \mu \circ \tilde{\varphi}$ for all $\mu \in M(G)$. Thus, elements of $N \simeq L^1(G)$ are supported in the closed subgroup $\tilde{\varphi} bG$ of S. We claim that $\tilde{\varphi}$ is one-to-one. Indeed, since the Silov boundary of $L^1(G)$ equals $\Delta L^1(G)$, every multiplicative linear functional on $L^1(G) \simeq N$ extends to a multiplicative linear functional on B. Therefore every continuous character on $\tilde{\varphi} G$ can be obtained by restricting an element of $\hat{S}$. Therefore $\hat{S}|_{\tilde{\varphi} G} = \hat{S}|_{\tilde{\varphi} bG} = (\tilde{\varphi} G)^\wedge = (bG)^\wedge$, where each of the preceding equality is one of groups, without taking into consideration their topologies. Therefore $\tilde{\varphi}\colon bG \to S$ is one-to-one. We now pass to the maximal group containing $\tilde{\varphi} bG$, apply (i) and obtain the required conclusion. The Proposition is proved. □

The proof of the next theorem is difficult; the best approach to the proof is through Taylor [1] followed by Taylor [6].

5.3.3. Theorem. *Let B be a semisimple CCMA and ψ a critical point for B. Then there exists a locally compact abelian group G and an L-subalgebra M of $M(G)$ such that the following hold.*

(i) $L^1(G) \subseteq M \subseteq \text{Rad } L^1(G)$.
(ii) $\Gamma(\psi)$ *is the dual group and* $K(\psi)$ *is the Bohr compactification of* G.
(iii) *There is an injective L-homomorphism* $\varphi: M \to B$ *such that* $\varphi M = \{\mu \in B: \text{support } \mu \subseteq K(\psi)\}$.
(iv) *If* $B = M(H)$ *for some locally compact abelian group* H, *then* $\psi B = M(H_\tau)$ *where* H_τ *is a refinement of* H *and* $M = \text{Rad } L^1(H_\tau)$.

We now turn to some applications of 5.3.3 that use what we call *critical point induction*, which allows us to prove for μ a property satisfied by $\psi\mu$ for all $0 \leq \psi < 1$. We begin with some notation. We let $\Delta^+ = \Delta^+ M(G)$ be the set of those $\psi \in \Delta M(G)$ such that $\psi_\mu \geq 0$ a.e. $d\mu$ for all $\mu \in M(G)$. Note that by 5.1.2, Δ^+ and $\hat{S}^+$ are the same, where S is the structure semigroup of $M(G)$. We say that $\psi \geq \rho$ for ψ and $\rho \in \Delta^+$ if $\psi_\mu \geq \rho_\mu$ a.e. $d\mu$ for all $\mu \in M(G)$.

5.3.4. *A Proof of the Idempotent Theorem.* Let $H^0(G)$ be the additive subgroup of $M(G)$ generated by the idempotents of $M(G)$. Let $H_1^0(G)$ be the additive subgroup of $M(G)$ generated by measures $\gamma\mu$ where $\gamma \in \Gamma$, the dual group of G, and μ is normalized Haar measure on a compact subgroup of G. If G_τ is any refinement of G, then $H^0(G_\tau) \subseteq H_1^0(G)$ and $H_1^0(G_\tau) \subseteq H_1^0(G)$. If $\mu \in H^0(G)$, then $\hat{\mu}$ is integer-valued. Therefore if μ, $\nu \in H^0(G)$ and $\mu \neq \nu$, $\|\mu - \nu\| \geq \|\hat{\mu} - \hat{\nu}\|_\infty \geq 1$.

We claim that if $\mu \in H^0(G)$ and $\psi\mu \in H_1^0(G)$ for all $\psi \in \Delta^+$ with $\psi \neq 1$, then $\mu \in H_1^0(G)$. Let $\psi \in \Delta^+$. Then

$$\|\mu - \psi\mu\| + \|\psi\mu\| = \|\mu - \psi^t\mu\| + \|\psi^t\mu\|$$

for $0 < t \leq 1$, since $\hat{\mu}$ and $(\psi^t\mu)^\wedge$ are integer-valued. Letting $t \to 0$, we see that $\|\mu - \psi\mu\| + \|\psi\mu\| = \|\mu - \psi^0\mu\| + \|\psi^0\mu\|$. Since $\psi^0 = \lim_{t\to 0} \psi^t$ is idempotent, $(\mu - \psi\mu) \perp \psi\mu$. Therefore $\|\mu\| = \|\mu - \psi\mu\| + \|\psi\mu\|$. Therefore for all $\psi \in \Delta^+$ with $\psi \neq 1$ either $\psi\mu = 0$, or $\|\mu - \psi\mu\| \leq \|\mu\| - 1$. By induction and our hypothesis, $\mu = \nu + \omega$ where $\nu \in H_0^1(G)$ and $\psi\omega = 0$ for all $\psi \in \Delta^+ \setminus \{1\}$. But then for all $\rho \in \Delta \setminus \Gamma$, $|\hat{\omega}(\rho)| \leq \| |\rho| \omega\| = 0$, so $\omega \in \text{Rad } L^1(G) \subseteq M_0(G)$. Since $\omega \in H^0(G)$ also, the support of the Fourier-Stieltjes transform of ω is a compact-open subset of Γ. But every compact-open subset of Γ is a finite union of cosets of compact-open subgroups of Γ (Rudin [1, 2.3.4(c)]). Therefore $\omega \in H_1^0(G)$. The claim is established.

Let $\mu \in H^0(G)$ be fixed and let

$$\Lambda = \{\psi \in \Delta^+ M(G): \psi\mu \notin H_1^0(G)\}.$$

We claim that Λ is open and closed in the strong topology. Indeed, $H_1^0(G)$ is a closed subgroup of the discrete group $H^0(G)$. Therefore $H_1^0(G)$ is both open and closed and $H^0(G) \setminus H_1^0(G)$ is open and closed in $H^0(G)$. Since $\psi \to \psi\mu$ is strongly continuous, Λ is open and closed. Therefore Λ has a minimal element ρ. By 5.3.1, ρ is a critical point. By 5.3.3(iv) $\rho M(G) = M(G_\tau)$

for some refinement G_τ of G. Since $M(G_\tau)$ is a prime L-subalgebra of $M(G)$, $\Delta M(G_\tau) \subseteq \Delta M(G)$ and $\Delta^+ M(G_\tau) \subset \Delta^+ M(G)$, the identifications being made via the map $v \to \hat{v}(\rho\psi)$, where $v \in M(G)$ and either $\psi \in \Delta M(G_\tau)$ or $\psi \in \Delta^+ M(G_\tau)$. Therefore, if $0 \leq \psi < 1$ and $\psi \in \Delta M(G_\tau)$, then $0 \leq \psi < \rho$, so, by the minimality of ρ, $\psi\rho\mu \in H_1^0(G)$. But then $\psi\rho\mu \in \rho H_1^0(G) = H_1^0(G_\tau)$. Since the preceding holds for all $\psi \in \Delta^+ M(G_\tau)$ with $\psi < 1$, $\rho\mu \in H_1^0(G_\tau) = \rho H_1^0(G) \subseteq H_1^0(G)$. Therefore Λ cannot have a minimal element. Therefore $\Lambda = \emptyset$. The Idempotent Theorem is proved. □

The *spine* $\mathscr{L}(B)$ of a CCMA B is the L-algebra generated by the group algebras contained in B. When $B = M(G)$, we write $\mathscr{L}(G)$ in place of $\mathscr{L}(M(G))$. A more elaborate version of the argument used in 5.3.4 proves the next result. See Taylor [6] for the details, as well as for the proof of the corollary that follows.

5.3.5. Theorem. *Let G be a locally compact abelian group and $\mu \in M(G)$ an invertible measure. Then*

$$\mu = v * \exp(2\pi i\omega)$$

where $\omega \perp \mathscr{L}(G)$ and $v \in \mathscr{L}(G)$.

5.3.6. Corollary. *If $\mu \in M(G)$ is invertible, then $\mu = \mu_1 * \cdots * \mu_n * \exp(2\pi i\omega)$ where $\omega \perp \mathscr{L}(G)$, $\mu_j \in [L^1(G_{\tau_j}) + C\delta(0)]$ and the G_{τ_j} are refinements of G for $1 \leq j \leq n$.*

Remarks and Credits. When $G = R$, the spine is just $M_d(R) + L^1(R)$. The maximal ideal space of $L^1(R) + C\delta(0)$ is $R \cup \{0\}$, a circle. By 5.3.6, an invertible measure μ in $L^1(R) + C\delta(0)$ has a logarithm if and only if $\hat{\mu}$ has winding number zero around the origin. If μ is an invertible discrete measure, then $\mu = \delta(x) * \exp(2\pi i\omega)$, for some unique $x \in R$ and $\omega \in M_d(R)$ by a result of H. Bohr [1, p. 55]. (See Taylor [1, 8.3.3] for a modern proof.) Thus each invertible measure $\mu \in M(R)$ has the form

$$\mu = \eta^k * \delta(x) * \exp(2\pi i\omega)$$

where $x \in R$, $\eta \in L^1(R) + C\delta(0)$, $k \in Z$ and $\omega \in M(R)$. The numbers k and x are unique. That result has applications to Wiener-Hopf operators. See Douglas and Taylor [1] and Taylor [1].

If G is a locally compact abelian group, $(\hat{G})_\infty$ will denote $\hat{G}$ if $\hat{G}$ is compact, and the one-point (Alexandroff) compactification if $\hat{G}$ is not compact. Theorems 1.1.1 and 5.3.5 characterize the zeroth and first cohomology groups of $\Delta M(G)$. Taylor [5] has proved the following general theorem.

5.3.7. Theorem. *Let B be a CCMA with identity. Then the pth Čech cohomology group $H^p(\Delta B, H)$ for any coefficient group H is given by*

$$H^p(\Delta B, H) = \oplus \sum_{\psi} H^p(\hat{G}_\psi, H),$$

where ψ ranges over the critical points of ΔB. Thus, the cohomology of ΔB is the same as the cohomology of $\Delta \mathscr{L}(B)$.

The results of this section are due to J. L. Taylor [1, 4, 5, 6]. Another proof of the Idempotent Theorem is in Host and Parreau [1].

It is unknown whether $\mu \in M(R)$ and $|\psi_\mu| = 0$ or 1 for all $\psi \in \Delta M(R)$ imply that μ belongs to the spine. When G is an infinite product group, there exists $\mu \in M(G)$ such that $\mu^n \perp \mathscr{L}(G)$, for $n = 1, 2, \ldots$ and $\psi_\mu = 0$, or 1 for all $\psi \in \Delta(G)$ (Izuchi [2]).

5.4. A Guide for Further Study

(i) Idempotent semigroups. The structure semigroup is idempotent if and only if $\hat{\mu}(\Delta B) \subseteq [0, \infty)$ for all $\mu \in B$, $\mu \geq 0$ (S. E. Newman [1]). Functions of bounded variation on idempotent semigroups are related to measure algebras in S. E. Newman [1, 2]. See Lawson, Luikkonen, and Mislove [1] for some idempotent semigroups S such that $M(S)$ is symmetric.

(ii) ΔB is discrete in the norm topology if and only if ΔB is a union of groups and ΔB is discrete in the Gel'fand topology if and only if the spectral radius closure of B in $M(S)$ is an ideal in $M(S)$. (Here S is the structure semigroup of B). (Baker [1, 2])

(iii) Host and Parreau [1] have used CCMA theory to show that $\mu * L^1(G)$ is closed in $L^1(G)$ if and only if $\mu = \nu * \tau$ where $\nu = \nu^2$ and τ is invertible in $M(G)$.

(iv) Continuity by translation. For a locally compact topological semigroup, S, let $\tilde{L}(S)$ denote the set of measures $\mu \in M(S)$ such that the mapping $x \to \delta(x) * |\mu|$ and $x \to |\mu| * \delta(x)$ are both weakly continuous. Then $\tilde{L}(S)$ has properties analogous to those of $L^1(G)$, for a group G. See Baker and Baker [1], Dzinotyiweyi [1], Saka [1], and Sleijpen [1].

(v) Convolution algebras that are not CCMA's. An important property of a CCMA is this: if $\mu, \nu \geq 0$ and $\mu' \ll \mu$, $\nu' \ll \nu$, then $\mu' * \nu' \ll \mu * \nu$. Let B be the set of rotation invariant measures in R^n. Then B is not a convolution measure algebra. B can be identified with $M([0, \infty))$ but the multiplication is *not* convolution. See Ragozin [1] for results pertaining to B, and Jewett [1], Dunkl [1, 2], Spector [1, 2] for some of the general theory of such algebras (called "convos" or "measure algebras on hypergroups").

(vi) In addition to the things already mentioned, Taylor [1] contains the following: a brief survey of topics in which CCMAs are involved; some examples of CCMAs not given here; and results on the Šilov boundary and Gleason parts for CCMAs (see Chapter 8).

Chapter 6

Independent Power Measures

6.1. Introduction and Initial Results

Let G be a locally compact abelian group with dual group Γ and $\mu \in M(G)$. The measure μ has *independent powers*, or μ is *i.p.*, if $\mu^m \perp \mu^n$ whenever $0 \leq m < n < \infty$. The measure μ has *strongly independent powers*, or μ is *strongly i.p.*, if $\delta(x)*\mu^m \perp \mu^n$ for $0 \leq m < n < \infty$ and all $x \in G$. The measure μ is *tame* if for each $\psi \in \Delta M(G)$ there exist $a \in C$ and $\gamma \in \Gamma$ such that $\psi_\mu = a\gamma$ a.e. $d\mu$. The measure μ is *monotrochic* if for each $\psi \in \Delta M(G)$ there exists $r \in [0, 1]$ such that $|\psi_\mu| = r$ a.e. $d\mu$. The measure μ is *strongly tame* or *strongly monotrochic* if the preceding holds for all $\psi \in \Delta N$ where N is the L-subalgebra of $M(G)$ generated by μ. The measure μ is *Hermitian* if $\tilde{\mu} = \mu$.

Let D be a countable subgroup of G. Then $\mu \in M(G)$ is *D-ergodic* if for all Borel sets E such that $E + D = E$, either $|\mu|(E) = 0$ or $|\mu|(E) = \|\mu\|$.

In the remainder of this Section we calculate the spectra of most i.p. measures and give results that relate the definitions of the two preceding paragraphs to each other. In Sections 6.2 and 6.3 measures on independent sets and generalizations of independent sets are discussed. In Section 6.4 infinite product measures are studied and used to give examples of tame Hermitian i.p. measures. Sections 6.5 to 6.8 are devoted to aspects of infinite convolutions: general results in Section 6.5, Bernoulli convolutions in Section 6.6, coin tossing in Section 6.7, and, in Section 6.8, a proof that $M_o(G)$ contains a tame Hermitian i.p. probability measure if G is non-discrete.

For applications of i.p. measures, see Chapters 8 and 9. For Riesz products, see Chapter 7.

Let N be a closed subalgebra of $M(G)$ and let $\mu \in N$. The spectrum of μ as an element of N is denoted by $\sigma_N(\mu)$. We write $\sigma(\mu)$ for $\sigma_{M(G)}(\mu)$. If μ is i.p. and N is the closed subalgebra generated by μ, then N is isometrically isomorphic to the convolution algebra of sequences $\{a_n\}_1^\infty$ for which $\sum_1^\infty |a_n| \|\mu^n\| < \infty$. Every z in the disc of radius $r = \lim \|\mu^n\|^{1/n}$ determines a multiplicative linear functional on N: $\sum a_n \mu^n \to \sum a_n z^n$. Thus $\sigma_N(\mu)$ is that disc. Its boundary $\{z : |z| = r\}$ is of course contained in $\sigma(\mu)$. (See Rudin [3, 10.18] or Rickart [1, 3.3.19].)

6.1.1. Theorem. *Let G be a locally compact abelian group and μ an i.p. probability measure in $M(G)$.*

(i) *If μ is Hermitian, then $\sigma(\mu) = \{z:|z| \leq 1\}$.*
(ii) *If μ is strongly i.p., then $\sigma(\mu) = \{z:|z| \leq 1\}$.*
(iii) *If $\sigma(\mu) \neq \{z:|z| \leq 1\}$, and μ is continuous, then G contains an infinite compact subgroup K such that G/K is not a torsion group.*
(iv) *If $G = T$ or $G = R$ and μ is continuous, then $\sigma(\mu) = \{z:|z| \leq 1\}$.*

Remark. If G and K are as in (iii), then there exists a continuous probability measure, namely $\mu = \delta(x) * m_K$, with independent powers and spectrum $\{0\} \cup \{z: |z| = 1\}$. Here m_K denotes Haar measure on K, and $x \in G$ is such that $x + K$ has infinite order in G/K.

6.1.2. Lemma. *Let μ be an i.p. probability measure in $M(G)$. Then for each $\theta \in [0, 2\pi)$, there exists $\psi \in \Delta M(G)$ such that $\psi_\mu = e^{-i\theta}$ a.e. $d\mu$.*

Proof. Since $e^{i\theta}$ is in the boundary of $\sigma_N(\mu)$, $e^{i\theta}$ is in the boundary of $\sigma(\mu)$. Therefore there exists $\psi \in \Delta M(G)$ such that $\int \overline{\psi_\mu(x)}\, d\mu(x) = e^{i\theta}$, so $1 = \int e^{-i\theta}\overline{\psi}_\mu(x)\, d\mu$. Since $1 = \|\mu\| = \int d\mu$ and $|\psi_\mu(x)| \leq 1$ a.e. $d\mu$, we have $e^{-i\theta}\overline{\psi}_\mu = 1$ a.e. $d\mu$. The Lemma is proved. □

Remark. It follows at once from the Lemma that the spectrum of an i.p. probability measure is rotation-invariant; that is, $e^{i\theta}\sigma(\mu) = \sigma(\mu)$ for all $\theta \in [0, 2\pi]$.

6.1.3. Lemma. *Let μ be a probability measure in $M(G)$. Let $r \in (0, 1)$ be such that $\sigma(\mu) \cap \{z: |z| = r\} = \varnothing$. Then there exist idempotents $e_1, e_2 \in M(G)$ such that $e_1 + e_2 = \delta_0$, and such that for all $\psi \in \Delta M(G)$, $|\hat{\mu}(\psi)| > r$ if and only if $\hat{e}_2(\psi) = 1$, and $|\hat{\mu}(\psi)| < r$ if and only if $\hat{e}_1(\psi) = 1$.*

Proof. Let F be a function that is one in a neighborhood U of $\sigma(\mu) \cap \{z: |z| < r\}$ and zero in a neighborhood V of $\sigma(\mu) \cap \{z: |z| > r\}$. Then F is analytic on the neighborhood $U \cup V$ of $\sigma(\mu)$. Therefore $e_1 = F \circ \mu \in M(G)$ (see Rudin [3, 10.31] or Rickart [1, 3.5.3]). Set $e_2 = \delta_0 - e_1$. Now the conclusion follows easily. □

6.1.4. Lemma. *Let μ be a Hermitian probability measure in $M(G)$. Suppose that there exists $c \in \sigma(\mu)$ with $|c| = 1$ and* Im $c \neq 0$. *Then for all $r \in (0, 1)$,*

(1) $$\sigma(\mu) \cap \{z: |z| = r\} \neq \varnothing.$$

Proof. We argue by contradiction. Suppose that (1) were false for some $r \in (0, 1)$. Let e_1 and e_2 be idempotents given by 6.1.3. Then $e_1 + e_2 * \mu$ is invertible. Let $\nu = (e_1 + \mu * e_2)^{-1} * e_2$. Then $\mu * \nu = e_2$. By the proof of the Idempotent Theorem 1.1.1,

(2) $$\mu * \nu = e_2 = \sum_{j=1}^{N} \lambda_j$$

and

(3) $$\lambda_j = \sum_{k=1}^{m(j)} n_{j,k}\gamma_{j,k} m_{K(j)} \quad \text{for } 1 \le j \le N,$$

where the $m_{K(j)}$ are Haar measures on compact subgroups $K(1), \ldots, K(N)$ of G, the $n_{j,k}$ are integers, and $\gamma_{j,k} \in \Gamma$. We may assume that the λ_j are mutually singular and all non-zero. Let $\psi \in \Delta M(G)$ be such that $\hat{\mu}(\psi) = c$. Then $\psi_\mu = \bar{c}$.

Since $\hat{e}_2(\psi) = 1$, we have $(\gamma_{j,k} m_{K(j)})\hat{}(\psi) \neq 0$ for some (j, k). Let such a (j, k) be fixed and write $\gamma_0 = \gamma_{j,k}$, $K = K(j)$. Since $(\gamma_0 m_K)\hat{}(\psi) \neq 0$, $\psi_{m_K} = \bar{\gamma}_0$ a.e. dm_K. Since $L^1(K)$ is an ideal in $M(K)$, each element $\Delta L^1(K)$ has a unique extension to $M(K)$ and thence to $M(G)$ by 5.2.3. Of course, in the present case, we may take the extension of ψ to be evaluation at a continuous character (still called γ_0) of G. In particular, for $\tau \in M(K)$,

(4) $$\hat{\tau}(\psi) = \hat{\tau}(\gamma_0).$$

By (2) and (3)

$$\int \mu(K - x)|v|(x) = \mu * |v|(K) \ge |\lambda_j|(K) > 0.$$

Therefore $\mu(K - x_0) > 0$ for some $x_0 \in G$, and $\mu_0 = \delta(-x_0) * \mu|_K$ satisfies

(5) $$0 \neq \mu_0 \in M(K),\ \mu_0 \ge 0, \quad \text{and } \delta(x_0) * \mu_0 \ll \mu.$$

Then (4) and (5) imply that

(6) $$\bar{c}\|\mu_0\| = (\delta(x_0) * \mu_0)\hat{}(\psi) = \hat{\mu}_0(\psi)\delta(x_0)\hat{}(\psi) = \hat{\mu}_0(\gamma_0)\delta(x_0)\hat{}(\psi).$$

Since $\delta(x_0)^{\sim} * \tilde{\mu}_0 \ll \tilde{\mu} = \mu$, we have by (4) and (5) again,

$$\begin{aligned}\bar{c}\|\mu_0\| &= (\delta(x_0)^{\sim} * \tilde{\mu}_0)\hat{}(\psi) = \tilde{\mu}_0\hat{}(\psi)\delta(x_0)^{\sim}\hat{}(\psi)\\ &= (\tilde{\mu}_0)\hat{}(\gamma_0)\delta(x_0)^{\sim}\hat{}(\psi) = [\hat{\mu}_0(\gamma_0)\delta(x_0)\hat{}(\psi)]^{-}.\end{aligned}$$

Therefore $c\|\mu_0\| = \bar{c}\|\mu_0\|$. Since c is not real, and $\mu_0 \neq 0$, we have a contradiction and the Lemma is proved. □

Proof of 6.1.1. (i) By 6.1.2 there exists $\psi \in \Delta M(G)$ such that $\psi_\mu = c$ where $|c| = 1$ and $\operatorname{Im} c \neq 0$. By 6.1.2, $\sigma(\mu)$ is rotation-invariant. That and (1) show that $\sigma(\mu) \supseteq \{z : |z| = r, r \in (0, 1)\}$. Since $\sigma(\mu)$ is compact, $\sigma(\mu) = \{z : |z| \le 1\}$.

(ii) Whenever $|c| = 1$, $n \geq 1$, and $x_1, \ldots, x_k$ are distinct elements of $G \backslash \{0\}$,

$$\left\| \left[(\delta(0) + \bar{c}\mu) * \left(\delta(0) + \sum_{j=1}^{k} \delta(x_j) \right) \right]^n \right\| = 2^n(1 + k)^n.$$

A simple weak* limit argument now shows that there exists $\psi \in \Delta M(G)$ such that

(7) $$\hat{\mu}(\psi) = c \text{ and } \delta(x)\hat{\ }(\psi) = 1 \text{ for all } x \in G.$$

We suppose that $\{z : |z| = r\} \cap \sigma(\mu) = \varnothing$ for some $r \in (0, 1)$ and repeat the proof of 6.1.4 through (4) word for word using the ψ of (7). By (3) and (7), $\gamma_0 = 1$ on K, and hence

$$c\|\mu_0\| = (\mu_0 * \delta(x_0))\hat{\ }(\psi) = \hat{\mu}_0(\psi) = \|\mu_0\| \neq 0,$$

which contradicts the choice of $c \neq 1$. We have established (ii).

(iii) We continue the notation of the above paragraphs, assuming further that $c^n \neq 1$ for all $n = 1, 2, \ldots$. If $n(K + x_0) \neq K$ for all integers $n > 1$, then we have proved (iii). Otherwise we may suppose that $n(K + x_0) = K$ for some $n > 1$. Then $(\mu^n)\hat{\ }(\psi) = c^n$ and (2), (3), (4), and (5) hold with μ^n in place of μ, μ_0^n in place of μ_0 and nx_0 in place of x_0. Then by (4), $\hat{\mu}_0^n(\gamma_0) = \hat{\mu}_0^n(\psi) = c^n\|\mu_0\|$. Therefore μ_0^n is concentrated on $K' = \{x \in K : \langle x, \gamma_0 \rangle = c^n\}$. Since $c^m \neq 1$ for all $m \neq 0$, K' is a coset of a compact subgroup K'' of K such that K/K'' is not a torsion group. That proves (iii).

(iv) is immediate from the fact that neither T nor R contains an infinite compact proper subgroup. That ends the proof of 6.1.1. □

6.1.5. Theorem. *Let G be a locally compact abelian group and μ a probability measure in $M(G)$.*

- (i) *If μ is Hermitian and i.p., then $\mu \perp$ Rad $L^1(G_\tau)$ for all refinements G_τ of G.*
- (ii) *If μ is monotrochic, then either $\mu \perp$ Rad $L^1(G_\tau)$ for all τ or $\mu \in$ Rad $L^1(G_\tau)$ for some τ.*
- (iii) *If μ is monotrochic and $\mu \perp$ Rad $L^1(G_\tau)$ for every τ, then for each z with $|z| \leq 1$, there exists $\psi \in \Delta M(G)$ such that $\psi_\mu = z$ a.e. $d\mu$, and μ is strongly i.p.*

The following Corollary is immediate from 6.1.5(i) and 6.1.5(iii).

6.1.6. Corollary. *Let μ be a Hermitian, tame, i.p. probability measure on the locally compact abelian group G. Then*

$$\{\psi_\mu : \psi \in \Delta M(G)\} = \{a\gamma : \gamma \in \Gamma, a \in C, |a| \leq 1\}.$$

Proof of 6.1.5. (i) If $\mu \not\perp \text{Rad } L^1(G_\tau)$ for some τ, then we may assume that $\mu \in \text{Rad } L^1(G_\tau)$ and that $\|\mu\| = 1$. Since μ is i.p., the spectral radius of $\mu^2 - \mu^4$ is 2. Since $\mu \in \text{Rad } L^1(G_\tau)$, the spectral radius of $\mu^2 - \mu^4$ is the supremum of $\hat{\mu}(\gamma)^2 - \hat{\mu}(\gamma)^4$ as γ ranges over Γ. Because μ is Hermitian, $0 \le \hat{\mu}(\gamma)^2 \le 1$, so $0 \le \hat{\mu}(\gamma)^2 - \hat{\mu}(\gamma)^4 \le \frac{1}{2}$. That contradiction proves that $\mu \perp \text{Rad } L^1(G_\tau)$.

(ii) Suppose that $\mu \not\perp \text{Rad } L^1(G_\tau)$. Let $\psi \in \Delta M(G)$ be the element such that $\psi_\omega = 1$ a.e. $d\omega$ for all $\omega \in M(G_\tau)$ and $\psi_\mu = 0$ a.e. for all $\nu \perp M(G_\tau)$. (See Section 5.2 for the construction of ψ.) Since μ is monotrochic, $\psi_\mu \equiv 1$ a.e. $d\mu$. Therefore, $\mu \in M(G_\tau)$. Let $\mu = \mu_1 + \mu_2$, where $\mu_1 \in \text{Rad } L^1(G_\tau)$ and $\mu_2 \perp \text{Rad } L^1(G_\tau)$. Then $\mu_1 \ne 0$ by assumption. If $\mu_2 \ne 0$, there exists $\varphi \in M(G_\tau)\backslash(G_\tau)\hat{}$ such that $\hat{\mu}_2(\varphi) \ne 0$. Of course $\varphi_{\mu_1} = 0$ a.e. $d\mu_1$. That contradicts the monotrochicity of μ. Therefore $\mu_2 = 0$ and $\mu \in \text{Rad } L^1(G_\tau)$.

(iii) We argue by contradiction and first suppose that if $\psi \in \Delta M(G)$, then either $\psi_\mu = 0$ or $|\psi_\mu| = 1$. Then $I = \{\psi \in \Delta M(G)\colon \psi \ge 0, \psi_\mu \ne 0\}$ is both open and closed. Let ρ be a minimal element of I. By 5.3.1 and 5.3.3 (iv), $\rho = \psi^{(\tau)}$, where $\psi^{(\tau)}$ is the idempotent associated with a refinement G_τ of G. Then $\psi_\mu^{(\tau)} = 1$. By replacing G by G_τ, we see that we may assume $G = G_\tau$ and $\rho = 1$. But $\mu \perp \text{Rad } L^1(G)$. Therefore there exists $\psi \in \Delta M(G)\backslash\Gamma$ such that $\psi_\mu = 1$. But then $\psi < 1 = \rho$, which contradicts the minimality of ρ. Therefore there exists $r \in (0, 1)$ and $\psi \in \Delta M(G)$ such that $\psi_\mu = r$. On replacing ψ by $|\psi|^z$ where $\text{Re } z > 0$, and taking limits, (iii) follows. To show that μ is strongly i.p., note that

$$|\psi|_{\delta(x) * \mu^m} = r^m \quad \text{for all } m \ge 1 \quad \text{and } x \in G.$$

(That follows, for example, by passing to the structure semigroup, where ψ is continuous.) Therefore $\delta(x) * \mu^m \perp \mu^n$ whenever $x \in G$ and $0 \le m < n < \infty$. That ends the proof of 6.1.5. □

Remark. When G is compact and $\mu \in M_o(G)$ is tame, a more elementary argument yields 6.1.6 as follows.

We argue by contradiction and first suppose that for every $\psi \in \Delta M(G)$, either $|\psi_\mu| \equiv 0$ a.e. $d\mu$ or $|\psi_\mu| \equiv 1$ a.e. $d\mu$. Then $\sigma(\mu)$ would be contained in the set $\{a\hat{\mu}(\gamma)\colon \gamma \in \Gamma, |a| = 1\} \cup \{0\}$, which is disconnected by the assumptions on G and μ, and which therefore cannot equal $\{z\colon |z| \le 1\}$. That would contradict 6.1.1(i). Therefore there exists $\psi \in \Delta M(G)$ such that $|\psi_\mu| = r$ a.e. $d\mu$ for some $r \in (0, 1)$. The proof now continues as before.

6.1.7. Proposition. *Let G and H be locally compact abelian groups. Let μ and ν be probability measures in $M(G)$ and ω a probability measure in $M(H)$.*

(i) *If μ and ν are monotrochic, then $\mu * \nu$ is monotrochic.*
(ii) *If μ and ω are tame, then $\mu \times \omega$ is tame.*
(iii) *If ω is tame and H is a closed subgroup of G, then ω is tame in $M(H)$.*
(iv) *If μ is tame and $x \in G$, then $\delta(x) * \mu$ is tame.*
(v) *If D is a countable subgroup of G and μ and ν are D-ergodic, then $\mu * \nu$ is D-ergodic.*

Proof. (i). We pass to the structure semigroup S of $M(G)$. Then every element $\psi \in \Delta M(G)$ is identified with a continuous semicharacter on S. We identify each measure in $M(G)$ with its image in $M(S)$. See Section 5.1 for details.

Let $\psi \in \Delta M(G)$ and suppose that $|\psi_\mu| = r$ a.e. $d\mu$ and $|\psi_\nu| = s$ a.e. $d\nu$ where $r, s \in [0, 1]$. Then, considered as a continuous function on S, $|\psi| = r$ on the support X of μ and $|\psi| = s$ on the support Y of ν. Since ψ is multiplicative on S, $|\psi| = rs$ on XY. Since XY contains the support of $\mu * \nu$, $|\psi_{\mu*\nu}| = rs$ a.e. $d(\mu * \nu)$. Therefore $\mu * \nu$ is monotrochic if μ and ν are monotrochic.

(ii)–(iv) are routine. To prove (v), note first that for every Borel subset E of G, $\mu(E + x) = f(x)$ is a Borel function. (Indeed f is the supremum of the continuous functions $g(x) = \int_E k(y)d\delta(-x) * \mu(y)$ as k varies over the continuous functions with $0 \le k \le 1$.) Therefore if $E + D = E$, then $E_1 = \{x: \mu(E - x) = 1\}$ is a Borel set that has $E_1 + D = E_1$. If $\nu(E_1) = 0$, then

$$\mu * \nu(E) = \int \mu(E - x)d\nu(x) = 0;$$

if $\nu(E_1) = 1$, then $\mu * \nu(E) = 1$. That proves (v) and the Proposition is proved. $\square$

6.1.8. Theorem. *Let D be a countable subgroup of G and $\mu \in M(G)$ a D-ergodic probability measure.*

(i) *If $\delta(d) * \mu \approx \mu$ for all $d \in D$, then μ is strongly monotrochic.*
(ii) *Either μ is strongly i.p., or some power μ^n ($n \ge 1$) of μ belongs to $L^1(G_\tau)$ for some refinement G_τ of G.*
(iii) *If $\nu \in M(G)$ is such that $\delta_d * \nu \ll \nu$ for all $d \in D$, then either $\mu \perp \nu$ or $\mu \ll \nu$.*

Proof. (i) Let N be the L-subalgebra of $M(G)$ that is generated by μ. We may apply 5.1 (8): $|\psi_\mu(x + d)| = |\psi_\mu(x)|$ a.e. $d\mu$ for all $d \in D$ and $\psi \in \Delta N$. Therefore $\psi_\mu(x + d) = \psi_\mu(x)\psi_{\delta(d)}(d)$ a.e. $d\mu$ for all $x \in G$.

Let $E = \{x \in \text{supp}\, \mu: |\psi_\mu(x)| < \hat{\mu}(|\psi|)\}$. Then for all d, $E + d$ differs from E by a μ-null set and $H = \bigcup\{E + d: d \in D\}$ is D-invariant, that is $H + D = H$. Although E may not be a Borel set, E differs from a Borel set E' by a μ-null set. Because $\delta_x * \mu \approx \mu$ for all $x \in D$, $H' = E' + D$ has μ-measure equal to the μ-measure of H. By the D-ergodicity of μ, either $\mu(H') = \|\mu\|$ or $\mu(H') = 0$. But $\mu(H) = \mu(H') = \|\mu\|$ is absurd, by the definition of E. Therefore $\mu(H') = 0 = \mu(E)$, which implies that $|\psi_\mu(x)| \ge \hat{\mu}(|\psi|)$ a.e. $d\mu$. Similarly, $|\psi_\mu| \le \hat{\mu}(|\psi|)$, and (i) is proved.

(ii) We suppose that μ is not strongly i.p. Let $0 < m < n$ and $x \in G$ be such that $\mu^m \not\perp \delta(x) * \mu^n$. Let $\tau \in M(G)$ be any discrete probability measure such that $\{y: \tau(\{y\}) \ne 0\} = Gp(\{x\} \cup D)$. Then $\tau * \mu$ is $[D + Gp(\{x\})]$-ergodic and $(\tau * \mu)^m \not\perp (\tau * \mu)^n$. We have reduced to the case that $x = 0$ and $\delta(d) * \mu \approx \mu$ for all $x \in D$. Of course we may further assume that the support of μ generates a dense subgroup of G. Note that 6.1.5(iii) shows that $\mu \in \operatorname{Rad} L^1(G_\tau)$ for some τ, but we must prove more. Let N be the L-subalgebra generated by μ and $\psi \in \Delta N$. By (i) $|\psi_\mu|$ is constant a.e. $d\mu$. Since μ is not i.p. we must have $|\psi_\mu| = 1$ a.e. $d\mu$, so ΔN is a group. By Theorem 5.3.7, $L^1(H) \subseteq N \subseteq \operatorname{Rad} L^1(H)$ for some locally compact abelian group H. The natural injection $\varphi: N \to M(G)$ induces a continuous mapping $\beta: \Delta M(G) \to \Delta N \cup \{0\} = H\hat{} \cup \{0\}$. (Here "0" is the function identically equal to zero.) Because the support of μ generates a dense subgroup of G, the set of transforms $N\hat{}$ separate elements of Γ. Therefore β maps Γ one-to-one into $H\hat{}$. Furthermore, $\beta(\Gamma)$ must be dense in $H\hat{}$, because $L^1(H) \subseteq N$ and φ is one-to-one. Therefore the adjoint map α of β, from H into G, is a one-to-one continuous homomorphism. Let G_τ be the refinement of G which makes H an open subgroup of G and under which $\alpha: H \to \alpha(H)$ is a homeomorphism. Then, for all $\nu \in N$, $\nu = \varphi\nu = \nu \circ \alpha^{-1}$, so $L^1(G_\tau)|_{\alpha(H)} \subseteq N \subseteq \operatorname{Rad} L^1(G_\tau)$. Therefore $\mu \in \operatorname{Rad} L^1(G_\tau)$. Then $\varphi\mu^n = \mu^n \not\perp L^1(G_\tau)$ for some $n \geq 1$. By 6.1.7(v), μ^n is D-ergodic. Let E be a Borel subset of zero Haar measure. Then $\mu^n(D + E) < \|\mu^n\| = 1$ since $\mu^n \not\perp L^1(G_\tau)$. By the D-ergodicity of μ^n, $\mu^n(D + E) = 0$. Therefore $\mu^n \in L^1(G_\tau)$.

(iii). Suppose that μ is not absolutely continuous with respect to ν. Let E be a Borel subset of G such that $\mu(E) > 0$ and $|\nu|(E) = 0$. Since $\delta(d) * \nu \ll \nu$ for all $d \in D$, $|\nu|(E + d) = 0$ for all $d \in D$. Therefore $|\nu|(E + D) = 0$, while $\mu(E + D) = 1$ by the D-ergodicity of μ. Therefore $\nu \perp \mu$. That ends the proof of 6.1.8. □

Remarks and Credits. Theorem 6.1.1(iv) was proved by Taylor [1, 4], using critical points. The remainder of 6.1.1 is due to Bailey, Brown, and Moran [1]. Lemma 6.1.4 was communicated to us by Saeki. The analogue of 6.1.1 fails for joint spectra. See Brown [5] for examples and positive results.

Theorem 6.1.5(i) goes back to Williamson [3]; 6.1.5(ii) is Lemma 3 of Brown and Moran [9]; 6.1.5(iii) a variant of Brown and Moran [4, Proposition 2]. Brown [2] proves 6.1.6; the elementary proof given in the Remark for the special case G compact, $\mu \in M_o(G)$ is new.

Proposition 6.1.7(i) is a special case of 6.5.2, which is from Brown and Moran [9, Lemma 2]. Parts (ii)–(iv) of 6.1.7 are due to Brown [2] and Moran [2]. Part (v) is Brown and Moran [4, Lemma 1].

The convolution of two tame measures may not be tame: let $\mu = \delta(0) + \delta(1)$ and $\nu = \delta(0) + \delta(\sqrt{2})$. Then μ and ν are both tame measures in $M(R)$ (that's obvious), but $\mu * \nu$ is not tame, as an application of 6.5.7 shows. By using 6.7.1 (iii) one can construct a pair of tame i.p. probability measures in $M_o(R)$ whose convolution product is not tame.

Theorem 6.1.8(i) and (ii) are from Brown and Moran [4]; (iii) may be new.

One property of i.p. probability measures μ that is often used is this: the spectral radius of $\delta_0 - \mu^2$ is 2. Therefore there exists $\psi \in \Delta M(G)$ such that $\hat{\mu}(\psi) = i$. When μ is Hermitian, that yields a proof of the asymmetry of $M(G)$. See Chapter 8 for generalizations of that idea.

6.2. Measures on Algebraically Scattered Sets

A subset E of G is *algebraically scattered* if for each $x \in GpE$, $x \neq 0$, there exists a countable subset Y of E such that $x \notin Gp(E \setminus Y)$. Every independent set is algebraically scattered, as is every set of the form $\bigcup_{n=-\infty}^{\infty} \{nx: x \in E\}$ where E is independent. The last assertion is easily proved.

For a subset E of G we define $0E = \{0\} = (0) * E$, $nE = E + (n-1)E$ and $(n) * E = \{nx: x \in E\}$ for $n = 1, 2, \ldots$.

6.2.1. Theorem. *Let E be an algebraically scattered subset of the locally compact abelian group G. Then for every non-negative integer n, every non-empty countable subset F of G, every Borel subset $K \subseteq nE + F$, and every set $\{\mu_1, \ldots, \mu_{n+1}\} \subseteq M_c(G)$,*

(i) $\mu_1 * \cdots * \mu_n((K + x) \cap (K + y)) = 0$ *if* $x, y \in G, \quad x - y \notin F - F$;

and

(ii) $\mu_1 * \cdots * \mu_{n+1}(K + x) = 0$ *for all* $x \in G$.

Proof. If $n = 0$, then $\mu_1 * \cdots * \mu_n = \delta(0)$, $nE = \{0\}$, and $K \subseteq F$. Then (i)–(ii) follow easily.

We now shall prove that for each $n \geq 1$, (i) $\Rightarrow$ (ii). We apply (i) with F replaced by $H = GpF$. Then (i) has the form

$$\mu_1 * \cdots * \mu_n((K + x) \cap (K + y)) \quad \text{if } x - y \notin H.$$

We write G as a union $\bigcup_w (K + H + w)$ where w ranges over a set of distinct representatives of cosets of H. If $w \neq w'$, then by (i),

$$\mu_1 * \cdots * \mu_n((K + H + w) \cap (K + H + w')) = 0.$$

Since $\mu_1 * \cdots * \mu_n$ is a finite measure, there exists a countable set W of distinct representatives of cosets of H such that $\mu_1 * \cdots * \mu_n(K + x) \neq 0$ implies $x \in W$. Therefore

$$(1) \quad \mu_1 * \cdots * \mu_{n+1}(K + x) = \int \mu_1 * \cdots * \mu_n(K + x + w) d\mu_{n+1}(w) = 0,$$

since μ_{n+1} is continuous. Thus (ii) holds.

To establish (i) we shall use induction on n, so (ii) for $n - 1$ will be available to us. By replacing μ_1 by $\delta(-x) * \mu_1$, we may assume that $x = 0$. There is nothing to prove unless $K \cap (K + y) \neq 0$. But

$$K \cap (K + y) \subseteq (nE + F) \cap (nE + F + y) \subseteq \bigcup_{s,t \in F} (nE + s) \cap (nE + t + y).$$

If $(nE + s) \cap (nE + y + t) \neq 0$, then $y + t - s \in GpE$. If $y \notin F - F$, then $y + t - s \neq 0$. Hence, there exists a countable subset $Y_{s,t} \subseteq E$ such that $y + t - s \notin Gp(E \setminus Y_{s,t})$. We claim that

$$n(E + s) \cap (nE + y + t) \subseteq [(n - 1)E + Y_{s,t}] \cup [(n - 1)E + Y_{s,t} + y + t - s].$$

Indeed, let $u \in (nE) \cap (nE + y + t - s)$; then $u = \sum_{j=1}^{n} e'_j + y + t - s = \sum_{j=1}^{n} e_j$, where e_j and $e'_j \in E$ for $j \geq 1$. Since $y + t - s \notin Gp(E \setminus Y_{s,t})$, at least one element of the set $\{e_1, \ldots, e_n, e'_1, \ldots, e'_n\}$ is in $Y_{s,t}$. Thus, either $\sum_1^n e_j \in (n - 1)E + Y_{s,t}$ or $\sum_1^n e'_j + y + t - s \in (n - 1)E + Y_{s,t} + y + t - s$. That establishes the claim. We set $W = \bigcup_{s,t \in F} (Y_{s,t} + s) \cup (Y_{s,t} + t + y)$. Then W is countable and

$$K \cap (K + y) \subseteq (n - 1)E + W. \tag{2}$$

If $n = 1$, then $K \cap (K + y) \subseteq W$ and (i) follows. If $n > 1$, then the $(n - 1)$-case of (ii) with W in place of F and $L = K \cap (K + y)$ in place of K shows that $\mu_1 * \cdots * \mu_n(L) = 0$. That ends the proof of 6.2.1. □

In the results that follow, the hypothesis "E is σ-compact" is used to ensure that nE is a Borel set.

Since $\bigcup_{r=1}^{n} (k(r) \in Q) \cup (j(r) * Q)$ is algebraically scattered when $Q = E \cup -E$ and E is independent, 6.2.2 is immediate from 6.2.1.

6.2.2. Corollary. *Let E be a σ-compact independent subset of the locally compact abelian group G and $Q = E \cup -E$. Let $n \geq 1$ and $k(r)$, $j(r)$ be non-negative integers for $1 \leq r \leq n$ and let $\mu_1, \ldots, \mu_{n+1}$ be continuous regular Borel measures in G. Then*

(i) $\mu_1 * \cdots * \mu_n([\sum_{r=1}^{n} (k(r)) * Q + x] \cap [\sum_{r=1}^{n} (j(r)) * Q + y]) \neq 0$ *implies $x = y$; and*
(ii) $\mu_1 * \cdots * \mu_{n+1}(\sum_{r=1}^{n} (k(r)) * Q + y) = 0$ *for all $y \in G$.*

6.2.3. Corollary. *Let E be a σ-compact algebraically scattered Borel subset of the locally compact abelian group G. Let $\mu \in M(G)$ be a continuous measure concentrated on E. Then μ is strongly i.p. and, in fact,*

$$\delta_x * \mu^m \perp \delta_y * \mu^n \quad \text{for } x, y \in G,$$

unless $m = n$ and $x = y$.

Proof. $\delta_x * \mu^m$ is concentrated on $mE + x$ and $\delta_y * \mu^n$ is concentrated on $nE + y$. If $m \neq n$, then 6.2.1(ii) gives the result; if $m = n$ and $x \neq y$ then 6.2.1(i) with $F = \{0\}$ gives the result. □

6.2.4. Corollary. *Let G be a locally compact abelian group, and let E be a σ-compact algebraically scattered subset of G, and let H be a σ-compact non-discrete locally compact abelian group that is continuously embedded in G. Let m_H denote Haar measure on H. Then $m_H(GpE + y) = 0$ for $y \in G$.*

Proof. Since H is σ-compact, the structure theorem implies the existence of an open subgroup H_0 of H of the form $H_0 = R^n \times H_1$, where H_1 is compact and H/H_0 is countable. Let $\nu_1 \in M(H_0)$ be defined by

$$\int f\, d\nu_1 = \int [f(x_1, \ldots, x_n, y)/(1 + x_1^2) \cdots (1 + x_n^2)] dx_1, \ldots, dx_n\, dy,$$

where $(x_1, \ldots, x_n) \in R^n$ and $y \in H_1$. Let $\nu = \sum 2^{-j}\delta(z_j) * \nu_1$ where $\{z_j\}_1^\infty + H_0 = H$. Then $\nu \in L^1(H)$, $\nu \geq 0$, ν is continuous, and $\nu^k \approx \nu^j$ for $1 \leq j, k < \infty$. Furthermore, $\omega \ll \nu$ for all $\omega \in L^1(H)$.

By 6.2.1(ii), $\nu^{n+1}(n(E \cup -E) + y) = 0$ for all $y \in G$ and $n \geq 1$. Therefore $n(E \cup -E) + y$ has zero m_H-measure for $n \geq 1$ and $y \in G$. Since $GpE = \bigcup n(E \cup -E)$ we see that $GpE + y$ has zero m_H-measure for all $y \in G$. □

6.2.5. Theorem. *Let E be a maximal independent subset of R. Then one of the sets $n(E \cup -E)$, $n = 1, 2, \ldots$ is not Lebesgue-measurable.*

Proof. Suppose that $n(E \cup -E)$ were Lebesgue measurable for all integers $n = 1, 2, \ldots$. Let $(1/k)E = \{x/k : x \in E\}$ for integers $k = 1, 2, \ldots,$ Then $(1/k)E$ is independent and

$$R = \bigcup_{k=1}^{\infty} Gp((1/k)E). \tag{3}$$

Of course, each set $n((1/k)E \cup -(1/k)E) = Q(k, n)$ would be Lebesgue measurable. Now (3) and the Lebesgue measurability of the sets $Q(k, n)$ imply that some set $Q(k, n)$ would have nonzero Lebesgue measure.

The argument of 6.2.1 can now be used to show that if $G = R$, the measures μ_j are all in $L^1(R)$, and the sets K are Lebesgue measurable, then the conclusions of 6.2.1 hold.

But there exists a measure μ (for example $d\mu/dx = (1 + x^2)^{-1}$) in $L^1(R)$ such that Lebesgue measure is absolutely continuous with respect to μ^n for $n = 1, 2, \ldots$. By the "Lebesgue" version of 6.2.1, $\mu^{n+1}(Q(k, n)) = 0$. That contradiction proves Theorem 6.2.5. □

We now prove a "multivariable" form of 6.2.3.

6.2.6. Theorem. *Let E be a σ-compact algebraically scattered subset of the locally compact abelian group G. Suppose that $\mu_1, \ldots, \mu_n$ are pairwise singular* *continuous measures that are concentrated on E. Let* $a = (a(1), \ldots, a(n))$ *and* $b = (b(1), \ldots, b(n))$ *be n-tuples of non-negative integers and* $y \in G$. *If either* $a \neq b$ *or* $y \neq 0$, *then*

$$\lambda \perp \rho \tag{4}$$

where $\lambda = \mu_1^{a(1)} * \cdots * \mu_n^{a(n)}$ *and* $\rho = \delta(y) * \mu_1^{b(1)} * \cdots * \mu_n^{b(n)}$.

Proof. We may suppose that the μ_j are non-negative and that they are concentrated on pairwise disjoint compact subsets $K_1, \ldots, K_n$ of E. We may also suppose that $m = \sum_1^n a(j) \geq q = \sum_1^n b(j)$.

Case I, $a \neq b$. If $m > q$, then, by 6.3.1(ii), $\lambda(qE + w) = 0$ for all $w \in G$ and, therefore,

$$\lambda\left(\sum_1^n b(j)K_j + w\right) = 0 \quad \text{for all } w \in G. \tag{5}$$

Since ρ is concentrated on $\sum_1^n b(j)K_j + y$, (4) follows. We may therefore assume that $m = q$. We shall use induction on m to establish (5). Clearly, (5) implies (4).

For $m = 1$, we may suppose that $a(1) = b(2) = 1$. If $w \neq 0$, then

$$\lambda(b(2)K_2 + w) = \mu_1(K_1 \cap (K_2 + w)) = 0$$

by 6.2.1(i). If $w = 0$, then $K_1 \cap K_2 = \varnothing$ and (5) also follows. We now may assume that (5) holds whenever $\sum b(j) \leq m - 1$, $\lambda = \mu^{a'(1)} * \cdots * \mu^{a'(n)}$, and $\sum a'(j) = m - 1$.

Let $\sum_1^n a(j) = m \geq 2$ and $\sum_1^n b(j) = m$. We may assume that $a(n) > b(n)$. For $1 \leq j < n$, let $a'(j) = a(j)$, and let $a'(n) = a(n) - 1$. Let

$$\lambda' = \mu_1^{a'(1)} * \cdots * \mu_n^{a'(n)}, \quad \text{and} \quad m' = m - 1.$$

We claim that for each $u \in K_n \backslash \{0\}$, the set

$$Q_u = \left[\sum_{j=1}^n a'(j)K_j\right] \cap \left[\sum_{j=1}^n b(j)K_j - u\right]$$

has zero λ'-measure. Let $u \in K_n \backslash \{0\}$ be fixed. Since E is algebraically scattered, there exists a countable set $Y \subseteq E$ such that $u \notin Gp(E \backslash Y)$. For $1 \leq j \leq n$, set $Y_j = Y \cap K_j$ and let $I = \{j\colon 1 \leq j \leq n, b(j) \geq 1\}$. Then a straightforward calculation shows that

$$Q_u \subseteq [(m' - 1)E + Y] \cup \bigcup_{k \in I} \left[\sum_{j \neq k} b(j)K_j + (b(k) - 1)K_k + Y_k - u\right]. \tag{6}$$

By 6.2.1(ii), $\lambda'((m' - 1)E + Y) = 0$. If $k \in I$ and $k < n$, then $0 \notin Y_k - u$ and

$$\sum_{j \neq k} b(j)K_j + (b(k) - 1)K_k + Y_k - u \subseteq (m - 1)E + Y_k - u.$$

Since λ' is concentrated on $(m - 1)E$, 6.2.1(i) with $F = \{0\}$ implies that $\lambda'((m - 1)E + Y_k - u) = 0$. If $n \in I$, then

$$\lambda'\left(\sum_{j=1}^{n-1} b(j)K_j + (b(n) - 1)K_n + Y_n - u\right) = 0,$$

by (5) applied to λ'. Therefore $\lambda'(Q_u) = 0$ for all non-zero $u \in K_n$. But then

$$\lambda(\sum b(j)K_j) = \int \lambda'(Q_u)d\mu_n(u) = 0,$$

so (5) follows if $w = 0$. If $w \neq 0$, then (5) follows from 6.2.1(i).

Case II, $a = b$. Since λ is concentrated on mE and ρ on $mE + y$, it will suffice to show that $mE \cap (mE + y)$ has zero λ-measure. And that follows from 6.2.1(i) with $F = \{0\}$, $n = m$, $x = 0$, and $K = mE$. That ends the proof of 6.2.6. □

6.2.7. Corollary. *Let E be a σ-compact algebraically scattered subset of the locally compact abelian group G, and let $\mu_1, \ldots, \mu_n$ be continuous measures concentrated on E. Then $\|\mu_1 * \cdots * \mu_n\| = \|\mu_1\| \cdots \|\mu_n\|$.*

Proof. Let $\varepsilon > 0$, $m \geq 1$. Let $g_1, \ldots, g_n$ be simple Borel functions on G such that $\|g_j|\mu_j| - \mu_j\| < \varepsilon$ for each j. Then $g_j|\mu_j| = \sum c_{j,k}\mu_{j,k}$ where $\mu_{j,k} \geq 0$. By replacing the g_j if necessary, we may assume that for all j, k, j', k' either $\mu_{j,k} \perp \mu_{j',k'}$ or $\mu_{j,k} = \mu_{j',k'}$, and that $\|\mu_{j,k}\| \leq \|\mu_j\|/m$. Then

$$(g_1|\mu_1|) * \cdots * (g_n|\mu_n|) = \sum\{\bigstar(c_{j,k}\mu_{j,k})^{\alpha(j,k)}: 0 \leq \alpha(j,k) \leq 1, \sum \alpha(j,k) = n\}.$$

The convolutions on the right hand side are mutually singular except when some $\mu_{j,k} = \mu_{j',k'}$. A straightforward combinatorial argument shows that

$$\sum\{\|\bigstar(c_{jk}\mu_{jk})^{\alpha(j,k)}\|\} - \sum\{\|\bigstar c_{jk}\mu_{jk}\|: \mu_{j,k} \neq \mu_{j',k'}\} = o(m) \text{ as } m \to \infty.$$

Similarly,

$$\|g_1|\mu_1|\| \cdots \|g_n|\mu_n|\| - \sum\{\|\bigstar(c_{jk}\mu_{jk})^{\alpha(j,k)}\|: \mu_{jk} \neq \mu_{j',k'}\}| = o(m).$$

Therefore $\|g_1|\mu_1|\| \cdots \|g_n|\mu_n|\| = \|(g_1|\mu_1|) * \cdots * (g_n|\mu_n|)\|$. Since

$$\|g_1|\mu_1| * \cdots * g_n|\mu_n| - \mu_1 * \cdots * \mu_n\| = o(\varepsilon)$$

as $\varepsilon \to 0$ and $\left| \|g_1|\mu_1|\| \cdots \|g_n|\mu_n|\| - \|\mu_1\| \cdots \|\mu_n\| \right| = o(\varepsilon)$, the Corollary follows. □

6.2.8. Corollary. *Let E be a σ-compact algebraically scattered subset of the locally compact abelian group G. Let f be a continuous linear functional on $M_c(E)$ of norm $\|f\| \leq 1$. Then there exists $\psi \in \Delta M(G)$ such that $\hat{\mu}(\psi) = \langle f, \mu \rangle$ for each $\mu \in M_c(E)$.*

Proof. Suppose that for each finite subset α of $M_c(E)$ there exists $\psi^{(\alpha)} \in \Delta M(G)$ such that $\hat{\mu}(\psi^{(\alpha)}) = \langle f, \mu \rangle$ for all $\mu \in \alpha$. Let ψ be any cluster point of the net $\{\psi^{(\alpha)}\}$, where α ranges over all finite subsets of $M_c(E)$ and $\alpha' \geq \alpha$ if $\alpha' \supseteq \alpha$. Then $\hat{\mu}(\psi) = \langle f, \mu \rangle$ for all $\mu \in M_c(E)$. It remains to find $\psi^{(\alpha)}$ given the finite set $\alpha \in M_c(E)$.

Let $\alpha = \{\mu_1, \ldots, \mu_n\} \subseteq M_c(E)$. By the linearity and continuity of f, we may assume that $1 = \|\mu_1\| = \cdots = \|\mu_n\|$ and that the μ_j are mutually singular.

For $1 \leq j \leq n$, let $z_j = \langle f, \mu_j \rangle$. Since $|z_j| \leq 1$, there exist complex numbers v_j, w_j such that

$$z_j = \tfrac{1}{2}(v_j + w_j) \quad \text{and} \quad |v_j| = |w_j| = 1 \quad \text{for } 1 \leq j \leq n. \tag{7}$$

Since the measures μ_j are continuous, there exist continuous measures $\sigma_j, \tau_j \in M_c(E)$ such that

$$\sigma_j \perp \tau_j,\ \mu_j = \tfrac{1}{2}(\sigma_j + \tau_j) \quad \text{and} \quad \|\sigma_j\| = \|\tau_j\| = 1 \quad \text{for } 1 \leq j \leq n. \tag{8}$$

We set

$$\omega = \delta_0 + \sum_{j=1}^{n} (\bar{v}_j \sigma_j + \bar{w}_j \tau_j).$$

Then 6.2.6 and 6.2.7 imply that $\|\omega^k\| = (1 + 2n)^k$, for $k = 1, 2, \ldots$. Therefore, there exists $\psi \in \Delta M(G)$ such that

$$|1 + \textstyle\sum \bar{v}_j \hat{\sigma}_j(\psi) + \bar{w}_j \hat{\tau}_j(\psi)| = 1 + 2n.$$

By (7) and (8), $\hat{\sigma}_j(\psi) = v_j$ and $\hat{\tau}_j(\psi) = w_j$. Using (7) and (8) again, we conclude that $\hat{\mu}_j(\psi) = z_j$ for $1 \leq j \leq n$. The Corollary is proved. □

6.2.9. Corollary. *Let E be a σ-compact algebraically scattered subset of the locally compact abelian group G. Let U denote the closed unit ball of $M_c(E)^*$, and N the L-subalgebra of $M(G)$ that is generated by $M_c(E)$ and $M_d(G)$. Then*

$$\Delta N = \partial N = b\Gamma \times U.$$

Proof. We first show that $b\Gamma \times U \subset \Delta N$ as sets. Any $\mu \in N$ has the form

$$\mu = \sum_{0}^{\infty} \nu_j * \mu_j \quad \text{with} \quad \|\mu\| = \sum_{j=0}^{\infty} \|\mu_j\|, \tag{9}$$

where for $j \geq 0$, $\nu_j \in M_d(g)$ and μ_j is absolutely continuous with respect to a product of j continuous measures that are each concentrated on E. Then 6.2.6 and 6.2.7 imply that the decomposition (9) is unique and that the norm equality holds.

Let $\rho = (\gamma, f) \in b\Gamma \times U$. We shall show that ρ defines a unique element of ΔN. Corollary 6.2.8 tells us that there exists an extension ψ of f to all $M(G)$, where $\psi \in \Delta M(G)$. If $\mu_1, \ldots, \mu_j \in M_c(E)$, then

$$(\mu_1 * \cdots * \mu_j)\hat{\ }(\psi) = \langle f, \mu_1 \rangle \cdots \langle f, \mu_j \rangle.$$

Then the restriction of ψ to N is uniquely determined by f. We now define $\hat{\mu}(\rho)$ by

$$\hat{\mu}(\rho) = \sum \hat{\nu}_j(\gamma)\hat{\mu}_j(\psi), \tag{10}$$

where μ has the form (9). It follows from the pairwise mutual singularity of the $\nu_j * \mu_j$ that (10) defines a linear functional on N. We must show that $\hat{\mu}(\rho)\hat{\nu}(\rho) = (\mu * \nu)\hat{\ }(\rho)$ for all μ, $\nu \in N$. Without loss of generality we may assume that $\mu = \delta(x) * \mu_1$ and $\nu = \delta(y) * \nu_1$ where μ_1 is the product of j measures from $M_c(E)$ and ν_1 is the product of k measures from $M_c(E)$. Then $\hat{\mu}(\rho) = \langle \gamma, x \rangle \hat{\mu}_1(\psi)$ and $\hat{\nu}(\rho) = \langle \gamma, y \rangle \hat{\nu}_1(\psi)$. And $\mu * \nu = \delta(x + y) * \mu_1 * \nu_1$. Therefore

$$(\mu * \nu)\hat{\ }(\rho) = \langle \gamma, x + y \rangle (\mu_1 * \nu_1)\hat{\ }(\psi) = \langle \gamma, x \rangle \hat{\mu}_1(\psi) \langle \gamma, y \rangle \hat{\nu}_1(\psi) = \hat{\mu}(\rho)\hat{\nu}(\rho).$$

Thus, $b\Gamma \times U \subseteq \Delta N$.

We now show that $b\Gamma \times U \supseteq \Delta N$. Let $\varphi \in \Delta N$. Define $\gamma \in b\Gamma$ by $\langle \gamma, x \rangle = \delta(x)\hat{\ }(\varphi)$ and $f \in U$ by $f = \varphi|_{M_c(E)}$. Then it is easy to see that $\rho_\varphi = (\gamma, f) \in b\Gamma \times U$ and $\hat{\mu}(\rho_\varphi) = \hat{\mu}(\varphi)$. Thus $b\Gamma \times U = \Delta N$.

It should be obvious that the correspondence between φ and ρ_φ is a homeomorphism when U is given the weak$*$ topology and $b\Gamma \times U$ product topology. It is obvious also that $|\varphi_\mu| = 1$ a.e. $d\mu$ for all $\mu \in N$ if and only if $|f_\nu| = 1$ a.e. $d\nu$ for all $\nu \in M_c(E)$. The set of all $(\gamma, f) \in b\Gamma \times U$ such that $|f_\nu| = 1$ a.e. $d\nu$ for all $\nu \in M_c(E)$ is dense in $b\Gamma \times U$. By 5.1.3, each such (γ, f) is identified with some $\varphi \in \partial N$. Therefore $\partial N = \Delta N = b\Gamma \times U$. The Corollary is proved. ☐

6.2.10. Lemma. *Let F be a countable subset of the algebraically scattered set E. Then there exists a countable subset Y such that $F \subseteq Y \subseteq E$ and $(GpY) \cap Gp(E \setminus Y) = \{0\}$.*

Proof. Let $Y_0 = F$. For $n = 1, 2, \ldots$, there exists a countable subset Y_n of E such that $GpY_{n-1} \cap Gp(E \setminus Y_n) = \{0\}$. Then $Y = \bigcup_0^\infty Y_n$ will do. □

We remind the reader that an L-subalgebra N of $M(G)$ is *prime* if $N^\perp$ is an ideal. See Section 5.2.

6.2.11. Lemma. *Let E and N be as in* 6.2.9. *Let* $\mathscr{S}_0 = M_d(G)$. *For each* $n \geq 1$, *let* $\mathscr{S}_n$ *be the set of measures* μ *such that there exists a countable union of translates of* nE *on which* μ *is concentrated and such that* $|\mu|((n-1)E + x) = 0$ *for all* x. *Then:*

(i) $\sum_0^\infty \mathscr{S}_n$ *is a prime L-subalgebra of* $M(G)$;
(ii) $N \supseteq \sum_0^\infty \mathscr{S}_m$, *and* $\mathscr{S}_m \perp \mathscr{S}_n$ *for* $0 \leq m \neq n < \infty$;
(iii) $\mathscr{S}_m * \mathscr{S}_n \subseteq \mathscr{S}_{m+n}$ *for* $0 \leq m, n < \infty$; *and*
(iv) $[\mathscr{S}_m \cap M(mE)] * [\mathscr{S}_n \cap M(nE)] \subseteq \mathscr{S}_{n+m} \cap M((m+n)E)$.

Proof. It is easy to see that $\sum_0^\infty \mathscr{S}_n$ is the L-algebra of measures concentrated on the Raĭkov system (See 5.2) generated by E, so (*i*) follows from 5.2.1. Part (ii) is obvious.

For (iii) and (iv) we may assume that $m, n \geq 1$. Let $\mu \in \mathscr{S}_m$ and $\nu \in \mathscr{S}_n$. Then we may assume that $\mu \geq 0$, $\mu \in M(mE)$, $\nu \geq 0$ and $\nu \in M(nE)$. Let $p \geq 0$ be a minimal integer such that

$$\mu * \nu(pE + v) \neq 0 \tag{11}$$

for some $v \in G$. Of course $1 \leq p \leq m + n$, since $\mu * \nu \in M((m+n)E$. It will suffice to show that $v = 0$ and $p = m + n$.

Suppose that $v \neq 0$. Let $Y \subseteq E$ be a countable subset such that $v \notin Gp(E \setminus Y)$. By the minimality of p, $\mu * \nu((p-1)E + Y + v) = 0$. Therefore (11) implies that

$$\mu * \nu(pE + v) = \mu * \nu(p(E \setminus Y) + v) > 0. \tag{12}$$

Since μ is concentrated on $m(E \setminus Y)$ and ν on $n(E \setminus Y)$, $\mu * \nu$ is concentrated on $(m+n)(E \setminus Y)$. Because $v \notin Gp(E \setminus Y)$,

$$[(m+n)(E \setminus Y)] \cap [p(E \setminus Y) + v] = \varnothing. \tag{13}$$

But (12) and (13) are contradictory. Therefore $v = 0$. Then (11) may be rewritten in the form

$$\iint \chi_{pE}(x + y)d\mu(x)d\nu(y) > 0. \tag{14}$$

Note that formulas such as (14) are valid because pE is a Borel set and μ, ν, $\mu \times \nu$ and $\mu * \nu$ are regular Borel measures.

Because $\mu \in \mathscr{S}_m$, (14) implies that there exists $x \in mE \setminus \bigcup_{j=0}^{m-1} jE$ such that

$$\int \chi_{pE}(x + y)dv(y) > 0. \tag{15}$$

By Lemma 6.2.10, there exists a countable subset $Y \subseteq E$ such that $x \in GpY$ and $GpY \cap Gp(E \setminus Y) = \{0\}$. Because $v \in \mathscr{S}_n$, there exists, by (15), $y \in [n(E \setminus Y) \setminus \bigcup_{j=0}^{n-1} jE]$ such that $x + y \in pE$. We write $x + y = w_1 + \cdots + w_r + v_1 + \cdots + v_s$, where $\{w_j\}_1^r \subseteq Y$ and $\{v_j\}_1^s \subset E \setminus Y$ and $r + s = p$. Since

$$GpY \cap Gp(E \setminus Y) = \{0\}, x = w_1 + \cdots + w_r \text{ and } y = v_1 + \cdots + v_s.$$

Since $x \in mE \setminus \bigcup_0^{m-1} jE$, $r \geq m$, and similarly, $s \geq n$. Therefore $p = m + n$, and (iii)–(iv) hold. That ends the proof of the Lemma. □

6.2.12. Theorem. *Let E be σ-compact algebraically scattered subset of the locally compact abelian group G. Let N be the L-subalgebra of $M(G)$ that is generated by $M_c(E) \cup M_d(G)$. Then N is prime.*

Proof. Let $\mu \in M(G)$ and $v \in N^\perp$. We must show that $\mu * v \in N^\perp$. We may assume that $\mu \geq 0$ and $v \geq 0$ by A.6.1. Let $\mathscr{S}_j$ be as in the statement of 6.2.11 and $N_1 = \sum_{j=0}^{\infty} \mathscr{S}_j$. Then $N_1 \supseteq N$ and $N_1^\perp \subseteq N^\perp$. Lemma 6.2.11(i) shows that $N_1^\perp$ is an ideal. We may therefore assume that $\mu \in N_1$, that $v \in N_1 \cap N^\perp$, and thus that $\mu \in \mathscr{S}_m$ and $v \in \mathscr{S}_n$ for some integers $m, n \geq 1$. To prove the Theorem, it will suffice to show that $\delta(w) * \tau^k \perp \mu * v$ for all $w \in G$, $k \geq 1$ and non-negative measures $\tau \in M_c(E)$. Let $\tau \geq 0$, $\tau \in M_c(E)$, $w \in G$, and $k \geq 1$. By 6.2.1(ii), $\delta(w) * \tau^k \in \mathscr{S}_k$. Therefore we may assume that $k = m + n$. If $w \neq 0$, let Y be a countable subset of E such that $w \notin Gp(E \setminus Y)$. Then $\mu \in \mathscr{S}_m$ implies $|\mu|(m(E \setminus Y)) = \|\mu\|$. Similarly $|v|(n(E \setminus Y)) = \|v\|$. Of course, then $\mu * v(kE + w) = 0$. We may therefore assume that $w = 0$.

Let F be a σ-compact subset of nE such that $v(nE \setminus F) = 0$ and $\tau^n(F) = 0$. Then the set $F^* = \{(x_1, \ldots, x_n) \in E \times \cdots \times E : \sum x_j \in F\}$ has zero $\tau^{(n)} = \tau \times \cdots \times \tau$ (n times) measure. Let F_k consist of all k-tuples of elements $x_1, \ldots, x_k$ of E such that for some choice of integers $1 \leq j(1) < \cdots < j(n) \leq k$, $\sum x_{j(i)} \in F$. By Fubini's Theorem, F_k has zero $\tau^{(k)}$-measure. Let

$$Q = \{(x_1, \ldots, x_k) \in E \times \cdots \times E \setminus F_k : \sum x_j \in mE + F\}.$$

Since $\mu * v$ is concentrated on $mE + F$, it will suffice to show that $\tau^k(mE + F) = 0$. That is equivalent to showing that Q has zero $\tau^{(k)}$-measure. Suppose that $\tau^{(k)}(Q) > 0$. Then there exists $x_1 \in E$ such that

$$Q_1 = \left\{(x_2, \ldots, x_k) \in E \times \cdots \times E : (x_1, \ldots, x_k) \notin F_k, \sum_1^k x_j \in mE + F\right\}$$

has $\tau^{(k-1)}(Q_1) > 0$. By 6.2.10, there is a countable set Y_1 such that $x_1 \in Y_1 \subseteq E$ and $Gp Y_1 \cap Gp(E \setminus Y_1) = \{0\}$. By induction, there exist $x_2, \ldots, x_k \in E$ and countable subsets $Y_2, \ldots, Y_k$ of E such that

(16) $$(x_1, \ldots, x_k) \in E \times \cdots \times E \setminus F_k, \sum_1^k x_j \in mE + F;$$

(17) $$\begin{cases} x_i \in Y_i \subseteq E \setminus \bigcup_{j=1}^{i-1} Y_j \text{ and } Gp\, Y_i \cap Gp(E \setminus Y_i) = \{0\} \text{ for} \\ \qquad 1 \leq i \leq k-1 \text{ and } Y_1 \subseteq \cdots \subseteq Y_{k-1}; \end{cases}$$

and such that

(18) $$x_k \in E \setminus Y_{k-1}.$$

By (16), $\sum_1^k x_j = \sum_1^m y_j + \sum_1^n y_{m+j}$ where $y_j \in E$ for all j and $\sum y_{m+j} \in F$. Therefore $\sum_1^k x_j = \sum_1^{k-1} \sum \{y_j : y_j \in Y_k\} + \sum \{y_j : y_j \notin Y_{k-1}\}$. Then (17) and (18) imply that $\{y_1, \ldots, y_k\}$ is a permutation of $\{x_1, \ldots, x_k\}$. Therefore, for an appropriate choice of distinct $j(i)$, $\sum_{i=1}^n x_{j(i)} \in F$, a contradiction. That proves Theorem 6.2.12. □

The proof of the next corollary shows that the algebras N of 6.1.12 are distinct from all algebras on Raĭkov systems.

6.2.13. Corollary. *Let G be a non-discrete locally compact abelian group. Then there exists $\psi \in \Delta M(G)$ such that $\psi_\mu^2 = \psi_\mu$ a.e. $d\mu$ for all $\mu \in M(G)$ and such that for each Raĭkov system $\mathscr{R}$ on G either there exists a non-zero measure μ concentrated on an element of $\mathscr{R}$ with $\psi_\mu \equiv 0$, or there exists a non-zero measure μ such that $\mu(X) = 0$ for all $X \in \mathscr{R}$ and $\psi_\mu \equiv 1$.*

Proof. Let E be a perfect compact metrizable independent subset of G. Let N be the L-subalgebra of $M(G)$ that is generated by $M_c(E \cup -E)$ and $M_d(G)$. Let p be the projection of $M(G)$ onto N. Since $E \cup -E$ is algebraically scattered, 6.2.12 implies that p is multiplicative. We define ψ by $\hat{\mu}(\psi) = (p\mu)\hat{}(0) = \int d(p\mu)$. It is easy to see that $\psi_\mu^2 = \psi_\mu$ a.e. $d\mu$ for all $\mu \in M(G)$ (see 5.2.2).

Let $\mathscr{R}$ be any Raĭkov system on G. Suppose that $\mu(X) = 0$ whenever $X \in \mathscr{R}$ and $\mu \in M_c(E \cup -E)$. Then any non-zero measure $\mu \in N \cap M_c(G)$ will have $\psi_\mu \equiv 1$ a.e. $d\mu$ and $\mu(X) = 0$ for all $X \in \mathscr{R}$. Thus, we may assume that there exists $\mu \in M_c(E \cup -E)$ and $X \in \mathscr{R}$ such that $\mu(X) \neq 0$. We may assume that X is compact since μ is regular. Also $X = X_1 \cup X_2$ where X_2 is countable and X_1 is perfect and compact. (That is straightforward.) We may therefore assume that $X = X_1 \subseteq E$.

Let $f: X \to X$ be any continuous one-to-one function such that $f(x) \neq x$ and $f(f(x)) \neq x$ for all $x \in X$. (Such an f can be obtained by identifying X with the infinite product $Z_3 \times Z_3 \times \cdots$ and letting $f(x) = x + y$, where $y \neq 0$ is fixed.) Let $Y = \{x + f(x): x \in X\}$. Then $Y \in \mathscr{R}$. Let v be any continuous measure that is concentrated on Y.

We claim that $\psi_v \equiv 0$ a.e. dv. Indeed, v is singular with respect to $\delta(y) * \mu^k$ for all $\mu \in M_c(E \cup -E)$ and $k \geq 1$, as is easily seen. Therefore $v \perp N$, and $\psi_v \equiv 0$. The Corollary is proved. □

6.2.14. Corollary. *Let G be a locally compact abelian group, $E \subseteq G$ a compact, perfect metrizable totally disconnected independent subset of G, and N the L-subalgebra of $M(G)$ generated by $M_c(E \cup -E)$ and $M_d(G)$. Let $H = \{\gamma \in b\Gamma: \gamma$ is μ-measurable for all $\mu \in N\}$. Let $N_1 = \{\mu: \gamma$ is μ-measurable for all $\gamma \in H\}$. Then $N_1 \neq N$.*

Proof. As in the proof of 6.2.13, we left $f: E \to E$ be any homeomorphism such that $f(x) \neq x$ and $f(f(x)) \neq x$ for all $x \in E$. Let $\mu \in M_c(E)$, $\mu \neq 0$, and let μ' be the measure on $Y = \{x + f(x): x \in E\}$ such that

$$\text{(19)} \qquad \int g(u)d\mu'(u) = \int g(x + f(x))d\mu(x) \quad \text{for } g \in C(E + E).$$

We claim that $\gamma \circ f$ is μ-measurable for all $\mu \in M(E)$. To see that, let $\check{f}$ be the map of measures defined by $\int g\, d(f\mu) = \int g \circ f\, d\mu$ for all $g \in C(E)$ and $\mu \in M(E)$. Then f is an automorphism of $M(E)$, so γ is $f\mu$-measurable for all μ. Therefore $\gamma \circ f$ is μ-measurable for all $\mu \in M(E)$. Therefore

$$x \to \langle x + f(x), \gamma \rangle = \gamma \circ f(x)\gamma(x)$$

is μ-measurable for all $\mu \in M(E \cup -E)$. But the (homeomorphic) map $x \to x + f(x)$ sends μ-measurable sets to μ'-measurable sets and μ-measurable functions to μ'-measurable functions. Therefore γ is μ'-measurable. But $\mu' \notin N$ by the proof of 6.2.13. The Corollary is proved. □

It is obvious from 6.2.8 that a monotrochic measure cannot be concentrated on an independent set. The next result strengthens that observation. Whether a continuous *tame* measure on T can be concentrated on a proper Borel subgroup remains an open question. (Lemma 6.5.2 shows that $\mu = \bigstar[\frac{1}{2}\delta(0) + \frac{1}{2}\delta((n!)^{-1})]$ is monotrochic. Clearly, μ is concentrated on a proper Borel subgroup of T.)

6.2.15. Theorem. *Let E_1 be a σ-compact independent subset of the locally compact abelian group G and $E = E_1 \cup -E_1$. Let μ be a continuous measure on G. If $\mu(GpE + y) \neq 0$ for some $y \in G$, then there exists $\psi \in \Delta M(G)$ such that $\{x: \psi_\mu(x) = 1\}$ and $\{x: \psi_\mu(x) = 0\}$ both have non-zero μ-measure.*

Proof. If the assertion of the theorem holds, we may assume that $\psi_\nu^2 = \psi_\nu$ for all $\nu \in M(G)$. We say in that case that ψ_μ is a non-trivial idempotent. We may assume that $\mu \geq 0$. We have several steps.

(A) Suppose that ψ_ν is idempotent for all $\nu \in M(G)$. Then ψ_μ is a nontrivial idempotent if and only if $\psi_{\delta_x * \mu}$ is a nontrivial idempotent for all $x \in G$. Indeed, $\psi(\delta_x) = 1$ so $\psi_{\delta_x * \mu}(x + y) = \psi_{\delta_x}(x)\psi_\mu(y)$ a.e. $d\mu$ by 5.1(8). The assertion now follows.

(B) Since the integers are well ordered and μ is continuous, there exists a minimal integer $m > 0$ such that

(20) $$\text{there is an } x \in G \text{ with } \mu(x + mE) = 0$$

and

(21) $$\text{for all } y \in G \text{ and } 0 < j < m,\ \mu(y + jE) = 0.$$

By using (A) we see we may assume that both

(22) $$\mu(mE) \neq 0$$

and (21) hold. We leave m fixed. We may further assume (by replacing μ with a measure absolutely continuous with respect to μ) that μ has support contained in mE.

(C) Let $x_1, \ldots, x_m \in E$ with $x_1 + \cdots + x_m = x$ belonging to the support of μ. Let $X = \{x_j, -x_j : 1 \leq j \leq m\}$. Then

(23) $$mE = \bigcup_{j=0}^{m} [j(E \setminus X) + (m - j)X].$$

From (21) and (23), we see that

(24) $$\mu(mE) = \mu(m(E \setminus X)).$$

The regularity of μ, (21), and the continuity of addition in G imply that

(25) $$\mu(mE) = \sup\{\mu(m(E \setminus W))\},$$

where the supremum in (27) is taken over neighborhoods $W = -W$ of X. But if W is any open neighborhood of X, then $(E \setminus W)$ misses an entire (relative) neighborhood of $x_1 + \cdots + x_m = x$ in the support of μ. Therefore

(26) $$\mu(mE) > \mu(m(E \setminus W)).$$

Of course for a sufficiently small neighborhood W of X, (25) combined with (26) shows that

(27) $$\mu(mE) > \mu(m(E\setminus W)) > \tfrac{1}{2}\mu(mE) > 0.$$

(D) Assume that a neighborhood $W = -W$ of X has been chosen such that (27) holds. Then

(28) $$mE = \bigcup_{j=0}^{m} j(E\setminus W) + (m - j)(E \cap W),$$

and (27) and (29) together imply that for some $0 \le j_1 < m$,

(29) $$\mu(j_1(E\setminus W) + (m - j_1)(E \cap W)) > 0.$$

(E) Let $\mathscr{R}$ be the Raĭkov system in G which is generated by $F = E\setminus W$. Let p be the map from $M(G)$ to $A_{\mathscr{R}}$, the algebra of measures ω on G that are concentrated on sets in $\mathscr{R}$.

Then the definition of p and (27) together imply that

$$\|p\mu\| \ge \mu(mF) = \mu(m(E\setminus W)) > 0.$$

Let ψ be the maximal ideal of $M(G)$ defined by $\psi(\omega) = (p\omega)\hat{}(0)$ for all $\omega \in M(G)$. As in the proof of 6.2.13, ψ_ω is idempotent for all $\omega \in M(G)$. In particular $\psi(\mu) = \|p\mu\| > 0$, so $\psi_\mu = 1$ on a set of nonzero μ-measure.

(F) Let $r = m - j_1$, $L = E \cap W$, and $K = j_1E + rL$. We claim that

(30) $$\mu(K \cap (y + GpF)) = 0 \text{ for all } y \in G.$$

Suppose that (30) holds. Then $p\mu(K) = 0$, while (29) implies that $\mu(K) > 0$. Therefore $\psi_\mu = 0$ for μ-almost all x in K. Thus the Theorem will follow. It remains to establish (30).

So suppose that $K \cap (y + GpF) \ne \varnothing$ for some $y \in G$. The definition of K implies that there exist $x \in rL$ and $a \in GpF$ such that $y = x + a$. If $z \in K \cap (y + GpF)$, then for some $b \in GpF$, $c \in j_1F$, and $d \in rL$,

$$z = c + d = y + b = (a + b) + x.$$

Since E_1 is independent, $(GpF) \cap (GpL) = \{0\}$. Therefore $c = a + b$ and $d = x$. Therefore $z = x + c \in [x + j_1F]$ and $K \cap (y + GpF) \subseteq x + j_1F$. The last set has zero μ-measure by (21). The Theorem is proved. □

Let G be a locally compact abelian group. A closed subset E of $\Delta M(G)$ is an *interpolation set for* $M(G)$ if $M(G)\hat{}|_E = C(E)$. The following result shows that 2.1.2 does not generalize from $L^1(G)$ to $M(G)$.

6.2.16. Theorem. *There exist interpolation sets E_1 and E_2 for $M(T)$ whose union is not an interpolation set.*

Proof. Let $n_j = j!$, for $j \geq 1$. It is not hard to show that for each sequence $\{z_j\}$ of complex numbers of modulus one there exists $x \in T$ such that

$$|z_j - \langle x, n_j \rangle| < \tfrac{1}{3} \quad \text{for } j \geq 1. \tag{31}$$

Let E be a perfect compact independent subset of T, and let $E_1, E_2, \ldots$ be a sequence of disjoint perfect compact subsets of E. For $j \geq 1$, let $\mathscr{R}_j$ be the Raĭkov system generated by $\bigcup_{m=j}^{\infty} E_m$, A_j the algebra of measures concentrated on sets in $\mathscr{R}_j$, and $\rho_j, \psi_j \in \Delta M(G)$ be the multiplicative linear functionals, defined by the following properties.

$$(\psi_j)_\mu = 0 \text{ a.e. } d\mu \quad \text{for } \mu \in A_j^{\perp}; \tag{32}$$

$$(\rho_j)_\mu = 0 \quad \text{for } \mu \in A_{j+j}^{\perp}; \tag{33}$$

$$(\psi_j)_\mu(x) = \langle x, n_j \rangle \quad \text{for } \mu \in A_j; \tag{34}$$

and

$$(\rho_j)_\mu(x) = \langle x, n_j \rangle \quad \text{for } \mu \in A_{j+1}. \tag{35}$$

For $j = 1, 2, \ldots$ let q_j denote the projection on A_j and $\gamma_j(x) = \langle x, n_j \rangle$. Then

$$\left\{ \begin{array}{l} \hat{\mu}(\psi_j) = (q_j\mu)\hat{}(n_j) \quad \text{and} \quad \hat{\mu}(\rho_j) = (q_{j+1}\mu)\hat{}(n_j) \quad \text{for all } j \geq 1 \\ \text{and all } \mu \in M(G). \end{array} \right. \tag{36}$$

Since $M_d(T) \subseteq A_j$ for all $j = 1, 2, \ldots$, (31) and (36) imply that the closures E_1 of $\{\psi_j\}$ and E_2 of $\{\rho_j\}$ are interpolation sets in $\Delta M(T)$. It is easy to see that $\psi_j \neq \rho_k$ for all $1 \leq j, k < \infty$. We claim that $E_1 \cup E_2$ is not an interpolation set in $\Delta M(T)$. To prove the claim it will be sufficient to prove that if $\mu \in M(T)$, $n \geq 1$ and

$$\hat{\mu}(\psi_{2j}) = 1, \hat{\mu}(\rho_{2j}) = 0 \quad \text{for } j = 1, \ldots, n, \tag{37}$$

then $\|\mu\| \geq n$. Let μ be any measure and n any integer such that (37) holds. Then

$$\|\mu\| \geq \sum_{j=1}^{\infty} \|(q_{2j} - q_{2j+1})\mu\|, \quad \text{and} \quad (q_{2j} - q_{2j+1})\mu \perp (q_{2k} - q_{2k+1})\mu \tag{38}$$
$$\text{for } 1 \leq j \neq k < \infty.$$

Now (36) and (37) imply that

$$1 = (q_{2j} - q_{2j+1})\mu^\wedge(n_j), \quad \text{so } \|(q_{2j} - q_{2j+1})\mu\| \geq 1 \quad \text{for } 1 \leq j \leq n.$$

By (38), $\|\mu\| \geq n$, as required. Theorem 6.2.16 is proved. □

Remarks and Credits. The first perfect compact independent subset of R was constructed by von Neumann [1]; his set is algebraically independent. Independent sets E such that every unimodular function on E is the uniform limit on E of continuous characters are called *Kronecker sets* or *K-sets.* Many constructions of such sets and their finite order analogues, the K_p*-sets,* are known. See Rudin [1, Section 5.2] or Hewitt and Ross [1, vol. II, pp. 554–558], for the standard Cantor-type dissection process. Lindahl and Poulsen [1, Section I.3] give the category method of Kaufman [2, 6]. For the Körner construction, see Section 4.7 or Lindahl and Poulsen [1, Section XIII.3].

Theorem 6.2.1, Corollary 6.2.3, Theorem 6.2.6 and Corollaries 6.2.7 and 6.2.8 were proved by Hewitt [1] and Hewitt and Kakutani [1] for the case $E = E_1 \cup -E_1$, where E_1 is independent. The versions here are due to Saeki [16]. Theorem 6.2.2 is from Rago [1] and Saeki [16]; they extended results of Salinger and Varopoulos [1] and Hartman and Ryll-Nardzewski [4]. The first published version of 6.2.4 is in Rudin [1, Theorem 5.3.6], but the proof there appears to be incomplete. Proofs of 6.2.4 appeared later in Graham [4] and Rago [1]. Theorem 6.2.5 is, of course, well-known. The theorem has to be phrased in the manner given because of the possible existence of Lebesgue measurable sets E and F of R such that $E + F$ is not measurable. For another proof of the existence of non-measurable sets, see Halmos [1, p. 69].

Corollary 6.2.9 for independent sets is due to Simon [3]; Theorem 6.2.12 (for independent sets) is from Simon [2]; the argument given here was suggested by Saeki. Corollaries 6.2.13 and 6.2.14 appear to be new. Theorem 6.2.15 and its proof are taken from Graham [9]. Theorem 6.6.6(ii) will show that $\mu = \bigstar_{j=1}^{\infty}[\frac{1}{2}\delta(0) + \frac{1}{2}\delta(3^{-j})]$ cannot be concentrated on a proper Raĭkov system, although μ is i.p. Theorem 6.2.16 appears in Graham [5], where the generalization to all non-discrete locally compact abelian groups is also proved.

6.3. Measures on Dissociate Sets

Let H be a subset of the abelian group G. A subset E of $G\backslash H$ is *dissociate mod H* if $n \geq 1$, $\{x_1, \ldots, x_n\} \subseteq E$, $\{m_1, \ldots, m_n\} \subseteq \{\pm 2, \pm 1, 0\}$ and $\sum m_j x_j \in H$ imply $m_1 x_1 \in H, \ldots, m_n x_n \in H$. If $H = \{0\}$, E is *dissociate.* The set E is *independent mod H* if the preceding holds with $\{\pm 2, \pm 1, 0\}$ replaced by Z. Since many results of this section are analogous to those of Section 6.2, we omit many proofs. We begin by giving examples that show that "dissociate" and "algebraically scattered" are distinct notions. If E is independent, then $E \cup -E$ is algebraically scattered but not dissociate. If $E \subseteq T$ is independent and uncountable and if $x \in T \backslash E$ is such that $E \cup \{x\}$ is independent, then $E' = E \cup (\{3y: y \in E\} + x)$ is dissociate but not algebraically scattered. To see that, note first that if Y is a countable subset of E', then $x \in Gp(E' \backslash Y)$. That E' is dissociate is a simple argument which we leave to the reader.

For $n = 1, 2, \ldots$ and $E \subseteq G$ we let

$$E(n) = \left\{\sum_1^n \pm x_j : x_j \in E, \pm x_j \neq x_k \quad \text{for } 1 \leq j \neq k \leq n\right\}.$$

6.3.1. Theorem. *Let E be a dissociate subset of the locally compact abelian group G. Then for every integer $n = 1, 2, \ldots$, every non-empty countable subset F of G, every Borel subset $K \subseteq E(n) + F$ and all sets $\{\mu_1, \ldots, \mu_{n+1}\} \subseteq M_c(G)$,*

(i) $\mu_1 * \cdots * \mu_n((K + x) \cap (K + y)) = 0$ *if* $x - y \notin F - F$; *and*
(ii) $\mu_1 * \cdots * \mu_{n+1}(K + x) = 0$ *for all* $x \in G$.

Proof. The case $n = 0$ and the fact that (i) $\Rightarrow$ (ii) for each $n \geq 1$ follow as in the proof of 6.2.1. For (i), there is nothing to prove unless $K \cap (K + y) \neq \varnothing$. (As in the proof of 6.2.1(i) we may assume $x = 0$.) We may assume that $y \notin F - F$. Then

$$K \cap (K + y) \subseteq \bigcup_{s, t \in F} (E(n) + s) \cap (E(n) + y + t).$$

Fix $s, t \in F$. Suppose that $(E(n) + s) \cap (E(n) + y + t) \neq \varnothing$. Since $y + t - s \neq 0$, $y + t - s \in E(n) - E(n)$, so there exist distinct $x_1, \ldots, x_n \in E$ and distinct $y_1, \ldots, y_n \in E$ such that $y + t - s = \sum \pm x_j \mp y_j$. Let $Y_{s,t} = \{x_j, y_j\}_1^\infty$, and $W = \bigcup_{s,t \in F} (Y_{s,t} + s) \cup (Y_{s,t} + t + y)$. Then $K \cap (K + y) \subseteq E(n - 1) + W$. By (ii) for $n - 1$, $|\mu_1| * \cdots * |\mu_n|(E(n - 1) + w) = 0$, so $\mu_1 * \cdots * \mu_n(K \cap (K + y)) = 0$. That proves the Theorem. □

6.2.2 has the following analogue. If G and $\{\mu_j\}$ are as in 6.2.2 and E is a σ-compact dissociate subset of G, then $E(n) + x$ has zero $\mu_1 * \cdots * \mu_{n+1}$-measure for every integer $n \geq 1$ and every $x \in G$. The proof is essentially the same as the proof of 6.2.2. However, we can have $\mu_1 * \cdots * \mu_{n+1}(GpE) \neq 0$. Here is an example. Let H be the infinite product Z_3 and let $G = D \times H$. Let $\varphi: D \to H$ be a homeomorphism of D onto a set of type K_3. Then $E = \{(x, \varphi(x)): x \in D\}$ is dissociate, since the projection on H is independent. But $(3) * E = D \times \{0\}$, so $(m_{D \times \{0\}})^4((3) * E) = 1$. We do not know whether a compact dissociate set can generate T.

6.3.2. Lemma. *Let $\nu_1, \ldots, \nu_n$ be continuous measures concentrated on the Borel sets $E_1, \ldots, E_n$ of the locally compact abelian group G. Then $\nu_1 * \cdots * \nu_n$ is concentrated on the set $L = \{\sum_1^n x_j \colon x_j \in E_j$ and $x_j \neq \pm x_k$ for $1 \leq j \neq k \leq n\}$.*

Proof. We may assume that the ν_j are probability measures. By Fubini's Theorem,

$$F = \{(x_1, \ldots, x_n) \in E_1 \times \cdots \times E_n \colon x_j = \pm x_k \quad \text{for some } j \neq k\}$$

has zero $\nu_1 \times \cdots \times \nu_n$-measure. Therefore there exists a σ-compact subset L' of $E_1 \times \cdots \times E_n \backslash F$ such that $\nu_1 \times \cdots \times \nu_n(L') = 1$. Then $L'' = \{\sum x_j \colon (x_1, \ldots, x_n) \in L'\}$ is a σ-compact subset of L and $\nu_1 * \cdots * \nu_n(L'') = 1$. That proves the Lemma. □

6.3.3. Corollary. *Let E be a σ-compact dissociate subset of the locally compact abelian group G and $\mu \in M_c(G)$ a measure concentrated on E. Then μ is strongly i.p., and in fact*

$$\delta(x) * \mu^m \perp \delta(y) * \mu^n$$

unless $m = n$ and $x = y$.

Proof. By 6.3.2, $\delta(x) * \mu^m$ is concentrated on $E(m) + x$ and $\delta(y) * \mu^n$ is concentrated on $E(n) + y$. The proof of 6.2.3 now applies. □

The proof of the next Corollary is like that of 6.2.4.

6.3.4. Corollary. *Let G be a locally compact abelian group and E be a σ-compact dissociate subset of G. Let H be a σ-compact non-discrete locally compact abelian group that is continuously embedded in G. Then $m_H(E(n) + x) = 0$ for all $n = 0, 1, \ldots,$ and $x \in G$.*

If H is a Borel subset of G, we let $\mathscr{S}(H)$ be the set of regular Borel measures that are concentrated on countable unions of translates of H. Then $\mathscr{S}(H)^\perp = \{\mu \in M(G) \colon |\mu|(x + H) = 0 \text{ for all } x \in G\}$ is an L-ideal. If H is a subsemigroup, then $\mathscr{S}(H)$ is the set of measures concentrated on the Raĭkov system generated by H. The following lemma, a variant of 6.2.11, will be used in the proof of 6.3.6, the "dissociate" version of 6.2.6. We let $E(\infty) = \bigcup_{n=0}^{\infty} E(n)$.

6.3.5. Lemma. *Let H be a σ-compact subgroup of the locally compact abelian group G, and let E be a compact subset of G that is dissociate* mod H. *For*

$n \geq 1$, let $\mathscr{S}_n = \mathscr{S}(E(n) + H) \cap (\mathscr{S}(E(n-1) + H))^{\perp}$, and let $\mathscr{S}_0 = \mathscr{S}(H)$. Then:

(i) *$\mathscr{S}_n \perp \mathscr{S}_m$ for $0 \leq m \neq n < \infty$;*
(ii) *$\mathscr{S}_n * \mathscr{S}_m \subseteq \mathscr{S}_{n+m}$ for $0 \leq n, m < \infty$;*
(iii) *$\mathscr{S}(E(\infty) + H) = \sum_0^\infty \mathscr{S}_n$;*
(iv) *if $x \in G \setminus H$, $1 \leq n < \infty$, and $\mu \in \mathscr{S}_n$, then*

$$\mu([H + E(n)] \cap [x + H + E(n)]) = 0; \text{ and}$$

(v) *$\mathscr{S}(E(\infty) + H)$ is a prime L-subalgebra of $M(G)$.*

Proof. (i) We will first prove that if $0 \leq m < n$, then $E(m)$ is covered by finitely many translates of $E(n)$. That is obvious for $m = 0$. Suppose $1 \leq m < n$ and that $E(m-1)$ is covered by finitely many translates of $E(j)$ for all $j > m - 1$. Let $x_1, \ldots, x_{n-m}$ be distinct elements of E. Then

$$E(m) \subseteq \left[\bigcup_{j=1}^{n-m} (E(m-1) + \{x_j, -x_j\})\right] \cup [E(n) - (x_1 + \cdots + x_{n-m})].$$

The inductive hypotheses now establishes the claim. Therefore

$$\mathscr{S}(H + E(n-1))^{\perp} \supseteq \mathscr{S}(H + E(n))^{\perp} \quad \text{for } n = 1, 2, \ldots,$$

which implies (i).

(ii) Let $\mu \in \mathscr{S}_m$, $\nu \in \mathscr{S}_n$. Since $\mathscr{S}_m$ and $\mathscr{S}_n$ are translation-invariant L-subspaces contained, respectively, in $\mathscr{S}(E(m) + H)$ and $\mathscr{S}(E(n) + H)$, we may assume that $\mu, \nu \geq 0$ and $\mu \in M(E(m) + H)$, $\nu \in M(E(n) + H)$. We claim that $\mu * \nu \in M(E(m + n) + H)$ and that $\mu * \nu \perp \mathscr{S}(E(n + m - 1) + H)$. From those claims (if established), (ii) follows at once.

Let K be a Borel set of non-zero $\mu * \nu$ measure. Then

$$\text{(1)} \qquad \iint \chi_K(x + y) d\mu(x) d\nu(y) > 0.$$

Since $\mu \in M(E(m) + H)$, there exists $x \in E(m) + H$ such that

$$\text{(2)} \qquad \int \chi_K(x + y) d\nu(y) > 0.$$

Let F be any finite subset of E such that $x \in F(m) + H$. Since

$$\nu \in M(E(n) + H) \cap \mathscr{S}(H + E(n-1))^{\perp},$$

(2) implies that there exists $y \in (E \backslash F)(n) + H$ such that $x + y \in K$. Then

$$x + y \in [F(m) + H + (E \backslash F)(n) + H] \subseteq H + E(m + n).$$

Therefore, $\mu * \nu(K) \neq 0$ implies $K \cap (E(m + n) + H) \neq \varnothing$. Therefore $\mu * \nu$ is carried by $E(m + n) + H$.

We now show that $\mu * \nu \perp \mathscr{S}(E(n + m - 1) + H)$. Let p denote the smallest integer such that $\mu * \nu(w + E(p) + H) \neq 0$ for some $w \in G$. Then $p \leq m + n$. Also

$$\mu * \nu([E(m + n) + H] \cap [w + E(p) + H]) > 0,$$

since $\mu * \nu \in M(E(m + n) + H)$. Therefore

$$(E(m + n) + H) \cap (w + E(p) + H) \neq \varnothing.$$

Let E_1 be a finite subset of E such that $w \in H + E_1(m + n) - E_1(p)$.

Since p is minimal, $\mu * \nu(w + E_1(1) + E(p - 1) + H) = 0$. Let $K = (w + E(p) + H) \backslash (w + E_1(1) + E(p - 1) + H)$. Then $K \subseteq w + (E \backslash E_1)(p) + H$ and K satisfies (2). Therefore there exists a finite subset F of $E \backslash E_1$ and $x \in F(m) + H$ such that $\int \chi_K(x + y) d\nu(y) > 0$. There also exists

$$y \in [(E(n) + H) \backslash ((E_1 \cup F)_{(1)} + E(n - 1) + H)]$$

with $x + y \in K$. Let $w = h_1 + \sum_1^{m+n} \pm w_j - \sum_{m+n+1}^{m+n+p} \pm w_j$ where the w_j are in E and $w_j \neq \pm w_k$ if either $1 \leq j \neq k \leq m + n$ or $m + n + 1 \leq j \neq k \leq m + n + p$. By our choices of E_1, F, x, and y, there are distinct elements $x_1, \ldots, x_m \in F \subseteq E \backslash E_1$, distinct elements $y_1, \ldots, y_n \in E \backslash (E_1 \cup F)$, distinct elements $u_1, \ldots, u_p \in E \backslash K$, and $h_1, \ldots, h_4 \in H$ such that $x + y = w + \sum_1^p \pm u_j + h_2$, $x = \sum_1^m \pm x_j + h_3$ and $y = \sum_1^n \pm y_j + h_4$. Then

$$\text{(3)} \qquad \sum_1^m \pm x_j + \sum_1^n \pm y_j - \sum_1^{m+n} \pm w_j + \sum_{m+n+1}^{m+n+p} \pm w_j - \sum_1^p \pm u_j$$
$$= h_1 + h_2 - h_3 - h_4.$$

Since no u_j equals $\pm$ a w_k and no x_j or y_j equals $\pm$ a w_k and no y_j equals $\pm$ an x_k, the left side of (3) has the form $\sum \varepsilon_j v_j + \sum \varepsilon_j' v_j$ where all $|\varepsilon_j| \leq 2$, $|\varepsilon_j'| \leq 2$, and the v_j are distinct elements of E. Since E is dissociate mod H, $\varepsilon_j v_j \in H$ and $\varepsilon_j' w_j \in H$ for all j. Thus $w \in H$, that is, we may assume that $w = 0$. But then (3) reduces to $\sum_1^m x_j + \sum_1^n y_j - \sum_1^p u_j \in H$. Since no x_j is a y_k, we must have $p = m + n$, again because E is dissociate mod H.

(iii) is immediate from (ii) and the fact, established in the proof of (i), that finite unions of translates of $E(n)$ cover $E(m)$ for $0 \leq m < n < \infty$.

(iv) This is a variant of 6.2.1. If $(x + E(n) + H) \cap (E(n) + H) = \varnothing$, the conclusion of (iv) certainly holds. Otherwise, there exists a finite subset F of E such that $x \in [H + F(n) - F(n)]$. Then $[x + (E\backslash F)(n) + H] \cap [E\backslash F)(n) + H] = \varnothing$ since E is H-dissociate. Therefore

$$(4) \qquad \begin{aligned}&(x + E(n) + H) \cap (E(n) + H)\\ &\subseteq (x + F + E(n-1) + H) \cup (F + E(n-1) + H).\end{aligned}$$

Since both sets on the right side of (4) are finite unions of translates of $E(n-1) + H$ and $\mu \in \mathscr{S}_n$, (iv) holds.

(v) Since $\mathscr{S}(X)^{\perp}$ is an L-ideal for any σ-compact set X, (v) follows from (ii) and (iii). That proves the Lemma. □

6.3.6. Theorem. *Let H be a σ-compact subgroup of the locally compact abelian group G. Suppose that E is a compact subset of G that is dissociate* mod *H. Let $\nu_1, \dots, \nu_n$ be pairwise mutually singular measures in $M_c(E)$ and $\mu, \nu \in \mathscr{S}(H)$. If $a = (a(1), \dots, a(n))$ and $b = (b(1), \dots, b(n))$ are distinct n-tuples of nonnegative integers, then*

$$(5) \qquad \lambda \perp \rho$$

where

$$\lambda = \mu * \nu_1^{a(1)} * \cdots * \nu_n^{a(n)} \quad \text{and} \quad \rho = \nu * \nu_1^{b(1)} * \cdots * \nu_n^{b(n)}.$$

Proof. Let $m = a(1) + \cdots + a(n)$ and $q = b(1) + \cdots + b(n)$. By 6.3.2, λ is concentrated on $H + E_1(a(1)) + \cdots + E_n(a(n))$ and ρ is concentrated on $H + E_1(b(1)) + \cdots + E_n(b(n))$, where the E_j are σ-compact disjoint subsets of E with ν_j concentrated on E_j. If $[H + E_1(a(1)) + \cdots + E_n(a(n))] \cap [H + E_1(b(1)) + \cdots + E_n(b(n))] \neq \varnothing$ then

$$h \pm x_{1,1} \cdots \pm x_{1,a(1)} \pm \cdots \pm x_{n,a(n)} = h' \pm y_{1,1} + \cdots \pm y_{n,b(n)}$$

for some $h, h' \in H$, $x_{1,1}, \dots, x_{1,a(1)}, y_{1,1}, \dots, y_{1,b(1)} \in E_1, \dots, y_{n,b(n)} \in E_n$. Therefore $\pm x_{11} \pm \cdots \pm x_{n,a_n(n)} \mp y_{11} \mp \cdots \mp y_{n,b(n)} \in H$. The x's are distinct, the y's distinct, $x_{j,k} \neq y_{j',k'}$ if $j \neq j'$ for all k, k', and E is dissociate mod H. Therefore

$$\pm x_{j,1} \pm \cdots \pm x_{j,a(j)} \mp \cdots \mp y_{j,1} \mp \cdots \mp y_{j,b(j)} \in H \text{ for } 1 \leq j \leq n.$$

Since each E_j is dissociate mod H, $a(j) = b(j)$ for each j. That proves (5). □

6.3.7. Corollary. *Let E, H, μ, $a(1), \ldots, a(n)$, and $\nu_1, \ldots, \nu_n$ be as in 6.3.6. Then*

$$\|\mu * \nu_1^{a(1)} * \cdots * \nu_n^{a(n)}\| = \|\mu\| \, \|\nu_1\|^{a(1)} \cdots \|\nu_n\|^{a(n)}. \tag{6}$$

Proof. We use the argument of 6.2.7, combined with the observation that if $\varepsilon > 0$, there exists a neighborhood U of 0 such that $\|\omega * \mu\| \geq (1 - \varepsilon)\|\mu\|$ for all probability measures ω supported in U. The details are left to the reader. □

The proofs of the next three results are modifications of the proofs of 6.2.8, 6.2.9 and 6.2.14 and are omitted.

6.3.8. Corollary. *Let H be a σ-compact subset of the locally compact abelian group G, and E a σ-compact subset of G that is dissociate* mod H. *Let f be a continuous linear functional on $M_c(E)$ of norm $\|f\| \leq 1$ and $\rho \in \Delta\mathscr{S}(H)$. There exists $\psi \in \Delta M(G)$ such that $\hat{\nu}(\psi) = \langle f, \nu \rangle$ for each $\nu \in M_c(E)$ and $\hat{\mu}(\psi) = \hat{\mu}(\rho)$ for each $\mu \in \mathscr{S}(H)$.*

6.3.9. Corollary. *Let H be a σ-compact subgroup of the locally compact abelian group G. Let E be a perfect compact subset of G that is dissociate* mod H. *Let N be the L-subalgebra generated by $M_c(M)$ and N' the L-subalgebra generated by $M_c(E) \cup \mathscr{S}(H)$. Then*

$$\Delta N' = \Delta\mathscr{S}(H) \times \partial N = \Delta\mathscr{S}(H) \times U,$$

where U is the closed unit ball of $M_c(K)^*$.

6.3.10. Corollary. *Let G, E, and H be as in 6.3.9. Let $N_1 = \mathscr{S}(E(\infty) + H)$. Let $\Lambda = \{\gamma \in b\Gamma : \gamma$ is μ-measurable for all $\mu \in N_1\}$. Let $N_2 = \{\mu \in M(G) : \gamma$ is μ-measurable for all $\gamma \in \Lambda\}$. Then $N_1 \neq N_2$.*

6.3.11. Theorem. *Let G be a locally compact abelian group and H a Borel subgroup of G that has zero Haar measure. Let E be a σ-compact subset of G that is dissociate* mod H, *and let N be the L-subalgebra of $M(G)$ that is generated by $M_d(G) \cup M_c(H \cup E)$. Then N is prime.*

Proof. We mimic the proof of 6.2.12, using 6.3.5 in place of 6.2.11. We indicate the modifications necessary, leaving the details to the reader.

It will suffice to show the following. For all integers $k \geq 0$, $m, n \geq 1$, all non-negative measures $\tau \in M_c(E)$, $\sigma \in M_c(H)$, $\mu \in \mathscr{S}(H + E(m)) \cap N$ and $\nu \in \mathscr{S}(H \cap E(n)) \cap N^{\perp}$ and all elements $w \in G$,

$$\delta(w) * \sigma * \tau^k \perp \mu * \nu. \tag{7}$$

(If $k = 0$, then (7) holds, since $\nu \in \mathscr{S}(H)^{\perp}$, which is an ideal.) We argue by contradiction, and assume that $k \geq 1$ is a minimal integer such that (7) fails for some choice of τ, μ, ν and w. By 6.3.5(ii), $k = m + n$. Changing w if necessary, we may assume that $\mu \in M(H + E(m))$ and $\nu \in M(H + E(n))$. A routine application of Fubini's Theorem shows that $\mu * \nu \in M(H + E(k))$. As in the proof of 6.2.12, we reduce to the case that $w = 0$, which will follow upon reduction to the case that $w \in H$.

We choose a σ-compact subset $F \subseteq H + E(n)$ such that ν is concentrated on F and $\tau^n(F) = 0$. The argument now proceeds as in 6.2.12, being slightly complicated by the addition of a term h_0 in $(h_0, x_1, \ldots, x_k)$ and slightly simplified by the fact that the choices $Y_j = \{x_j\}$ will suffice. □

For a set $K \subseteq G$, $Gp'K$ is the set of sums of the form $\pm x_1 + \sum_2^n m_j x_j$ where $x_1, \ldots, x_n$ are distinct elements of K and $m_2, \ldots, m_n \in Z$.

6.3.12. Theorem. *Let G be a locally compact abelian group. Let W be a σ-compact subset of G and μ a probability measure concentrated on W. Let K be a compact independent subset of G such that $\mu(Gp'K + W) = 0$. Suppose that $K_1, \ldots, K_n$ are pairwise disjoint σ-compact subsets of K and that $\nu_1, \ldots, \nu_n$ are continuous measures that are concentrated on $K_1, \ldots, K_n$, respectively.*

(i) *If $(a(1), \ldots, (a(n)) \neq (b(1), \ldots, b(n))$ are distinct n-tuples of non-negative integers and $y \in GpK$, then*

(8) $$\lambda \perp \rho,$$

where

$$\lambda = \delta_y * \mu * (|\nu_1| + |\tilde{\nu}|)^{a(1)} * \cdots * (|\tilde{\nu}_n| + |\nu_n|)^{a(n)}$$

and

$$\rho = \mu * (|\nu_1| + |\tilde{\nu}_1|)^{b(1)} * \cdots * (|\nu_n| + |\tilde{\nu}_n|)^{b(n)}.$$

(ii) *If $\omega \ll \mu$, then*

$$\|\omega * \nu_1^{a(1)} * \cdots * \nu_n^{a(n)}\| = \|\omega\| \, \|\nu_1\|^{a(1)} \cdots \|\nu_n\|^{a(n)}$$

for all n-tuples $(a(1), \ldots, a(n))$ of non-negative integers.

Proof. (i) We may suppose that $a(1) < b(1)$ and that $y \in Gp(F)$, where F is a finite subset of K such that $F \cap K_j = \varnothing$ for $1 \leq j \leq n$. That uses the fact that the ν_j are continuous measures. By 6.3.2, $\delta_y * \nu_1^{a(1)} * \cdots * \nu_1^{a(n)} * \mu$ is concentrated on $y + K_1(a(1)) + \cdots + K_n(a(n)) + W$. But

(9)
$$\rho(y + K_1(a(1)) + \cdots + K_n(a(n)) + W)$$
$$= \int [\mu(y + K_1(a(1)) + \cdots + K_n(a(n)) + W + z)] d\nu_1^{b(1)} * \cdots * \nu_n^{b(n)}(z).$$

Of course the variable of integration z in (9) runs over $K_1(b(1)) + \cdots + K_n(b(n))$. But then, since $a(1) < b(1)$,

$$y + K_1(a(1)) + \cdots + K_n(a(n)) + z \subset Gp'K + D$$

for all but a set of z's of $\nu^{b(1)}$-measure zero. Therefore the integral in (9) is zero, and (8) is established.

(ii) We argue exactly as in 6.3.7. That ends the proof of 6.3.12. □

6.3.13. Proposition. *Let G be a metrizable locally compact abelian group. Let μ be a probability measure on G. Let W be a σ-compact subset of G of zero Haar measure. Then there exists a strongly independent compact M_o-subset E of G such that $GpE + W$ has zero Haar measure, $\mu(Gp'E + W) = 0$, and $0 \notin Gp'E + W$.*

Proof. The idea is to modify the Körner construction of an independent compact M_o-set. Here is how to modify the argument given in Section 4.7 for $G = T$. For all the details, and for the general case, see Saeki [18, Lemma 2]. We write $W = \cup W_n$ where the W_n are compact and $W_{n+1} \supseteq W_n$ for all n.

At the mth stage in the construction, we shall require that $E_m \cap W_m = \varnothing$. Since W_m has zero Haar measure and is compact, there is no difficulty caused by adding that requirement. We shall also require that the set X in (2) of 4.7 satisfies the following: if $|u_j| \le L_m$ for $1 \le j \le N_m$ and some $|u_j| = 1$, then $(W_n + \sum u_j x_j) \cap W_n = \varnothing$. Of course, that last intersection remains void when the x_j are replaced by x'_j, where $x_j - x'_j \in V$, a sufficiently small neighborhood of 0. □

6.3.14. Proposition. *Let H be a σ-compact subgroup of G having zero Haar measure.*

(i) *There exists a compact perfect subset E of G that is independent* mod H.
(ii) *If G is metrizable, E can be chosen to be of type M_o.*
(iii) *If H is such that $\{px : x \in U\} \subseteq H$ for all neighborhoods U of 0 and all $1 \le p < q(G)$, then E may be chosen (of type M_o in case G is metrizable) such that $(GpE) \cap H = \{0\}$.*

Proof. (i) is a routine modification of the standard Cantor set construction. See Rudin [1, 5.2.4] for the original method, or any of Williamson [2], Graham [5], Brown and Moran [15] or Saeki and Sato [1, Lemma 5] for the general case. Part (ii) is a routine modification of the Körner construction. See Section 4.7 or Saeki [18, Lemma 2]. For (iii), see Saeki and Sato [1, Lemma 5]. □

A Raĭkov system $\mathscr{R}$ on the locally compact abelian group is *symmetric* if $E \in \mathscr{R}$ implies $-E \in \mathscr{R}$. If $\mathscr{R}$ is countably generated, then $\mathscr{R}$ is singly generated. If $\mathscr{R}$ is symmetric and singly generated, then $\mathscr{R}$ is generated by a Borel subgroup. Those last two facts are easily established.

We now point out that the phenomenon of Wiener-Pitt (see 8.2.6) can be generalized. An element $\rho \in \Delta M(G)$ is *symmetric* if $\hat{\mu}(\rho)^{-} = (\tilde{\mu})^{\wedge}(\rho)$ for all $\mu \in M(G)$. See Section 8.2 for more about symmetric elements of $\Delta M(G)$.

6.3.15. Corollary. *Let $\mathscr{R}$ be a symmetric Raĭkov system on the locally compact metrizable abelian group G that is generated by a σ-compact set. Let $\mathscr{R}'$ be a strictly larger symmetric Raĭkov system. Then there exists a measure μ concentrated on a set in $\mathscr{R}'$ such that*

(i) $\|\mu\| = \sup\{|\hat{\mu}(\rho)| : \rho \in \Delta M(G)\} = 1$;
(ii) $|\hat{\mu}(\rho)| \leq \frac{1}{2}$ *for all symmetric* $\rho \in \Delta M(G)$; *and*
(iii) $\mu^k * \nu \perp \mu^j * \omega$ *if* $0 \leq k < j < \infty$ *and* ν, ω *are concentrated on sets in* $\mathscr{R}$.

Outline of Proof. Since $\mathscr{R}$ is properly contained in $\mathscr{R}'$, there exists a compact set $E' \in \mathscr{R}'$ such that $E' \cap E$ is of first category in E' for all $E \in \mathscr{R}$. (We omit the proof. See Williamson [2].)

Let H be a σ-compact subgroup of G that generates $\mathscr{R}$. Then it is not hard to show that E' contains a compact perfect subset that is dissociate mod H. Then if $\nu \in M_c(E)$, it is easy to show (along the lines of 6.3.7) that $\mu = \frac{1}{2}((\nu * \tilde{\nu})^2 - (\nu * \tilde{\nu})^4)$ has spectral radius one. Then (i)–(ii) follow at once, and (iii) is 6.3.6. □

Remarks and Credits. The example of a set that is dissociate but not algebraically scattered was communicated to us by G. S. Shapiro. Theorem 6.3.1 is perhaps new in this context but hardly surprising, given 6.2.1; a similar remark applies to 6.3.4, 6.3.8, 6.3.9, 6.3.10 and 6.3.12. Lemma 6.3.2 is a generalization of a standard result (for example, Rudin [1, p. 109]); it was brought to our attention by Saeki. Corollaries 6.3.3 and 6.3.15 are from Williamson [2] Lemma 6.3.5 is taken from Saeki and Sato [1] and Brown and Moran [15]. Theorem 6.3.6 implicit in Williamson [2] and Saeki [18] and explicit in Saeki and Sato [1]. Corollary 6.3.7 is in Saeki [18]. Theorem 6.3.12 is from Brown, Graham and Moran [1, Lemma 2.6].

Here follow some results related to those of this Section. The first two are from Saeki [19]; the third is from Saeki [22].

6.3.16. Theorem. *Let $f_1, \ldots, f_r$ be non-negative functions on the locally compact abelian group G. Suppose that for all $1 \leq j \leq r$, $f_j \in L^1(G) \cap L^\infty(G)$. Let $\varepsilon > 0$. Then there exist singular measures $\nu_1, \ldots, \nu_r$ such that for $1 \leq j, k \leq r$,*

(i) $\nu_j * \nu_k \in L^1(G)$;
(ii) *support* ν_j *is a compact subset of support* f_j;
(iii) $\|\nu_j * \nu_k - f_j * f_k\|_1 < \varepsilon$;
(iv) $\|\hat{\nu}_j \hat{\nu}_k - \hat{f}_j \hat{f}_k\|_1 < \varepsilon$ *if* $1 \leq j \neq k \leq r$.

6.3.17. Theorem. *Let $n \geq 1$ and let $S \subseteq (Z^+)^n$ be a set such that $(m_1, \ldots, m_n) \in S$ and $0 \leq k_j \leq m_j$ for $1 \leq j \leq n$ implies $(k_1, \ldots, k_n) \in S$. Let G be a non-discrete locally compact abelian group. Then there exist probability measures $\mu_1, \ldots, \mu_n \in M(G)$ such that*

$$\mu_1^{m_1} * \cdots * \mu_n^{m_n} \perp L^1(G) \quad \textit{if } (m_1, \ldots, m_n) \in S$$

and

$$\mu_1^{m_1} * \cdots * \mu_n^{m_n} \in L^1(G) \quad \textit{if } (m_1, \ldots, m_n) \notin S.$$

In particular, there exist $\mu, \nu \in M(G)$, $\mu, \nu \geq 0$ such that $\mu^2 \in L^1(G)$, $\nu^2 \in L^1(G)$, but $\mu * \nu \perp L^1(G)$.

6.3.18. Theorem. *There exists a singular probability measure $\mu \in M(T)$ such that the Fourier-Stieltjes series of μ^2 converges uniformly.*

That one cannot replace the hypothesis "H is a σ-compact subgroup" by "H is a σ-compact semigroup" in the results of this section is illustrated by the following result of Haight [1].

6.3.19. Theorem. *There exists a σ-compact subsemigroup $E \subseteq R$ of zero Haar measure such that $E - E = R$.*

The next result, from Connolly and Williamson [1], is the measure analogue of 6.3.19. Whether one can find $\mu \geq 0$, $\mu \in M(R)$ such that $\mu^n \perp L^1(R)$ for all $n \geq 1$ with $\mu * \tilde{\mu} \in L^1(R)$ is unknown; the measure of Connolly and Williamson has $\mu^3 \in L^1(R)$ (Ludvik [2]). Also unknown is whether there exists $\mu \in M(R)$ such that $\mu \geq 0$, $\mu^2 \in L^1(R)$ and $\mu * \tilde{\mu} \perp L^1(R)$.

6.3.20. Theorem. *There exists $\mu \in M(R)$, $\mu \geq 0$, such that $\mu^2 \perp L^1(R)$ and $\mu * \tilde{\mu} \in L^1(R)$.*

6.4. Infinite Product Measures

For $j \geq 1$, let G_j be a compact abelian group and let μ_j be a probability measure in $M(G_j)$. In this section we shall consider infinite product measures of the form $\mu = \bigtimes_{j=1}^{\infty} \mu_j$ on the infinite product group $G = \prod_{j=1}^{\infty} G_j$. By making the appropriate choice of the μ_j, one can ensure that μ is tame, i.p., and/or in $M_o(G)$. That is done in Theorem 6.4.2. Related results are given. Some of the methods used here find applications also in the study of infinite convolutions (Section 6.5 to 6.8) and of Riesz products (Chapter 7).

We begin by giving two criteria for the mutual singularity of infinite product measures. The first is due to Kakutani [1]. For an exposition of

Kakutani's proof, see Hewitt and Stromberg [1, pp. 453–455]; for another proof, see Brown and Moran [11].

6.4.1. Kakutani's Criterion. *For $j \geq 1$, let μ_j, v_j be probability measures on a measure space (Ω_j, Σ_j) with $v_j \ll \mu_j$. Let $v = \times_{j=1}^{\infty} v_j$ and $\mu = \times_{j=1}^{\infty} \mu_j$. Then either*

(i) $v \ll \mu$ *and* $\prod_{j=1}^{\infty} \int (dv_j/d\mu_j)^{1/2} d\mu_j > 0$; *or*
(ii) $v \perp \mu$ *and* $\prod_{j=1}^{\infty} \int (dv_j/d\mu_j)^{1/2}\, d\mu_j = 0$.

If $v_j \approx \mu_j$ for all j, then either $v \approx \mu$ or $v \perp \mu$.

6.4.2. Lemma. *For each $j = 1, 2, \ldots$ let (Ω_j, Σ_j) be measurable spaces and μ_j, v_j probability measures on Σ_j, and suppose that θ_j is a Σ_j-measurable function on Ω_j. Let $\mu = \times_{j=1}^{\infty} \mu_j$ and $v = \times_{j=1}^{\infty} v_j$. If*

(1) $$\sup_j \int |\theta_j|^2\, d(\mu_j + v_j) < \infty$$

and

(2) $$\sum_{j=1}^{\infty} \left| \int \theta_j\, d\mu_j - \int \theta_j dv_j \right|^2 = \infty,$$

then

$$\mu \perp v.$$

Proof. Let $a_j = \int \theta_j\, d\mu_j$ and $b_j = \int \theta_j\, dv_j$, for $j = 1, 2, \ldots$. Let $\{c_j\}_{j=1}^{\infty}$ be a sequence of complex numbers and $\{n(k)\}_{k=1}^{\infty}$ a strictly increasing sequence of positive integers such that

(3) $$\sum_{j=1}^{\infty} |c_j|^2 < \infty \quad \text{and} \quad \sum_{j=n(k)+1}^{n(k+1)} c_j(a_j - b_j) = 1 \quad \text{for } k \geq 1.$$

The existence of $\{c_j\}$ and $\{n(k)\}$ is immediate from (1) and (2).

Each function θ_j defines a function (which we also call θ_j) on $\Omega = \times_{j=1}^{\infty} \Omega_j$ by $\theta_j(x_1, x_2, \ldots) = \theta_j(x_j)$. We define f_k on Ω by $f_k = \sum_{j=n(k)+1}^{n(k+1)} c_j(\theta_j - b_j)$. Then (3) and (1) imply that

(4)

$$\int |f_k|^2\, dv = \sum |c_j|^2 \int |\theta_j - b_j|^2\, dv_j + \sum_{j \neq l} c_j \bar{c}_l \int (\theta_j - b_j) dv \int (\bar{\theta}_j - \bar{b}_l) dv$$
$$= o(1) \quad \text{as } k \to \infty.$$

An expansion similar to (4) reveals that $\int |f_k - 1|^2\, d\mu = o(1)$ as $k \to \infty$. Since we may replace $\{f_k\}$ by a subsequence, we may assume that $f_k \to 0$ a.e. dv and $f_k \to 1$ a.e. $d\mu$. It follows that $\mu \perp v$. The Lemma is proved. □

6.4.3. Theorem. *For $j \geq 1$, let G_j be a non-trivial compact abelian group with Haar measure m_j and identity e_j. Let $0 < \alpha_j < 1$, and $\omega_j = \alpha_j \delta(e_j) + (1 - \alpha_j)m_j$. Let $G = \prod_1^\infty G_j$, and $\omega = \mathsf{X}_{j=1}^\infty \omega_j$.*

(i) *Then $\omega \in M_o(G)$ if and only if $\lim \alpha_j = 0$ and each G_j is finite.*

(ii) $\Sigma\{1 - \alpha_j : \alpha_j > \frac{1}{2}\} + \Sigma\{\alpha_j^k : \alpha_j < \frac{1}{2}\} = \infty$ *for all $k > 0$ if and only if ω is i.p.*

(iii) *If the groups G_j are finite, then ω is tame if and only if $\limsup \alpha_j < 1$.*

Proof. For $j \geq 1$, let Γ_j be the dual group of G_j, and let elements of $\Gamma = \Sigma\Gamma_j$ be written $\gamma = (\gamma_1, \gamma_2, \ldots)$. Then

$$\hat{\omega}(\gamma) = \prod_{j=1}^\infty \hat{\omega}_j(\gamma_j) = \Pi\{\alpha_j : \gamma_j \neq 0\}$$

Now (i) follows easily.

(ii) Suppose that $\Sigma\{1 - \alpha_j : \alpha_j \geq \frac{1}{2}\} + \Sigma\{\alpha_j^m : \alpha_j < \frac{1}{2}\} < \infty$ for some $m \geq 1$. Let $n > m$, and let

$$\lambda_k = \left(\mathop{\mathsf{X}}_{j=1}^{k-1} \omega_j^m\right) \times \left(\mathop{\mathsf{X}}_{j=k}^{\infty} \omega_j^n\right) \quad \text{for } k \geq 1.$$

Then

$$\begin{aligned}
\|\omega^m - \lambda_k\| &\leq \sum_{j=k}^\infty \|\omega_j^m - \omega_j^n\| \\
&\leq \sum_{j=k}^\infty \{|\alpha_j^m - \alpha_j^n| + |(1 - \alpha_j^m) - (1 - \alpha_j^n)|\} \\
&= 2 \sum_{j=k}^\infty |\alpha_j^m - \alpha_j^n| = o(1) \quad \text{as } k \to \infty.
\end{aligned}$$

Since $\lambda_k \ll \omega^n$ for all k and since λ_k converges to ω^m in norm, $\omega^m \ll \omega^n$. Therefore ω is not i.p.

We now prove the converse. For each $j \geq 1$, define functions on G by

$$\theta_j(x) = \begin{cases} 1 & \text{if } x = e_j \\ -(1 - \alpha_j)^{-1/2} & \text{if } x \neq e_j, \end{cases}$$

and let $\beta_j = 1/\#G_j$ if $\#G_j < \infty$ and $\beta_j = 0$ if $\#G_j = \infty$. Then, for $j, m = 1, 2, \ldots,$

$$(5) \qquad \int |\theta_j|^2 \, d\omega_j^m = \alpha_j^m + (1 - \alpha_j)^{-1}(1 - \alpha_j^m)(1 - \beta_j) + (1 - \alpha_j^m)\beta_j.$$

The function $(1 - \alpha^m)/(1 - \alpha)$ is increasing on $[0, 1)$ with

$$\lim_{\alpha \to 1} (1 - \alpha^m)/(1 - \alpha) = m.$$

That, (5), and the fact that the α_j are in $(0, 1)$ show that $\int |\theta_j|^2 \, d\omega_j^m \leq 2 + m$. Setting $\mu_j = \omega_j^m$ and $\nu_j = \omega_j^n$, we see that (1) holds for all $m, n \geq 1$. To apply 6.4.2, we must show that (2) holds when $m < n$ and $\Sigma\{(1 - \alpha_j): \alpha_j > \frac{1}{2}\} + \Sigma\{\alpha_j^{2m+2}: \alpha_j \leq \frac{1}{2}\} = \infty$. But then

$$\begin{aligned} \left| \int \theta_j \, d\omega_j^m - \int \theta_j \, d\omega_j^n \right| &= (\alpha_j^m - \alpha_j^n)(1 - \beta_j)((1 - \alpha_j)^{-1/2} - 1) \\ &\geq (\alpha_j^m - \alpha_j^n)(1 - \beta_j)(\tfrac{1}{2}\alpha_j)(1 - \alpha_j)^{-1/2} \\ &\geq \alpha_j^{m+1}(1 - \alpha_j)^{1/2}/4. \end{aligned}$$

Therefore

$$\Sigma \left| \int \theta_j \, d\omega_j^m - \int \theta_j \, d\omega_j^n \right|^2 \geq \frac{1}{16} \sum_{\alpha_j > 1/2} 2^{-2m-2}(1 - \alpha_j) + \frac{1}{32} \sum_{\alpha_j < 1/2} \alpha_j^{2m+2} = \infty.$$

Therefore (2) holds with $\mu_j = \omega_j^m$ and $\nu_j = \omega_j^n$. By 6.3.2, $\omega^m \perp \omega^n$. Now (ii) follows.

(iii) Suppose that $\alpha = \limsup \alpha_j < 1$. We write ω_j' for the measure on $G = \prod_1^\infty G_j$ given by $\omega_j' = \delta(0) \times \cdots \times \delta(0) \times \omega_j \times \delta(0) \times \cdots$. Then $\omega = \bigstar\omega_j'$. (In doing that, we observe that infinite products "are" infinite convolutions.)

Let $\psi \in \Delta M(G)$ and suppose that $\psi_\omega \not\equiv 0$. Then there exists $\gamma_0 \in \Gamma$, the dual group of G, such that $\hat{\omega}(\gamma_0 \psi) \neq 0$. Let $\varphi = \gamma_0 \psi$.

Since each $\omega_j \in L^1(G_j)$ and $\varphi_\omega \not\equiv 0$, the restriction $\varphi_{\omega_j'}$ is given by a character γ_j of G_j. If $\gamma_j \neq 0$, then $\hat{\omega}_j(\varphi) = \hat{\omega}_j(\gamma_j) = \alpha_j$. Let $J = \{j: \gamma_j \neq 0\}$. Then

$$0 \neq \hat{\omega}(\varphi) \leq \Pi\{\alpha_j: j \in J\} \leq \alpha^m,$$

where $m = \text{card } J$. Since $\alpha < 1$, $m < \infty$. Therefore $\gamma_j = 0$ for all $j > j_0$. Let $\gamma_\infty = (\gamma_1, \gamma_2, \ldots) \in \Gamma = \sum_1^\infty \Gamma_j$. Let $\sigma_n = \bigstar_{j=1}^n \omega_j'$ and $\tau_n = \bigstar_{j=n+1}^\infty \omega_j'$ for all $n \geq 1$. Let $a = \lim_{m \to \infty} \hat{\tau}_m(\varphi)$. We claim that $\varphi_\omega = \bar{a}\gamma_\infty$ a.e. $d\omega$ and thus, that $\psi_\omega = \bar{a}\bar{\gamma}_0\gamma_\infty$ a.e. $d\omega$.

To show that $\varphi_\omega = \bar{a}\gamma_\infty$, it will suffice to show that for all $k \geq 1$,

$$a \int \overline{\gamma\gamma_\infty} \, d\omega = \int \bar{\gamma}\overline{\varphi}_\omega \, d\omega \quad \text{for all } \gamma \in \sum_{j=1}^k \Gamma_j,$$

where $\sum_{j=1}^{k} \Gamma_j$ is regarded as a subset of $\Gamma = \sum_{j=1}^{\infty} \Gamma_j$. Let $\gamma \in \sum_1^k \Gamma_j$. Then

$$\int \bar{\gamma}\varphi_\omega \, d\omega = \hat{\sigma}_m(\gamma\varphi)\hat{\tau}_m(\gamma\varphi) \quad \text{for } n = 1, 2, \ldots.$$

Now $\gamma\gamma_\infty = (\lambda_1, \ldots, \lambda_{m_0}, 0, 0, \ldots)$. Let $m \geq m_0$. Then

$$\hat{\sigma}_m(\gamma\varphi) = \hat{\sigma}_m(\gamma\gamma_\infty) \quad \text{and} \quad \hat{\tau}_m(\gamma\varphi) = \hat{\tau}_m(\varphi) = a.$$

Therefore $\int \bar{\gamma}\varphi_\omega \, d\omega = a \int \bar{\gamma}\gamma_\infty \, d\omega$. That proves the claim. Therefore ω is tame.

Now suppose that $\limsup \alpha_j = 1$. Let $\alpha_{j(1)}, \alpha_{j(2)}, \ldots$ be such that $\prod_1^\infty \alpha_{j(k)} > 0$. Let $\gamma_{j(k)} \in \Gamma_{j(k)} \backslash \{0\}$ for all k. We leave to the reader the task of showing that $\lim \sum_{k=1}^{m} \gamma_{j(m)}$ converges weak* in $L^\infty(\omega)$. It will follow that there exists $\varphi \in \Gamma^- \backslash \Gamma$ such that $\varphi_{\omega'_{j(k)}} = \gamma_{j(k)}$ for all k, and therefore that $\varphi_\omega \neq a\gamma$ for all $\gamma \in \Gamma$. □

The next result is immediate from 6.4.3.

6.4.4. Corollary. *Let $G = \prod_{j=1}^{\infty} G_j$ be an infinite product of non-trivial finite abelian groups. Then there exists a tame Hermitian i.p. probability measure ω in $M_o(G)$.*

6.4.5. Proposition. *For $j = 1, 2, \ldots$, let ω_j be a monotrochic probability measure on the compact abelian group G_j. Then $\omega = \times_{j=1}^{\infty} \omega_j$ is a monotrochic measure on $G = \prod_{j=1}^{\infty} G_j$.*

Proof. Let c_j be the constant value of $|\psi_{\omega_j}|$ (a.e. $d\omega_j'$). Then $|\psi_\omega(x_1, x_2, \ldots)| = (\prod_{j=1}^{n} c_j)|\psi_{\tau_n}(x_{n+1}, \ldots)|$ for all $n \geq 1$, where $\tau_n = \times_{j=1}^{n} \delta(0) \times \times_{j=n+1}^{\infty} \omega_j$. Let $c = \hat{\omega}(|\psi|)$. Then $|\psi_\mu(x_1, x_2, \ldots)| < c$ is a tail-event as is

$$|\psi_\mu(x_1, x_2, \ldots)| > c.$$

If $|\psi_\infty(x_1, x_2, \ldots)| < c$ occurs with non-zero probability, then by the zero-one law (see Kahane [1, p. 6] or Hewitt and Stromberg [1, p. 443]) $|\psi_\omega(x_1, x_2, \ldots)| < c$ a.e. $d\omega$ and $\hat{\omega}(|\psi|) < c$, a contradiction, so

$$|\psi_\omega(x_1, x_2, \ldots)| \geq c \text{ a.e. } d\omega.$$

Similarly $|\psi_\omega| \leq c$ a.e. $d\omega$. Therefore $|\psi_\omega| = c$ a.e. $d\omega$. That proves the Proposition. □

6.4.6. Proposition. *For $j = 1, 2, \ldots$, let μ_j be a discrete probability measure on the compact abelian group G_j. Let D be the subgroup of $G = \prod_{j=1}^{\infty} G_j$ that is generated by all the elements $(0, \ldots, 0, x_j, 0, \ldots)$, where $\mu_j(\{x_j\}) \neq 0$ and $1 \leq j < \infty$. Then $\mu = \times_{j=1}^{\infty} \mu_j$ is D-ergodic.*

Proof. Let D_j be the subgroup of G_j generated by those x with $\mu_j(\{x\}) \neq 0$. Let $H_m = D_1 \times \cdots \times D_m \times \prod_{j=m+1}^{\infty} G_j$. Then $\mu|_{H_m} = \mu$ and $\mu(E) = \mu(H_m \cap E)$ for all Borel sets E of G. Let E be a Borel subset of E with $E + D = E$, and $\mu(E) > 0$. Then $E \cap H_m = D_1 \times \cdots \times D_m \times \pi_m E$ where π_m is the projection of E on $\prod_{j=m+1}^{\infty} G_j$. It follows at once that E is a tail event for μ. Therefore $\mu(E) = 1$. □

Credits. Lemma 6.4.2 is standard. Theorem 6.4.3 is due to Varopoulos [1] (part (i)), Kaufman [1] (part (ii)) and J. L. Taylor [2] (part (iii)). Corollary 6.4.4 is from Brown [2] and Moran [2], and 6.4.5 is from Brown and Moran [9], as is 6.4.6.

6.5. General Results on Infinite Convolutions

In this section we shall study measures of the form $\mu = \bigstar_{j=1}^{\infty} \mu_j$ where the μ_j are probability measures and the convergence is weak*. Most of our results will be for discrete μ_j's.

6.5.1. Lemma. *Let G be a metrizable locally compact abelian group and $\{U_j\}$ a sequence of compact neighborhoods of the identity of G such that*

(1) $$U_{j+1} + U_{j+1} \subseteq U_j \quad \textit{for } j = 1, 2, \ldots, \textit{and } \bigcap_{j=1}^{\infty} U_j = \{0\}.$$

Let $\mu_1, \mu_2, \ldots$ be probability measures on G such that

(2) $$\operatorname{supp} \mu_j \subseteq U_j \quad \textit{for } j = 1, 2, \ldots.$$

Then the infinite convolution $\mu = \bigstar_{j=1}^{\infty} \mu_j$ converges weak to a probability measure μ such that* $\operatorname{supp} \mu \subseteq U_1 + U_1$, *and*

$$\mu^n = \bigstar_{j=1}^{\infty} \mu_j^n \quad \text{for } 1 \leq n < \infty.$$

Proof. For $1 \leq n < m$, let $\sigma_n = \bigstar_{j=1}^{n} \mu_j$ and $\sigma_{n,m} = \bigstar_{j=n+1}^{m} \mu_j$. Then for each continuous function f on G and $\varepsilon > 0$ there exists $n_0 \geq 1$ such that

$$|f(x) - f(y)| < \varepsilon \quad \text{if} \quad x - y \in U_{n_0}.$$

Therefore $\sup\{|f(x) - \int f(x + y)d\sigma_{n,m}(y)| : x \in U_1 + U_1\} < \varepsilon$ if $n_0 < n < m < \infty$. But if $n_0 < n < m$, then

$$\left| \int f \, d\sigma_n - \int f \, d\sigma_m \right| < \varepsilon.$$

Thus σ_n converges weak$*$ to a measure μ. That μ is supported in $U_1 + U_1$, that μ is a probability measure and that the equalities $\mu^n = \bigstar_{j=1}^{\infty} \mu_j^n$ hold are easily established. That ends the proof of 6.5.1. $\square$

6.5.2. Lemma. *Let G be an infinite metrizable locally compact abelian group. Let $\mu = \bigstar_{j=1}^{\infty} \mu_j$ be an infinite convolution on G, obtained as in Lemma* 6.5.1. *If $\mu_1, \mu_2, \ldots$ are monotrochic, then μ is monotrochic.*

Proof. When the infinite convolution is actually an infinite product, 6.5.2 is 6.4.5. We will reduce to that case.

For each j, let H_j denote the Bohr compactification of $G_j = G$. Then μ_j is monotrochic in $M(H_j)$ for $j = 1, 2, \ldots$. Let $\mathscr{R}$ denote the Raĭkov system on $H = \prod_{j=1}^{\infty} H_j$ that is generated by the compact set $U = \prod_{j=1}^{\infty} U_j$, and let K denote the σ-compact subgroup of H generated by U. Here the U_j are as in (1)–(2). Note that $\mathscr{R}$ is generated by K.

We will construct a map $\check{p}$: $M(K) \to M(G)$. If $x = (x_1, x_2, \ldots) \in kU$, then $p(x) = \sum x_j$ exists in G and the map $x \to p(x)$ is continuous from kU to G. Hence p extends to a Borel map from K to G, and the formula $\check{p}(\mu)(E) = \mu(p^{-1}E)$, E Borel, defines an L-homomorphism of convolution measure algebras (see Section 5.1).

If $\psi \in \Delta M(G)$, then $\psi \circ \check{p} \in \Delta M(K)$. Suppose that we have extended $\psi \circ p$ to all of $M(H)$. Let $\omega = \times \mu_j$. Then $|(\psi \circ p)_\omega|$ is constant a.e. $d\omega$, and $\check{p}\omega = \mu$. Therefore $\psi_\mu(px) = (\psi \circ \check{p})_\omega(x)$. Thus $|\psi_\mu|$ will be constant a.e. $d\mu$ and μ will be monotrochic. It remains to extend $\varphi = \psi \circ \check{p}$ from $M(K)$ to $M(H)$.

If $\omega \in A_{\mathscr{R}}$, the L-algebra of measures concentrated on sets in $\mathscr{R}$, then

$$\omega = \sum_{1}^{\infty} \delta(x_j) * \omega_j, \quad \text{where } \omega_j \in M(K), \tag{3}$$

and

$$(x_j + K) \cap (x_k + K) = \varnothing \quad \text{for } 1 \le j \ne k < \infty. \tag{4}$$

Since $\varphi \in \Delta M(K)$, φ defines a character γ on K by $\langle \gamma, x \rangle = \delta(x)\hat{\ }(\psi)$, $x \in K$. Let λ be any extension of γ to H. We define φ_1 on $A_{\mathscr{R}}$ by $\varphi_1(\omega) = \Sigma \langle \lambda, x_j \rangle \hat{\omega}_j(\varphi)$, where ω, ω_j and x_j are as in (3)–(4). Straightforward calculations show that φ_1 defines an element of $\Delta A_{\mathscr{R}}$. By 5.2.2, $\Delta A_{\mathscr{R}}$ is embedded as a subset of $\Delta M(H)$. Therefore φ_1 is the required extension of φ to $M(H)$. The Lemma is proved. $\square$

6.5.3. Lemma. *Let G be an infinite metrizable locally compact abelian group. Let $\mu = \bigstar_{j=1}^{\infty} \mu_j \in M(G)$ be obtained as in Lemma* 6.5.1. *Suppose that μ_1, $\mu_2, \ldots$ are discrete and that D is the subgroup of G generated by the supports of the μ_j. Then $\mu = \bigstar_{j=1}^{\infty} \mu_j$ is D-ergodic.*

Proof. We use the method of proof of 6.5.2 and the notation of that proof. By 6.4.6, $\omega = \bigstar_{j=1}^{\infty} \mu'_j$ is D'-ergodic in $M(H)$, where D' is the subgroup of H generated by the supports of the measures μ'_j. (Here μ'_j denotes $\delta(0) \times \cdots \times \mu_j \times \delta(0) \times \cdots$, where μ'_j appears in the jth place.)

If $E \subseteq G$ is any D-invariant Borel set, then $p^{-1}(E)$ is a D'-invariant Borel subset of $K \subseteq H$. Therefore $\omega(p^{-1}E) \in \{0, 1\}$. Since $\omega(p^{-1}E) = \check{p}\omega(E) = \mu(E)$, $\mu(E) \in \{0, 1\}$ so μ is indeed D-ergodic. The Lemma is proved. □

6.5.4. Corollary. *If $\mu = \bigstar \mu_j$ is an infinite convolution of discrete measures, then either μ is strongly i.p., or for some $n \geq 1$, $\mu^n \in L^1(G_\tau)$ for some refinement G_τ of G.*

The proof of 6.5.4 is immediate from 6.1.8 and 6.5.3.

The next result is a strengthening of the classical *Jessen-Wintner Purity Theorem* (Jessen and Wintner [1]), which states that an infinite convolution of discrete measures on R or T is purely discrete, purely continuous and singular, or in L^1.

6.5.5. The Generalized Purity Theorem. *Let $\mu = \bigstar_{j=1}^{\infty} \mu_j$ be an infinite convolution of discrete measures on $G = R$ or T. Then exactly one of the following holds.*

(i) *μ is discrete.*
(ii) *$\mu \in L^1(G)$.*
(iii) *For some $k \geq 1$, $\mu, \mu^2, \ldots, \mu^k$ are singular, continuous, and mutually singular, whereas $\mu^{k+1} \in L^1(G)$.*
(iv) *μ is continuous and strongly i.p.*

Proof. Let D be the group generated by those x such that $\mu_j(\{x\}) \neq 0$ for some j. Then μ is D-ergodic by 6.5.3. Theorem 6.1.5(ii) and Theorem 6.1.8(ii) shows that either (iv) holds or $\mu^n \in L^1(G_\tau)$ for some n and a refinement of G. We may assume that $\mu^n \in L^1(G_\tau)$. But G and G_d are the only refinements of G. Therefore either $\mu \in M_d(G)$, $\mu \in M_c(G)$, or $\mu^n \in L^1(G)$ for some n. Suppose the last holds. Let k be the smallest integer such that $\mu^k \in L^1(G)$. If $1 \leq m < n \leq k$, then $\mu^m * \mu^{k-n}$ is singular while $\mu^n * \mu^{k-n} = \mu^k \in L^1(G)$. Suppose that $\mu^m \not\perp \mu^n$. Then A.6.1 would imply that $\mu^m * \mu^{k-n} \not\perp \mu^n * \mu^{k-n} = \mu^k$ which would contradict the singularity of μ^{k+m-n}, that is, the minimality of k. That proves the Theorem. □

In general, it is difficult to decide which of the alternatives (i)–(iv) of 6.5.5 holds for a particular infinite convolution. But for certain types of infinite convolutions there are detailed results as we shall see in the next two sections. In particular, we will use 6.5.5 and 6.5.7 to construct tame Hermitian i.p. measures.

6.5.6. Lemma. *Let G be a compact abelian group. Let* $\mu = \bigstar_{j=1}^{\infty} \mu_j$ *where each* μ_j *is a discrete probability measure charging exactly the set* S_j. *For* $n \geq 1$, *let* $\tau_n = \bigstar_{j=n+1}^{\infty} \mu_j$. *Then:*

(i) *the linear hull of* $\{\delta(x) * \tau_n : x \in S_1 + \cdots + S_n\}$ *is dense in* $L^1(\mu)$;
(ii) *if* $\{\gamma_\alpha\} \subseteq \Gamma$, *the dual group of G, if* $\lim\langle \gamma_\alpha, x\rangle = 1$ *for all* $x \in \bigcup_{j=1}^{\infty} S_j$, *and if* $\lim \hat{\mu}(\gamma_\alpha) = z$, *then*

$$\psi \in \{\gamma_\alpha\}^{-}\backslash\Gamma \quad \textit{implies} \quad \psi_\mu = \bar{z} \text{ a.e. } d\mu;$$

(iii) *if* $\psi \in \Delta M(G)$ *and* $\psi_\mu \not\equiv 0$, *then* $\prod_{j=1}^{\infty} \hat{\mu}_j(\psi)$ *converges; and*
(iv) *if* $\psi \in \Delta M(G)$ *and* $\delta(x)\hat{\ }(\psi) = 1$ *for all* $x \in \bigcup_1^{\infty} S_j$, *then* ψ_μ *is constant a.e.* $d\mu$.

Proof. Let $\sigma_n = \bigstar_{j=1}^{n} \mu_j$. Then $\mu = \sigma_n * \tau_n$ for $n = 1, 2, \ldots$, so μ is a sum of the form $\Sigma\{\alpha(x)\delta(x) * \tau_n : x \in S_1 + \cdots + S_n\}$. Thus, $\delta(x) * \tau_n \ll \mu$ for all $x \in S_1 + \cdots + S_n$ and $n = 1, 2, \ldots$. It is easy to see that if $\gamma \in \Gamma$, the dual group of G, then

$$\lim_{n\to\infty} \|\gamma\mu - (\gamma\sigma_n) * \tau_n\| = 0.$$

To prove (i), it will therefore be sufficient to show that each $(\gamma\sigma_n) * \tau_n$ is in the linear hull of $\delta(x) * \tau_n : x \in S_1 + \cdots + S_n$. But that is obvious.

(ii) Since $\hat{\mu}(\gamma_\alpha) = \hat{\sigma}_n(\gamma_\alpha) \int \bar{\gamma}_\alpha \, d\tau_n$, the hypotheses of (ii) imply that

$$z = \lim_\alpha \hat{\mu}(\gamma_\alpha) = \lim_\alpha \hat{\tau}_n(\gamma_\alpha) \quad \text{for } n = 1, 2, \ldots. \tag{5}$$

If $\psi = \lim_\beta \gamma_{\alpha(\beta)}$, then (5) and (i) imply that $\psi_\mu = \bar{z}$ a.e. $d\mu$.

(iii) Using (i) we conclude from $\psi_\mu \not\equiv 0$ that there exists $m \geq 1$ and $x \in S_1 + \cdots + S_m$ with $\delta(x)\hat{\ }(\psi)\hat{\tau}_m(\psi) \neq 0$. But

$$\int \bar{\psi}_\mu \, d\tau_m = \prod_{j=m}^{n} \hat{\mu}_j(\psi) \int \bar{\psi}_\mu \, d\tau_n \neq 0 \tag{6}$$

for $m < n < \infty$. Therefore $\lim_{n\to\infty} \prod_{j=m}^{n} \hat{\mu}_j(\psi)$ exists and (iii) is proved.

(iv) We may assume that $\psi_\mu \not\equiv 0$. Then (6) implies that

$$\int \bar{\psi}_\mu \, d(\delta(x) * \tau_n) = \int \bar{\psi}_\mu \, d\tau_n \text{ for each } x \in S_1 + \cdots + S_n \text{ and } n \geq 1.$$

Thus, by (i), $\bar{\psi}_\mu = \hat{\mu}(\psi)$ a.e. $d\mu$. The Lemma is proved. □

6.5.7. Corollary. *Let* $\mu = \bigstar_{j=1}^{\infty} \mu_j$ *be the infinite convolution of discrete probability measures. Let D denote the (discrete) group generated by the* *sets charged by the μ_j. Then μ is tame if the restriction to $M(D)$ of each $\psi \in \Delta M(G)$ with $\psi_\mu \not\equiv 0$ agrees with the restriction of a continuous character of G.*

Proof. Let $\psi \in \Delta M(G)$, and let φ be the character on D induced by ψ: $\langle \varphi, -x \rangle = \psi(\delta(x)) = \delta(x)^\wedge(\psi)$. If $\psi_\mu \not\equiv 0$ then Lemma 6.5.6(iii) and the hypothesis show that $\varphi = \gamma$ for some $\gamma \in \Gamma$. Then $(\bar{\gamma}\psi)_\mu$ is constant a.e. $d\mu$ by 6.5.6(iv). Therefore $\psi = a\gamma$, where $a = (\bar{\gamma}\psi)_\mu$ a.e. $d\mu$. The Corollary is proved. □

Credits. 6.5.1 goes back at least as far as Wintner's notes [1]. The purity theorem is in Jessen and Wintner [1]. Lemma 6.5.6 and its corollary are taken from Brown [2], though its roots lie in Taylor [2] and B. E. Johnson [2]. The remaining results of this section are due to Brown and Moran [4, 7, 9]. The implication in Lemma 6.5.6(iii) can not be reversed, as the following example shows.

There are measures $\mu \in M(R)$ and characters $\gamma \in D \backslash R$ such that $\prod \hat{\mu}_j(\gamma)$ converges, yet for no $\psi \in \Delta M(R)$ does ψ agree with γ on $M(D)$ and also have $\hat{\mu}(\psi) \neq 0$. To see that note that if $\varepsilon_j \to 0$ sufficiently fast, then $\bigstar\, [\frac{1}{2}\delta(0) + \frac{1}{2}\delta(2^{-j} + \varepsilon_j)] \in L^1(R)$. Of course, we can choose $\varepsilon_j \to 0$ that fast, and yet such that $\{2^{-j} + \varepsilon_j : 1 \leq j < \infty\}$ is independent. The choice of γ is now easy.

6.6. Bernoulli Convolutions

We work on R. An *antisymmetric Bernoulli convolution* is an infinite convolution of the form

(1) $$\mu = \bigstar_{j-1}^{\infty} [\tfrac{1}{2}\delta(0) + \tfrac{1}{2}\delta(r_j)],$$

where

(2) $$0 < r_j < \infty \quad \text{for} \quad 1 < j < \infty \quad \text{and} \quad \sum_{j=1}^{\infty} r_j < \infty.$$

A *symmetric Bernoulli convolution* is an infinite convolution of the form

(3) $$\nu = \bigstar_{j=1}^{\infty} [\tfrac{1}{2}\delta(-r_j) + \tfrac{1}{2}\delta(r_j)]$$

where

(4) $$\sum_{j=1}^{\infty} r_j^2 < \infty.$$

We shall show below in 6.6.1 that (4) implies the convergence of (3).

The two cases of Bernoulli convolutions that have been most thoroughly studied are

(5) $$r_j = \theta^{-j} \quad \text{for } j = 1, 2, \ldots$$

where $1 < \theta < \infty$, and

(6) $$r_j = (a_1 \cdots a_j)^{-1}, \quad \text{for } j = 1, 2, \ldots$$

where the a_j are integers ≥ 2.

In the case (5), μ is the usual Cantor-Lebesgue L-measure on the Cantor set with dissection ratio $1/\theta$. A famous theorem of Salem and Zygmund asserts that if μ is of the form (1) where (5) holds, then $\mu \notin M_o(R)$ if and only if θ is a Pisot number. The reader is referred to Meyer [1] or Salem [1] for a proof of that result and for related matters. For an extension, see Kaufman [15]. We shall discuss (6) below.

6.6.1. Proposition. *Let $r_j > 0$ for $j = 1, 2, \ldots$. Then $\bigstar_{j=1}^{\infty} [\frac{1}{2}\delta(-r_j) + \frac{1}{2}\delta(r_j)]$ converges weak$*$ to a probability measure in $M(R)$ if and only if $\sum r_j^2 < \infty$.*

Proof. Suppose that $\sum r_j^2 < \infty$. Taking Fourier-Stieltjes transforms, we find that

$$\left(\bigstar_{j=1}^{N} [\tfrac{1}{2}\delta(-r_j) + \tfrac{1}{2}\delta(r_j)]\right)^{\wedge}(t) = \prod_{j=1}^{N} \cos r_j t.$$

Since $1 - u^2/2 \leq \cos u \leq 1 - u^2/4$ (if $|u| \leq \frac{1}{2}$), we see that $\prod_1^N \cos r_j t$ converges uniformly on compact subsets of R. Of course, that implies the weak$*$ convergence of $\bigstar_1^N [\frac{1}{2}\delta(-r_j) + \frac{1}{2}\delta(r_j)] = v_N$ since each v_N is a probability measure.

Now suppose that $\bigstar_1^{\infty} [\frac{1}{2}\delta(-r) + \frac{1}{2}\delta(r_j)]$ converges weak$*$ to a probability measure v. Suppose also that $\{r_j\}$ is bounded, say $r_j \leq M < \infty$ for $j = 1, 2, \ldots$. Then for $|t| \leq \frac{1}{4}M$,

(7) $$1 - (r_j t)^2/2 \leq \cos r_j t \leq 1 - (r_j t)^2/4, \quad \text{for } j = 1, 2, \ldots.$$

Of course $\prod_1^N \cos r_j t$ decreases monotonically as $N \to \infty$. Therefore, if $\bigstar_1^{\infty}[\frac{1}{2}\delta(-r_j) + \frac{1}{2}\delta(r_j)]$ is not to converge weak$*$ to the zero measure, there must exist some t in $(0, \frac{1}{4}M)$ such that

(8) $$\lim_{N \to \infty} \left| \prod_1^N \cos r_j t \right| > 0.$$

Putting (7) and (8) together, we conclude that $\sum r_j^2 < \infty$.

It remains to show that $\{r_j\}$ is bounded if the convolution converges. Note that the weak topology $\sigma(M(R), C(R))$ agrees with the weak$*$ topology $\sigma(M(R), C_o(R))$ on the set of probability measures. Therefore $\hat{v}(t) = \prod_{j=1}^{\infty} \cos r_j t$ for all $t \in R$. If $\{r_j\}$ is not bounded, then for all $\varepsilon > 0$ there exists $0 < t < \varepsilon$ and $j \geq 1$ such that $|\cos r_j t| < \varepsilon$. It follows that $\liminf_{t \to 0} \prod_{j=1}^{\infty} \cos r_j t = 0$, which contradicts the assumption that v is a probability measure on R. Therefore $\sup r_j < \infty$ and the Proposition is proved. □

The proof of the next result is left to the reader.

6.6.2. Proposition. *Let $r_j \geq 0$ for $j = 1, 2, \ldots$. Then $\bigstar_{j=1}^{\infty} [\frac{1}{2}\delta(0) + \frac{1}{2}\delta(r_j)]$ converges weak$*$ to a probability measure if and only if $\sum_{j=1}^{\infty} r_j < \infty$.*

By making $r_j \to 0$ very rapidly, it is easy to construct Bernoulli convolutions (symmetric and antisymmetric) with every power singular with respect to $L^1(R)$, and hence, by the Generalized Purity Theorem 6.5.5, with mutually singular powers. Much more can be said. The reader is referred to Brown and Moran [2] and [2, Corrigendum] or to Lin and Saeki [1] for a proof of the following theorem.

6.6.3. Theorem. *Let $E = \{r = (r_j): r_j \geq 0, \sum r_i^2 < \infty\}$, with metric $d((r_j), (s_j)) = (\sum (r_j - s_j)^2)^{1/2}$. For $r = (r_j) \in E$, let $v_r = \bigstar_{j=1}^{\infty} [\frac{1}{2}\delta(-r_j) + \frac{1}{2}\delta(r_j)]$. Then, except for a set of first category in E, every v_r is such that there exists, for all $t \in (-1, 1)$, $\psi \in R^- \backslash R \subseteq \Delta M(R)$ with $\psi_{v_r} = t$ a.e. dv_r.*

For antisymmetric Bernoulli convolutions, we have the analogous result with a simpler proof.

6.6.4. Theorem. *Let $F = \{r = (r_j): r_j \geq 0, \sum_{j=1}^{\infty} r_j < \infty\}$, with metric $d((r_j), (s_j)) = \sum |r_j - s_j|$. For $r \in F$, let $\mu_r = \bigstar_{j=1}^{\infty} [\frac{1}{2}\delta(0) + \frac{1}{2}\delta(r_j)]$. Then, except for a set of first category in F, every μ_r is such that for all $z \in \mathbb{C}$ with $0 < |z| < 1$, there exists $\psi \in R^- \backslash R$ with $\psi_{\mu_r} = z$ a.e. $d\mu_r$.*

The proof of the Lemma below is a simple exercise.

6.6.5. Lemma. *Let $|z| < 1$ and $\varepsilon > 0$. Then there exists a set of complex numbers $z_1, \ldots, z_n$, all having unit modulus, such that*

$$\left| z - \prod_{j=1}^{n} [\tfrac{1}{2} + \tfrac{1}{2} z_j] \right| < \varepsilon.$$

Proof of Theorem 6.6.4. Since the products $\prod [\frac{1}{2} + \frac{1}{2}\delta(r_n)\hat{}\,]$ converge uniformly on finite subsets of the integers Z, the map $r \to \hat{\mu}_r(k)$ is a continuous function from F to $\mathbb{C}$ for each $k \in Z$.

For each $z \in \mathbb{C}$ with $|z| < 1$, $\varepsilon > 0$, and $n > 0$, it is clear that the subset of F defined by

(9) $$X = X(z, \varepsilon, n) = \bigcap_{k=1}^{\infty} \left\{ r \in F: \sum_{j=1}^{n} |1 - \exp(ir_j k)| + |\hat{\mu}_r(k) - \bar{z}| \geq \varepsilon \right\}$$

is closed. We claim that X has no interior in F. Indeed, let $r \in X$ and $\alpha > 0$. We must find $r' \in F$ with $\|r - r'\| = \sum |r_j - r'_j| < \alpha$ and $k \geq 1$ such that

$$\sum_{1}^{n} |1 - e^{ir'_j k}| + |\hat{\mu}_{r'}(k) - \bar{z}| < \varepsilon.$$

Let $z_1, \ldots, z_m \in T$ be such that

$$\left| z - \prod_{j=1}^{m} (\tfrac{1}{2} + \tfrac{1}{2} z_j) \right| < \varepsilon.$$

Let $J \geq n + m$ be such that $\sum_{j \geq J} r_j < \alpha/2$. Let $r'_1, \ldots, r'_j$ be such that

(10) $$|r_j - r'_j| < \alpha/2J \quad \text{for } 1 \leq j \leq J,$$

(11) $$e^{ir'_j} \text{ is a root of unity for } 1 \leq j \leq J - m, \text{ and}$$

(12) $$\{e^{ir'_j}: J - m < j \leq J\} \text{ is an independent set.}$$

We set $r'_j = 0$ if $j > J$. Then $\|r - r'\| < \alpha/2$. By (11) there exists $K \geq 1$ such that

(13) $$e^{ir'_j K} = 1 \quad \text{for } 1 \leq j \leq J - m,$$

and a positive multiple k of K (by (12)) such that

(14) $$\left| z - \prod_{j=m+1}^{J} (\tfrac{1}{2} + \tfrac{1}{2} e^{ikr'_j}) \right| < \varepsilon.$$

Now (13) and (14) show that $|\hat{\mu}_{r'}(k) - \bar{z}| < \varepsilon$ and $\sum_1^n |1 - e^{ir'k}| = 0$, which show that $r' \notin X$. Since α and r are arbitrary, X has no interior.

Let $\{z_m: m = 1, 2, \ldots\}$ be a dense sequence in $\{z: |z| < 1\}$. Then the union

$$Y = \bigcup_{m=1}^{\infty} \bigcup_{k=1}^{\infty} \bigcup_{n=1}^{\infty} X(z_m, 1/k, n)$$

is first category in F. If $r \notin Y$, then for each m, k, n, there exists q such that

$$\sum_{1}^{n} |\exp(ir_j q) - 1| + |\hat{\mu}_r(q) - z_m| < 1/k.$$

By varying k and m, we see that for each z with $|z| < 1$, there exists a sequence $q(1), q(2), \ldots,$ in Z such that $\lim_{n\to\infty} \langle r_j, q(n)\rangle = \lim_{n\to\infty} \exp(ir_j q(n)) = 1$ for $j = 1, 2, \ldots$ and $\lim \hat{\mu}_r(q(n)) = \bar{z}$. By Lemma 6.5.6(ii) there exists $\psi \in Z^- \backslash Z \subseteq R^- \backslash R$ such that $\psi_{\mu_j} = \bar{z}$ a.e. du_r. The Theorem is proved. □

Remarks and Credits. Šreĭder [2] showed that for $\mu = * [\frac{1}{2}\delta(0) + \frac{1}{2}\delta(3^{-j})]$, the Cantor-Lebesgue measure, there is an element $\psi \in \bar{Z}\backslash Z$, with $\psi_\mu \equiv a$ a.e. $d\mu$ where $0 < a < 1$. That was the first example of a generalized character with $|\psi_\mu|^2 \neq |\psi_\mu|$. Šreĭder's example was generalized by Hewitt and Kakukani [2] and Kaufman [4].

Let $\mathscr{B}$ denote the set of antisymmetric Bernoulli convolutions μ_r, where $r = (r_j)$ has the form (6). The measure μ_r is *fine* if $\limsup a_j = \infty$, and *coarse* otherwise.

6.6.6. Proposition. *Let* $\mu_r \in \mathscr{B}$.

(i) *Then* μ_r *is fine if and only if for all* z *with* $|z| \leq 1$, *there exists* $\psi \in Z^- \backslash Z$ *such that* $\psi_{\mu_r} = z$ *a.e.* du_r.
(ii) *If* μ *is coarse and* E *is a compact set with* $\mu(E) > 0$, *then* $E + E$ *has positive Lebesgue measure.*

The article of Brown and Moran [7] contains the proof of Theorem 6.6.6(i), and provides a deep and detailed study of the maximal ideal spaces and structure semi-groups of L-algebras generated by Bernoulli convolutions. Theorem 6.6.6(ii) is due to Talagrand [1] when $a_j = 3$ for all j; J.-F. Méla pointed out to us that (ii) holds in general.

For a more detailed version of 6.6.3 and its extension to general locally compact abelian groups, see Lin and Saeki [1].

6.7. Coin Tossing

We now turn to consideration of another type of infinite convolution on R and T:

$$\mu = \mathop{\ast}_{j=1}^{\infty} [(1 - r_j)\delta(0) + r_j\delta(2^{-j})].$$

Such a measure may be interpreted in the following way. A countably infinite set of coins is tossed, with r_n the probability of the nth coin being heads. Let X_n be the random variable which is one if the nth coin is heads and zero otherwise. Then μ is the distribution of $X = \sum_1^\infty 2^{-n}X_n$. If we carry out k independent trials of this experiment, then μ^k is the distribution of the sum of the k (independent) X's. The case of unbiased coins, $r_n = \frac{1}{2}$ for all n, yields Lebesgue measure on [0, 1].

6.7.1. Theorem. *Let $0 \le r_j \le 1$ for $j = 1, 2, \ldots$. Let*

$$\nu = \mathop{\bigstar}_{j=1}^{\infty} [(1 - r_j)\delta(0) + r_j\delta(2\pi/2^j)].$$

Let $b = \limsup |1 - 2r_j|$. Then:

(i) $2b/\pi \le \lim\sup_{|t|\to\infty} |\hat{\nu}(t)| \le b$;
(ii) $\nu^k \in L^1(R)$ *if* $\sum_{j=1}^{\infty} |1 - 2r_j|^{2k} < \infty$;
(iii) $(\nu * \tilde{\nu})^k \perp L^1(R)$ *if* $\sum_{j=1}^{\infty} |1 - 2r_j|^{4k} = \infty$; *and*
(iv) $\nu \in M_d(R)$ *if and only if* $\sum r_j < \infty$.

In the lemmas that follow we shall write ν for a measure of the form $\nu = \bigstar_{j=1}^{\infty} [(1 - r_j)\delta(0) + r_j\delta(2\pi/2^j)]$ where $0 \le r_j \le 1$, and μ for a measure of the form $\mu = \bigstar_{j=1}^{\infty} [(1 - s_j)\delta(0) + (s_j/2)(\delta(2\pi/2^j) + \delta(-2\pi/2^j))]$, where $s_j = 2(r_j - r_j^2)$ for all j. Then $\nu * \tilde{\nu} = \mu$.

6.7.2. Lemma. *Let k, j and q be positive integers and $0 \le r_j \le 1$. Let $\nu_j = (1 - r_j)\delta(0) + r_j\delta(2\pi/2^j)$. Then*

$$\sum_{m=0}^{1} |\hat{\nu}_j(q + m2^{j-1})|^{2k} \le 1 + (1 - 2r_j)^{2k}. \tag{1}$$

Proof. Straightforward calculations using $\cos x = 1 - 2\sin^2(x/2)$ show that

$$|\hat{\nu}_j(m)|^2 = 1 - 2s_j \sin^2(\pi m/2^j) \quad \text{for all } m.$$

(Here $s_j = 2(r_j - r_j^2)$, as indicated above.) Then the left hand side of (1) equals

$$(1 - 2s_j \sin^2 \pi q/2^j)^k + (1 - 2s_j + 2s_j \sin^2 \pi q/2^j)^k.$$

Differential calculus shows that $0 \le 2s_j \le 1$, and (differentiate with respect to x) that $(1 - 2sx)^k + (1 - 2s + 2sx)^k \le 1 + (1 - 2s)^k$ for $0 \le 2s, x \le 1$. Therefore the left hand side of (1) is at most

$$1 + (1 - 2s_j)^k = 1 + (1 - 2r_j)^{2k},$$

which establishes (1). □

6.7.3. Lemma. *If $\Sigma(1 - 2r_j)^{2k} < \infty$, then $\nu^k \in L^1(R)$.*

Proof. It will suffice to show that $\int |\hat{v}|^{2k}\, d\gamma < \infty$. Since v is a probability measure supported in $[0, 2\pi]$, the integral $\int |\hat{v}|^{2k}\, d\gamma$ is finite if and only if $\sum_{m=0}^{\infty} |\hat{v}(m)|^{2k} < \infty$. For each $n = 0, 1, 2, \ldots$

$$(2)\qquad \sum_{m=0}^{2^n-1} |\hat{v}(m)|^{2k} = \sum_{m(1)=0}^{1} \cdots \sum_{m(n)=0}^{1} \prod_{j=1}^{\infty} \left| \hat{v}_j\left(\sum_{p=1}^{n} m(p) 2^{p-1} \right) \right|^{2k}$$

where $v_j = (1 - r_j)\delta(0) + r_j \delta(2\pi/2^j)$ for all j. By (1), (2) and induction, we conclude that

$$\sum_{0}^{2^n-1} |\hat{v}(m)|^{2k} \leq \prod_{j=1}^{n} [1 + (1 - 2r_j)^{2k}].$$

Since $\prod_1^\infty [1 + (1 - 2r_j)^{2k}] < \infty$ if and only if $\sum (1 - 2r_j)^{2k} < \infty$, we see that $\sum_0^\infty |\hat{v}(m)|^{2k} < \infty$, and the Lemma follows. □

The proof of the next result is straightforward.

6.7.4. Lemma. *In the preceding notation,*

$$\hat{\mu}(2^m) = (1 - 2s_{m+1}) \prod_{j=m+2}^{\infty} [1 - s_j + s_j \cos(\pi 2^{m+1-j})] > 0.$$

6.7.5. Lemma. *If* $\lim_{j\to\infty} r_j = r \neq \frac{1}{2}, 0, 1$, *then* μ *is strongly i.p.*

Proof. By 6.7.4, $\lim_{k\to\infty} \hat{\mu}(2^{j(k)}) = 1 - 2s$, where $s = 2(r - r^2)$. Since $r \neq 0, \frac{1}{2}, 1$, $1 - 2s \neq 0, 1$. Of course, the characters $2^{j(k)}$ tend to one on the elements $2\pi/2^j$ of R. Therefore 6.5.6 applies: there exists ψ in the closure of $\{2^m\}_{m=1}^{\infty}$ such that $\psi_\mu = 1 - 2s$ a.e. $d\mu$. Therefore μ is strongly i.p. □

6.7.6. Lemma. *Let ω and v be probability measures on the measurable space X. Let $\{\psi_n\}$ be a sequence of real-valued measurable functions on X such that*

$$(3)\qquad \sup \int |\psi_m|^2\, dv < \infty;$$

$$(4)\qquad \int \psi_m \psi_n\, dv = 0 \quad \textit{for } m \neq n;$$

$$(5)\qquad \sum \left(\int \psi_m\, d\omega \right)^2 = \infty;$$

and there is a constant $C > 0$ such that

$$(6)\qquad \left| \int \psi_n \psi_m \, d\omega - \int \psi_n \, d\omega \int \psi_m \, d\omega \right| \le C 2^{m-n} \quad \textit{for } 1 \le m \le n.$$

Then ω is not absolutely continuous with respect to ν.

Proof. For $m = 1, 2, \ldots$, let $\alpha_m = \int \psi_m \, d\omega$. Then $\{\alpha_m\} \notin l^2$, so there exist integers $0 = m(1) < m(2) < \ldots$, and numbers $c_1, c_2, \ldots$ such that

$$(7)\qquad \sum_{j=m(k)+1}^{m(k+1)} c_j \alpha_j = 1 \quad \text{for } k \ge 1 \text{ and } \sum_{j=1}^{\infty} c_j^2 < \infty.$$

We let $f_k = \sum_{j=m(k)+1}^{m(k+1)} c_j \psi_j$ for $k \ge 1$. Then

$$\int |f_k|^2 \, d\nu \le \sum |c_j|^2 \sup \int |\psi_m|^2 \, d\nu \to 0 \text{ as } k \to \infty.$$

By passing to a subsequence of $\{f_k\}$, we may assume that

$$(8)\qquad f_k \to 0 \quad \text{pointwise a.e. } d\nu.$$

We claim that

$$(9)\qquad \sup \int |f_k|^2 \, d\omega < \infty.$$

To see (9), note that (6) implies that

$$\left| \int |f_k|^2 \, d\omega - \left(\int f_k \, d\omega \right)^2 \right| = \left| \sum c_m c_n \left(\int \psi_m \psi_n \, d\omega - \int \psi_m \, d\omega \int \psi_n \, d\omega \right| \right.$$
$$\le 4C \sum |c_m c_n|.$$

Since $(\int f_k \, d\omega)^2 = 1$ for all k, (9) now follows.

Let us suppose that $\omega \ll \nu$. Then (9) and (8) would imply that $\int f_k \, d\omega \to 0$, contradicting (7). [To see that $\int f_k \, d\omega \to 0$, fix $\varepsilon > 0$ and let Y be a measurable set of μ-measure $1 - \varepsilon$ such that $f_k \to 0$ uniformly on Y. Such a Y exists by Egorov's Theorem (Halmos [1, p. 88]). Then

$$\int f_k \, d\omega = \int f_k \chi_Y \, d\omega + \int f_k \chi_{X \setminus Y} \, d\omega.$$

Since $|\int f_k \chi_{X \setminus Y} \, d\omega| \le (\int |f_k|^2 \, d\omega)^{1/2} (\int \chi_{X \setminus Y} \, d\omega)^{1/2} \le \varepsilon^{1/2} \sup (\int |f_k|^2 \, d\omega)^{1/2}$, and $\int f_k \psi_Y \, d\omega \to 0$, $\int f_k \, d\omega \to 0$.] The Lemma is proved. $\square$

6.7.7. Lemma. *If* $\sum_{j=1}^{\infty}(1-2s_j)^{2k}=\infty$, *then* $\mu^k \perp L^1(T)$.

Proof. We shall apply 6.7.6 with $\omega=\mu^k$, $v=m_T$, and $\psi_m(t)=\exp(2\pi i 2^m t)+\exp(-2\pi i 2^m t)$. That (3) and (4) hold is obvious. For (5), observe that

$$(10)\quad \int \psi_m\, d\mu^k = 2(1-2s_m)^k \prod_{j=m+2}^{\infty}[1-s_j+s_j\cos(\pi 2^{m-j})]^k, \quad \text{for } m \geq 1.$$

Since the infinite products in (10) are uniformly bounded away from zero, our hypothesis and (10) imply (5). To obtain (6) we first observe that

$$(11)\qquad \psi_m(s+t)=\psi_m(s) \quad \text{if } t=m/2^m,$$

and

$$(12)\qquad |\psi_m(s+t)-\psi_m(t)| \leq 4\pi 2^m|s| \quad \text{for all } t.$$

For $1\leq m\leq n$, $\mu^k=\sigma_m^k * \sigma_{m,n}^k * \tau_n^k$, where $\sigma_m = \bigstar_{j=1}^{m}\mu_j$, $\sigma_{m,n}=\bigstar_{j=m+1}^{n-1}\mu_j$ and $\tau_n=\bigstar_{m}^{\infty}\mu_j$. Then (11) implies that

$$(13)\qquad \int \psi_m\psi_n\, d\mu^k = \int \psi_m\psi_n\, d\tau_m^k \quad \text{and} \quad \int \psi_m\, d\mu^k = \int \psi_m\, d\tau_m^k$$

for $1\leq m\leq n<\infty$. And (12) implies that

$$(14)\qquad \iint |\psi_m(s+t)-\psi_m(s)|\,|\psi_n(t)|\, d\sigma_{m,n}^k(s)d\tau_n^k(t) \leq 8\pi k 2^{m-n+1}$$

and

$$(15)\qquad \iint |\psi_m(s+t)-\psi_m(s)|\, d\sigma_{m,n}^k(s)d\tau_n^k(t) \leq 4\pi k 2^{m-n+1}.$$

Then (13), (14), and (15) show that

$$\begin{aligned}
&\left|\int \psi_m(s)\psi_n(s)\, d\mu^k - \int \psi_m(s)d\mu^k \int \psi_n(t)d\mu^k\right| \\
&\quad = \left|\iint \psi_m(s+t)\psi_n(t)d\sigma_{m,n}^k(s)d\tau_n^k(t)\right. \\
&\qquad \left. - \iint \psi_m(s+t)d\sigma_{m,n}^k(s)d\tau_{m,n}^k(t) \int \psi_n(t)d\tau_n^k(t)\right| \\
&\quad \leq 16\pi k 2^{m-n+1},
\end{aligned}$$

which establishes (6). We now apply 6.7.6 to conclude that $\mu^\kappa \notin L^1(T)$. By the Generalized Purity Theorem 6.5.5, $\mu^k \perp L^1(T)$. The Lemma is proved. □

Proof of 6.7.1. (i) We shall use the identity

$$(16) \qquad (\sin u)/u = \prod_{j=1}^{\infty} \cos(u/2^j),$$

which follows from the identities

$$\sin u = 2\cos(u/2)\sin(u/2) = 2^n \sin(u/2^n) \prod_{j=1}^{n} \cos(u/2^j),$$

that $\lim_{x\to\infty} x \sin(u/x) = u$, and that the infinite product in (16) converges.

We claim that for all $t \in R$,

$$(17) \qquad |\hat{v}(t)|^2 = \prod_{j=1}^{\infty} [\cos^2 \pi t 2^{-j} + 4|r_j - \tfrac{1}{2}|^2 \sin^2 \pi t 2^{-j}].$$

Indeed, $|\hat{v}(t)|^2 = \prod_{j=1}^{\infty} [1 - 2r_j + 2r_j^2 + (2r_j - 2r_j^2)\cos 2\pi t 2^{-j}]$. We apply the identity $\cos x = 1 - 2\sin^2 x/2$ to the preceding and (17) follows. Suppose that $0 < b' < b$. Let $1 \le j(1) < j(2) < \cdots$ be such that $|1 - 2r_{j(k)}| > b'$ for $k = 1, 2, \ldots$. Then (17) and (16) imply that

$$\begin{aligned} |\hat{v}(2^{j(k)})|^2 &\ge (b')^2 \prod_{j=j(k)+1}^{\infty} \cos^2(\pi 2^{j(k)-j}) \\ &= (b')^2 \prod_{j=2}^{\infty} \cos^2 \pi 2^{-j} = 4(b')^2/\pi^2 \end{aligned}$$

Thus $\limsup |\hat{v}| \ge 2b'/\pi$ for all $b' < b$. That proves the first half of (i).

When $b = 1$, then $|\hat{\mu}|$ is certainly bounded by b, since μ is a probability measure. Now suppose that $b < 1$ and that $b^2 < b' < 1$. Then there exists an integer J such that $4|r_j - \tfrac{1}{2}|^2 < b'$ whenever $j \ge J$. Then (17) implies that

$$(18) \qquad |\hat{\mu}(t)| \le \prod_{j=J+1}^{\infty} [\cos^2 \pi t 2^{-j} + b' \sin^2 \pi t 2^{-j}].$$

Each term in every factor of the right hand side of (18) is non-negative. Let $f(t, b') = \prod_{J+1}^{\infty} [\cos^2 \pi t 2^{-j} + b' \sin^2 \pi t 2^{-j}]$. Then

$$f(t, b')/b' = (1/b') \prod_{J+1}^{\infty} \cos^2 \pi t 2^{-j} + f_1(t) + b' f_2(t) + \cdots$$

where $f_n(t)$ is the coefficient of $(b')^n$ in the expansion of $f(t, b')$. Using (16) with $u = \pi t 2^J$, we have $\prod_{J+1}^{\infty} \cos^2 \pi t 2^{-j} = \sin^2 \pi t 2^{-J}/(\pi t 2^{-J})^2 \to 0$ as $t \to \infty$. Therefore

$$(1/b')\limsup_{t\to\infty} f(t, b') = \limsup_{t\to\infty}\{f_1(t) + b' f_2(t) + \cdots\}$$

$$\leq \limsup \sum_1^{\infty} f_n(t),$$

since all summands in the preceding are non-negative and $0 < b' < 1$. But $f(t, 1) = 1$ and $\sum_1^{\infty} f_n(t) = f(t, 1) - \sin^2 \pi t 2^{-J}/(\pi t 2^{-J})^2$. Therefore

$$\limsup_{t\to\infty} f(t, b') \leq b',$$

or $\limsup |\hat{v}(t)|^2 \leq b'$. Since that holds for all $b' > b^2$, we have

$$\limsup_{t\to\infty} |\hat{v}(t)| \leq b.$$

(ii) is just 6.7.3. (iii) follows from 6.7.7, the observation that $(1 - 2s_j) = (1 - 2r_j)^2$ and 6.7.5. (iv) is routine. That ends the proof of 6.7.1. □

6.7.8. Theorem. *Let* $\mu = \bigstar_{j=1}^{\infty} [(1 - s_j)\delta(0) + (s_j/2)(\delta(-2\pi/2^j) + \delta(2\pi/2^j))]$. *If* $\liminf s_j = s > 0$, *then* μ *is a tame measure in* $M(R)$.

Proof. Let $\psi \in \Delta M(R)$ and suppose that $\psi_\mu \not\equiv 0$. Then 6.5.6(iii) implies that for some $J \geq 1$,

$$\prod_{j=J}^{\infty} |1 - s_j + s_j \operatorname{Re} \delta(2\pi/2^j)\hat{\ }(\psi)| > 0. \tag{19}$$

Let D denote the subgroup of R generated by $\{2\pi/2^j : 1 \leq j \leq \infty\}$, and let γ denote the character on D that ψ induces. (That is, $\langle -x, \gamma\rangle = \delta(x)\hat{\ }(\psi)$ for all x.) Then (19) becomes

$$\prod_{j=J}^{\infty} |1 - s_j + s_j \operatorname{Re}\langle 2^{-j}, \gamma\rangle| > 0. \tag{20}$$

But (20) holds if and only if

$$\sum_{j=J}^{\infty} |s_j - s_j \operatorname{Re}\langle 2^{-j}, \gamma\rangle| < \infty. \tag{21}$$

Since $\liminf s_j > 0$, (21) holds if and only if $\sum |1 - \operatorname{Re}\langle 2^{-j}, \gamma\rangle| < \infty$, or if and only if

$$\sum_{j=1}^{\infty} |1 - \langle 2^{-j}, \gamma\rangle| < \infty.$$

Let $x = \sum_1^\infty \varepsilon_j 2^{-j} \in R$ where $\varepsilon_j = 0, \pm 1$. Then for $1 \leq N < M < \infty$,

$$\left|\left\langle \sum_1^N \varepsilon_j 2^{-j}, \gamma\right\rangle - \left\langle \sum_1^M \varepsilon_j 2^{-j}, \gamma\right\rangle\right| = \left|1 - \left\langle \sum_{N+1}^M \varepsilon_j 2^{-j}, \gamma\right\rangle\right|$$
$$\leq \sum_{N+1}^M |1 - \langle 2^{-j}, \gamma\rangle| \to 0.$$

Therefore γ has a unique continuous extension to $\{x \in R: |x| \leq 1\}$, and of course that extension extends in turn to a continuous character on R. An application of 6.5.7 completes the proof. □

6.7.9. Theorem. *$M_o(R)$ and $M_o(T)$ both contain tame Hermitian i.p. probability measures.*

Proof. Set $\mu = \bigstar_1^\infty [(1 - s_j)\delta(0) + (s_j/2)(\delta(-2\pi/2^j) + \delta(2\pi/2^j))]$ where $0 < s_j \leq \frac{1}{2}$ for all j, $\lim s_j = \frac{1}{2}$, and $\Sigma(1 - 2s_j)^k = \infty$ for $k = 1, 2, \ldots$. Then μ has the required properties by 6.7.1, 6.7.8 and the Generalized Purity Theorem 6.5.5. □

Remarks and Credits. Theorem 6.7.1(i) is due to Blum and Epstein [1]. Theorem 6.7.1(ii)–(iii) is due to Brown and Moran [12]. They proved a more general result involving convolutions of the form

$$\omega = \bigstar_{n=1}^{\infty} \left[\sum_{j=0}^{a(n)-1} b(j, n)\delta(jd_n)\right] \tag{22}$$

where $d_n^{-1} = a(1) \cdots a(n)$, $\{a(j)\}$ is a bounded sequence of integers greater than or equal to two, and the $b(j, k)$ are non-negative with $\sum_j b(j, n) = 1$ for all n. Then ω is singular if and only if $\sum_n \max |b(j, n) - a(n)^{-1}|^{2k} = \infty$. The proof of 6.7.1(ii)–(iii) contains the essential ideas of Brown and Moran's arguments. Part (iv) of 6.7.1 is standard.

Brown [4] has constructed a family $\{\mu_t: t > 0\}$ of strongly i.p. probability measures in $M_o(R)$ such that $\mu_t * \mu_s = \mu_{t+s}$ for all $0 < t, s < \infty$.

6.7.9 was proved by Brown [2] and Moran [2] independently. Saeki [20] gives a more general version of some of the results of this section, using L^p-martingales.

Blum and Epstein [2] identify the sets on which $\hat{\omega} \to 0$ where ω is as in (22).

6.8. $M_o(G)$ Contains Tame i.p. Measures

6.8.1. Theorem. *Let G be a non-discrete locally compact abelian group. Then there exists a tame Hermitian i.p. probability measure in $M_o(G)$.*

A *Taylor-Johnson measure* is a tame, Hermitian i.p. probability measure.

6.8.2. Lemma. (*i*) *Let H and G be locally compact abelian groups. If $M_o(H)$ contains a Taylor-Johnson measure, then $M_o(G \times H)$ contains a Taylor-Johnson measure.*

(*ii*) *If H is an open subgroup of the locally compact abelian group G, and $M_o(H)$ contains a Taylor-Johnson measure, then $M_o(G)$ contains a Taylor-Johnson measure.*

(*iii*) *If H is a compact subgroup of the locally compact abelian group G, and $M_o(G/H)$ contains a Taylor-Johnson measure, then $M_o(G)$ contains a Taylor-Johnson measure.*

(*iv*) *If $M_o(G)$ contains a Taylor-Johnson measure whenever G is an infinite compact abelian group, then $M_o(G)$ contains a Taylor-Johnson measure whenever G is a non-discrete locally compact abelian group.*

Proof. (i) follows from 6.1.7(ii), with μ Haar measure restricted to a relatively compact open subset of G and ν a Taylor–Johnson measure in $M_o(H)$, and (ii) is 6.1.7(iii).

(iii) Let $\nu \in M_o(G/H)$ be a Taylor-Johnson measure. Let ω denote normalized Haar measure on H. Then $p: G \to G/H$ induces an isomorphism $\check{p}$ from $\omega * M(G)$ to $M(G/H)$. Let $\mu \in \omega * M(G)$ be such that $\mu \to \nu = \check{p}\mu$ by this isomorphism. Then μ is an i.p. Hermitian probability measure.

We claim that if $\psi \in \Delta M(G)$, then either $\psi_\omega \equiv 0$ a.e. $d\omega$ or $|\psi_\omega| \equiv 1$ a.e. $d\omega$. We claim, further, that there exists φ in $\Delta M(G/H)$ such that for some $\gamma \in \Gamma$,

$$\psi_\mu = \gamma(p^*\varphi)_\mu, \tag{1}$$

where p^* is the mapping of $\Delta M(G/H)$ to $\Delta M(G)$ induced by p. Indeed if $\psi_\omega \not\equiv 0$, then $\psi_\omega \equiv \gamma$ a.e. $d\omega$ for some $\gamma \in \Gamma$. Let $\lambda = \bar{\gamma}\psi$, and let $\varphi = \lambda \circ \check{p}^{-1}$. (We obviously restrict λ to $\omega * M(G)$.) Then it is easy to see that $\varphi \in \Delta M(G/H)$, and $\lambda = p^*\varphi$. We have established (1).

Of course, the tameness of ν implies that $\varphi_\nu = a\lambda$ for some $a \geq 1$ and $\lambda \in (G/H)^\wedge \subseteq \Gamma$. Then (1) implies that $\psi_\mu = a\gamma\lambda$ a.e. $d\mu$. Therefore μ is tame.

(iv) By (ii), we may assume (using the structure theorem) that $G = R^n \times K$, where K is compact. If $n = 0$, there is nothing to prove. If $K \neq \{0\}$, the result follows from the hypothesis and 6.8.2(i). That $M_o(R)$ contains a Taylor-Johnson measure follows from the hypothesis and an application of A.7.1. An application of (i) now completes the proof in the case that $n > 0$. □

Proof of 6.8.1. By 6.8.2(iv), we may assume that G is compact.

Let Γ be the dual group of G. If Γ contains an element of infinite order, then G has compact subgroup H such that G/H is isomorphic to the circle group T. Then the existence of Taylor-Johnson measures in $M_o(G)$ follows from 6.8.2(iii) and 6.7.9. We may therefore assume that Γ is a torsion group.

We now apply some results from the theory of abelian groups. A group Λ is *divisible* if for each $\lambda \in \Lambda$ and integer $n \geq 1$, $n\gamma = \lambda$ has a solution $\gamma \in \Lambda$. It is well-known that a divisible abelian group is a direct summand of every abelian group that contains it and that every non-trivial divisible abelian torsion group contains a subgroup that is isomorphic either to the rationals Q or to $Z(p^\infty)$. (We remind the reader that, for a fixed prime p, $Z(p^\infty) = \{x \in T: p^k x = 0 \text{ for some } k\}$.) For proofs of those facts, see Fuchs [1, Volume 1, Section 20, 21, and 23]. We conclude from the preceding that we may assume that either $\Gamma = Z(p^\infty)$ for some prime p, or that Γ contains no divisible subgroups.

Suppose that Γ contains no divisible subgroups. Then Γ contains a subgroup of the form $\oplus \sum_{j=1}^{\infty} Z(m_j)$. If Γ is uncountable, that is straightforward. If Γ is countable, that follows from Fuchs [1, Volume II, Proposition 7.7.5.] An appeal to 6.8.2(iii) and 6.4.4 shows that $M_o(G)$ contains Taylor-Johnson measures. We have reduced the proof of 6.8.1 to the case that $\Gamma = Z(p^\infty)$ for some prime p. That case will be settled in Lemma 6.8.3, and the proof of 6.8.1 will be complete. □

Before stating 6.8.3, we have a number of preliminaries. We begin with some comments on $Z(p^\infty)$ and its dual group Δ_p. The compact group Δ_p may be identified with the set of all formal power series $\sum_0^\infty a_j p^j$, with integer coefficients satisfying $0 \leq a_j < p$ for all j. Addition in Δ_p is then defined as follows:

$$\sum a_j p^j + \sum b_j p^j = \sum c_j p^j,$$

where integers t_j and c_j are obtained inductively: $t_{-1} = 0$ and t_j, c_j are integers satisfying the inequalities $0 \leq t_j \leq 1$, $0 \leq c_j \leq p$ and the equation $a_j + b_j + t_{j-1} = t_j p + c_j$. A neighborhood basis at the identity of Δ_p is formed by the sets $V_n = p^n \Delta_p$, $n \geq 1$. In that representation of Δ_p, elements of $Z(p^\infty)$ have the form kp^{-m} where $0 < k < p$ and $0 < m < \infty$. The value of kp^{-m} at $\sum a_j p^j$ is given by $\prod_{j=1}^{m-1} \exp(2\pi i a_j k p^{j-m})$. For more about Δ_p and $Z(p^\infty)$ the reader may consult Hewitt and Ross [1, Volume I, pages 108ff], or Fuchs [1], or Kurosh [1, Volume 1].

Let $0 < t < (2p)^{-1}$. For $n \geq 1$ we set

$$\beta_n = t(\delta(0) - \delta(p^{n-1})) + p^{-1} \sum_{j=0}^{p-1} \delta(jp^{n-1}).$$

Then β_n is a discrete probability measure with support contained in V_n. Let $\alpha_n = \beta_n * \tilde{\beta}_n$.

6.8.3. Lemma. *There exists a sequence of integers, $1 \le m(1) < m(2), \ldots,$ with $\lim_{j\to\infty} m(j) = \infty$ such that*

$$\nu = \mathop{\bigstar}_{j=1}^{\infty} \alpha_j^{m(j)}$$

is a Taylor–Johnson measure in $M_o(\Delta_p)$, where the α_n are as above.

Proof. We shall first show that ν is tame, whatever the sequence of positive integers, that $\nu \in M_o(\Delta_p)$ if $\lim m(n) = \infty$, and finally, that for some choice of $1 < m(1) < m(2) < \cdots$ with $\lim m(j) = \infty$, ν is i.p. That ν is Hermitian is obvious. The Lemma will have been proved.

ν is tame. We note first if $D = \sum \{a_j p^j : a_j = 0 \text{ for } j \ge J,\ a_j < p \text{ for } j \ge J\}$, then D is an isomorphic image of Z under the map $1 \to 1 \cdot p^0 = 1$ and ν is D-ergodic.

Let $\varphi \in \hat{D} = T$. If $\prod \hat{\alpha}_j^{m(j)}(\varphi)$ converges then $\lim |\hat{\beta}_j(\varphi)| = 1$. Let $|\hat{\beta}_j(\varphi)| = \theta_j \hat{\beta}_j(\varphi)$, where $|\theta_j| = 1$. Then by the Cauchy-Schwarz inequality

$$\begin{aligned}\left(\int |\theta_j - \varphi|\, d\beta_j\right)^2 &\le \int |\theta_j - \varphi|^2\, d\beta_j \\ &= (1 - \theta_j \overline{\hat{\beta}_j(\varphi)})(1 - \theta_j \hat{\beta}(\varphi)) + (1 - |\hat{\beta}_j(\varphi)|^2) \\ &\le (1 - |\hat{\beta}_j(\varphi)|)^2 + (1 - |\hat{\beta}_j(\varphi)|^2).\end{aligned}$$

Therefore $\lim \int |\theta_j - \varphi|\, d\beta_j = 0$. But

$$\int |\theta_j - \varphi|\, d\beta_j \ge (2p)^{-1} \sum_{k=0}^{p-1} |\theta_j - \varphi(kp^{j-1})|.$$

Considering the terms corresponding to $k = 0$ and $k = 1$, we see that $\theta_j \to 1$ and $\varphi(p^{j-1}) \to 1$. Since $p^{j-1} \in \Delta_p$ is the image of $p^{j-1} \in Z$ and $D = Z$, we have $\varphi(p^{j-1}) = \exp(2\pi i x p^{j-1})$. Since $\lim \varphi(p^{j-1}) = 1$, x must have the form $x = kp^{-m}$ for non-negative integers k and m. It follows at once that φ extends to an element of $\Delta_p{}^\wedge = Z(p^\infty)$. Then ν is tame by 6.5.7.

ν belongs to $M_o(\Delta_p)$. We first observe that for all integers $j = 1, 2, \ldots$ and $k = 0, \ldots, p^j - 1$, with $p \nmid k$,

$$\hat{\alpha}_j(kp^{-j}) = t^2 |1 - \exp(2\pi i k p^{-1})|^2 \le 4t^2 < 1.$$

Therefore $|\hat{\nu}(\gamma)| \le (4t^2)^{m(j)}$ for γ outside of the finite set $\{kp^{-n} : 0 \le n \le j,\ 0 \le k < p\}$. Since $4t^2 < 1$, and $\lim m(j) = \infty$, $\hat{\nu} \in C_o(Z(p^\infty))$.

ν can be chosen to be i.p. We shall begin by considering the infinite convolutions

$$\tau_n = \mathop{\bigstar}_{j=n+1}^{\infty} \alpha_j, \text{ for } n \ge 0.$$

We claim that each τ_n is strongly i.p. The standard geometric series calculation applied to $\hat{\alpha}_k(p^{-j-1})$ shows that

$$(2)\qquad \hat{\tau}_0(p^{j-1}) = \prod_{q=1}^{j} |p^{-1}(1 - \exp(2\pi i p^{q-j}))/(1 - \exp(2\pi i p^{q-j-1})) + t(1 - \exp(2\pi i p^{q-j-1}))|^2.$$

Straightforward derivative estimates show that

$$(3)\quad |p^{-1}(1 - \exp(2\pi i p^{-q}))/(1 - \exp(2\pi i p^{-q-1}))| \leq C|1 - \exp(2\pi i p^{-q})|$$

and

$$(4)\qquad |1 - \exp(2\pi i p^{-q})| \leq 2\pi p^{-q}.$$

Using (3) and (4), we conclude that

$$\lim_{j\to\infty} \hat{\tau}_0(p^{j-1}) = a$$

exists. That $1 > a > 0$ follows from our choice of t between 0 and $\frac{1}{2}p$. Since $p^j = \gamma_j \to 1$ pointwise on $D(\tau_0)$, τ_0 is strongly i.p. Similar arguments show that τ_n is strongly i.p. for $n \geq 1$.

We now construct the sequence $m(j)$ as follows. Since τ_0 is strongly i.p., there exists a finite set $E(1)$ of continuous characters on Δ_p such that for all complex numbers c_1, c_2,

$$\left\| \sum_{j=1}^{2} c_j \hat{\tau}_0^j \Big|_{E(1)} \right\|_{A(E(1))} \geq (1 - 2^{-1}) \sum_{j=1}^{2} |c_j|.$$

Because Δ_p is totally disconnected, there exists a neighborhood $U(1)$ of the identity of Δ_p, such that $U(1)$ is a compact-open subgroup of Δ_p and such that each $\gamma \in E(1)$ is constant (with value one, obviously) on $U(1)$.

Choose $j(1) \geq 2$ so large that α_j is supported in $U(1)$ for all $j \geq j(1)$. Then for any probability measure ω' supported in $U(1)$, $\omega = \omega' * \bigstar_1^{j(1)} \alpha_j$ satisfies

$$(5)\qquad \left\| \sum_{j=1}^{2} c_j \hat{\omega}^j \Big|_{E(1)} \right\|_{A(E(1))} \geq (1 - 2^{-1}) \sum_{j=1}^{2} |c_j|$$

for all complex numbers c_1, c_2.

In particular, (5) holds with $\omega' = \tau_{j(1)}^2$. We set $m(1) = \cdots = m(j(1) - 1) = 1$. That begins the induction.

Suppose that for some integer $k \geq 1$ we have the following: finite subsets $E(1) \subseteq \cdots \subseteq E(k)$ of $Z(p^\infty)$; integers $1 \leq j(1) < \cdots < j(k)$; integers $1 \leq m(1) \leq \cdots \leq m(j(k))$ with $m(j(i)) = i$ for $1 \leq i \leq k$; and subgroups $U(1) \supseteq \cdots \supseteq U(k)$ in Δ_p such that for all $1 \leq i \leq k$,

$$\text{(6)} \qquad \langle \gamma, x \rangle = 1 \quad \text{for} \quad \gamma \in E(i), x \in U(i);$$

$$\text{(7)} \begin{cases} \text{for all probability measures } \omega' \text{ in } M(U(i)) \text{ and complex numbers} \\ c_1, \ldots, c_{j(i)+1}, \text{ if } \omega = \omega' * \bigstar_1^{j(i)} \alpha_j^{m(j)} \text{ then} \\ \left\| \sum_1^{i+1} c_j \hat{\omega}^j \Big|_{E(i)} \right\|_{A(E(i))} \geq (1 - 2^{-i}) \sum_{j=1}^{i+1} |c_j| \end{cases}$$

$$\text{(8)} \qquad \alpha_j \text{ is supported in } U(i) \text{ for } j > j(i).$$

We set $\sigma_k = \bigstar_{j=1}^{j(k)-1} \alpha_j^{m(j)}$. Since $\tau_{j(k)}$ is strongly i.p., there exists a finite subset $E(k+1)$ of $Z(p^\infty)$ such that, for $\omega = \sigma_k * \tau_{j(k)}^{k+1}$,

$$\left\| \sum_{j=1}^{k+2} c_j \hat{\omega}^j \Big|_{E(k+1)} \right\|_{A(E(k+1))} \geq (1 - 2^{-k-1}) \sum_{j=1}^{k+2} |c_j|.$$

for all complex numbers $c_1, \ldots, c_{k+2}$. There is no loss of generality in assuming that $E(k+1) \supseteq E(k)$.

Let $U(k+1) = \{x \in \Delta_p : \langle \gamma, x \rangle = 1 \text{ for } \gamma \in E(k+1)\}$. Let $j(k+1) > j(k)$ be so large that α_j is supported in $U(k+1)$ for all $j > j(k+1)$. Let $m(j(k)+1) = \cdots = m(j(k+1)) = k+1$. Routine calculations will show that the new $m(j)$'s, $j(k+1)$, $E(k+1)$, and $U(k+1)$ have the desired properties. That ends the induction.

That $\lim m(j) = \infty$ is obvious. That ν is i.p. follows from (6), (7) and (8). That ends the proof of Lemma 6.8.3. □

In course of the proof of 6.8.1, we proved the following result.

6.8.4. Theorem. *Let $\mathscr{B}$ be a class of locally compact abelian groups that has the following properties:*

(i) $R, T \in \mathscr{B}$;
(ii) $\Delta_p \in \mathscr{B}$ *for all primes* p; $(Z_p)^\infty \in \mathscr{B}$ *for all primes* p;
(iii) $G \in \mathscr{B}$, H *compact imply* $G \times H \in \mathscr{B}$;
(iv) $H \in \mathscr{B}$ *and* $G \supseteq H$ *as an open subgroup imply* $G \in \mathscr{B}$; *and*
(v) $G \supseteq H$ *a compact subgroup and* $G/H \in \mathscr{B}$ *imply* $G \in \mathscr{B}$.

Then $\mathscr{B}$ contains all non-discrete locally compact abelian groups.

Remarks and Credits. An alternative proof to 6.8.1 can be given through the use of Riesz products: that many Hermitian Riesz products are Taylor-Johnson measures is a principal result of the next chapter. The proof of 6.8.3 is adapted from Brown [2]. Theorem 6.8.1 is due to Brown [2] and Moran [2]. Theorem 6.8.4 is standard; in it (iii) may be replaced by "(iii)′ $G, H \in B$ imply $G \times H \in B$."

Chapter 7

Riesz Products

7.1. Introduction and Initial Results

This chapter is devoted to a study of Riesz product measures. For elementary results on Riesz products on the circle see A.1. The class of measures that we shall call "Riesz products" is defined in Section 7.1. The existence and elementary properties of such measures are established also in 7.1. In Section 7.2 we discuss the orthogonality of pairs of Riesz products and show that it is often the case that two Riesz products are either equivalent or mutually singular. In Section 7.3 we give a criterion for a Riesz product to be tame and use that criterion to show that every non-discrete locally compact abelian group supports a tame Hermitian i.p. probability measure in $M_o(G)$, a result proved by other methods in Section 6.8 and used in Section 8.2 to show that the Šilov boundary of $M(G)$ is not all of the maximal ideal space. Section 7.3 also contains an application to the symbolic calculus for $M(G)$, a topic covered in detail in Chapter 9.

In Section 7.4 we construct a singular probability measure $\mu \in M_o(G)$ such that $\mu^2 \approx \mu$. In Section 7.5 we show that if a set $X \subseteq M_c(G)$ is compact in the pseudonorm topology, then there exists an i.p. probability measure not in $M_o(G)$ such that $\rho * X \subseteq A(G)$, and that the support of a singular measure's transform is large in a certain sense.

In Section 7.6 we prove, in two different ways, that there are continuous measures μ on T such that $\{\gamma : \hat{\mu}(\gamma) \geq 1\}$ is dense in bZ. In Section 7.7 we construct non-trivial idempotents in $B(E)$, where $E \subseteq Z$.

We now define Riesz products and give some preliminary results. *For abstract dual groups Γ we shall use multiplicative notation in this chapter only. For Z we shall use additive notation.*

In 1918, Frédéric Riesz [1] gave the first example of a continuous measure μ on T with Fourier-Stieltjes coefficients *not* tending to zero. Riesz's example was the weak$*$ limit of the nonnegative absolutely continuous measures μ_n where

(1) $$d\mu_n = \prod_{j=1}^{n} (1 + \cos 4^j x)dx.$$

Direct calculations show that if $m \in Z$, then

$$(2)\quad \hat{\mu}_n(m) = \begin{cases} 1 & \text{if } m = 0; \\ 2^{-k} & \text{if } m = \pm 4^{j(1)} \pm \cdots \pm 4^{j(k)},\ 1 \le j(1) < \cdots < j(k) \le n; \\ 0 & \text{otherwise.} \end{cases}$$

Formula (2) and the boundedness of $\{\|\mu_n\|\}$ imply that the sequence μ_n does indeed possess a weak$*$ limit point μ and that

$$(3)\quad \hat{\mu}(m) = \begin{cases} 1 & \text{if } m = 0; \\ 2^{-k} & \text{if } m = \pm 4^{j(1)} \pm \cdots \pm 4^{j(k)},\ 1 \le j(1) < \cdots < j(k); \\ 0 & \text{otherwise.} \end{cases}$$

The salient features of Riesz's example are two. For the sake of generality we express them in terms of a compact abelian group G with dual group Γ and indicate in parentheses the correspondence with Riesz's example. We remind the reader that Γ will be written *multiplicatively.*

The first feature is that we have a subset Θ $(=\{4^j : j = 1, 2, \ldots\})$ of $\Gamma(=Z)$ such that every element ω of Γ may be expressed in *at most* one way as a product (sum)

$$(4)\qquad \omega = \prod_{j=1}^{n} \theta_j^{\varepsilon_j}.$$

where the θ_i are distinct elements of Θ, and ε_i is allowed to be 1 or -1 if $\theta_i^2 \neq 1$ but must equal $+1$ if $\theta_i^2 = 1$. A set Θ with that property is called *dissociate.*

The second feature of Riesz's example is the choice of a function $a\colon \Theta \to \mathbb{C}$ that defines for each $\theta \in \Theta$ a polynomial q_θ on G (Riesz chose $a(4^j) = \frac{1}{2}$ for all j). The function a is restricted as follows.

Let $q_\theta = 1 + a(\theta)\theta$ if $\theta^2 = 1$, and $q_\theta = 1 + a(\theta)\theta + \overline{a(\theta)\theta}$ if $\theta^2 \neq 1$. We restrict $a(\theta)$ in such a way that $q_\theta \geq 0$. It is customary to require, as we shall, that $-1 \le a(\theta) \le 1$ if $\theta^2 = 1$ and $|a(\theta)| \le \frac{1}{2}$ if $\theta^2 \neq 1$. (If $\theta^3 = 1$, then it is easy to see that $a(\theta) = -\frac{1}{2} + i\sqrt{3}/2$ also gives $q_\theta \geq 0$, but such special choices are rarely needed.)

We define for each finite subset Φ of Θ the partial product

$$(5)\qquad P_\Phi = \prod_{\theta \in \Phi} q_\theta.$$

Then P_Φ is a non-negative element of $L^1(G)$. Since Θ is dissociate, the Fourier coefficients of P_Φ are given as follows, where $\Phi = \{\theta_1, \ldots, \theta_n\}$.

$$(6)\qquad \hat{P}_\Phi(\omega) = \begin{cases} 1 & \text{if } \omega = 1; \\ \displaystyle\prod_{i=1}^{n} a(\theta_i)^{(\varepsilon_i)} & \text{if (4) and (5) hold;} \\ 0 & \text{otherwise.} \end{cases}$$

Here $a(\theta_i)^{(\varepsilon_i)} = a(\theta_i)$ if $\varepsilon_i = 1$ and $a(\theta_i)^{(\varepsilon_i)} = \overline{a(\theta_i)}$ if $\varepsilon_i = -1$. It follows from (6) and the non-negativity of P_Φ that $\|P_\Phi\|_1 = 1$. It is now clear that $\mu \equiv \lim P_\Phi$ exists weak$*$ in $M(G)$. Here the limit is taken over increasing finite subsets Φ of Θ. Then μ is a probability measure on G and the Fourier-Stieltjes coefficients of μ are given by

(7) $$\hat{\mu}(\omega) = \begin{cases} 1 & \text{if } \omega = 1; \\ \prod_{i=1}^{n} a(\theta_i)^{(\varepsilon_i)} & \text{if (4) holds;} \\ 0 & \text{otherwise.} \end{cases}$$

We say that μ is the *Riesz product based on* Θ *and* a. Each product ω of the form (4) will be called a *word* with *letters* $\theta_1, \ldots, \theta_n$. The set of all words (4) using letters from Θ will be denoted $\Omega(\Theta)$. It is a useful fact that if ω_1 and $\omega_2 \in \Omega(\Theta)$ have no letters in common, then $\omega_1\omega_2 \in \Omega(\Theta)$.

If $a(\theta) = 1, -1$, or 0, then q_θ equals $2\chi_{\ker\theta}$ or $2\chi_{G\backslash\ker\theta}$, or 1 respectively. Thus the presence of q_θ in (5) adds nothing of interest. We shall therefore assume that

(8) $$\begin{aligned} &-1 < a(\theta) < 1, \quad \text{for all } \theta \in \Theta \text{ with } \theta^2 = 1; \quad \text{and} \\ &0 \le |a(\theta)| \le \tfrac{1}{2} \quad \text{for all } \theta \in \Theta \text{ with } \theta^2 \ne 1. \end{aligned}$$

We let $R(G)$ be the set of all measures μ on G such that μ is the Riesz product based on an infinite dissociate set Θ and a function a on Θ that satisfies (8). While ruling out $a(\theta) = 0$ would not restrict the class $R(G)$, it would complicate some arguments in Section 7.2. For example, m_G is a Riesz product, one which is convenient to include in the class of measures we study.

Let $\mu \in R(G)$ be based on Θ and a. For a finite subset Φ of Θ we let μ_Φ be the Riesz product based on $\Theta\backslash\Phi$ and the restriction of a to $\Theta\backslash\Phi$. Then a straightforward computation shows that

(9) $$\mu = P_\Phi \mu_\Phi = \sum\{\hat{\mu}(\omega)\omega\mu_\Phi : \omega \in \Omega(\Phi)\}.$$

It is easy to see from the definition of μ_Φ that Haar measure on G is the weak$*$ limit of the measures μ_Φ. Thus if μ is equivalent to μ_Φ for all finite subsets Φ of Θ, then the support of μ is all of G. That occurs if, for example, $G = T$ or $|a(\theta)| < \frac{1}{2}$ for all $\theta \in \Theta$ that have order not equal to two. The reader is referred to Brown [3, Proposition 3 and its proof] for details. For a qualitative result along those lines, see Peyrière [2, p. 135]. The next Lemma is proved by a simple induction. See 11.3.2 for a quantitative version.

7.1.1. Lemma. *Let Θ be an infinite subset of the abelian group Γ.*

(i) *If $\{\theta^2: \theta \in \Theta\}$ is infinite, then Θ contains an infinite dissociate subset.*
(ii) *If $\{\theta^2: \theta \in \Theta\}$ is finite, then there exist an infinite subset $\{\theta_j\} \subseteq \Theta$ such that both $\{\theta_1\theta_j: j = 2, 3, \ldots\}$ and $\{\theta_{2j}\theta_{2j+1}: j = 1, 2, \ldots\}$ are dissociate.*

We now obtain some elementary consequences of the definition of Riesz products. Let $\mu \in R(G)$ be based on Θ and a. A net $\{\omega_\alpha\} \subseteq \Omega(\Theta)$ is *tail-dissociate (relative to* Θ) if for every $\theta \in \Theta$ there exists an α_0 such that if $\alpha \geq \alpha_0$, then θ is not one of the letters of ω_α. We recall that $\bar{\Gamma}$ denotes the closure of Γ in $\Delta M(G)$. The weak$*$ topology on $L^\infty(\mu)$ is that induced by $L^1(\mu)$.

7.1.2. Proposition. *Let G be an infinite compact abelian group and let $\mu \in R(G)$ be based on Θ and a.*

(i) *Let $\psi \in \bar{\Gamma} \setminus \Gamma$, and $\hat{\mu}(\psi) \neq 0$. Then ψ_μ is a constant a.e. $d\mu$ if and only if there exists a tail-dissociate sequence $\{\omega_n\} \subseteq \Omega(\Theta)$ such that* $\lim \hat{\mu}(\omega_n) = \hat{\mu}(\psi)$, *in which case $\omega_n \to \psi_\mu$ weak$*$ in $L^\infty(\mu)$.*
(ii) *If $\sum_{\theta \in \Theta}(1 - |a(\theta)|) = \infty$, then there exists a tail-dissociate sequence $\{\omega_n\} \subseteq \Omega(\Theta)$ such that $\omega_n \to 0$ weak$*$ in $L^\infty(\mu)$.*

Proof. (i) Suppose that $\{\omega_\alpha\}$ is a tail-dissociate sequence in $\Omega(\Theta)$ such that $c = \lim \hat{\mu}(\omega_\alpha)$. If $\omega \in \Omega(\Theta)$, then eventually ω and ω_α have no letters in common. Therefore $\hat{\mu}(\omega\omega_\alpha) = \hat{\mu}(\omega)\hat{\mu}(\omega_\alpha)$ eventually. Therefore for all $\omega \in \Omega(\Theta)$,

$$\lim \int \bar{\omega}\bar{\omega}_\alpha \, d\mu = c \int \bar{\omega} \, d\mu. \tag{10}$$

To show that $\lim \omega_\alpha = c$, a.e. $d\mu$, it suffices to show that (10) holds when $\omega \notin \Omega(\Theta)$, that is, to show that

$$\lim \int \bar{\omega}\bar{\omega}_\alpha \, d\mu = 0 \quad \text{for } \omega \notin \Omega(\Theta). \tag{11}$$

If $\omega\omega_\alpha \notin \Omega(\Theta)$ for all $\alpha \geq \alpha_0$, then (11) holds. Thus we may assume that $\omega\omega_\beta \in \Omega(\Theta)$ for a cofinal subset $\{\omega_\beta\}$ of $\{\omega_\alpha\}$.

Since $\{\omega_\beta\}$ is tail-dissociate, we may inductively choose a sequence $\{\omega_k\}$ from $\{\omega_\beta\}$ and a sequence $\{\theta_k\}$ of distinct elements of Θ such that the letters in ω_n are those in the set $\{\theta_k: M(n) \leq k \leq N(n)\}$, the letters in $\omega\omega_n$ are included in $\{\theta_k: 1 \leq k \leq N(n)\}$, and $N(n) < M(n + 1)$. Thus for appropriate $\varepsilon(j, k)$, $\delta(j, k) \in \{-1, 0, 1\}$,

$$\omega\omega_j = \prod_{k=1}^{N(j)} \theta_k^{\varepsilon(j,k)} \tag{12}$$

and

$$\omega_j = \prod_{k=M(j)}^{N(j)} \theta_k^{\delta(j,k)}. \tag{13}$$

Thus

$$\omega = \prod_{k=1}^{M(j)-1} \theta_k^{\varepsilon(j,k)} \prod_{k=M(j)}^{N(j)} \theta_k^{\varepsilon(j,k)-\delta(j,k)} \quad \text{for } j = 1, 2, \ldots.$$

Since $\omega \notin \Omega(\Theta)$, there exists $k = k(j)$ between $M(j)$ and $N(j)$ such that $\theta_k^2 \neq 1$, and $\varepsilon(j,k) - \delta(j,k) = \pm 2$. By (12), $\omega\omega_j$ contains the letters $\theta_{k(1)}, \ldots, \theta_{k(j)}$. Since $|\hat{\mu}(\omega\omega_j)| \leq \prod_{m=1}^{j} |\hat{\mu}(\theta_{k(m)})| = 2^{-j}$, $\hat{\mu}(\omega\omega_j) \to 0$. Therefore $\omega_j \to \psi_\mu = c$ weak$*$ in $L^\infty(\mu)$. That completes one direction of (i).

Now suppose that ψ_μ is constant a.e. $d\mu$, $\psi \in \bar{\Gamma}\backslash\Gamma$, and $\hat{\mu}(\psi) \neq 0$. Let $\{\gamma_\alpha\} \subseteq \Gamma$ be a net with $\lim \gamma_\alpha = \psi$ weak$*$ in $L^\infty(\mu)$. Since $c = \hat{\mu}(\psi) \neq 0$, we may assume that $\{\gamma_\alpha\} \subseteq \Omega(\Theta)$. We claim that we may assume that $\{\gamma_\alpha\}$ is tail-dissociate. Indeed, otherwise there would exist $\theta \in \Theta$ and a cofinal subset $\{\omega_\beta\}$ of $\{\gamma_\alpha\}$ such that θ is a letter of each ω_β. We may assume then that each $\omega_\beta = \lambda\lambda_\beta$ where either $\lambda = \theta$ for all β or $\lambda = \bar{\theta}$ for all β and θ is not a letter of any λ_β. Then $\hat{\mu}(\lambda\lambda_\beta) = \hat{\mu}(\lambda)\hat{\mu}(\lambda_\beta) \to c$. Therefore $d = \lim \hat{\mu}(\lambda_\beta)$ exists and $\hat{\mu}(\lambda)d = c$. Since $\omega_\beta \to \psi$, $\bar{\lambda}\omega_\beta = \lambda_\beta \to \bar{\lambda}\psi$. Therefore $\hat{\mu}(\bar{\lambda}\omega_\beta) \to \hat{\mu}(\bar{\lambda}\psi) = \hat{\mu}(\bar{\lambda})\hat{\mu}(\psi) = \hat{\mu}(\bar{\lambda})c$. Thus $d = \hat{\mu}(\bar{\lambda})c = \hat{\mu}(\bar{\lambda})\hat{\mu}(\lambda)d$, so $|\hat{\mu}(\lambda)| = 1$, that is $\lambda = \theta = 1$, a contradiction. Therefore $\{\gamma_\alpha\}$ is tail-dissociate.

To replace $\{\gamma_\alpha\}$ by a sequence, we let $\omega_1 = \gamma_{\alpha(1)}$ where $\alpha(1)$ is arbitrary. After $\omega_1, \ldots, \omega_n$ are chosen, we let $\gamma_{\alpha(n+1)}$ be such that $\gamma_{\alpha(n+1)}$ has no letters in common with $\omega_1, \ldots, \omega_n$ and such that $|\hat{\mu}(\psi) - \hat{\mu}(\gamma_{\alpha(n+1)})| < 1/n$, and set $\omega_{n+1} = \gamma_{\alpha(n+1)}$. Then $\omega_n \to \psi$ weak$*$ in $L^\infty(\mu)$ by the first part of the proof of (i).

(ii) Since $\sum(1 - |a(\theta)|) = \infty$, there exists, for each finite set $\Phi \subseteq \Theta$ and each $\varepsilon > 0$ a finite set $\Phi' \subseteq \Theta$ such that $\prod\{|a(\theta)| : \theta \in \Phi\} < \varepsilon$. A simple induction completes the proof. □

7.1.3. Corollary. *Let G be a nondiscrete locally compact abelian group. Then $M(G)$ is not symmetric, that is, there exists a Hermitian measure $\mu \in M(G)$ such that $\hat{\mu}$ is not everywhere real-valued on $\Delta M(G)$.*

Proof. It is easy to see that if H is a quotient group of a closed subgroup of G and if $M(H)$ is not symmetric, then $M(G)$ is not symmetric. By the structure theorem, G has an open subgroup of the form $R^n \times D$, where D is compact and $n \geq 0$. Let us take the quotient H of $R^n \times D$ over $Z^n \times \{0\}$. It will be sufficient to show that $H = T^n \times D$ is such that $M(H)$ is non-symmetric.

Let Θ be any infinite dissociate subset of $\hat{H}$ and let μ be the Riesz product based on Θ and a where $a(\theta) = \frac{1}{2}$ for all $\theta \in \Theta$. Let $\psi \in \bar{\Gamma}\backslash\Gamma$ be any cluster point of Θ. Then, by Proposition 7.1.2, $\psi_\mu = \frac{1}{2}$ a.e. $d\mu$ since Θ is tail-dissociate.

Let $\rho = |\psi|^\alpha$ where $\alpha = 1 - i\pi/(2\log 2)$. Then $\hat{\mu}(\rho) = i/2$. Since $\hat{\mu}$ is real on Γ, μ is Hermitian, while $\hat{\mu}$ is not everywhere real-valued on $\Delta M(G)$. That proves 7.1.3. □

7.1.4. Proposition. *Let G be an infinite compact abelian group. Let $\mu \in R(G)$ be based on Θ and a.*

(i) *If $\sum_{\theta\in\Theta}(1 - |a(\theta)|) = \infty$, then μ is continuous.*
(ii) *If $\sum_{\theta\in\Theta}(1 - |a(\theta)|) < \infty$, then μ is either purely discrete or purely continuous, and both cases occur.*
(iii) *If* Gp Θ *has finite index in Γ and $\sum(1 - |a(\theta)|) < \infty$, then μ is discrete.*

Proof. (i) Suppose that $\sum(1 - |a(\theta)|) = \infty$. By 7.1.3 (ii), there exists $\psi \in \Gamma^- \backslash \Gamma$ such that $\psi_\mu = 0$ a.e. $d\mu$. But if $\mu_d \neq 0$, then $|\psi_{\mu_d}| = 1$ a.e. $d\mu_d$, so $\psi_\mu \neq 0$. Therefore $\mu_d = 0$ and μ is continuous.

(ii) and (iii). Suppose that

$$\sum_{\Theta}(1 - |a(\theta)|) < \infty. \tag{14}$$

The definition of $R(G)$ and (14) imply that $\Phi = \{\theta \in \Theta : \theta^2 \neq 1\}$ is finite and that Θ is countable. Since $\mu \ll \mu_\Phi$, it will be sufficient to prove that μ_Φ is either purely discrete or purely continuous, that is, we may assume that $\Phi = \varnothing$.

We argue by contradiction. Suppose that the subgroup $\Lambda = \text{Gp}\,\Theta$ has infinite index in Γ. Then the idempotent ω in $M(G)$ such that $\hat{\omega}$ is the characteristic function of Λ would be Haar measure on an infinite compact subgroup of G, and ω would be a continuous measure. Then $\omega * \mu = \mu$ would be continuous. That shows that the case of continuous μ can occur. We now assume that Λ has finite index in Γ. We set $H = \{x : \langle x, \lambda\rangle = 1, \lambda \in \Lambda\}$. Then H is finite and $p: G \to G/H$ maps μ to $\check{p}\mu$, which is the Riesz product on G/H based on Θ and a. Since H is finite, μ is continuous if and only if $\check{p}\mu$ is. We may therefore assume that $H = \{0\}$ and $\Lambda = \Gamma$. Since Θ consists only of elements of order two, we may write $\Theta = \{\theta_k : k = 1, 2, \ldots\}$ and $G = \prod_{k=1}^\infty Z_2(k)$, where $Z_2(k)$ is isomorphic to the two element group $\{0, x_k\}$ and the dual of $Z_2(k)$ is $\{1, \theta_k\}$ for $k = 1, 2, \ldots$. Also $\mu = \mathsf{X}_1^\infty[(1 + a_k)\delta(0) + (1 + a_k)\delta(x_k)]$.

Let $G_n = \{0\}^n \times \prod_{k=n+1}^\infty Z_2(k)$. Then for all $y \in G$,

$$\mu(\{y\}) = \lim_{n\to\infty} \mu(y + G_n) = \lim_{n\to\infty} \prod_{k=1}^n \frac{1}{2}(1 + a(\theta_k)\langle\theta_k, y\rangle).$$

Now, by (14), the above product converges to a non-zero number whenever $y \in G$ is such that $\langle\theta_k, y\rangle = \text{sgn}\, a(\theta_k)$ for all but a finite number of integers k.

Hence, μ is not purely continuous. Theorem 6.5.5(i) shows that μ is discrete. That ends the proof of the Proposition. □

The next result is a variant of 7.2.2. Its proof is simpler.

7.1.5. Proposition. *Let G be an infinite locally compact abelian group. Let $\mu \in R(G)$ be based on Θ and a. Then μ has strongly independent powers and every power of μ is singular with respect to every measure ω in $M_o(G)$ if either one of these conditions holds.*

(i) $\{|a(\theta)|: \theta \in \Theta\}$ *has a cluster point in* (0, 1).
(ii) $\sum\{1 - |a(\theta)|: \theta^2 = 1, |a(\theta)| > \frac{1}{2}\} = \infty$.

Proof. We leave to the reader the simple task of showing that in both cases (i) and (ii) there exists a tail-dissociate net $\omega_j \subseteq \Omega(\Theta)$ such that $c = \lim \hat{\mu}(\omega_j)$ and $0 < c \leq \frac{1}{2}$. Let $\psi \in \Gamma^- \backslash \Gamma$ be a cluster point of $\{\omega_j\}$. Then $\psi_\mu = c$ a.e. $d\mu$ by 7.1.2. Of course, $|\psi_{\delta(x)*\mu^m}| = c^m$ a.e. $d(\delta(x) * \mu^m)$, and therefore $\delta(x) * \mu^m$ and μ^n are mutually singular if $0 \leq m \neq n < \infty$. If $\omega \in M_o(G)$, then $\psi_\omega = 0$ a.e. $d\omega$. Therefore μ^n and ω are mutually singular for $n = 0, 1, 2, \ldots$. The Proposition is proved. □

Remarks and Credits. For classical results concerning Riesz products on the circle group, the reader may consult Zygmund [1, Volume 1, Section V.7]. Hewitt and Zuckerman [2] coined the term "dissociate" and proved 7.1.1. Proposition 7.1.2 is essentially Brown [3, 4.2]; the proof here being a modification of Brown's that was suggested by Saeki, the modification being necessitated by the following example of Saeki. Let $\Gamma = R_d$ and $\Theta = \{(\frac{3}{2})^n\}_1^\infty$. Then $-3 + (\frac{3}{2})^n \in \Omega(\Theta)$ for all n, while $-3 \notin \Omega(\Theta)$. Thus we can have $\gamma\omega_n \in \Omega(\Theta)$ for all n, where $\gamma \notin \Omega(\Theta)$ and ω_n is a tail-dissociate sequence in $\Omega(\Theta)$.

7.1.3 is due to Šreĭder [1], Hewitt [1] and Williamson [1]. Wiener and Pitt [1] proved a result implied by 7.1.3: there exists $\mu \in M(T)$ such that $\hat{\mu} \geq 1$ on Z and $1/\hat{\mu} \notin B(Z)$. The proof of 7.1.3 given here is due to Y. Meyer (See Padé [1, page 1454].) For other proofs of 7.1.3, see Section 8.2. Proposition 7.1.4 is a version of Brown [3, 3.4], and 7.1.5 is implicit in Brown [3, 3.4].

We conclude this section with references to some results related to Riesz products. For applications of Riesz products to Sidon sets, see Lòpez and Ross [1, pages 24–30]. A bounded function analogue of 7.2.2 is known: if $\Theta \subseteq \Gamma$ is dissociate and $a: \Theta \to R$, then $\prod [1 + ia(\theta)(\theta + \bar{\theta})]$ is a bounded function on G if $\sum a(\theta)^2 < \infty$; see A.1.3. If μ is a Riesz product on T based on a and $\{3^n\}$ and $|a(3^n)| \sum_1^n |a(3^k)| = O(1)$, then $\|\sum_{j=-m}^m \hat{\mu}(j)e^{ijx}\| = O(1)$; such a μ may be singular; see M. Weiss [1], Hewitt and Zuckerman [3], and Katznelson [5], the latter two for related results. If G is compact, $\varepsilon > 0$, and f is a trigonometric polynomial on G, then $f = \mu * \nu$ where $\|\mu\|\|\nu\| \leq (1 + \varepsilon)\|f\|$ and μ and ν are singular continuous measures. That is not difficult; see Graham and A. MacLean [1]. For the size of sets on which Riesz products may be concentrated, see Peyrière [2].

7.2. Orthogonality Relations for Riesz Products

Let $a: \Theta \to \mathbb{C}$ be any function on a set $\Theta \subseteq \Gamma$. We define $a'(\theta) = a(\theta)$ if $\theta^2 = 1$ and $a'(\theta) = 2a(\theta)$ if $\theta^2 \neq 1$. We use the convention that $0/0 = 0$.

7.2.1. Theorem. *Let G be an infinite compact abelian group. Let $\mu, \nu \in R(G)$ be based on Θ, a, and on Ψ, b. Extend a to be zero on $\Psi \backslash \Theta$ and b to be zero on $\Theta \backslash \Psi$. Then the following hold:*

(i) $\mu \perp \nu$ *if* $\sum_{\theta \in \Theta \cap \Psi} |a(\theta) - b(\theta)|^2 = \infty$.
(ii) $\nu \ll \mu$ *if* $\sum_{\theta \in \Theta \cup \Psi} |a(\theta) - b(\theta)|^2/(2 - |a'(\theta) + b'(\theta)|) < \infty$ *and* $|a(\theta)| < \frac{1}{2}$ *for all θ with $\theta^2 \neq 1$.*

Before proving 7.2.1, we shall draw a number of corollaries. The first was proved by Zygmund (see Zygmund [1, p. 209]) for $G = T$ and extended to more general compact abelian groups by Hewitt and Zuckerman [2].

7.2.2. Theorem. *Let G be an infinite compact abelian group and let $\mu \in R(G)$ be based on Θ and a. Then either*

(i) $\mu \perp L^1(G)$ *and* $\sum_{\theta \in \Theta} |a(\theta)|^2 = \infty$; *or*
(ii) $\mu \in L^1(G)$, $d\mu/dx \in L^2(G)$, *and* $\sum_{\theta \in \Theta} |a(\theta)|^2 < \infty$.

Proof of 7.2.2. Let us note that if $\Theta = \{\theta_n\}$, and $b(\theta_n) = 0$ for $n = 1, 2, \ldots$ then m_G is the Riesz product based on Θ and b. If $\sum |a(\theta)|^2 = \infty$, then $\sum |a(\theta) - b(\theta)|^2 = \infty$, so $\mu \perp \nu$ by 7.2.1(i). Since ν is Haar measure, 7.2.2(i) holds.

Now suppose that $\sum |a(\theta)|^2 < \infty$. Then

$$\sum_{\gamma} |\hat{\mu}(\gamma)|^2 = |\hat{\mu}(1)|^2 + \sum_{\substack{\omega \in \Omega(\Theta) \\ \omega \neq 1}} |\hat{\mu}(\omega)|^2$$

$$= 1 + \sum_{n=1}^{\infty} \frac{1}{n!} \sum \left\{ \prod_{j=1}^{n} |a(\theta_j)|^2 : \theta_i \neq \theta_j \in \Theta, 1 \leq i \neq j \leq n < \infty \right\}$$

$$\leq 1 + \sum_{n=1}^{\infty} \frac{1}{n!} \left(\sum_{\theta \in \Theta} |a(\theta)|^2 \right)^n = 1 + \exp(\sum |a(\theta)|^2) < \infty.$$

Therefore $\hat{\mu} \in L^2(\Gamma)$. Since $\hat{\mu} \in L^2(\Gamma)$ and $\hat{\mu} \in B(\Gamma)$, $\mu \in L^1(G)$. By the Plancherel Theorem, $d\mu/dx \in L^2(G)$. That ends the proof of 7.2.2. □

7.2.3. Corollary. *Let G be an infinite compact abelian group. Let $\mu \in R(G)$ be based on Θ and a. Then μ has independent powers if and only if*

$$(1) \qquad \sum\{|a(\theta)|^{2n}: |a(\theta)| \le \tfrac{1}{2}\} + \sum\{1 - |a(\theta)|: |a(\theta)| > \tfrac{1}{2}\} = \infty$$

for all integers $n \ge 1$.

Proof. If $\sum\{1 - |a(\theta)|: |a(\theta)| > \frac{1}{2}\} = \infty$, then μ is i.p. by 7.1.5. We may therefore assume that $\sum\{1 - |a(\theta)|: |a(\theta)| > \frac{1}{2}\} < \infty$. Then for all $m, n \ge 1$,

$$(2) \qquad \sum\{|a(\theta)^m - a(\theta)^n|^2: |a(\theta)| > \tfrac{1}{2}\} < \infty.$$

Suppose that (1) holds for some $n \ge 1$ and $0 \le m < n$. Then by (1) and (2),

$$\sum_{\theta \in \Theta} |a(\theta)^n - a(\theta)^m|^2 \ge \sum\{|a(\theta)|^{2m}(1 - 2^{n-m})^2: |a(\theta)| \le \tfrac{1}{2}\} = \infty.$$

Therefore $\mu^m \perp \mu^n$ by 7.2.1(i).

Now suppose that

$$\sum\{1 - |a(\theta)|): |a(\theta)| > \tfrac{1}{2}\} + \sum\{|a(\theta)|^{2n}: |a(\theta)| \le \tfrac{1}{2}\} < \infty$$

for some integer $n \ge 1$. We shall show that $\mu^{2n} \approx \mu^n$ by applying 7.2.1(ii). For each $\theta \in \Theta$ let

$$f(\theta) = |a(\theta)^{2n} - a(\theta)^n|^2/(2 - |a'(\theta)^{2n} + a'(\theta)^n|).$$

If $\theta^2 \ne 1$, $f(\theta) \le |a(\theta)|^{2n}(1 - (\frac{1}{2})^n)^2/(2 - (\frac{1}{2} + 1))$. Suppose that $\theta^2 = 1$ and $a(\theta)^n \ge 0$. Let $x = a(\theta)^n$. Then $f(\theta) = (1 - x)^2x^2/(2 - (1 + x)x) \le 1 - x$, as an elementary application of differential calculus will show. Similarly, if $\theta^2 = 1$ and $a(\theta)^n < 0$, we set $x = -a(\theta)^n$, and conclude that $f(\theta) \le 1 - x$. In both cases, we have $f(\theta) \le 1 - |a(\theta)|^n$.

That $\sum f(\theta) < \infty$ is now easily shown, thereby justifying the application of 7.2.1(ii). That ends the proof of 7.2.3. □

The reader may easily construct examples of Riesz products with independent powers. For example, if $\Theta = \{\theta_j\}_{j=1}^{\infty}$ is dissociate, let a be defined by $a(\theta_j) = (\log(j + 7))^{-1}$ for $j = 1, 2, \ldots$. Then the measure based on Θ and a is in $M_o(G)$ and has independent powers (Brown [3]).

7.2.4. Corollary. *Let G be a non-discrete locally compact abelian group, and $n \ge 1$. Then there exists a set*

$$\{\mu_t: 0 \le t \le 1\}$$

of pairwise singular probability measures such that

$$\mu_s^k * \mu_t^l \perp L^1(G) \quad \textit{for} \quad s, t \in [0, 1], k \geq 0, l \geq 0 \quad \textit{and} \quad k + l \leq n$$

and

$$\mu_s^k * \mu_t^l \in L^1(G) \quad \textit{for} \quad s, t \in [0, 1], k \geq 0, l \geq 0 \quad \textit{and} \quad k + l \geq m + 1.$$

Proof. An application of the structure theorem and A.7.1 shows that it will be sufficient to prove 7.2.4 for compact groups G.

Let $\{\theta_k\}_{k=1}^{\infty}$ be an infinite countable dissociate subset of the dual group Γ of G. Let $\{a_k\}$ be any real-valued sequence such that $0 < a_k < \frac{1}{2}$ for $k = 1, 2, \ldots, \sum_{k=1}^{\infty} a_k^{2n} = \infty$ and $\sum_{k=1}^{\infty} a_k^{2n+2} < \infty$. Let μ be the Riesz product based on $\{\theta_k\}$ and a where $a(\theta_k) = a_k$ for $k = 1, 2, \ldots$. An application of 7.2.2 shows that $\mu^n \perp L^1(G)$ and $\mu^{n+1} \in L^1(G)$. An application of 8.3.3 shows that $H = \{y \in G: \delta_y * \mu \not\perp \mu\}$ is a σ-compact Borel subgroup of zero Haar measure. By 6.3.14(i), G/H has cardinality at least c. Let $\{y(t): t \in [0, 1]\}$ be a subset of G such that $y(t) - y(s) \notin H$ whenever $0 \leq s < t \leq 1$. It is easy to see that

$$\{\mu_t = \delta_{y(t)} * \mu : 0 \leq t \leq 1\}$$

has the required properties. The Corollary is proved. □

Proof of 7.2.2(i). We assume that $\sum_{\Theta \cap \Psi} |a(\gamma) - b(\gamma)|^2 = \infty$, so there exists a countable subset $\{\theta_n\}_{n=1}^{\infty}$ of $\Theta \cap \Psi$ such that

$$\sum_{n=1}^{\infty} |a(\theta_n) - b(\theta_n)|^2 = \infty. \tag{3}$$

Let $\{c_n\}_{n=1}^{\infty}$ be a sequence of complex numbers and $1 \leq n(1) < n(2) \cdots$ be a strictly increasing sequence of natural numbers such that

$$c_n(a(\theta_n) - b(\theta_n)) \geq 0 \quad \text{for} \quad n = 1, 2, \ldots, \tag{4}$$

$$\sum |c_n|^2 < \infty, \tag{5}$$

and

$$\sum_{j=n(k)+1}^{n(k+1)} c_j(a(\theta_j) - b(\theta_j)) = 1 \quad \text{for} \quad k \geq 1. \tag{6}$$

That such sequences $\{c_j\}$ exists follows easily from (3). We define polynomials

$$f_k = \sum_{j=n(k)+1}^{n(k+1)} c_j(\bar{\theta}_j - b(\theta_j)). \tag{7}$$

Straightforward calculations show that

$$\int |f_k|^2\, d\nu = \sum_{n(k)+1}^{n(k+1)} |c_j|^2(1 - |b(\theta_j)|^2)$$

and

$$\int |f_k|^2\, d\mu = 1 + \sum_{n(k)+1}^{n(k+1)} |c_j|^2(1 - |a(\theta_j)|^2)$$

for all k. Thus

(8) $$\lim \int |f_k|^2\, d\nu = 0 \quad \text{and} \quad \lim \int |f_k|^2\, d\mu = 1.$$

We claim that

(9) $$f_k \to 1 \text{ in the norm of } L^1(\mu).$$

To see (9), note that

$$\left(\int |f_k - 1|\, d\mu\right)^2 \leq \int |f_k - 1|^2\, d\mu = \int |f_k|^2\, d\mu - 2\mathrm{Re} \int f_k\, d\mu + 1$$
$$= 1 + o(1) - 2\,\mathrm{Re} \sum_{n(k)+1}^{n(k+1)} c_j(a(\theta_j) - b(\theta_j)) + 1$$

by the Cauchy–Schwarz inequality and (8). Applying (6), we have $\int |f_k - 1|\, d\mu = o(1)$ as $k \to \infty$. That establishes (9).

By (8) there exists a Borel set F and a subsequence $f_{k(j)}$ with $\nu(F) = 1$ and $f_{k(j)} \to 0$ pointwise on F. Since $f_{k(j)} \to 1$ in $L^1(\mu)$, $\mu(F) = 0$. Therefore $\mu \perp \nu$. That completes the proof of 7.2.1(i). □

7.2.5. Lemma. *Let ω be a regular Borel measure on G and let $f, g \in L^1(\omega)$. Suppose that $f\omega$, $g\omega$, and ω are all probability measures. Then*

$$\left(\int |f - g|\, d\omega\right)^2 \leq 4\left(1 - \left(\int f^{1/2} g^{1/2}\, d\omega\right)^2\right).$$

Proof. By the Cauchy–Schwarz inequality,

$$\left(\int |f - g|\, d\omega\right)^2 \leq \left(\int |f^{1/2} - g^{1/2}|^2\, d\omega\right)\left(\int |f^{1/2} + g^{1/2}|^2\, d\omega\right)$$
$$= \left(2 - 2\int f^{1/2} g^{1/2}\, d\omega\right)\left(2 + 2\int f^{1/2} g^{1/2}\, d\omega\right)$$
$$= 4\left(1 - \left(\int f^{1/2} g^{1/2}\, d\omega\right)^2\right). \quad \square$$

For the next lemma and for the proof of 7.1.2(ii), we shall need some *ad hoc* notation. We suppose that $\Theta = \Psi = \{\theta_n\}_{n=1}^{\infty}$. For $n \geq 1$, let $a_n = a(\theta_n)$ and $b_n = b(\theta_n)$, and let

$$q_n = 1 + a_n\theta_n,\ q'_n = 1 + b_n\theta_n \quad \text{if} \quad \theta_n^2 = 1, \tag{10}$$

$$q_n = 1 + a_n\theta_n + \overline{a_n\theta_n},\ q'_n = 1 + b_n\theta_n + \overline{b_n\theta_n} \quad \text{if } \theta_n^2 \neq 1, \tag{11}$$

$$p_n = \prod_{j=1}^{n} q_j, \quad \text{and} \quad p'_n = \prod_{j=1}^{n} q'_j. \tag{12}$$

Then, of course, μ is the weak$*$ limit of $p_n m_G$ and ν is the weak-$*$ limit of $p'_n m_G$.
Let $n \geq 1$, $r \geq 0$ and $s \geq 0$ be integers. We set

$$h(n, r, s) = p'_n \prod_{k=n+r+1}^{n+r+s} q_k \tag{13}$$

and

$$I(n, r, s) = \int \prod_{k=n+1}^{n+r} (q_k q'_k)^{1/2} h(n, r, s) dy. \tag{14}$$

(The empty product is taken to be one.) Then $h(n, r, s)$ is a non-negative function on G. By the Cauchy-Schwarz inequality, $0 \leq I(n, r, s) \leq 1$ for all n, r, s.

7.2.6. Lemma. *Suppose that* $\lim_{n\to\infty}(\inf_{r,s} I(n, r, s)) = 1$. *Then* ν *is absolutely continuous with respect to* μ *and* $\lim p'_n/p_n = d\nu/d\mu$ *in the norm of* $L^1(\mu)$.

Proof. Because $|p_m| > 0$ everywhere for all m (by the hypothesis on $a(\theta)$), $(p'_n/p_n)\mu$ is the weak$*$ limit of the measures $(p'_n/p_n)p_{n+s}m_G$ as s tends to infinity. Therefore

$$\begin{aligned}\int |p'_n/p_n - p'_{n+r}/p_{n+r}| d\mu &= \lim_{s\to\infty} \int |p'_n/p_n - p'_{n+r}/p_{n+r})| p_{n+r+s}\, dm_G \\ &= \lim_{s\to\infty} \int \left| \prod_{k=n+1}^{n+r} q_k - \prod_{k=n+1}^{n+r} q'_k \right| p'_n \prod_{k=n+r+1}^{n+r+s} q_k\, dm_G.\end{aligned}$$

Let $\omega = h(n, r, s)m_G$. Then Lemma 7.2.5 implies that

$$\int \left[\left| \prod_{k=n+1}^{n+r} q_k - \prod_{k=n+1}^{n+r} q'_k \right| p'_n \prod_{k=n+r+1}^{n+r+s} q_k \right] dm_G \leq 4(1 - I(n, r, s)^2). \tag{15}$$

(To apply 7.2.5, note that $\omega = h(n, r, s)$ is a Riesz product, as are $\prod_{n+1}^{n+r} q_k\omega$ and $\prod_{n+1}^{n+r} q'_k\omega$.) By the hypothesis, the last term in (15) tends to zero as n increases. Therefore p'_n/p_n is a Cauchy sequence in $L^1(\mu)$. Since $(p'_n/p_n)\mu$ converges to ν in the weak* topology, we see that $\lim(p'_n/p_n)\mu = \nu$ in $M(G)$ norm. Therefore $\nu \ll \mu$, and the Lemma is proved. □

Proof of 7.2.1(ii). For $n, r \geq 1$, straightforward computations show that

$$(16)\quad \prod_{n+1}^{n+r}(q_k q'_k)^{1/2} = \prod_{k=n+1}^{n+r}[(\tfrac{1}{2}q_k + \tfrac{1}{2}q'_k)(1 - (q_k - q'_k)^2/(q_k + q'_k)^2)^{1/2}]$$

$$\geq \prod_{n+1}^{n+r}[(\tfrac{1}{2}q_k + \tfrac{1}{2}q'_k)(1 - (q_k - q'_k)^2/(q_k + q'_k)^2)]$$

$$\geq \left[\prod_{n+1}^{n+r}(\tfrac{1}{2}q_k + \tfrac{1}{2}q'_k)\right]\left[1 - \prod_{n+1}^{n+r}(q_j - q'_j)^2/(q_j + q'_j)^2\right]$$

$$\geq \prod_{k=n+1}^{n+r}(\tfrac{1}{2}q_k + \tfrac{1}{2}q'_n) - \sum_{j=n+1}^{n+r}\left[\left(\prod_{\substack{i=n+1\\ i\neq j}}^{n+r}(\tfrac{1}{2}q_i + \tfrac{1}{2}q'_i)\right)(q_j - q'_j)^2\Big/\inf_x(q_j(x) + q'_j(x))\right].$$

But $\inf_x(q_j(x) + q'_j(x)) \geq 2 - |a'(\theta_j) + b'(\theta_j)|$ and

$$(17)\quad \int \prod_{\substack{i=n+1\\ i\neq j}}^{n+r}(\tfrac{1}{2}q_i + \tfrac{1}{2}q'_i)(q_j - q'_j)^2 h(n, r, s)dy = |a(\theta_j) - b(\theta_j)|^2.$$

Therefore

$$I(n, r, s) \geq \int \prod_{n+1}^{n+r}(\tfrac{1}{2}q_k + \tfrac{1}{2}q'_k)h(n, r, s)dy$$

$$- \sum_{j=n+1}^{n+r}(2 - |a'(\theta_j) + b'(\theta_j)|)^{-1}$$

$$\times \int \prod_{\substack{i=n+1\\ i\neq j}}^{n+r}(\tfrac{1}{2}q_i + \tfrac{1}{2}q'_i)(q_j - q'_j)^2 h(n, r, s)dy$$

$$= 1 - \sum_{j=n+1}^{n+r}|a(\theta_j) - b(\theta_j)|^2/(2 - |a'(\theta_j) + b'(\theta_j)|).$$

It is now clear from the hypothesis of 7.2.1(ii) that $\lim_{n\to\infty}\inf_{r,s} I(n, r, s) = 1$. By 7.2.6, $\nu \ll \mu$. (The reader will note that in the case that $2 - |a'(\theta_j) + b'(\theta_j)| = 0$, there is no contribution to the summands in (16) or (17), so our use of the convention $0/0 = 0$ is justified.)

When Θ and Ψ are equal but uncountable, then it is easy to see that $a = b$ except on a countable subset $\{\theta_n\}_{n=1}^{\infty}$ and the arguments given above apply in this case without change.

Now suppose that $\Theta \neq \Psi$. Set $\Theta_1 = \Theta \cap \Psi$ and let a_1, b_1 be the restrictions of a and b respectively to Θ_1, and μ_1 the Riesz product based on Θ_1, a_1. Then the proof for the case $\Theta = \Psi$ shows that $\nu \ll \mu_1$ and $\mu \approx \mu_1$. Therefore $\nu \ll \mu$ and the proof of Theorem 7.2.1 is finished. $\square$

Remarks and Credits. Brown and Moran [8] essentially proved 7.2.1(i); a weaker version appears in Peyrière [1]. The proof of 7.2.1(i) given here is Brown and Moran's, as modified by a suggestion of Saeki. Brown and Moran [8] proved a weaker version of 7.2.1(ii), the one requiring $\sum|a(\theta) - b(\theta)|^2/(2 - |a'(\theta) + b'(\theta)|)^2 < \infty$. The version here appears in Ritter [1, (4.5)]; our proof is Brown and Moran's, except that we use the estimate (6) of Ritter [1, p. 262] at our formula (16). Brown and Moran and Ritter claim that $\mu \approx \nu$, a claim that is false, for example on $\prod Z_4$ (arrange things so that $q_1' = 0$ on a large set, while $q_j = q_j'$ for $j \geq 2$). The convergence condition in 7.2.1(ii) is the best possible, as can be deduced from Brown and Moran [8, formula (25)]. When G is an infinite product group and the elements of Θ are in distinct summands of Γ, then two Riesz products based on Θ are either mutually singular or equivalent. That follows from 6.4.1. The conditions for mutual singularity depend both on the coefficient functions a and b and on the factors of G. See Brown and Moran [8, Section 5] for a discussion.

Whether two Riesz products on T are either equivalent or mutually singular is unknown. Ritter [2, p. 88] shows that that dichotomy holds for products μ based on $\Theta = \{r^n\}_1^{\infty}$. Here is a proof of that. For $n \geq 1$, let H_n be the group of (r^n)th roots of unity, and let $D = \bigcup H_n$. Then μ is D-ergodic by Brown [3, Theorem 5], and $\delta(x) * \mu \approx \mu$ for all $x \in D$, by an application of 7.2.1(ii). Now 6.1.8(iii) yields the result.

Corollary 7.2.3 is from Brown [3]. Corollary 7.2.4 is a variant of Stromberg [2]. Wiener and Wintner [1] proved that there exists a singular probability measure $\mu \in M(T)$ with $\mu * \mu \in L^1(T)$; Hewitt and Zuckerman [2] showed that every non-discrete abelian group supports such a measure. Varopoulos [3] contains examples of i.p. Riesz products. For orthogonality of generalized Riesz products, see 7.5.6 and the Remarks and Credits paragraph of Section 7.5.

7.3. Most Riesz Products Are Tame

In this section we give a necessary and sufficient condition for a Riesz product $\mu \in R(G)$ to be tame and applications of that result.

Let G be an infinite compact abelian group. We let $R_t(G)$ be the set of measures $\mu \in R(G)$ such that $\limsup_{\gamma \to \infty} |\hat{\mu}(\gamma)| < 1$.

7.3.1. Theorem. *Let G be an infinite compact abelian group. Let $\mu \in R(G)$. Then μ is tame if and only if $\mu \in R_t(G)$.*

7.3.2. Corollary. *Let G be a non-discrete locally compact abelian group. Then there exists a tame, Hermitian independent power probability measure in $M_o(G)$.*

Proof of Corollary. A *Taylor–Johnson measure* is a tame Hermitian independent power probability measure. It is easy to see that if H is an open subgroup of G and if $M_o(H)$ contains a Taylor–Johnson measure, then $M_o(G)$ does also. (See 6.8.2 for the proof.) By the structure theorem, there exists an open subgroup H of G of the form $R^n \times K$ where $n \geq 0$ and K is compact. If $n = 0$, then there exist Riesz products on K that have the required properties. (We simply choose $\mu \in R(K) \cap M_o(K)$ such that μ has independent powers and μ is Hermitian. Then μ is tame by 7.3.1.)

If $n \geq 1$, let $\mu_1 \in M_o(T^n \times K)$ be a Taylor–Johnson measure, which exists by 7.3.1. Let μ be a lifting of μ_1 to $R^n \times K$ given by A.7.1. Then μ has the required properties. The proof of the Corollary is finished. □

7.3.3. Lemma. *Let Θ be an infinite dissociate subset of the abelian group Γ. Then*

$$\bigcap \{\Omega(\Theta\backslash\Phi)\cdot\Omega(\Theta\backslash\Phi)\colon \Phi \textit{ is finite}\} = \{1\}. \tag{1}$$

Proof. Let $\gamma = \omega_1\omega_2 \in \Omega(\Theta)\cdot\Omega(\Theta)$, where $\omega_1, \omega_2 \in \Omega(\Theta)$. Let Φ be the set of letters of ω_1 and ω_2. We claim that if $\gamma \in \Omega(\Theta\backslash\Phi)\cdot\Omega(\Theta\backslash\Phi)$, then $\gamma = 1$. That will be sufficient to prove the Lemma.

Indeed, if $\omega_3, \omega_4 \in \Omega(\Theta\backslash\Phi)$ are such that $\gamma = \omega_3\omega_4$, then

$$\omega_1\omega_3^{-1} = \omega_4\omega_2^{-1}.$$

Now ω_1 and $\omega_2^{-1} \in \Omega(\Phi)$ and ω_3^{-1} and $\omega_4 \in \Omega(\Theta\backslash\Phi)$; therefore ω_1 and ω_3^{-1} (and ω_2^{-1} and ω_4) have no letters in common. The uniqueness of expression of elements of $\Omega(\Theta)$ now implies that the letters and exponents of ω_1 are exactly those of ω_2^{-1}; that is, $\omega_1 = \omega_2^{-1}$ and therefore $\gamma = 1$. The Lemma is proved. □

For the next Lemma and the proof of 7.3.1, we remind the reader that $\hat{\mu}(\varphi) = \int \overline{\varphi}_\mu \, d\mu$ for all $\varphi \in \Delta M(G)$, $\mu \in M(G)$.

7.3.4. Lemma. *Let Θ be an infinite dissociate subset of the discrete group Γ. Let G be the dual group of Γ and let $\mu \in R_t(G)$ be a Riesz product based on Θ. Let $\varphi \in \Delta M(G)$ with $\hat{\mu}(\varphi) \neq 0$.*

(i) *For each finite subset Φ of Θ there exists a unique $\omega_\Phi = \omega_{\Phi,\varphi} \in \Omega(\Phi)$ such that $\hat{\mu}_\Phi(\overline{\omega}_\Phi\varphi) \neq 0$ and $\hat{\mu}(\varphi) = \hat{\mu}(\omega_\Phi)\hat{\mu}_\Phi(\overline{\omega}_\Phi\varphi)$.*

(ii) *There exists a finite subset Φ_o of Θ and $\omega_o \in \Omega(\Phi_0)$ such that $\omega_\Phi = \omega_o$ for all finite subsets Φ of Θ that contain Φ_o.*

(iii) *For ω_o as in (ii) and $\psi = \bar{\omega}_o \varphi$, we have*

$$\hat{\mu}_\Phi(\psi) = \hat{\mu}(\psi)$$

for all finite subsets Φ of Θ.

(iv) *For ψ as in (iii), we have $\hat{\mu}(\gamma\psi) = \hat{\mu}(\gamma)\hat{\mu}(\psi)$ for all $\gamma \in \Gamma$.*

Proof. (i) From 7.1, formula (9), we obtain

$$\hat{\mu}(\varphi) = \sum\{\hat{\mu}(\omega)\hat{\mu}_\Phi(\bar{\omega}\varphi): \omega \in \Omega(\Phi)\}.$$

It is easy to see that if ω, ω' are distinct elements of $\Omega(\Phi)$, then

$$[\omega \cdot \Omega(\Theta \backslash \Phi)] \cap [\omega' \cdot \Omega(\Theta \backslash \Phi)] = \varnothing.$$

Therefore $(\omega\mu_\Phi) * (\omega'\mu_\Phi) = 0$ for all pairs $\omega \neq \omega'$ of elements of $\Omega(\Phi)$. Thus $\hat{\mu}(\omega)\hat{\mu}_\Phi(\bar{\omega}\varphi)$ is non-zero for exactly one element $\omega = \omega_\Phi$ of $\Omega(\Phi)$ and therefore $\hat{\mu}(\varphi) = \hat{\mu}(\omega_\Phi)\hat{\mu}_\Phi(\bar{\omega}_\Phi \varphi)$.

(ii) For $\omega \in \Omega(\Theta)$, let $l(\omega)$ denote the number of letters of ω. Since $\mu \in R_t(G)$, $\sup\{|\hat{\mu}(\theta)|: \theta \in \Theta\} = d < 1$. Now (i) implies that

$$0 < |\hat{\mu}(\varphi)| \leq |\hat{\mu}(\omega_\Phi)| \leq d^{l(\omega_\Phi)}$$

for all finite subsets Φ of Θ. Therefore there exists a finite subset Φ_o of Φ such that

$$l(\omega_{\Phi_o}) = \sup\{l(\omega_\Phi): \Phi \subseteq \Theta, \Phi \text{ finite}\}.$$

Let us set $\omega_o = \omega_{\Phi_o}$. We shall show that ω_o and Φ_o have the required property.

Suppose that Φ is a finite subset of Θ containing Φ_o. By part (i) applied to $\mu_{\Phi_o}, \bar{\omega}_o\varphi$, and $\Phi\backslash\Phi_o$,

$$0 \neq \hat{\mu}_{\Phi_o}(\bar{\omega}_o \varphi) = \hat{\mu}_{\Phi_o}(\omega)\hat{\mu}_\Phi(\bar{\omega}\bar{\omega}_o \varphi)$$

for some $\omega \in \Omega(\Phi\backslash\Phi_o)$. Since $\omega\omega_o \in \Omega(\Phi)$, $\omega_\Phi = \omega\omega_o$ by the uniqueness of ω_Φ. But $l(\omega_\Phi) = l(\omega) + l(\omega_o) = l(\omega_o)$, and therefore $l(\omega) = 0$; that is, $\omega = 1$. Therefore $\omega_\Phi = \omega_o$. That proves (ii).

(iii) Let Φ be a finite subset of Θ. We claim that $\hat{\mu}_\Phi(\bar{\omega}_o \varphi) = \hat{\mu}_{\Phi_o}(\bar{\omega}_o \varphi)$. If $\Phi \supseteq \Phi_o$ that is obvious from (ii). Otherwise, set $\Phi_1 = \Phi \cup \Phi_o$. The only element ω of $\Omega(\Phi_1 \backslash \Phi)$ such that $\hat{\mu}_{\Phi_1}(\bar{\omega}\bar{\omega}_o \varphi) \neq 0$ is $\omega = 1$ by (ii) and therefore (using 7.1 (9) again) $\hat{\mu}_{\Phi_1}(\bar{\omega}_o \varphi) = \hat{\mu}_\Phi(\bar{\omega}_o \varphi) = \hat{\mu}_{\Phi_o}(\bar{\omega}_o \varphi)$. We may apply those equalities even to the empty set $\Phi = \varnothing$, thus obtaining $\hat{\mu}(\bar{\omega}_o\varphi) = \hat{\mu}_\Phi(\bar{\omega}_o\varphi)$ for all finite subsets Φ of Θ. That proves (iii).

(iv) Let $\omega \in \Omega(\Theta)$ be fixed. Let Φ be a finite subset of Θ such that $\omega \in \Omega(\Phi)$. Then 7.1 (9) implies that $\hat{\mu}(\bar{\omega}\psi) = \hat{\mu}(\bar{\omega})\hat{\mu}_\Phi(\psi)$. Applying (iii), we have $\hat{\mu}(\bar{\omega}\psi) = \hat{\mu}(\bar{\omega})\hat{\mu}(\psi)$. Thus (iv) holds for elements $\gamma = \bar{\omega}$ of $\Omega(\Theta)$.

Suppose that $\gamma \notin \Omega(\Theta)$. We claim that $\hat{\mu}(\gamma\psi) = 0$. Otherwise (i)–(iii), applied to $\varphi = \gamma\psi$, show that there exists $\omega_1 \in \Omega(\Theta)$ such that $\hat{\mu}_\Phi(\bar{\omega}_1\gamma\psi) \neq 0$ for all finite subsets Φ of Θ. Set $\tau = \omega_1\bar{\gamma}$. Then

$$[(\bar{\tau}\mu_\Phi) * \mu_\Phi]\hat{\ }(\psi) = \hat{\mu}_\Phi(\tau\psi)\hat{\mu}_\Phi(\psi) \neq 0.$$

Therefore $\bar{\tau}\mu_\Phi * \mu_\Phi \neq 0$ for all finite subsets Φ of Θ. Therefore $[\bar{\tau} \cdot \Omega(\Theta\backslash\Phi)] \cap [\Omega(\Theta\backslash\Phi)] \neq \varnothing$ for all finite subsets Φ of Θ, that is, $\bar{\tau} \in \Omega(\Theta\backslash\Phi) \cdot \Omega(\Theta\backslash\Phi)$ for all finite subsets Φ of Θ. By 7.3.3, $\tau = 1$; that is $\bar{\omega}_1\gamma = \tau = 1$ and $\gamma \in \Omega(\Theta)$. That proves that $\hat{\mu}(\gamma\psi) = 0$ for all $\gamma \notin \Omega(\Theta)$. Now $\hat{\mu}(\gamma\psi) = \hat{\mu}(\gamma)\hat{\mu}(\psi)$ follows when $\gamma \notin \Omega(\Theta)$. The proof of Lemma 7.3.4 is finished. □

Proof of Theorem 7.3.1. Suppose that $\mu \in R_t(G)$, $\varphi \in \Delta$ and $\varphi_\mu \not\equiv 0$ a.e. $d\mu$. Then there exists $\gamma_o \in \Gamma$ such that $\hat{\mu}(\bar{\gamma}_o\varphi) \neq 0$. By 7.3.4, there exists $\omega_o \in \Omega(\Theta)$ such that $\hat{\mu}(\gamma\bar{\omega}_o\bar{\gamma}_o\varphi) = \hat{\mu}(\gamma)\hat{\mu}(\bar{\omega}_o\bar{\gamma}_o\varphi)$ for all $\gamma \in \Gamma$. It follows at once that

$$\omega_o\gamma_o\bar{\varphi} = \hat{\mu}(\bar{\omega}_o\bar{\gamma}_o\varphi) = b \quad \text{a.e. } d\mu;$$

that is, $\varphi = \bar{b}\omega_0\gamma$ a.e. $d\mu$. One direction of the proof of 7.3.1 is finished. Here is the other.

Implicit in the hypothesis that $\mu \in R(G)\backslash R_t(G)$ is that there exists an infinite subset $\Psi = \{\theta_j\}_{j=1}^\infty$ of Θ such that every element of Ψ has order two and such that

$$\prod_{j=1}^{\infty} a(\theta_j) > 0. \tag{2}$$

Let $\gamma_n = \theta_1 \cdots \theta_n$, and $a_n = a(\theta_n)$ for $n \geq 1$. Then by the Cauchy-Schwarz inequality and the fact that $\theta_n^2 = 1$ for all n,

$$\left(\int |\gamma_n - \gamma_{n+p}| d\mu\right)^2 \leq \int 2(1 - \bar{\gamma}_n\gamma_{n+p})d\mu = 2\left(1 - \prod_{j=n+1}^{n+p} a_j\right), \tag{3}$$
$$\text{for} \quad 1 \leq n, p < \infty.$$

(2) and (3) show that $\{\gamma_n\}$ converges in $L^1(\mu)$ to a function φ. Therefore there exists an element (which we call φ also) of $\bar{\Gamma}$ such that $\lim \gamma_n = \varphi_\mu$ in $L^1(\mu)$. It follows at once that $|\varphi_\mu| = 1$ a.e. $d\mu$. Suppose that φ_μ were of the form $\alpha\gamma$ for some complex number α and $\gamma \in \Gamma$. Then $|\alpha| = 1$ and $\gamma \in \Omega(\Theta)$. Of

course, there exists $q \geq 1$ such that θ_n is not a letter of γ for all $n \geq q$ and θ_q is a letter of γ. Therefore $|\int \alpha\gamma\gamma_n \, d\mu| \leq a_q$ if $n \geq q$. Then

$$|\alpha\hat{\mu}(1)| = \left| \int \alpha\gamma\bar{\varphi} \, d\mu \right| \leq \lim_{n \to \infty} \left| \int \gamma\bar{\gamma}_n \, d\mu \right| < a_q < 1.$$

Therefore $\alpha\gamma \neq \varphi_\mu$. That proves that φ_μ is not of the form $\alpha\gamma$, contrary to our assumption. Theorem 7.3.1 is proved. □

7.3.5. Proposition. *Let G be an infinite compact abelian group and let $\mu \in R_t(G)$. Let B be the L-subalgebra of $M(G)$ that is generated by μ. Let $\varphi \in \Delta B$. Then there exists $\gamma \in \Gamma$ and a complex number a such that $\varphi_\mu = a\gamma$* a.e. $d\mu$.

Proof. In view of 7.3.1, it would be sufficient to show that $\mu_\Phi \in B$ for all finite subsets Φ of Θ, and then apply the arguments used above. But it may be the case that some μ_Φ is not absolutely continuous with respect to μ. We therefore let A be the L-subalgebra of $M(G)$ that is generated by the set $\{\mu_\Phi : \Phi$ is a finite subset of $\Theta\}$. Then $A * B \subseteq A$; that is, B is an L-ideal in A. Therefore, the maximal ideal space of B is a *subset* of the maximal ideal space of A. Now, the proof of 7.3.1 applies to A and the proof of 7.3.5 is finished. □

We now show how the methods of this Section may be applied to prove a symbolic calculus theorem. For another proof, see Section 9.5.

A complex-valued function F on $[-1, 1]$ *operates* on a subalgebra N of $M(G)$ if $F \circ \hat{\mu}$ is a Fourier-Stieltjes transform whenever $\mu \in N$ and $\hat{\mu}(\Gamma) \subseteq [-1, 1]$. Here Γ is the dual group of G.

7.3.6. Theorem. *Let G be an infinite compact abelian group and let μ be any Hermitian measure in $R_t(G) \backslash \mathrm{Rad}\ L^1(G)$. Suppose that $F : [-1, 1] \to \mathbb{C}$ operates on the subalgebra of $M(G)$ that is generated by μ and $L^1(G)$. Then F coincides with an entire function in a neighborhood of zero.*

7.3.7. Corollary. *Let G be a non-discrete locally compact abelian group. If F operates on $M_o(G)$, then F coincides with an entire function in a neighborhood of zero.*

Proof of 7.3.7. An application of the structure theorem shows that it will be sufficient to prove the Corollary in the case that G is of the form $R^n \times K$, where K is a compact abelian group.

It is easy to see that we may assume that $n = 0$ (we restrict ourselves to the restrictions of $B(R^n \times \hat{K})$ to $Z^n \times K$). We may thus assume that G is compact. But then $M_o(G)$ contains an element of $R_t(G) \backslash \mathrm{Rad}\ L^1(G)$ and the Corollary follows. □

Proof of 7.3.6. Let μ be based on Θ and a. We may assume that $F(0) = 0$. By 9.3.1 and 9.3.2, there exist numbers $\delta > 0$ and $b_1, b_2, \ldots$ such that

(4) $$F(t) = \sum_0^\infty b_n t^n \quad \text{for } |t| < 2\delta$$

and $\sum |b_n||t|^n < \infty$ for $|t| < 2\delta$. It will suffice to show that $b_n = o(K^{-n})$ for all $K > 0$. Let $K > 1$ be fixed. Let $j > 1$ be an integer such that $\nu = K(\mu^j - m_G)$ satisfies $\hat{\nu}(\Gamma) \subseteq [-\delta, \delta]$. Then ν belongs to the algebra generated by μ and m_G, and ν has spectral radius K. Therefore for each $\omega \in \Omega(\Theta)$, the sum $\sum_1^\infty b_n \hat{\nu}(\omega)^n K^{-n} \nu^n$ belongs to $M(G)$. For $\omega, \omega' \in \Omega(\Theta)\backslash\{1\}$ with no letters in common, $\hat{\nu}(\omega\omega') = \hat{\nu}(\omega)\hat{\nu}(\omega')/K$. We claim that if Φ is a finite subset of Θ and $\omega \in \Omega(\Phi)\backslash\{1\}$, then

(5)

$$(\bar{\omega}F \circ \nu) * \mu_\Phi = \left(\sum b_n \hat{\nu}(\omega)^n K^{-n} \nu^n\right) * \mu_\Phi + \left(F \circ \hat{\nu}(\omega) - \sum b_n \hat{\nu}(\omega)^n K^{-n}\right) m_G.$$

Indeed, if $\omega' \in \Omega(\Theta\backslash\Phi)\backslash\{1\}$, then

$$(\bar{\omega}F \circ \nu)\hat{}\,(\omega') = \sum_{n=1}^\infty b_n \hat{\nu}(\omega)^n \hat{\nu}(\omega')^n K^{-n}.$$

Therefore (5) holds.

Let $\gamma \in \Gamma\backslash\Omega(\Theta)$. Then the proof of 7.3.3 shows that there exists a finite subset $\Phi \subseteq \Theta$ such that

(6) $$(\bar{\gamma}F \circ \nu) * \mu_\Phi = 0.$$

Let $0 < s < \delta/K$. By 6.1.6 and 7.3.1, there exists $\psi \in \Delta M(G)$ such that $\psi_\mu = s^{1/j}$ a.e. $d\mu$. By replacing μ by μ^2, we may assume that $|a(\theta)| < \frac{1}{2}$ for all $\theta^2 \neq 1$, and, hence, that $\mu_\Phi \ll \mu$ for all finite sets $\Phi \subseteq \Theta$. Therefore $\psi_{\mu_\Phi} = s^{1/j} = \hat{\mu}_\Phi(\psi)$ for all Φ. Let $\omega \in \Omega(\Theta)\backslash\{1\}$, and let Φ be the set of letters of ω. Then (5) implies that

$$\begin{aligned}[(\bar{\omega}F \circ \nu) * \mu_\Phi]\hat{}\,(\psi) &= \sum_1^\infty b_n \hat{\nu}(\omega)^n K^{-n} \hat{\nu}^n(\psi) s^{1/j} \\ &= \sum_1^\infty b_n \hat{\nu}(\omega)^n s^{(n+1)/j}\end{aligned}$$

and (6) implies that for all $\gamma \in \Gamma\backslash\Omega(\Theta)$, there exists Φ with $(\bar{\gamma}\psi F \circ \nu) * \psi\mu_\Phi = 0$. Therefore

$$[(\bar{\gamma}F \circ \nu) * \mu_\Phi]\hat{}\,(\psi) = \sum_1^\infty b_n \hat{\nu}(\gamma)^n s^{(n+1)/j}, \ \gamma \in \Gamma\backslash\{1\}.$$

Since $\mu_\Phi{}^\wedge(\psi) = s^{1/j}$ for all Φ, we see that

$$(\bar{\psi}F \circ v)^\wedge(\gamma) = (\bar{\gamma}F \circ v)^\wedge(\psi) = \sum_1^\infty b_n \hat{v}(\gamma)^n s^n \quad \text{for all } \gamma \in \Gamma\backslash\{1\}.$$

Therefore, for some constants c and c',

$$\text{(7)} \qquad \bar{\psi}F \circ v = \sum_1^\infty b_n s^n v^n + cm_G = \sum_1^\infty b_n s^n K^n \mu^{nj} + c'm_G.$$

Since μ is i.p.,

$$\text{(8)} \qquad F \circ v = \sum a_n \mu^n + \tau$$

where $\tau \perp \sum_1^\infty \mu^n/n!$, and the a_n are unique. Of course $\bar{\psi}F \circ v = \sum a_n s^n \mu^n + \bar{\psi}\tau$. Therefore $a_n s^n K^n$ and $b_n = a_n K^{-n}$ for $n \geq 1$. Since $\sum |a_n| < \infty$, $b_n = o(K^{-n})$. That proves the Theorem. □

Credits. Theorem 7.3.1, Proposition 7.3.5 and Theorem 7.3.6 are due to Brown [3], though his proofs must be modified to apply to general compact groups. (The modification is not necessary for $G = T$; it is needed for $G = bR$ and $\Theta = \{(\frac{3}{2})^n\}_1^\infty$.) The modifications used here were provided by Saeki. Varopoulos [3] first proved 7.3.7 (and 7.3.6 for certain Riesz products); the basic ideas of his complicated proof are exposed in Varopoulos [2]. For another proof of 7.3.6, see Section 9.5. Corollary 7.3.2 is due to Brown [2] and Moran [2]; see Section 6.8 for another proof.

7.4. A Singular Measure in $M_o(G)$ That Is Equivalent to Its Square

7.4.1. Theorem. *Let G be a non-discrete locally compact abelian group. Then there exists a singular Hermitian probability measure $\mu \in M_o(G)$ such that $\mu \approx \mu^2$.*

Proof of 7.4.1 for compact groups. Let G be an infinite compact abelian group with dual group Γ. Let $\Theta = \{\theta_n\}_{n=1}^\infty$ be an infinite countable dissociate subset of Γ and let $a\colon \theta \to (0, \frac{1}{4}]$ be any function such that

$$\text{(1)} \qquad \sum_{n=1}^\infty |a(\theta_n)|^{2k} = \infty \quad \text{for } k \geq 1$$

and

$$\text{(2)} \qquad \lim_{n\to\infty} a(\theta_n) = 0.$$

Let Q^+ denote the set of positive rationals. For each $r \in Q^+$, let a_r be a function from Θ to $(0, \frac{1}{4}]$ such that there exists an integer $n(r)$ such that

(3) $$a_r(\theta_n) = a(\theta_n)^r \quad \text{for } n(r) < n < \infty.$$

Let v_r be the Riesz product based on Θ and a_r. Then v_r is singular by (1), and Hermitian, since $a_r \geq 0$. Also, $v_r \in M_o(G)$ by (2).

For each finite subset Φ of Θ and $r \in Q^+$, let $v_{r,\Phi}$ be the Riesz product based on $\Theta \backslash \Phi$ and a_r. Because $|a_r| \leq \frac{1}{4}$, formula (9) of Section 7.1 implies that $v_{r,\Phi}$ and v_r are equivalent measures.

Let $r, s \in Q^+$. Then (3) implies that there exists $N = N(r, s)$ such that

(4) $$a_r(\theta_n)a_s(\theta_n) = a_{rs}(\theta_n) \quad \text{for } N < n < \infty.$$

If we set $\Phi = \{\theta_n : n \leq N\}$, then (4) implies that

(5) $$v_{r,\Phi} * v_{s,\Phi} = v_{rs,\Phi}.$$

Since v_r and $v_{r,\Phi}$ are equivalent and v_s and $v_{s,\Phi}$ are equivalent, A.6.1 implies that

(6) $$v_r * v_s \approx v_{r,\Phi} * v_{s,\Phi}.$$

Using (5), (6) and the equivalence of v_{rs} with $v_{rs,\Phi}$ we see that

(7) $$v_r * v_s \approx v_{rs} \quad \text{for } r, s \in Q^+.$$

We now let $f: Q^+ \to (0, 1]$ be such that $\sum_{r \in Q^+} f(r) = 1$. Define μ by

$$\mu = \sum_{r \in Q^+} f(r) v_r.$$

It is obvious that μ is a Hermitian probability measure in $M_o(G)$. Also, μ is singular since every v_r is singular. Finally, μ and μ^2 are equivalent because of (7) and the fact that $f(r) > 0$ for all $r \in Q^+$.

The proof of 7.4.1 for compact groups is finished.

Proof of 7.4.1 for non-compact groups. Let G be a non-compact, non-discrete locally compact abelian group. Let H be an open subgroup. If $\mu \in M_o(H)$, then $\mu \in M_o(G)$ and therefore it is enough to show that 7.4.1 holds for some open subgroup H of G.

The structure theorem says that G has an open subgroup H of the form $R^n \times K$, where K is compact and $n \geq 0$. If $n = 0$, then, by what we have proved above, $H = K$ supports a singular Hermitian probability measure that is equivalent to its square.

We may therefore assume that $n \geq 1$. Let μ_1 be a singular, Hermitian probability measure in $M_o(T^n \times K)$ that is equivalent to μ_1^2. Let μ_2 be defined on $R^n \times K$ by $\mu_2(E) = \mu_1(E)$ if the Borel set $E \subseteq [0, 2\pi)^n \times K$, and $\mu_2(E) = 0$ if $E \cap [0, 2\pi)^n \times K = \varnothing$. Let $2\pi Z^n$ be the obvious subgroup of $R^n \times K$, and let μ be defined by

$$\mu = \sum\{f(m)\delta_m * \mu_2 : m \in 2\pi Z^n\} \tag{8}$$

where $f: 2\pi Z^n \to (0, 1]$ is such that $\sum_m f(m) = 1$. It is now easy to see that μ has the required properties. That ends the proof of 7.4.1. □

Credits. Theorem 7.4.1 was given by Makarov [1] for $G = R$, and Brown and Hewitt [1] for the general case.

7.5. A Multiplier Theorem and the Support of Singular Fourier–Stieltjes Transforms

7.5.1. Theorem. *Let G be a non-discrete locally compact abelian group. Let X be a subset of $M_c(G)$ that is compact in the pseudomeasure norm topology. Then there exists a singular Hermitian i.p. probability measure $\rho \in M(G)$ such that $\rho * \mu \in A(G)$ for all $\mu \in X$ and such that the map from X to A(G) given by $\mu \to \rho * \mu$ is continuous when X has the pseudomeasure norm.*

We shall draw some corollaries before proving 7.5.1.

7.5.2. Corollary. *Let G be a σ-compact locally compact abelian group. Let F be a perfect non-empty Borel subset of G. Then there exists a Borel subset S of G of zero Haar measure such that $F + S = G$.*

Proof. Since F is perfect and Borel there is a continuous probability measure μ concentrated on F. Let $X = \{\mu\}$, and let ν be a singular probability measure on G such that $(\nu * \mu)^\wedge \in L^1(\Gamma)$. Now ν is concentrated on a Borel set E of zero Haar measure. The following argument is necessitated by the fact that E^- may have positive Haar measure.

Let $g(x) = d(\mu * \nu)/dx$. Since $g \in A(G)$ and $g \neq 0$, the set $U = \{x: g(x) \geq \frac{1}{2}\|g\|_\infty\}$ is a compact set with non-empty interior. We claim that $Y = U \backslash (E + F)$ has zero Haar measure. If $x \in E$ and $y \in F$, then $x + y \notin Y$, that is $\chi_{Y-y}(x) = 0$ for all $x \in E$, $y \in Y$. Therefore

$$\mu * \nu(Y) = \int_F \mu(Y - y)d\nu(y) = \int_F \int_E \chi_{Y-y}(x)d\mu(x)d\nu(y) = 0.$$

But $\mu * \nu(Y) = \int_Y g(x)dx \geq \frac{1}{2}\|g\|_\infty m_G(Y)$. Therefore $m_G(Y) = 0$ as claimed. Let $\{x_n\}$ be a sequence in G such that $G = \bigcup(U + x_n)$. Then $G\backslash\bigcup_{n=1}^{\infty}(x_n + E + F)$ has zero Haar measure. Let $y_1 \in E$ be fixed and let

$$S = \left[\bigcup_{n=1}^{\infty}(x_n + F)\right] \cup \left[\left(G\backslash \bigcup_{n=1}^{\infty}(x_n + E + F)\right) - y_1\right].$$

Then S has zero Haar measure, while $S + F = G$. The Corollary is proved. □

7.5.3. Corollary. *Let G be a non-discrete locally compact abelian group. Then $\Delta M_c(G)$ is not σ-compact.*

Proof. If $\Delta M_c(G)$ were σ-compact, then there would exist a sequence of measures $\{\mu_1, \mu_2, \ldots\} \subseteq M_c(G)$ such that $\lim \mu_j = 0$, and such that for each $\psi \in \Delta M_c(G)$, there exists a finite $j \geq 1$ such that $\hat{\mu}_j(\psi) \neq 0$. Let $X = \{\mu_1, \mu_2, \ldots\} \cup \{0\}$. Then X satisfies the hypotheses of 7.5.1. Therefore there exists $\mu \in M_c(G)$ such that

$$(1) \qquad \mu * \mu_j \in L^1(G) \quad \text{for } j \geq 1,$$

and such that μ is a Hermitian independent power probability measure. Of course that last implies that there exists $\psi \in \Delta M_c(G)\backslash\Gamma$ such that $\hat{\mu}(\psi) = 1$. But (1) implies that $\hat{\mu}_j(\psi) = 0$ for $j \geq 1$. That proves the Corollary. □

7.5.4. Lemma. *Let G and X be as in 7.5.1. Let $K \subseteq \Gamma$ be compact and $\varepsilon > 0$. Then there exists $\lambda \in \Gamma\backslash K$ such that*

$$\sup\{|\hat{\mu}(\gamma\rho)| : \rho = \lambda \quad \text{or} \quad \rho = \lambda^{-1} \quad \text{and} \quad \gamma \in K, \mu \in X\} < \varepsilon.$$

Proof. Let $\{\mu_j\}_{j=1}^{n}$ be $\varepsilon/3$-dense in X. Since every $\hat{\mu}_j$ is uniformly continuous, there exists a neighborhood W of $1 \in \Gamma$ such that

$$\sup_{\gamma\in\Gamma}|\hat{\mu}_j(\gamma) - \hat{\mu}_j(\gamma\rho)| < \varepsilon/3, \quad \text{for } 1 \leq j \leq n \quad \text{and} \quad \rho \in W.$$

Let $\mu_0 \in M_c(G)$ be such that $\hat{\mu}_0 = 1$ on K, and let $F \subseteq \Gamma$ be a finite set such that $F + W \supseteq K$. Let

$$\tau = \mu_0 * \tilde{\mu}_0 + \sum_{\gamma\in F}\sum_{j=1}^{n}\bar{\gamma}(\mu_j * \tilde{\mu}_j).$$

Since $\tau + \bar{\tau} \in M_c(G)$, A.2.1 implies that there exists λ such that $\hat{\tau}(\lambda) + \hat{\tau}(\lambda^{-1}) < (\varepsilon/3)^2$. Clearly, $\lambda \notin K$, $|\hat{\mu}_j(\gamma\lambda)| < \varepsilon/3$, and $|\hat{\mu}_j(\gamma\lambda^{-1})| < \varepsilon/3$ for all $\gamma \in F$ and $1 \leq j \leq n$. If $\rho \in K$, then $\rho = \gamma\omega$ for some $\gamma \in F$ and $\omega \in W$, so

$$|\hat{\mu}_j(\rho\lambda^{\pm 1})| = |\hat{\mu}_j(\gamma\omega\lambda^{\pm 1})| < |\hat{\mu}_j(\gamma\omega\lambda^{\pm 1}) - \hat{\mu}_j(\gamma\lambda^{\pm 1})| + \varepsilon/3 < 2\varepsilon/3.$$

Since $\{\mu_j\}$ is $\varepsilon/3$-dense in X, the conclusion follows, and the Lemma is proved. □

An elaboration of the construction of Riesz products proves the next lemma.

7.5.5. Lemma. *Let σ be a probability measure in $M(G)$ such that $S = \{\gamma: \hat{\sigma}(\gamma) \neq 0\}$ has compact closure. Let $\Theta \subseteq \Gamma$ be such that whenever $n \geq 1, \theta_1, \ldots, \theta_n$ are distinct elements of Θ, $\varepsilon_1, \ldots, \varepsilon_n \in \{0, \pm 1, \pm 2\}$, and $\theta_1^{\varepsilon_1} \cdots \theta_n^{\varepsilon_n} \in SS^{-1}$, then $\hat{\sigma}(\theta_j^{\varepsilon_j}) = 1$ for all j. Let $a: \Theta \to \mathbb{C}$ be such that $-1 < a(\theta) < 1$ if $\sigma(\theta^2) = 1$ and $|a(\theta)| \leq \frac{1}{2}$ if $\hat{\sigma}(\theta^2) \neq 1$. Then there exists a probability measure $\rho \in M(G)$ such that the following hold.*

(i) *If $\gamma = \lambda\theta_1^{\varepsilon_1} \cdots \theta_n^{\varepsilon_n}$ where $n \geq 1$, $\theta_1, \ldots, \theta_n \in \Theta$ are distinct, $\varepsilon_1 = \pm 1, \ldots, \varepsilon_n = \pm 1$, and $\lambda \in S$, then $\hat{\rho}(\gamma) = \hat{\sigma}(\lambda)a(\theta_1)^{(\varepsilon_1)} \cdots a(\theta_n)^{(\varepsilon_n)}$.*
(ii) *If γ is not of the above form, then $\hat{\rho}(\gamma) = 0$.*

The measure ρ in 7.5.5 is called the *Riesz product based on σ, Θ, and a*. The proof of the next lemma is a trivial variation of the proofs of 7.2.1(i) and 7.2.2.

7.5.6. Lemma. *Let σ and S be as in 7.5.5. Let ρ and τ be Riesz products based on σ, Θ, a, and σ, Θ, b respectively. Then the following hold.*

(i) *ρ is absolutely continuous if and only if $\sum |a(\theta)|^2 < \infty$.*
(ii) *If $\sum |a(\theta) - b(\theta)|^2 = \infty$, then $\rho \perp \tau$.*

Proof of 7.5.1. By the structure theorem, G has an open subgroup H of the form $R^n \times L$, where L is compact and $n \geq 0$. Let $Y = X \cup (X - X)$. Let $\Lambda = \{\gamma \in \Gamma: \langle x, \lambda \rangle = 1, x \in H\}$, and let p be the natural projection of Γ on Γ/Λ. Let $\sigma \in M(L)$ be a Hermitian probability measure such that $\hat{\sigma}$ has compact support S. In the case that $n = 0$, we let σ be Haar measure on L. We begin the construction of ρ with the construction of a set Θ that satisfies the hypotheses of 7.5.5.

Case I, $n = 0$. We let $K_0 = S^2$. By 7.5.4, there exists θ_0 such that (2) below holds for $j = 0$. Suppose that $m \geq 1$ and that $\theta_0, \ldots, \theta_{m-1}$ have been found such that (2) below holds for $1 \leq j \leq m - 1$ where

$$K_j = \bigcup\{\theta_0^{\varepsilon(0)} \cdots \theta_{j-1}^{\varepsilon(j-1)}S^2: |\varepsilon(k)| \leq 2, \quad \text{for } 0 \leq k \leq j - 1\}.$$

$$\text{(2)} \qquad \theta_j \notin K_j \quad \text{and} \quad |\hat{\mu}(\lambda\theta_j^{\pm 1})| < (12)^{-j} \quad \text{for all } \lambda \in K_j \quad \text{and} \quad \mu \in Y.$$

By 7.5.4 there exists $\theta_m \in \Gamma \backslash K_m$ such that (2) holds for $j = m$. That completes the induction. We shall now modify the set $\{\theta_m\}_1^\infty$.

Suppose that $\{p\theta_m^2\}_0^\infty$ is finite. Then we may assume that $p(\theta_m\theta_0^{-1})^2 = 1$ for all $m \geq 1$. We claim that $\{\theta_m\theta_0^{-1}: m \geq 1\}$, S, and σ satisfy the hypotheses of 7.5.5. Indeed, suppose that $\gamma = (\theta_1\theta_0^{-1})^{\varepsilon(1)} \cdots (\theta_m\theta_0^{-1})^{\varepsilon(m)} \in S^2 = SS^{-1}$ for some choice of $\{\varepsilon(1), \ldots, \varepsilon(m)\} \subseteq \{0, \pm 1, \pm 2\}$. If $|\varepsilon(m)| = 2$, then $p(\theta_m \theta_0^{-1})^{\varepsilon(m)} = 1$ so $(\theta_m \theta_0^{-1})^{\varepsilon(m)} = 1$ on the support of σ. We may therefore assume that $|\varepsilon(m)| \neq 2$, which would imply that $\gamma \notin S^2$. That establishes the claim. We replace Θ by $\{\theta_m \theta_0^{-1}\}_1^\infty$, obtaining a set Θ that satisfies the hypotheses of 7.5.5.

Case II, $n > 1$. Let p' denote the projection of Γ onto R^n, via $\Gamma \to \Gamma/\Lambda = R^n \times \hat{L} \to R^n$. By 7.5.4, there exists θ_0 such that (2) holds for $j = 0$ and $K_0 = S^2$. Suppose that $\theta_0, \ldots, \theta_{j-1}, K_0, \ldots, K_{j-1}$ have been found. Let $K_j \subseteq \Gamma$ be a compact symmetric set such that $p'K_j$ is convex and

$$(3) \qquad p'K_j \supseteq \bigcup \{p'(\theta_0^{\varepsilon(0)} \cdots \theta_{j-1}^{\varepsilon(j-1)} K_{j-1}) \colon |\varepsilon_m| \leq 4, 1 \leq m \leq j-1\}.$$

We apply 7.5.4 to find $\theta_j \in \Gamma \backslash K_j$ such that (2) holds. Suppose that $\gamma = \theta_0^{\varepsilon(0)} \cdots \theta_m^{\varepsilon(m)} \in S^2$ for some choice of $|\varepsilon(j)| \leq 2$. If $|\varepsilon(m)| = 1$, we have an immediate contradication of (3). If $|\varepsilon(m)| = 2$, then the convexity of $p'K_m$ implies that $p'(\gamma \bar{\theta}_m^{\varepsilon(m)/2}) \in p'K_m$, another contradiction of (3). Thus $\{\theta_m\}_1^\infty$ and σ satisfy the hypotheses of 7.5.5.

We now let ρ be the Riesz product based on σ, Θ, and a, where $a \equiv \frac{1}{2}$ on Θ. By 7.5.6, ρ is i.p. By (2) and 7.5.5, we have, for all $\mu \in Y$,

$$\int_P |\hat{\rho}\hat{\mu}|\, d\gamma \leq \int_S |\hat{\mu}|\, d\gamma + \sum \left\{ \int_W |\hat{\mu}|\, d\gamma \colon W = \theta_1^{\varepsilon(1)} \cdots \theta_m^{\varepsilon(m)} S; \right.$$

$$\left. \varepsilon(j) = 0, \pm 1 \quad \text{for } 1 \leq j \leq m, \sum |\varepsilon_j| \neq 0; m = 1, 2, \ldots \right\}$$

$$\leq m_\Gamma(S) \left[3^k \|\hat{\mu}\|_\infty + \sum_{m=k+1}^{\infty} (3/12)^m \right] \quad \text{for} \quad k \geq 0.$$

Since $X \subseteq Y$, $\rho * X \subseteq A(G)$. If $\mu_1, \mu_2 \in X$, then $\mu_1 - \mu_2 \in Y$. Therefore $\mu \to \mu * \rho$ is continuous from X to $A(G)$. That ends the proof of 7.5.1. □

The above proof shows that sets $\Omega(\Theta)$, when Θ is dissociate, can be thin relative to the support of a Fourier-Stieltjes transform. That they can also be "thick" is shown by the following result.

7.5.7. Theorem. *Let G be a compact abelian group with dual group Γ, and let $\mu \in M(G) \backslash \text{Rad}\, L^1(G)$.*

(i) *There exists an infinite dissociate set $\Theta \subseteq \Gamma$ and $\gamma \in \Gamma$ such that*

$$\hat{\mu}(\gamma\omega) \neq 0 \quad \textit{for all } \omega \in \Omega(\Theta).$$

(ii) *If $\mu \notin M_o(G)$, then there exists $\varepsilon > 0$, $\gamma \in \Gamma$, and an infinite dissociate subset Θ of Γ such that $|\hat{\mu}(\gamma\omega)| > \varepsilon$ for all words $\omega \in \Omega(\Theta)$ of length two.*

Proof. (i) Since $\mu \notin \operatorname{Rad} L^1(G)$, there exist $\gamma' \in \Gamma$ and $h \in \Delta M(G)\backslash\Gamma$ such that $h = h^2$ and $\hat{\mu}(\gamma' h') \neq 0$. Let $\nu = (\gamma' h\mu) * (\gamma' h\mu)^- * (\gamma'\mu) * (\gamma'\mu)^-$. Then ν is a real-valued measure, and $\hat{\nu}(h) = |\hat{\mu}(\gamma' h)|^4 \neq 0$. Also for all $\lambda \in \Gamma$, $|\hat{\nu}(h)| \leq |\hat{\mu}(\gamma'\lambda)| \, \|\mu\|^3$. Thus it will suffice to prove (i) for ν.

Since $\hat{\nu}(h) \neq 0$, $\nu \notin L^1(G)$ and therefore there exists $\theta_1 \in \Gamma\backslash\{1\}$ such that $\hat{\nu}(\theta_1) = \hat{\nu}(\bar{\theta}_1)^- \neq 0$. Let $\nu_1 = (\theta_1 \nu) * (\bar{\theta}_1 \nu) * \nu$. Then

$$\hat{\nu}_1(h) = |\hat{\nu}(\theta_1 h)|^2 \hat{\nu}(h) = |\hat{\mu}(\gamma'\theta_1 h)\bar{\mu}^\wedge(\gamma'\theta_1 h)|^2 \hat{\nu}(h),$$

while $\hat{\nu}(\theta_1) = \hat{\mu}(\gamma'\theta_1 h)\bar{\mu}^\wedge(\bar{\gamma}'\theta_1 h)\hat{\mu}(\gamma'\theta_1)\bar{\mu}^\wedge(\gamma\theta_1) \neq 0$. Therefore $\nu_1(h) \neq 0$.

Now suppose that $n \geq 1$ and that distinct elements $\theta_1, \ldots, \theta_n$ of $\Gamma\backslash\{1\}$ have been found such that

$$\nu_n = \bigstar\{\omega\nu \colon \omega \in \Omega(\{\theta_1, \ldots, \theta_n\})\} \tag{4}$$

has $\hat{\nu}_n(h) \neq 0$. Then $\nu_n \notin L^1(G)$, and there exists $\theta_{n+1} \in \Gamma\backslash\{1, \theta_1, \ldots, \theta_n\}$ such that $\hat{\nu}_n(\theta_{n+1}) \neq 0$. Set $\nu_{n+1} = (\theta_{n+1}\nu_n) * (\bar{\theta}_{n+1}\nu_n) * \nu_n$. Then $\hat{\nu}_{n+1}(h) \neq 0$, and $\nu_{n+1} = \bigstar\{\omega\nu \colon \omega \in \Omega(\{\theta_1, \ldots, \theta_{n+1}\})\}$. That completes the induction. Let $\Theta' = \{\theta_j\}_{j=1}^\infty$. Then $\hat{\nu}(\omega) \neq 0$ for all $\omega \in \Omega(\Theta')\backslash\{1\}$. Set $\Theta'' = \{\theta_j\}_{j=2}^\infty$ and $\gamma = \gamma'\theta_1$. Then $\hat{\mu}(\gamma\omega) \neq 0$ for all $\omega \in \Omega(\Theta'')$. To replace Θ'' by a dissociate set, we apply 7.1.1. The details are left to the reader.

(ii) Let $\{\lambda_n\}_1^\infty \subseteq \Gamma$ be such that $|\hat{\mu}(\lambda_n)| \geq \delta > 0$ for all n. There is no loss of generality in assuming that $\{\lambda_n\}_1^\infty$ generates Γ; that is, we may assume that G is metrizable.

Then there exists $\psi \in \Gamma^-\backslash\Gamma$ such that $\hat{\mu}(\psi) \neq 0$. We replace ψ by $|\psi|^2$. Then $|\psi|^2 \in \Gamma^-$ and $|\psi|^8\mu \neq 0$. Therefore there exists $\gamma \in \Gamma$ such that $\hat{\mu}(\gamma|\psi|^8) \neq 0$. Let $\nu = \gamma\mu * (\gamma\mu)^-$. Then $\hat{\nu}(|\psi|^8) = |\hat{\mu}(\gamma|\psi|^8)|^2 > 0$.

Since $|\psi|^2 \in \Gamma^-$, there exists a sequence $\{\gamma_j\} \subseteq \Gamma$ such that $\gamma_j \to |\psi|^2$, and therefore such that

$$|\hat{\nu}(\gamma_j|\psi|^6)| > \tfrac{1}{2}\hat{\nu}(|\psi|^8), \quad \text{for } j \geq j(1).$$

Similarly, there exist $1 \leq k(1) \leq k(2) < \cdots$ such that

$$|\hat{\nu}(\gamma_j\gamma_k^{\varepsilon_k}|\psi|^4)| > \tfrac{1}{2}\hat{\nu}(|\psi|^8) \quad \text{for } j \geq j(1),\ \varepsilon_k = \pm 1, 0 \quad \text{and} \quad k \geq k(j).$$

Proceding in this manner and passing to the usual diagonal subsequence, we find $\{\theta_j\}_1^\infty \subseteq \{\gamma_k\}_1^\infty$ such that

$$|\hat{\nu}(\theta_{j(1)}^{\varepsilon(1)} \cdots \theta_{j(4)}^{\varepsilon(4)})| > \hat{\nu}(|\psi^8)/2$$

for $1 \leq j(1) < \cdots < j(4) < \infty$ and $|\varepsilon(1)| = \cdots = |\varepsilon(4)| = 1$. We apply 7.1.1 as follows.

If $\{\theta_{j(k)}\}_{k=1}^{\infty}$ is an infinite dissociate subset of $\{\theta_j\}_1^{\infty}$, we set $\Theta = \{\theta_{j(2k+1)}\bar{\theta}_{j(2k)}: k = 1, 2, \ldots\}$. If $\{\theta_j^2\}_1^{\infty}$ is finite, we let $\{\theta_{j(k)}\}$ be dissociate and set $\Theta = \{\theta_{j(2k+1)}\bar{\theta}_{j(2k)}: k = 1, 2, \ldots\}$. That completes the proof of (ii). □

Remarks and Credits. Theorem 7.5.1 is due to Saeki (private communication); it extends a result proved for singleton X and compact G (Doss [3]) and more complicated X and G (Graham and A. MacLean [1]). Corollary 7.5.2 is due to Doss [3] for compact groups; the proof here is his. Talagrand [2] has shown by other methods that in case G is compact, S may be taken compact in 7.5.2; Körner [2, Lemma 6.10] shows that there exist probability measures μ, ν on Kronecker sets in T such that $\mu * \nu$ is a C^∞-function; Sato [1] has extended that result to general groups. Körner [2] also proves, as do Varopoulos [13] and Bernard and Varopoulos [1], that if F is a perfect Kronecker set in a compact metrizable abelian I-group G, then there exists another Kronecker set $E \subseteq G$ such that $E + F = G$. Corollary 7.5.3 is from Graham and A. MacLean [1]. For stronger versions of 7.5.6, see Host and Parreau [3] and Ritter [1, 2].

Theorem 7.5.6 is from Host and Parreau [2] and Graham, Host and Parreau [1]. In part (ii), "two" can be replaced by "n", for $n = 1, 2, \ldots$. The point of $n = 2$ is that the resulting set of ω's is not a Sidon set by 11.3.4.

7.6. Small Subsets of Z That Are Dense in bZ

When $\nu \in M_c(T)$, the sets $\{\gamma: |\hat{\nu}(\gamma)| \geq \varepsilon\}$ for $\varepsilon > 0$ have relative density zero in Z, as is shown in A.2.1. Those sets may be dense in bZ, however. In 7.6.1 we give a generalized Riesz product construction of such an example, and in 7.6.5 we show that continuous measures ν on Kronecker sets also have $\{\gamma: |\hat{\nu}(\gamma)| \geq (1 - \varepsilon)\|\nu\|\}$ dense in bZ for all $\varepsilon > 0$.

7.6.1. Theorem. *There exists a continuous measure ν on T such that $\{\gamma: \hat{\nu}(\gamma) \geq 1\}$ is dense in bZ.*

7.6.2. Lemma. *Let $\Lambda \subseteq Z$. Then the following conditions are equivalent.*

(i) *Λ is dense in bZ.*
(ii) *The probability measures on Λ are weak* dense in the probability measures on bZ.*
(iii) *Haar measure on bZ is in the weak* closure of the probability measures on Λ.*
(iv) *For every finite set $F \subseteq T\backslash\{0\}$ and every $\varepsilon > 0$, there exists a probability measure ν on Λ such that $|\hat{\nu}(x)| < \varepsilon$ for all $x \in F$.*
(v) *For all $k \geq 1$, $x_1, \ldots, x_k \in T\backslash\{0\}$ and $\varepsilon > 0$, the closed convex hull of $\{(\langle\lambda, x_1\rangle, \ldots, \langle\lambda, x_k\rangle): \lambda \in \Lambda\}$ contains the origin.*
(vi) *Whenever $\gamma \in Z$, $\varepsilon > 0$ and F is a finite subset of T, there exists $\lambda \in \Lambda$ such that $|\langle\lambda, x\rangle - \langle\gamma, x\rangle| < \varepsilon$ for all $x \in F$.*

Proof. (i) ⇒ (ii). Let μ be a probability measure on bZ, let $\varphi_1, \ldots, \varphi_k$ be continuous functions on bZ and let $\varepsilon > 0$. We must find a probability measure ν on Λ such that

$$\left| \int \varphi_j \, d\mu - \int \varphi \, d\nu \right| < \varepsilon \quad \text{for } 1 \le j \le k.$$

We may replace the integrals $\int \varphi_j \, d\mu$ by Riemann sums such that

$$(1) \qquad \left| \int \varphi_j \, d\mu - \sum_{i=1}^{n} \varphi_j(\gamma_i)\mu(X_i) \right| < \varepsilon \quad \text{for } 1 \le j \le k$$

where the same partition $\{X_i\}_1^m$ and choice of $\{\gamma_i\}_1^n$ is used for every j. Since Λ is assumed to be dense in bZ and the φ_j are continuous, we may assume that the γ_i are in Λ. That may be done while introducing an error so small that (1) still holds. Setting $\nu = \sum \mu(X_i)\delta_{\gamma_i}$, we see that $|\int \varphi_j \, d\mu - \int \varphi_j \, d\nu| < \varepsilon$ for $1 \le j \le k$. Obviously, ν is a probability measure on Λ.

The implications (ii) ⇒ (iii) ⇒ (iv) ⇒ (v) ⇒ (iii) ⇒ (i) are easily established. The proof of 7.6.2 is then complete. □

7.6.3. Lemma. *Let $\{n_j\}$ be any sequence in Z. For $m = 1, 2, \ldots$ let Λ_m be defined by*

$$\Lambda_m = \left\{ \lambda \colon \lambda = \sum_{j=m+1}^{\infty} \varepsilon_j n_j;\ \varepsilon_j = 0, 1;\ \sum \varepsilon_j < \infty \right\}.$$

Then the following are equivalent.

(i) *Λ_m is dense in bZ for all $m \ge 1$.*
(ii) *For each $x \in T \setminus \{0\}$,*

$$(2) \qquad \sum_{j=1}^{\infty} |\langle x, n_j \rangle - 1| = \infty.$$

Proof. If (ii) is not satisfied, then for some non-zero $x \in T$, there exists $m \ge 1$ such that

$$\sum_{j=m+1}^{\infty} |\langle x, n_j \rangle - 1| < \tfrac{1}{10}.$$

Thus $|\langle x, \lambda \rangle - 1| < \frac{1}{10}$ for all $\lambda \in \Lambda_m$, so $bZ \ne \Lambda_m^-$.

Now we shall show that (ii) implies that 7.6.2(iv) holds for each of the sets Λ_m. Fix m and $\varepsilon > 0$ and let $\{x_1, \ldots, x_k\}$ be an arbitrary finite subset of $T\backslash\{0\}$. Let r be such that $2^{-r} < \varepsilon$. By (2), there exists $\lambda_1 \in \Lambda_m$ such that $|\langle\lambda_1, x_1\rangle - 1| > \sqrt{3}$, from which it follows that $|1 + \langle\lambda_1, x_1\rangle| < 1$. Let $\lambda_1 = \sum_{j=m+1}^{m(1)} \varepsilon_j n_j$, and choose $\lambda_2 \in \Lambda_{m(1)}$ such that $|1 + \langle\lambda_2, x_1\rangle| < 1$. Proceeding inductively, we choose $\lambda_1, \ldots, \lambda_r$ from Λ_m such that $|1 + \langle\lambda_i, x_1\rangle| < 1$ for each i and such that no integer from the sequence $\{n_p\}$ occurs in more than one of the sums λ_i. We repeat the process k times so that for $1 \le j \le k$ and $1 \le i \le r$ we have $\lambda_{(j-1)r+i} \in \Lambda_m$ such that $|1 + \langle\lambda_{(j-1)r+i}, x_j\rangle| < 1$, and such that no integer from the sequence $\{n_p\}$ occurs in more than one of the sums λ_l. Let

$$\nu = 2^{-k} \mathop{\bigstar}_{l=1}^{kr} (\delta(0) + \delta(\lambda_l)).$$

Then ν is a probability measure on Λ_m and $|\hat{\nu}(x_j)| < 2^{-r} < \varepsilon$ for each j. Lemma 7.6.4 is proved. $\square$

Proof of 7.6.1. We define $n_1 = 1$, and set $n_{j+1} = jn_j + 1$ for $j = 1, 2, \ldots$. We define Λ_m as in the statement of 7.6.3. We shall show that (2) holds for $\{n_j\}$ and therefore that the sets Λ_m are all dense in bZ.

For $0 < x < 2\pi$, let $\varepsilon = |\exp(ix) - 1|$. Suppose that for some j,

$$|\exp(ixn_j) - 1| \le \varepsilon/2j.$$

Then $|\exp(ixjn_j) - 1| < \varepsilon/2$. Therefore

$$|\exp(ixn_{j+1}) - 1| = |\exp(ix)\exp(ixjn_j) - 1| \ge |\exp(ix) - 1| - \varepsilon/2 \ge \varepsilon/2.$$

Thus either $|\exp(ixn_j) - 1| \ge \varepsilon/2j$ for all sufficiently large j, or $|\exp(ixn_j) - 1| \ge \varepsilon/2$ for an infinite number of integers j. In either case (2) holds.

Let $f(t) = \max(1 - 6|t|, 0)$ for $-\infty < t < \infty$. Then $f * f(0) = \frac{1}{9}$ and $\int_{-\infty}^{\infty} f * f(t)dt = \frac{1}{36}$. Let $g(x) = 9f * f(x)$. Then g is decreasing on $[0, \infty)$, $\operatorname{supp} g \subseteq [-\frac{1}{3}, \frac{1}{3}]$, $g \in C^2(R)$, $g(0) = 1$, $g'(0) = 0$, g is positive-definite, and $g(x) \ge 1 - cx^2$ for some $c > 0$ and all x with $|x| < \frac{1}{4}$.

For $j = 1, 2, \ldots$ let $Q_j(x) = \sum_{m=-\infty}^{\infty} g(m/(j+3))\exp(imn_jx)$. Then $\hat{Q}_j$ is supported in $n_jZ \cap [-jn_j/3, jn_j/3]$, $Q_j \ge 0$ (because the restriction of g to $(j+3)^{-1}Z$ is positive-definite), $\hat{Q}_j(0) = 1$, and

$$\hat{Q}_j(n_j) \ge 1 - c/j^2. \tag{3}$$

Furthermore, because g is decreasing on $[0, \infty)$ and $g(x) = g(-x)$ for all x,

(4)

$$\sum_{-\infty}^{\infty} g(m/(j+3)) \le g(0) + (j+3)\int_{-\infty}^{\infty} g\,dx = 1 + (9/36)(j+3) = (j+7)/4.$$

Because $\hat{Q}_j$ is supported in $n_j Z \cap [-jn_j/3, jn_j/3]$, every integer $m \in Z$ has at most one expression as a sum $m = m_1 + \cdots + m_k$, where $0 \neq m_i \in$ support $\hat{Q}_{j_i}$ for $1 \leq i \leq k$ and $j_1 < j_2 < \cdots < j_k$. Therefore the weak $*$ limit μ of the products $\prod_{j=1}^{n} Q_j$ exists and is a probability measure on T. We claim that μ is continuous and that $\hat{\mu}$ is bounded away from zero on Λ_4. The second assertion is easy: if $\lambda \in \Lambda_4$, then

$$|\hat{\mu}(\lambda)| \geq \prod_{j=1}^{\infty} (1 - cj^{-2}) = d > 0.$$

We shall set $\nu = d^{-1}\mu$. If μ is shown to be continuous, then 7.6.1 will have been proved.

Some simple induction arguments (that we leave to the reader) will establish the following equations and estimates for all $k \geq 1$.

(5)
$$n_{k+1} = 2(k!) + (k!) \sum_{j=2}^{k} 1/j!.$$

(6)
$$\left\| \left(\prod_{j=1}^{k} \hat{Q}_j \right) \right\|_{l^1(Z)} \leq \prod_{j=1}^{k} (j + 7)/4 \leq (k + 7)! 4^{-k}.$$

(7) $\left(\prod_{j=1}^{k} Q_j \right)^{\wedge}$ is supported in $[-(n_{k+1}/2) - 1, (n_{k+1}/2) + 1]$.

(8)
$$\hat{\mu}(m) = \left(\prod_{1}^{k} Q_j \right)_{\wedge} (m) \quad \text{for} \quad |m| \leq \tfrac{1}{2} n_{k+1} + 1.$$

From (5)–(8) we see that, setting $m_k = (n_{k+1}/2) + 1$,

(9)
$$(2m_k + 1)^{-1} \sum_{j=-m_k}^{m_k} |\hat{\mu}(j)| \leq (k + 7)!/k! 4^k.$$

It is easy to see that the right hand side of (9) tends to zero as k tends to infinity. An application of A.2.1 (Wiener's Theorem) now shows that μ is continuous. Theorem 7.6.1 is proved. □

7.6.4. Lemma. *Let G be an abelian group and K an independent subset of G all of whose elements have infinite order. If $x_1, \ldots, x_n \in G$, then there exists $y_1, \ldots, y_m \in K$ such that*

$$\mathrm{Gp}(K \setminus \{y_j\}_1^m) \cap \mathrm{Gp}(\{x_j\}_1^n) = \{0\}.$$

Proof. We induct on n. If $n = 1$, and $0 \neq m_1 x_1 = \sum n_j y_j$ where the y_j are distinct and the n_j's are non-zero, then we claim $\mathrm{Gp}(K \setminus \{y_j\}) \cap \mathrm{Gp}\, x_1 = \{0\}$.

Indeed, suppose that $m_2 x_1 = \sum n'_j y'_j$ where the y'_j are distinct. Then $\sum m_2 n_j y_j = \sum m_1 n'_j y'_j$ so $m_2 n_j = 0$ for all j. Therefore $m_2 = 0$.

Now suppose that the conclusion holds for some $n \geq 1$. Let $x_1, \ldots, x_{n+1} \in G$. We may assume that the x_j are independent. Let $Y = \{y_1, \ldots, y_m\}$ be such that $\mathrm{Gp}(K \backslash Y) \cap \mathrm{Gp}\{x_j\}_1^n = \{0\}$. Let $H = \mathrm{Gp}\{x_j\}_1^n$. Then the image of $K \backslash Y$ is independent in G/H, so there exist cosets $y_{n+1} + H, \ldots, y_k + H$ where the y's are in $K \backslash Y$ such that $[\mathrm{Gp}\{x_j\}_1^{n+1}/H] \cap [\mathrm{Gp}(K \backslash \{y_j\}_1^k)/(H \cap \mathrm{Gp}(K \backslash \{y_j\}_1^k))] = \{0\}$. It is now clear that $y_1, \ldots, y_k$ suffice for $x_1, \ldots, x_{n+1}$. The Lemma is proved. □

7.6.5. Theorem. *Let $K \subseteq T$ be a compact perfect Kronecker set, $\mu \in M_c(K)$ and $\varepsilon > 0$. Then*

$$\{\gamma \in Z : |\hat{\mu}(\gamma)| > (1 - \varepsilon)\|\mu\|\}$$

is dense in bZ

Proof. Let $x_1, \ldots, x_n \in T$. Let $Y \subseteq K$ be a finite set such that $\mathrm{Gp}(K \backslash Y) \cap \mathrm{Gp}\{x_j\} = \{0\}$. The existence of such a set Y is guaranteed by Lemma 7.6.4.

We may assume that the x_j are independent, that x_1 has finite order $r > 1$, and that $d|\mu|/d\mu$ extends to a continuous unimodular function on K. Let $\gamma \in Z$ and $\delta > 0$.

Since $\varepsilon > 0$, there exists an open set $U \supseteq Y$ such that $|\mu|(K \backslash U) > (1 - \delta/2)\|\mu\|$. Since $\mathrm{Gp}\{x_j\} \cap \mathrm{Gp}(K \backslash U) = \{0\}$ and the x_j are independent, $\{x_j\}_2^n \cup (K \backslash U)$ is a Kronecker set. That follows by induction from a result in Bernard and Varopoulos [1]: if K is a totally disconnected compact Kronecker set and $x \notin K$, with $\{x\} \cup K$ independent, then $K \cup \{x\}$ is a Kronecker set.

Let $\gamma' \in Z$ be such that $|(\gamma')^r \gamma - d|\mu|/d\mu| < \delta\|\mu\|$ on $K \ \ U$ and $|\gamma' - 1| < \delta/r$ on $\{x_2, \ldots, x_n\}$. Such that a choice is possible because $(K \backslash U) \cup \{x_j\}$ is a Kronecker set. Then $|\int (\hat{\gamma}')^r \hat{\gamma}\, d\mu - \|\mu\|| < \delta\|\mu\|$ while $(\gamma')^r \gamma = \gamma$ at x_1, and $|(\gamma')^r \gamma - \gamma| < \delta$ on $x_2, \ldots, x_n$. The Theorem follows at once. □

Remarks and Credits. Theorem 7.6.1 is due to Katznelson [6]. Blum, Eisenberg and Hahn [1] point out that each of the following sets has closure of measure zero in bZ: $\{n!: n = 1, 2, \ldots\}$, $\{a^n: n = 1, 2, \ldots\}$ where $a \in Z$, $\{p: p \text{ is a prime}\}$, and $\{n^k: n = 1, 2, \ldots\}$ where $k \geq 2$. Theorem 7.6.4 is from Saeki [21] who gives a more general and stronger result; the basic idea of his proof is above.

7.7. Non-trivial Idempotents in $B(E)$ for $E \subset Z$

The following result is new, and answers a question of S. Hartman.

7.7.1. Theorem. *Let E be an infinite subset of Z. Then there exist infinite subsets E_0, E_1 of E and $\mu \in M(T)$ such that $\hat{\mu}(n) = j$ for all $n \in E_j$ and $j = 0, 1$.*

Proof. If no idempotent measure works, then for each $m = 1, 2, \ldots$ there exists k_m such that $E \backslash (mZ + k_m)$ is finite, with $0 \le k_m < m$. We may suppose that either

(1) $k_m = 0$ for infinitely many values of m, or else
(2) both $k_m \to \infty$ and $m - k_m \to \infty$.

For if (2) fails, then either k_m or $m - k_m$ takes on *some* finite value k_0 infinitely often; and it suffices to prove the result for whichever translate $E - k_0$ or $k_0 - E$ for which (1) holds. It suffices to find a dissociate sequence $\{\lambda_n\}_1^\infty \subseteq Z^+$ such that either λ_n or $-\lambda_n$ (or perhaps both) belong to E for each n, and such that

$$\left\{ \sum_{j=1}^{n} \varepsilon_j \lambda_j : \varepsilon_j = 0, 1 \text{ or } -1 \right\} \cap E \subseteq (\{\lambda_j\}_1^n \cup \{-\lambda_j\}_1^n) \quad \text{for all } n. \tag{3}$$

That suffices, because (replacing $\{\lambda_n\}$ by a subsequence if necessary) we may suppose that both $E_1 = E \cap [\{\lambda_n\} \cup \{-\lambda_n\}]$ and $E_0 = E \backslash E_1$ are infinite and let $\mu = 2$ weak$*$ $\lim_{n\to\infty} \prod_1^n (1 + \cos \lambda_j x)$. (We subtract $2m_T$ from μ if $0 \in E$.)

We select $\{\lambda_n\}$ by induction. Let λ_1 be any element of $Z^+ \cap (E \cup -E)$. When $\lambda_1, \ldots, \lambda_{n-1}$ have been suitably chosen, proceed as follows. Let $b > \sum_1^{n-1} \lambda_j$.

If (1) *holds*. Choose $m \ge b$ with $k_m = 0$. Then there exists $p > 0$ such that the set $[-pm - m + 1, -pm + m - 1] \cup [pm - m + 1, pm + m - 1]$ contains no points of the set $\{\sum_{j=1}^{n-1} \varepsilon_j \lambda_j : \varepsilon_j = 0, 1, \text{ or } -1\}$, and intersects E only in pm, or $-pm$, or both. Then let $\lambda_n = pm$.

If (2) *holds*. Choose m so that $m' \ge m \Rightarrow k_{m'} \ge b$ and $m' - k_{m'} \ge b$. There are only a finite number of points of E contained in $(mZ + k_m) \cap (2m + k_{2m})$. Therefore there exist p and q such that $b \le |q| \le m - b$ and such that the set $[-pm - m, -pm + m - 1] \cup [pm - m, pm + m - 1]$ contains no points of the set $\{\sum_{j=1}^{n-1} \varepsilon_j \lambda_j : \varepsilon_j = 0, -1, \text{ or } 1\}$, and intersects E in only one or two points: $\lambda^+ = pm + q$, $\lambda^- = pm + q$, or both. Then $|\lambda^+ + \lambda^-| \ge 2b$. Let $\lambda_n = \lambda^+$ if $\lambda^+ \in E$, or $-\lambda^-$ if $\lambda^- \in E$. Then λ_n (or $-\lambda_n$) is the only point in $E \cap ([-\lambda_n - b, -\lambda_n + b] \cup [\lambda_n - b, \lambda_n + b])$.

Thus, in either case, (1) or (2), λ_n can be found to satisfy (3).

Chapter 8

The Šilov Boundary, Symmetric Ideals, and Gleason Parts of $\Delta M(G)$

8.1. Introduction

In this chapter we shall study the algebra $M(G)$ of all regular Borel measures on the locally compact abelian group G, with particular emphasis on the maximal ideal space $\Delta M(G)$ of $M(G)$. We shall sometimes write Δ, Δ_o, ∂, ∂_o, Σ, and Σ_o, for $\Delta M(G)$, $\Delta M_o(G)$, $\partial M(G)$, $\partial M_o(G)$, $\Sigma M(G)$, and $\Sigma M_o(G)$. We remind the reader that $\Delta_o \supseteq \Delta$ and $\partial_o \supseteq \partial$, because $M_o(G)$ is an ideal in $M(G)$.

In Section 8.2 we show that $\Sigma M_o(G) \not\subseteq \partial M_o(G)$. From that two important facts about $M(G)$ follow: $\partial M(G) \neq \Delta M(G)$, and $\Gamma^- \neq \Sigma$.

In Section 8.3 some L-ideals in $M(G)$ are characterized in terms of the continuity of the map $x \mapsto \delta_x * \mu$ with respect to several topologies on $M(G)$ and the given topology on G. Those results have their origins in the theorem usually ascribed to Plessner [1] and Raïkov [1], which says that $\mu \in L^1(G)$ if and only if $\lim_{x \to 0} \|\delta_x * \mu - \mu\| = 0$. The results of Section 8.3 are used in Section 8.4 to show that $\Sigma M(G) \backslash \Gamma$ has no interior in $\Delta M(G) \backslash \Gamma$ and that $\partial M_o(G)$ is dense in $\partial M(G)$.

In Section 8.5 point derivations at certain elements of $\Delta M(G)$ are constructed. That construction is shown to eliminate certain candidates for strong boundary points for $M(G)$. At the end of the Section, some strong boundary points for $M(G)$ are identified.

In Section 8.6 the Gleason parts of $M(G)$ are identified.

8.2. The Šilov Boundary of $M(G)$

The *Šilov boundary* of a commutative Banach algebra B is the smallest closed subset ∂B of ΔB such that for all $\mu \in B$, $\|\hat{\mu}\|_\Delta = \sup\{|\hat{\mu}(\psi)| : \psi \in \partial B\}$, where $\|\hat{\mu}\|_\Delta = \sup\{|\hat{\mu}(\psi)| : \psi \in \Delta B\}$. A *strong boundary point* for B is an element $\psi \in \Delta B$ such that for each neighborhood U of ψ, there exists $\mu \in B$ such that

$$(1) \qquad \|\hat{\mu}\|_\Delta = 1, \ \hat{\mu}(\psi) = 1, \quad \text{and} \quad |\hat{\mu}(\rho)| \leq \tfrac{1}{2} \quad \text{for all } \rho \in \Delta B \backslash U.$$

The set of strong boundary points is a dense subset of ∂B. If $\psi \in \partial B$ and U is a neighborhood of ψ, then there exists $\mu \in B$ such that $\|\hat{\mu}\|_\Delta = 1$, and $|\hat{\mu}(\rho)| \leq \frac{1}{2}$

for all $p \in \Delta B \setminus U$. For proofs of those and other facts, see Gel'fand, Raïkov and Šilov [1, Section 11], Rickart [1, Section 3.3], or Gamelin [1, p. 10].

8.2.1. Proposition. *Let B be a commutative convolution measure algebra with maximal ideal space $\Delta B = \hat{S}$. Let $\psi \in \hat{S}$ be a strong boundary point. Then $|\psi|^2 = \psi$.*

Proof. Let $\psi = \lambda|\psi|$ be the polar decomposition of ψ (see Remark (ii) after 5.1.3). It will suffice to show that $\psi = \lambda|\psi|^2$. Let U be any neighborhood of ψ, and let $\mu \in B$ be such that (1) holds. Consider the function $z \mapsto \hat{\mu}(\lambda|\psi|^z)$. That function is analytic in the open right half-plane and is at most one in modulus by (1). Also, it attains the value 1 at $z = 1$, again by (1). Therefore $\hat{\mu}(\lambda|\psi|^z)$ is a constant function. In particular, $\hat{\mu}(\lambda|\psi|) = \hat{\mu}(\lambda|\psi|^2)$, so $\lambda|\psi|^2 \in U$. Since U was arbitrary, $\lambda|\psi|^2 = \psi$. That proves the Proposition. $\square$

We remind the reader that $\psi \in \Sigma$ if and only if $\hat{\mu}(\psi) = \hat{\tilde{\mu}}(\psi)^-$ for all $\mu \in M(G)$.

8.2.2. Proposition. *Let G be a locally compact abelian group. For each $\psi \in \Delta M(G)$, define $\tilde{\psi}$ by $\tilde{\psi}_\mu(y) = \overline{\psi_{\tilde{\mu}}(-y)}$. Then:*

(i) *$\tilde{\psi} \in \Delta M(G)$;*
(ii) *$\tilde{\tilde{\psi}} = \psi$;*
(iii) *$\psi \in \Sigma M(G)$ if and only if $\psi = \tilde{\psi}$;*
(iv) *$\psi\tilde{\psi} \in \Sigma M(G)$; and*
(v) *$\tilde{\mu}\hat{\ }(\psi) = \hat{\mu}(\tilde{\psi})^-$ for all $\mu \in M(G)$.*

Proof. (i) Straightforward computations show that $\tilde{\psi}$ has all the properties of a generalized character. (ii) is obvious. For (iii) first suppose that $\psi \in \Sigma$. Let $\mu \in M(G)$. Then

$$\hat{\mu}(\psi) = \int \overline{\psi_\mu(y)}\,d\mu(y) = \left(\int \overline{\psi_{\tilde{\mu}}(y)}\,d\tilde{\mu}(y)\right)^- = \hat{\tilde{\mu}}(\psi)^-$$

$$= \int \psi_{\tilde{\mu}}(y)\,d\overline{\tilde{\mu}(y)} = \int \psi_{\tilde{\mu}}(-y)\,d\mu(y) = \int \overline{\tilde{\psi}_\mu(y)}\,d\mu(y) = \hat{\mu}(\tilde{\psi}).$$

Therefore $\psi = \tilde{\psi}$. Now suppose that $\psi = \tilde{\psi}$. Then for each $\mu \in M(G)$,

$$\int \overline{\psi_\mu(y)}\,d\mu(y) = \int \psi_{\tilde{\mu}}(-y)\,d\mu(y) = \left(\int \overline{\psi_{\tilde{\mu}}(y)}\,d\overline{\mu}(-y)\right)^-$$

$$= \left(\int \overline{\psi_{\tilde{\mu}}(y)}\,d\tilde{\mu}(y)\right)^- = \hat{\tilde{\mu}}(\psi)^-.$$

Therefore $\psi \in \Sigma$. (iv) and (v) are obvious. That ends the proof of the Proposition. $\square$

8.2.3. Theorem. *Let G be a non-discrete locally compact abelian group. Then* $\Sigma M_o(G) \not\subseteq \partial M_o(G)$.

Proof. By either 6.8.1 or 7.3.2, there exists a tame Hermitian i.p. probability measure μ in $M_o(G)$. By 6.1.6, there exists $\rho \in \Delta M(G)$ such that $\rho_\mu = \frac{1}{2}$ a.e. $d\mu$. Because μ is Hermitian, $\psi = \rho\bar{\rho}$ has $\psi_\mu = \frac{1}{4}$ a.e. $d\mu$,so $\psi \in \Sigma_o$ by 8.2.2 (ii).

Now suppose that ψ were in ∂_o. Since the strong boundary points are dense in ∂_o, there would exist a net of them, $\{\psi^{(\beta)}\}$, such that $\lim \psi^{(\beta)} = \psi$. By 8.2.1, $|\psi^{(\beta)}|^2 = |\psi^{(\beta)}|$ for all β. Because μ is tame, $\psi_\mu^{(\beta)} = a_\beta \gamma^{(\beta)}$, where $a_\beta \in \{z : |z| \leq 1\}$ and $\gamma^{(\beta)} \in \Gamma$ for all β. Since $|\psi^{(\beta)}|^2 = |\psi^{(\beta)}|$, we have $|a_\beta|^2 = |a_\beta|$ for all β. Since $|\lim a_\beta \gamma^{(\beta)}| = \frac{1}{4}$, $|a_\beta| = 1$ for a cofinal subset of β's. We may assume therefore that $|a_\beta| = 1$ for all β.

Either $\gamma^{(\beta)}$ converges to some $\gamma \in \Gamma$, in which case $|\lim \psi_\mu^{(\beta)}| = |\lim a_\beta \gamma^{(\beta)}| = 1$ a.e. $d\mu$, or else $\gamma^{(\beta)} \to \infty$, and hence $\hat{\mu}(\gamma^{(\beta)}) \to 0$, in which case $\hat{\mu}(\psi) = \lim \hat{\mu}(a_\beta \gamma^{(\beta)}) = 0$. In either case, the fact that $\psi_\mu = \frac{1}{4}$ is contradicted. Therefore $\psi \notin \partial_o$ and Theorem 8.2.3 is proved. □

8.2.4. Corollary. *Let G be a non-discrete locally compact abelian group. Then* $\partial M(G) \neq \Delta M(G)$.

Proof. Let U be the set $\Delta_o \backslash \partial_o$, which is open in both $\Delta_o = \Delta M_o(G)$ and $\Delta = \Delta M(G)$, and, by the Theorem, non-empty. We claim that $U \cap \partial = \varnothing$. For if $U \cap \partial \neq \varnothing$, then there exists $\mu \in M(G)$ such that

$$1 = \sup\{|\hat{\mu}(\rho)| : \rho \in \Delta\} \quad \text{and} \quad \tfrac{1}{2} \geq \sup\{|\hat{\mu}(\rho)| : \rho \in \Delta \backslash U\}.$$

Let $\psi \in U$ be such that $|\hat{\mu}(\psi)| = 1$. There exists $\nu \in M_o(G)$ such that $\hat{\nu}(\psi) = 1$. Then $\hat{\mu}^n\hat{\nu}$ has modulus at most $2^{-n}\|\nu\|$ on $\Delta_o \backslash U$, and $(\hat{\mu}^n\hat{\nu})(\psi) = 1$. If n is sufficiently large, then $|\hat{\mu}^n\hat{\nu}|$ peaks on U and is at most one-half on $\Delta \backslash U$. Therefore $U \cap \partial_o \neq \varnothing$. That contradicts the definition of U. Therefore $U \cap \partial = \varnothing$, and so $\Delta \backslash \partial \supseteq U \neq \varnothing$. The Corollary is proved. □

The next two corollaries are immediate from 8.2.3.

8.2.5. Corollary. *Let G be a non-discrete locally compact abelian group with dual group* Γ. *Then* $\Sigma M(G) \neq \Gamma^-$.

8.2.6. Corollary. *Let G be a non-discrete locally compact abelian group with dual group* Γ. *Then there exists* $\mu \in M(G)$ *such that* $|\hat{\mu}(\gamma)| \geq 1$ *on* Γ *and* $1/\hat{\mu} \notin B(\Gamma)$.

Credits. Proposition 8.2.1 is from Taylor [1]. Proposition 8.2.2 is due to Johnson [1, Theorem 2], Drury [1], and Brown [2]. Theorem 8.2.3 is from

Brown [2]; that $\partial_o \neq \Delta_o$ was found independently by Moran [2]. Corollary 8.2.4 was proved by Taylor [2] for $G = D$ and Johnson [3] for arbitrary G. Corollary 8.2.5 is due to Simon [1] for $G = R$; the general case is due to Johnson [1] and Varopoulos [9]. Corollary 8.2.6 is, for $G = T$, a famous result of Wiener and Pitt [1]; the general version of 8.2.6 appears in Williamson [1]; for a stronger form of 8.2.6, see 6.3.15.

8.3. Some Translation Theorems

We discuss properties of the map $x \mapsto \delta_x * \mu$ from the locally compact abelian group G to the set of translates of the regular Borel measure $\mu \in M(G)$. Let us first recall that if $\mu \in L^1(G)$, then $y \mapsto \delta_y * \mu$ is a continuous map from G to $L^1(G)$:

(1) $$\lim_{y \to 0} \|\mu - \delta_y * \mu\| = 0.$$

(1) characterizes the ideal $L^1(G)$ as the next theorem shows.

8.3.1. Theorem. *Let G be a locally compact abelian group. For every $\mu \in M(G)$,*

(2) $$\limsup_{y \to 0} \|\mu - \delta_y * \mu\| = 2\|\mu_s\|.$$

8.3.2. Lemma. *Let $\mu \in M(G)$. Suppose that Y is a σ-compact subset of G with zero Haar measure. Then $\{y: \mu(y + Y) \neq 0\}$ is a Borel set with zero Haar measure.*

Proof of Theorem 8.3.1 assuming 8.3.2. Let Y be a σ-compact subset of G such that $|\mu_s|(Y) = \|\mu_s\|$ and such that Y has zero Haar measure. Then $\delta_y * \mu_s$ is concentrated on $y + Y$. By 8.3.2, $\{y: |\mu_s|(y + Y) = 0\}$ is dense. Therefore there exists a net $\{y_\beta\} \subseteq G$ such that $\lim y_\beta = 0$ and $|\mu_s|(y_\beta + Y) = 0$ for all β. Therefore for all β, μ_s and $\delta_{y_\beta} * \mu_s$ are mutually singular, and therefore $\limsup_\beta \|\delta_{y_\beta} * \mu_s - \mu_s\| = 2\|\mu_s\|$. Since $\lim_\beta \|\delta_{y_\beta} * (\mu - \mu_s) - (\mu - \mu_s)\| = 0$, formula (2) holds and Theorem 8.3.1 is proved. □

Proof of Lemma 8.3.2. It is easy to see that if Y is compact and $\nu \geq 0$, then the mapping $y \mapsto \nu(y + Y)$ is semicontinuous and therefore Borel. From that it follows at once that if Y is σ-compact, then $y \mapsto \mu(y + Y)$ is a Borel function. Therefore $\{y: \mu(y + Y) \neq 0\}$ is a Borel set.

We now reduce to the case that G is σ-compact. Let G_0 be an open σ-compact subgroup of G that contains both Y and the support of μ. If $y \notin G_0$, then $(y + Y) \cap G_0 = \varnothing$, and therefore $\mu(y + Y) = 0$. Thus we may assume that $G = G_0$.

Let U be any compact subset of G. Then by Fubini's theorem,

$$\int_U \mu(y + Y)dy = \iint \chi_U(y)\chi_Y(z - y)dy\, d\mu(z) = 0.$$

Since G is σ-compact, the Lemma follows. □

8.3.3. Corollary. *Let μ be a regular Borel measure on the locally compact abelian group G. Suppose that μ is singular with respect to Haar measure. Then $\{y: \delta_y * \mu \not\perp \mu\}$ is a Borel set of zero Haar measure.*

Proof. $\delta_y * \mu \perp \mu$ if and only if $\|\delta_y * |\mu| - |\mu|\| = 2\|\mu\|$. But

$$\|\delta_y * |\mu| - |\mu|\| = \sup\left\{\left|\int fd(\delta_y * |\mu| - |\mu|)\right| : f \in C_o(G), \|f\|_\infty \leq 1\right\}$$

is the supremum of continuous functions of y. Therefore

$$\{y: \|\delta_y * |\mu| - |\mu|\| < 2\|\mu\|\}$$

is a Borel subset of G.

Let Y be a σ-compact subset with zero Haar measure on which μ is concentrated. By Lemma 8.3.2, $\{y: |\mu|(y + Y) \neq 0\}$ has zero Haar measure. Thus $\{y: \delta_y * \mu \not\perp \mu\}$ has zero Haar measure. The Corollary is proved. □

Lemma 8.3.2 has another interesting consequence, the implication (ii) ⇒ (i) of the next theorem. The implication (i) ⇒ (ii) is a famous theorem of Steinhaus [1].

8.3.4. Theorem. *Let G be a locally compact abelian group and $\mu \in M(G)$. Then the following are equivalent.*

(i) $\mu \in L^1(G)$.
(ii) *Whenever A and B are Borel subsets of G such that $|\mu|(A) > 0$ and $|\mu|(B) > 0$, then $A - B$ has non-empty interior.*

Proof. (i) ⇒ (ii). Let A, B be Borel subsets of G such that $|\mu|(A) > 0$ and $|\mu|(B) > 0$. Since $\mu \in L^1(G)$, the Haar measures of A and B must be non-zero. We must show that $A - B$ has non-empty interior. There is no loss of generality in assuming that A and B are compact. We let f be the characteristic function of A and g be that of $-B$. Then $f * g$ is continuous and is supported in $A - B$. Because $\|f * g\|_1 = \|f\|_1\|g\|_1 > 0$, $f * g$ is not identically zero. Hence $A - B$ has non-empty interior.

(ii) ⇒ (i). It suffices to show that if K is a compact subset of G of zero Haar measure, then there exists a Borel subset A of K such that $|\mu|(A) = |\mu|(K)$

and $A - A$ has empty interior; for if μ were not in $L^1(G)$, there would exist such a K with $|\mu|(K) > 0$, and thus the Theorem follows.

By 8.3.2, $\{x: |\mu|(x + K) \neq 0\}$ is a Borel set of zero Haar measure. Let us assume that G is σ-compact and metrizable. Then there exists a dense sequence $\{x_1, x_2, \ldots\} \subseteq G$ such that $|\mu|(x_j + K) = 0$ for $j = 1, 2, \ldots$.

Let $S = \bigcup_{j=1}^{\infty}(K \cap (x_j + K))$. Then $|\mu|(S) = 0$. Set $A = K \backslash S$. Then $|\mu|(A) = |\mu|(K)$. We claim that $A - A$ has empty interior. For suppose that $V \subseteq A - A$ is a non-empty open set. Then V contains some x_j, so $x_j \in A - A$. Therefore there exists $a_1, a_2 \in A$ such that $x_j = a_1 - a_2$, or $a_1 = x_j + a_2 \in x_j + A$. Therefore $a_1 \in x_j + A$, or $A \cap (x_j + A) \neq \varnothing$, which contradicts the definition of A. That proves (ii) $\Rightarrow$ (i) when G is a σ-compact metrizable group.

The reduction of the general case to the σ-compact case is trivial: the measure μ is concentrated on an open σ-compact subgroup G_0, and there is no loss of generality in assuming G to be σ-compact.

We now prove (ii) $\Rightarrow$ (i) for σ-compact non-metrizable groups G. Let K be a compact subset of G that has zero Haar measure. Then there exists a compact subgroup H of G such that G/H is metrizable and $K + H$ has zero Haar measure. That is a standard result (Hewitt and Ross [1, Vol. I, p. 71]); this is the proof.

Let $U_1, U_2, \ldots$ be compact neighborhoods of K such that $\lim m_G(U_n) = 0$ and $U_1 \supseteq U_2 \supseteq \cdots$. Let $W_1, W_2, \ldots$ be compact symmetric neighborhoods of 0 such that

$$K + W_n \subseteq U_n \quad \text{and} \quad W_{n+1} + W_{n+1} \subseteq W_n, \quad \text{for } n = 1, 2, \ldots.$$

Let $H = \bigcap_{n=1}^{\infty} W_n$. It is easy to see that H is a compact group. Because H is a G_δ ($H = \bigcap \operatorname{Int} W_n$), G/H is metrizable. Since $K + H \subseteq K + W_n \subseteq U_n$ for all n, $m_G(K + H) = 0$.

Let p be the map from G to G/H. Then $p(K + H)$ has zero Haar measure. Let $\check{p}$ be the map of measures that p induces. Because G/H is metrizable and σ-compact, there exist $x_1, x_2, \ldots \in G$ such that $\{x_j + H: 1 \leq j < \infty\}$ is dense in G/H and

$$\check{p}|\mu|(p(K + H) + x_j + H) = \check{p}|\mu|(p(x_j + K + H)) = 0 \quad \text{for } j = 1, 2, \ldots.$$

Of course $\check{p}|\mu|(p(x_j + K + H)) = |\mu|(x_j + K + H)$. Let

$$S = (K + H) \cap \bigcup_{1}^{\infty} (x_j + K + H),$$

and let $A = K \backslash S$. Then $|\mu|(A) = |\mu|(K)$, but if $A - A$ contains an open set U, then for some finite $j \geq 1$, $(x_j + H) \in pU$, so $x_j + H \subseteq H + U \subseteq A - A + H$. But then there exist $a_1, a_2 \in A$ and $h_0, h_1 \in H$ such that $x_j + h_0 = a_1 - a_2 + h_1$, or

$$a_1 + h_1 = x_j + a_2 + h_0.$$

That leads to the contradiction: $(A + H) \cap (x_j + K + H) \neq \emptyset$. Theorem 8.3.4 is proved. □

Let $\psi \in \Delta$. Then the restriction of ψ to the discrete measures determines a character ψ^d of G. It is easy to see that the map $y \mapsto (\delta_y * \mu)\hat{}(\psi)$ from G to $\mathbb{C}$ is continuous at $y = 0$ if and only if either $\hat{\mu}(\psi) = 0$ or ψ^d is a continuous character.

8.3.5. Theorem. *Let G be a non-discrete locally compact abelian group with dual group Γ. Then*

$$\{\psi \in \Delta M(G) : \psi^d \text{ is not continuous}\}$$

is dense in $\Delta M(G) \backslash \Gamma$.

Proof. It suffices to show that for every $\rho \in \Delta \backslash \Gamma$ such that ρ^d is continuous and for every finite set of measures $\{\mu_1, \ldots, \mu_n\}$, there exists $\psi \in \Delta$ such that ψ^d is discontinuous and $\hat{\mu}_j(\rho) = \hat{\mu}_j(\psi)$ for $1 \leq j \leq n$.

Let $\omega = \exp(|\rho| \sum_{j=1}^{n} |\mu_j|)$. Then ω and ω^2 are mutually absolutely continuous. Because $\rho \in \Delta \backslash \Gamma$, $\rho\nu = 0$ for all $\nu \in L^1(G)$. Because $\rho\omega \approx \omega$, that implies $\omega \perp L^1(G)$.

We claim that

$$H = H(\omega) = \{y \in G : \delta_y * \omega \not\perp \omega\}$$

is a group. For suppose that $x, y \in H$. Then $\delta_x * \omega \not\perp \omega$, so $\delta_{-x} * \delta_x * \omega = \omega \not\perp \delta_{-x} * \omega$, by A.6.1. Thus $x \in H$ implies that $-x \in H$. Since $\omega \approx \omega^2$,

$$\delta_{x+y} * \omega \approx \delta_x * \omega * \delta_y * \omega \not\perp \omega * \omega \approx \omega.$$

Therefore $x + y \in H$. Therefore H is a group and the claim is proved. By 8.3.3, H is a Borel subset and therefore a Borel subgroup of zero Haar measure.

We claim that there exists a character γ' on the discrete group $(G/H)_d$ such that the character γ induced on G by $x \mapsto \langle \gamma', x + H \rangle$ is not continuous. For if H is closed, then G/H is a non-discrete locally compact abelian group and therefore has a discontinuous character γ'. Of course γ must be discontinuous on G, because γ is constant on cosets of H. If H is not closed, then some character γ' on G/H must have γ discontinuous or else H would be the intersection of the closed kernels of the γ as γ' ranged over the characters on G/H and H would be closed, a contradiction.

We now let γ be any discontinuous character on G which equals one on H. Let $N \subseteq M(G)$ be the L-subalgebra of measures generated by ω and $M_d(G)$. Then every $\nu \in N$ can be uniquely expressed in the form

$$\nu = \nu_0 + \sum_{j=1}^{\infty} \delta(x_j) * \nu_j,$$

where $\nu_0 \in M_d(G)$, $\nu_j \ll \omega_c$ for all j and the x_j belong to distinct cosets of H. Of course the definition of H implies that the $\delta(x_j) * \nu_j$ are mutually singular. Therefore the formula

$$F(\nu) = \sum_{1}^{\infty} \langle x_j, -\gamma \rangle \hat{\nu}_j(0) + \hat{\nu}_0(\gamma) \tag{3}$$

defines a bounded functional on N. It is easy to see that F is in fact multiplicative and that if ψ is the generalized character determined by F, then $|\psi_\nu| \equiv 1$ a.e. $d\nu$ for all $\nu \in N$ and $\psi_\omega \equiv 1$ a.e. $d\omega$.

Because $|\psi_\nu| \equiv 1$ for all $\nu \in N$, $\psi \in \partial N$ by 5.1.3. Therefore (Rickart [1, 3.3.26]) ψ may be extended to an element of $\partial M(G)$. We shall assume that ψ has been so extended.

Since $M_d(G) \subseteq N$, we have $\psi^d = \gamma$, so ψ^d is discontinuous. Since ρ^d is continuous, $(\rho\psi)^d = \rho^d\psi^d$ is discontinuous. We claim that

$$\hat{\mu}_j(\rho\psi) = \hat{\mu}_j(\rho) \quad \text{for } 1 \le j \le n.$$

Indeed, $\rho\mu_j \ll \omega$, and $\psi_\omega \equiv 1$. Therefore $(\psi\rho)_{\mu_j} \equiv \rho_{\mu_j}$ a.e. $d\mu_j$ for $1 \le j \le n$. That establishes the claim, and the proof of Theorem 8.3.5 is complete. □

The proof of the next result is left to the reader, who may consult Brown and Moran [5, pp. 308–309].

8.3.6. Theorem. *Let G and Γ be dual locally compact abelian groups. Let $\mu \in M(G)$. Then the following are equivalent.*

(i) $\mu \in M_o(G)$.
(ii) $y \to (\delta_y * \mu)\hat{}(\psi)$ *is continuous at* $y = 0$ *for all* $\psi \in \Gamma^- \backslash \Gamma$.
(iii) $\lim_{y \to 0} \sup_{\gamma \in \Gamma} |\langle y, \gamma \rangle \hat{\mu}(\gamma) - \hat{\mu}(\gamma)| = 0$.

Before giving a characterization of $M_c(G)$, we must make some preliminary remarks.

Let Γ be a σ-compact abelian group which is not compact. It is not hard to construct an increasing sequence $K_1 \subseteq K_2 \subseteq \cdots$ of open subsets of Γ, each of which has compact closure, such that

$$\Gamma = \bigcup_{j=1}^{\infty} K_j \tag{4}$$

and

$$\lim_{j \to \infty} \frac{m_\Gamma((\gamma + K_j) \cap K_j)}{m_\Gamma(K_j)} = 1 \quad \text{for all } \gamma \in \Gamma. \tag{5}$$

It is easy to see that if the K_j are as above, then

$$\lim_{j\to\infty}[m_\Gamma(K_j)]^{-1}\int_{K_j}\hat{\omega}(\gamma)d\gamma = \int_{b\Gamma}\hat{\omega}(x)dx \tag{6}$$

for all $\omega \in M_d(G)$, where G is the dual group of Γ. For a proof of (6), see Hewitt and Ross [1, Vol. I, pp. 255–256].

The next result is a variant of A.2.1.

8.3.7. Lemma. *Let G and Γ be dual locally compact abelian groups, and suppose that Γ is not compact. Let $\{K_j\}$ be as above, and $\mu \in M(G)$. Then the following are equivalent:*

(i) $\mu \in M_c(G)$.
(ii) $\lim_{j\to\infty}[m_\Gamma(K_j)]^{-1}\int_{K_j}|\hat{\mu}(\gamma)|^2\,d\gamma = 0$.
(iii) $\lim_{j\to\infty}[m_\Gamma(K_j)]^{-1}\int_{K_j}|\hat{\mu}(\gamma)|d\gamma = 0$.

Proof. For $j \geq 1$, let f_j be $1/m_\Gamma(K_j)$-times the indicator function of K_j. Then, by (6), $\lim_{j\to\infty}\hat{f}_j(y) = 0$ for all $y \in G\backslash\{0\}$, and $\hat{f}_j(0) = 1$ for all j.

If (i) holds, then $\lim_j[m_\Gamma(K_j)]^{-1}\int_{K_j}|\hat{\mu}(\gamma)|^2 = \lim_j \int \hat{f}_j\, d\mu * \tilde{\mu} = \mu * \tilde{\mu}(\{0\}) = 0$ by the dominated convergence theorem. Thus (i) implies (ii).

If (ii) holds, we apply the Cauchy-Schwarz inequality to obtain the following estimate.

$$\left[(m_\Gamma(K_j))^{-1}\int_{K_j}|\hat{\mu}(\gamma)|d\gamma\right]^2 \leq [m_\Gamma(K_j)]^{-1}\int_{K_j}|\hat{\mu}(\gamma)|^2\,d\gamma.$$

Therefore (ii) implies (iii).

Suppose that (iii) holds. Then

$$\lim_{j\to\infty}[m_\Gamma(K)]^{-1}\int_{K_j}|\langle y,\gamma\rangle\hat{\mu}(\gamma)|d\gamma \geq \lim_{j\to\infty}\left|\int \hat{f}_j\, d\delta_y * \mu\right| = \mu(\{-y\}) = 0$$

for all $y \in G$. Therefore (iii) implies (i). The proof of the Lemma is complete. □

8.3.8. Theorem. *Let G and Γ be dual locally compact abelian groups. Suppose that Γ is not compact and that $\{K_j\}$ is a sequence of compact subsets of Γ such that (4) and (5) hold. For $\mu \in M(G)$ let $N(\mu) = \sup_j[m_\Gamma(K_j)]^{-1}\int_{K_j}|\hat{\mu}(\gamma)|d\gamma$. Then $\mu \in M_c(G)$ if and only if*

$$\lim_{y\to 0} N(\delta_y * \mu - \mu) = 0. \tag{7}$$

Proof. Suppose that (7) holds. Let $\varepsilon > 0$, and let U be a symmetric compact neighborhood of the identity of G such that

$$N(\delta_y * \mu - \mu) < \varepsilon \quad \text{for all } y \in U.$$

Let φ be $(m_G(U))^{-1}$-times the characteristic function of U. Then $\hat{\varphi}$ is a real-valued function in $C_0(\Gamma)$. Let $K \subseteq \Gamma$ be a compact set such that $|\hat{\varphi}(\gamma)| \leq \frac{1}{2}$ for all $\gamma \in \Gamma \backslash K$. Then

$$[m_G(U)]^{-1} \int_U (1 - \langle \gamma, y \rangle) dy \geq \tfrac{1}{2} \quad \text{if } \gamma \in \Gamma \backslash K.$$

Therefore

$$(8) \qquad [m_G(U)]^{-1} \int_U |1 - \langle \gamma, y \rangle| dy \geq \tfrac{1}{2} \quad \text{for } \gamma \in \Gamma \backslash K.$$

Since $N(\delta_y * \mu - \mu) < \varepsilon$ for $y \in U$,

$$(9) \quad [m_\Gamma(K_j)]^{-1} \int_{K_j} |1 - \langle \gamma, y \rangle| |\hat{\mu}(\gamma)| d\gamma < \varepsilon \quad \text{for all } j \quad \text{and} \quad y \in U.$$

We multiply (9) by $[m_\Gamma(U)]^{-1}$ and integrate over U. A change of the order of integration then shows that

$$[m_G(U) m_\Gamma(K_j)]^{-1} \int_{K_j} |\hat{\mu}(\gamma)| \int_U |1 - \langle \gamma, y \rangle| dy\, d\gamma < \varepsilon, \quad \text{for } j \geq 1.$$

Therefore

$$(10) \quad [m_G(U) m_\Gamma(K_j)]^{-1} \int_{K_j \backslash K} |\hat{\mu}(\gamma)| \int_U |1 - \langle \gamma, y \rangle| dy\, d\gamma < \varepsilon, \quad \text{for } j \geq 1.$$

Then (8) and (10) imply that

$$[m_\Gamma(K_j)]^{-1} \int_{K_j \backslash K} |\hat{\mu}(\gamma)| d\gamma < 2\varepsilon \quad \text{for } j \geq 1.$$

But by (6) there exists an integer J so large that $j \geq J$ implies that

$$[m_\Gamma(K_j)]^{-1} \int_K |\hat{\mu}(\gamma)| d\gamma < \varepsilon.$$

Therefore

$$[m_\Gamma(K_j)]^{-1} \int_{K_j} |\hat{\mu}(\gamma)| \, d\gamma < 3\varepsilon \quad \text{if } j \geq J.$$

Now Lemma 8.3.8 shows that $\mu \in M_c(G)$.

If $\mu \in M_c(G)$, then Lemma 8.3.7 and an easy 2ε argument show that (7) holds. The proof of 8.3.8 is finished. □

Remarks and Credits. That (1) characterizes $L^1(G)$ is due to Plessner [1] (for $G = R$) and Raĭkov [1] (for general locally compact groups). Formula (2) seems to have been found first by Wiener and Young [1] for $G = R$, while 8.3.2 was discovered (for $G = R$) by Milicer-Grużewska [1], a year before the paper of Plessner. For another version of 8.3.1, see 12.1.4.

Rudin [8] proves that $\mu \in L^1(G)$ if and only if $x \mapsto |\mu|(x + E)$ is continuous for all Borel sets E. Glicksberg [2], extending results of Larsen and Tam, shows that if μ is a locally finite Borel measure on G and if there is a sequence $\{x_n\}_1^\infty \subseteq G$ such that $\{|\mu|(x_n + E)\}_1^\infty$ is dense in $\{|\mu|(x + E): x \in G\}$ for all compact sets E, then μ is absolutely continuous with respect to Haar measure.

The extension of 8.3.2 and 8.3.3 from R to general locally compact groups requires no new ideas, and we have followed the treatment of Wiener and Young.

The implication (ii) ⇒ (i) of 8.3.4 is due to Simmons [1], who uses a version of 8.3.2; an elegant proof that (i) ⇒ (ii) was found by Stromberg [1]. Theorem 8.3.4 is valid for non-abelian locally compact groups, and the proof given here applies without essential change. Theorem 8.3.5 provides a simple characterization, due to Brown and Moran [5], of Rad $L^1(G)$: $\mu \in \text{Rad } L^1(G)$ if and only if the mapping $y \mapsto (\delta_y * \mu)^\wedge(\psi)$ is continuous at $y = 0$ for all $\psi \in \Delta \backslash \Gamma$. The argument of Brown and Moran [5] was modified by Graham [8] to obtain 8.3.5. The proof of 8.3.5 shows that no singleton $\{\psi\} \subseteq \Delta \backslash \Gamma$ is a G_δ. Brown and Moran [5] proved 8.3.6, thereby strengthening Goldberg and Simon [1, Theorem B].

8.3.7 appears in Hewitt and Stromberg [2]. 8.3.8 is due to Goldberg and Simon [1].

For $\mu \in M(R)$, $H(\mu) = \{x: \delta_x * \mu \not\perp \mu\}$ can have various properties: it may contain only 0 (when μ is continuous and concentrated on an independent set—See 6.2.3); it may be countable (for example, if μ is Lebesgue measure on the Cantor middle third set, then $H(\mu)$ is the triadic rationals (Hille and Tamarkin [1], Wiener and Young [1])); it may be perfect (for example, for certain Riesz products—See Section 7.2). The function $\|\delta_x * \mu - \mu\|$ is usually badly behaved. When μ is Lebesgue measure on the middle-third set, $\|\delta_x * \mu - \mu\|$ is not of bounded variation (Milicer-Gruzewska [1]).

Saka [1] characterizes $M(G_\tau)$, G_τ a refinement of G (see Section 5.2), in terms of continuity of translations.

8.4. Non-symmetric Maximal Ideals in $M(G)$

For a Borel subset E of the locally compact abelian group G, let $\mathscr{A}(E)$ be the L-subalgebra of $M(G)$ that is generated by $M_c(E)$. For $\mu \in M(G)$, let $N(\mu)$ be the L-subalgebra generated by μ. For the definition of the projective tensor product $A \hat{\otimes} B$, see Section 11.1.

8.4.1. Lemma. *Let G be a metrizable non-discrete locally compact abelian group. Let $\mu \in M_s(G)$. Then there exists a compact independent M_o-set E of G such that the linear map $\varphi: \mathscr{A}(E) \hat{\otimes} N(\delta_0 + \omega) \to M(G)$ induced by $\varphi(\lambda \otimes \nu) = \lambda * \nu$ is an isomorphism whenever $0 \leq \omega^2 \ll \omega \ll \mu$.*

Proof. Let W be a σ-compact subset of zero Haar measure on which μ is concentrated, and let E be given by 6.3.13.

By 6.3.12 (i),

$$(1) \qquad \delta_y * \omega * \nu_1^{j(1)} * \cdots * \nu_m^{j(m)} \perp \omega * \nu_1^{k(1)} * \cdots * \nu_n^{k(n)}$$

whenever $\omega \ll \mu$, $1 \leq m \leq n$, $\nu_1, \ldots, \nu_n \in M_c(E)$ are mutually singular, $y \in Gp\, E$ and $(j(1), \ldots, j(m))$ and $(k(1), \ldots, k(m))$ are distinct m-tuples of non-negative integers. Let $\tau \in \mathscr{A}(E) \hat{\otimes} N(\delta_0 + \omega)$. Then τ has the form $\sum_1^\infty \tau_j$ where $\tau_j \in [\mathscr{A}(E) \cap M(jE)] \hat{\otimes} N(\delta_0 + \omega)$ for all j. By (1) $\|\tau\| = \sum \|\tau_j\| \geq \sum \|\varphi\tau_j\|$. Thus we may assume $\tau = \tau_j$ for some j, and by (1) again that $\tau = \sigma_1 * \cdots * \sigma_j * \nu'$ where $\nu' \ll \delta_0 + \mu$ and $\sigma_i \in M_c(E)$ for all i. An easy application of 6.3.12(ii) shows that $\|\sigma_1 * \cdots * \sigma_j * \nu'\| = \|\sigma_1\| \cdots \|\sigma_j\| \, \|\nu'\| = \|\varphi\tau\| = \|\tau\|$. That proves that φ is an isomorphism and ends the proof of 8.4.1. $\square$

8.4.2. Lemma. *Suppose that the interior U of $\Sigma \backslash \Gamma$ in $\Delta \backslash \Gamma$ is non-empty. Then U contains a strong boundary point for $M(G)$.*

Proof. Let $\psi \in U$, and let $V \subseteq U$ be any neighborhood of ψ. Since $M(G)^\wedge|_\Sigma$ is a symmetric separating subalgebra of $C(\Sigma)$, $M(G)^\wedge|_\Sigma$ is uniformly dense in $C(\Sigma)$. Therefore there exists $\nu_1 \in M(G)$ such that $\hat{\nu}_1(\psi) \geq \frac{3}{4}$, $0 \leq \hat{\nu}_1(\rho) \leq 1$ for all $\rho \in \Sigma$, and $\hat{\nu}_1(\rho) \leq \frac{1}{2}$ for $\rho \in \Sigma \backslash (V \cup \Gamma)$. Let $\mu = \delta_0 - [\hat{\nu}_1(\psi)\delta_0 - \nu_1]^2$. Then $\hat{\mu}(\psi) = 1$, $0 \leq \hat{\mu} \leq 1$ on Σ, and $\hat{\mu}(\rho) \leq \frac{15}{16}$ for $\rho \in \Sigma \backslash (V \cup \Gamma)$.

We apply the Local Peak Set Theorem 8.4.4 (below) to the case $B = M(G)/\text{Rad}\, L^1(G)$, $X = \{\psi : \hat{\mu}(\psi) = 1, \ \psi \in \Delta \backslash \Gamma\}$, $W = V$ and $f = \hat{\mu}|_{\Delta \backslash \Gamma}$: Thus there exists $\nu \in M(G)$ such that $\hat{\nu}(\psi) = 1$, $\sup\{|\hat{\nu}(\rho)| : \rho \in \Delta M(G) \backslash \Gamma\} = 1$, and $|\hat{\nu}| < \frac{1}{4}$ on $\Delta \backslash (\Gamma \cup V)$. (We take a power of the g given by 8.4.4.) By replacing ν by $\nu * \tilde{\nu}$ and using the fact that $\tilde{\nu}^\wedge = \hat{\nu}^-$ on Σ, we see that we may assume that $\hat{\nu} \geq 0$ on Σ. Let $c = \sup\{\hat{\nu}(\rho) : \rho \in \Sigma\}$. Then $\omega = \delta_0 - (\delta_0 - \nu)^2/c^2$ has $\hat{\omega}(\psi) = 1$, $|\hat{\omega}(\rho)| \leq 1 - (3c/4)^2$ for $\rho \in \Delta \backslash (V \cup \Gamma)$, and $0 \leq |\hat{\omega}(\rho)| \leq 1$ for all $\rho \in \Delta$.

Suppose that Γ is σ-compact. Then there exists $f \in L^1(G)$ such that $0 < \hat{f} < 1$ on Γ. We replace ω by $\omega_1 = \omega - f * \omega$. We claim that ω_1 peaks inside V. Indeed, $\hat{\omega}_1 < \hat{\omega}$ on Γ and $\hat{\omega}_1 = \hat{\omega}$ on $\Delta\backslash\Gamma$. That establishes the claim. Therefore ψ is itself a strong boundary point for $M(G)$.

If Γ is not σ-compact, we argue as follows. If $\psi \notin \Gamma^-\backslash\Gamma$, then the argument above (omitting the use of f) shows that ψ is again a strong boundary point. Thus we may assume that $\psi \in \Gamma^-\backslash\Gamma$. Then there exists a σ-compact open subgroup Λ of Γ such that for some sequence $\{\lambda_j\} \subseteq \Lambda$, $\lim \hat{\omega}(\lambda_j) = 1$. Let $\tau \in M(G)$ be the idempotent such that $\hat{\tau} = \chi_\Lambda$. Let $f \in L^1(G)$ be such that $0 < f \leq 1$ on Λ. Then $\omega_1 = \tau * (\omega - f)$ peaks on $\Gamma^-\backslash\Gamma$, so there exists a strong boundary point ψ' such that $\hat{\omega}_1(\psi') = 1 = \hat{\omega}(\psi')$. Therefore $\psi' \in U$. That completes the proof of the Lemma. $\square$

8.4.4. Rossi's Local Peak Set Theorem. *Let B be a commutative Banach algebra the following hold.*

(i) $\Sigma M(G)\backslash\Gamma$ *has no interior in* $\Delta M(G)\backslash\Gamma$.
(ii) $\Sigma M_o(G)\backslash\Gamma$ *has no interior in* $\Delta M_o(G)\backslash\Gamma$.
(iii) *Let* $\mu_1 \in M(G)$. *Then there exists a probability measure* $\omega \in M_o(G)$ *such that for all complex numbers* z *with* $|z| \leq 1$, *there are* 2^c *elements* $\psi \in \Delta M(G)\backslash\Sigma M(G)$ *such that* $\hat{\omega}(\psi) = z$ *and* $|\hat{\mu}_1(\psi)| = \sup\{|\hat{\mu}_1(\rho)|: \rho \in \Delta\backslash\Gamma\}$.
(iv) $\text{Rad } L^1(G) = \{\mu \in M(G): \hat{\mu}(\psi) = 0$ *for all* $\psi \in \Delta\backslash\Gamma\}$.
(v) $\partial M_o(G)^- = \partial M(G)$.

Proof. (iii) $\Rightarrow$ (i). (For the implication (iii) $\Rightarrow$ (ii), simply replace Δ by Δ_o and Σ by Σ_o in what follows.) We argue by contradiction, supposing that $\Sigma\backslash\Gamma$ has non-empty interior in $\Delta\backslash\Gamma$. An application of 8.4.2 then shows that $\Sigma\backslash\Gamma$ contains a strong boundary point ψ. Let $\mu \in M(G)$ be such that $\hat{\mu}(\psi) = 1 = \|\hat{\mu}\|_\Delta$ and $|\hat{\mu}| < 1$ off $\Sigma\backslash\Gamma$. By (iii), there exist $\omega \in M_o(G)$ and $\rho \in \Delta\backslash\Sigma$ such that $\hat{\omega}(\rho) = 1 = |\hat{\mu}(\rho)|$. Therefore $\rho \in \Sigma$. That contradiction completes the proof that (iii) implies (i).

(iii). We first suppose that G is metrizable. Let μ be the singular part of $\exp(|\mu_1|)$. Let E be as in the proof of 8.4.1. Let $\lambda \in \Delta\backslash\Gamma$ be such that $|\hat{\mu}_1(\lambda)| = \sup\{|\hat{\mu}_1(\rho)|: \rho \in \Delta\backslash\Gamma\}$. Then $|\lambda|^2 = |\lambda|$ by the proof of 8.2.1. Let $\omega = |\lambda|\mu$. Then $|\hat{\omega}(\lambda)| = \|\hat{\omega}\|_\Delta$, and ω and μ satisfy the hypotheses of 8.4.1. Let $\sigma \in M_o(E)$ be a fixed probability measure and let $\{\tau_\alpha\}$ be a fixed set of c mutually singular probability measures in $M_c(E)$ such that $\tau_\alpha \perp \sigma$ for all α. For each choice of $z = \{z_0, z_\alpha\}$, where $|z| = 1 = |z_\alpha|$ for all α, there exists $\psi(z) \in \Delta\mathscr{A}(E) \hat{\otimes} N(\delta_0 + \omega)$ such that

$$|\hat{\omega}(\psi^{(z)})| = |\hat{\omega}(\lambda)|, \hat{\sigma}(\psi^{(z)}) = z_0, \quad \text{and} \quad \hat{\tau}_\alpha(\psi^{(z)}) = z_\alpha$$

for all α. That follows at once from 6.3.12. Also by 6.3.12, and the fact that $\partial(A \hat{\otimes} B) = \partial A \times \partial B$ for commutative Banach algebras A and B, each $\psi^{(z)}$ belongs to the boundary of $\mathscr{A}(E) \hat{\otimes} N(\delta_0 + \omega)$. By 8.4.3, that algebra is mapped isometrically into $M(G)$. That and the standard extension theorem for elements of the Šilov boundary of a commutative Banach algebra

(Rickart [1, 3.3.27]) show that there exist elements $\psi^{(z)} \in \partial M(G)$ such that $\hat{\sigma}(\psi^{(z)}) = z_0$, $|\hat{\omega}(\psi^{(z)})| = |\hat{\omega}(\lambda)|$, and $\hat{\tau}_\alpha(\psi^{(z)}) = z_\alpha$ for all α. It remains to show that 2^c such $\psi^{(z)}$ can be found that are not in Σ. But on varying the z, one sees that there are 2^c possibilities for $\psi^{(z)}$ that have the same value at σ. It remains to show that those possibilities can be found in $\Delta \backslash \Sigma$. To that end, let us fix $\alpha(0)$. By dividing $\tau_{\alpha(0)}$ into successively two, four, ... parts, and taking weak$*$ limits, we see that there exist $\rho^{(z)} \in \partial$ such that $|\hat{\omega}(\rho^{(z)})| = |\hat{\omega}(\lambda)|$, $\hat{\sigma}(\rho^{(z)}) = z_0$, $\hat{\tau}_\alpha(\rho^{(z)}) = z_\alpha$ for all $\alpha \neq \alpha(0)$, and $(\rho^{(z)})_{\tau_{\alpha(o)}} = \frac{1}{2}$ a.e. $d\tau_{\alpha(0)}$. If $\rho^{(z)} \in \Sigma$, we let $\rho^{(z)} = |\rho^{(z)}|\gamma$ be the polar decomposition of $\rho^{(z)}$. Then $\gamma|\rho^{(z)}|^{(1+i)/2} \notin \Sigma$, as a simple application of 8.2.2 will show. The remaining details of the argument we leave to the reader. That ends the proof of 8.4.3(iii) for metrizable groups G.

Now suppose that G is not metrizable. Since $\mu = (\exp|\mu_1|)_s$ is singular, there exists a sequence of functions $\hat{f}_n \in A(G)$ such that $0 \le \hat{f}_n \le 1$, $\hat{f}_n \to 0$ a.e. dm_G, and $\int f_n \, d\mu = \|\mu\|$ for all n. Let Λ be the subgroup of the dual group of G that is generated by the union of the supports of the f_n. Then Λ is σ-compact. Let $H = \{x \in G : \langle x, \lambda \rangle = 1 \text{ for all } \lambda \in \Lambda\}$. Then G/H is metrizable, and the natural projection $p: G \to G/H$ induces a map $\check{p}$ of measures such that $\check{p}\mu$ is singular. We now use the result for the metrizable case, composing the elements of $\Delta M(G/H)$ with $\check{p}$ and using the fact that $\check{p}$ maps the non-negative measures in $M(G)$ onto those in $M(G/H)$. The remaining details are left to the reader. That ends the proof of 8.4.3(iii).

(iv) is obvious from (i). For (v), we argue as follows. Let $\psi \in \partial M(G)$ and let U be a neighborhood of ψ. It will suffice to find $\omega \in M_o(G)$ such that $\|\hat{\omega}\|_\Delta = 1$ and $\hat{\omega}(\rho) = 1$ for some $\rho \in U$, while $|\hat{\omega}| \le \frac{1}{2}$ on $\Delta \backslash U$. For, if we can find such a ω for each U, then each neighborhood U of ψ must meet ∂_o, so $\psi \in \partial_o^-$. Since $\partial_o \subseteq \partial$, we will have proved $\partial_o^- = \partial$.

It remains to find ω, given $\psi \in \partial$ and a neighborhood U of ψ. Because the strong boundary points for $M(G)$ are dense in ∂, we may assume that ψ is a strong boundary point. We may also assume that $\psi \in \partial \backslash \Gamma^-$. Let $\mu_1 \in M(G)$ be such that $\hat{\mu}_1(\psi) = \|\hat{\mu}_1\|_\Delta = 1$ and $|\hat{\mu}_1| \le \frac{1}{2}$ outside U. Let $\omega \in M_o(G)$ and $\rho \in \Delta$ be given by (iii) such that $\hat{\omega}(\rho) = 1$ and $|\hat{\mu}_1(\rho)| = 1$. Then $\rho \in U$. That completes the proof of (v) and ends the proof of 8.4.3. □

For a proof of the next result, consult Gamelin [1, p. 91], Gunning and Rossi [1, p. 62], or Rossi [1].

8.4.4. Rossi's Local Peak Set Theorem. *Let B be a commutative Banach algebra with maximal ideal space Δ. Suppose that B has an identity, that X is a compact subset of Δ, that W is a neighborhood of X, and that $f \in B$ is such that* $\sup\{|\hat{f}(w)| : w \in W\} = 1$, $f^{-1}(1) \neq \varnothing$, *and that* $|f| < \frac{1}{2}$ *on* $W \backslash X$. *Then there exists* $g \in B$ *such that* $\|\hat{g}\|_\Delta = 1$, $\hat{g}(x) = 1$ *if* $\hat{f}(x) = 1$, *and* $|\hat{g}| < 1$ *on* $\Delta \backslash X$.

Credits. Theorem 8.4.1 is a version of Brown, Graham and Moran [1, Theorem 2.1]. Parts (i)–(iii) of 8.4.3 are from Saeki [18]. Part (iv) of 8.4.3 is from Brown and Moran [6] and Graham [7]. Part (v) is from Brown and Moran [10]. Theorem 8.4.4 is from Rossi [1].

8.5. Point Derivations and Strong Boundary Points for $M(G)$

A linear functional d on a commutative Banach algebra B is a *point derivation at* $\psi \in \Delta B$ if

$$d(v\omega) = \hat{v}(\psi)d(\omega) + \hat{\omega}(\psi)d(v) \quad \text{for } v, \omega \in B.$$

A point derivation on the Banach algebra B may be B-norm bounded without being bounded in the spectral radius norm. For example, let $B = C^1(0, 1)$, the algebra of continuous and continuously differentiable functions on $[0, 1]$ with norm $\|f\|_B = \|f\|_\infty + \|f'\|_\infty$. Let $d(f) = f'(1)$. Then d is B-norm continuous ($|d(f)| \leq \|f\|_B$) but is not continuous in the spectral radius norm: $d(x^n) = n$ for $n \geq 1$ and $\|x^n\|_\infty = 1$. Since every point of $[0, 1]$ is a strong boundary point for $B = C^1(0, 1)$, the above discussion shows that the conclusion of the next proposition may be false if the norm on an algebra is not the spectral radius norm. We remind the reader that a *uniform algebra* (or *supnorm algebra*) is a commutative Banach algebra B whose norm is the spectral radius norm $\|\hat{f}\|_\Delta$.

8.5.1. Proposition. *Let B be a uniform algebra with maximal ideal space ΔB and identity 1. Let $\psi \in \Delta B$ be a strong boundary point. Then the only continuous point derivation at ψ is the zero functional.*

Proof. We claim that the ideal $N = \{f \in B: \hat{f}(\psi) = 0\}$ has a bounded approximate identity. To see that, fix $n \geq 1$, $f_1, \ldots, f_n \in N$, and $\varepsilon > 0$. Let $U \subseteq \Delta B$ be a neighborhood of ψ such that $\sup\{|\hat{f}_j(\rho)|: \rho \in U, 1 \leq j \leq n\} < \varepsilon$. Since ψ is a strong boundary point, there exists $g \in B$ such that $\hat{g}(\psi) = 1 = \|\hat{g}\|$, and $|\hat{g}(\rho)| < \varepsilon$ for $\rho \in \Delta B \backslash U$. Then

$$\|(1 - \hat{g})\hat{f}_j - \hat{f}_j\|_\Delta = \|\hat{g}\hat{f}_j\|_\Delta \leq \max\left(\sup_{\rho \in U}|\hat{g}(\sigma)\hat{f}_j(\rho)|, \sup_{\rho \notin U}|\hat{g}(\rho)\hat{f}_j(\rho)|\right) < \varepsilon\|\hat{f}_j\|_\Delta.$$

Of course, $1 - g \in N$ and $\|1 - \hat{g}\|_\Delta \leq 2$. That proves the claim.

Let $f \in N$, let $\varepsilon > 0$, and let $h \in N$ be such that $\|\hat{f} - \hat{f}\hat{h}\|_\Delta < \varepsilon$. Then

$$|d(f - hf)| < \varepsilon\|d\| \quad \text{and} \quad d(hf) = \hat{h}(\psi)d(f) + \hat{f}(\psi)d(h) = 0.$$

Hence $|d(f)| \leq \varepsilon\|d\|$ for all ε, so $d \equiv 0$ on N. Let $f \in B$. Then $f = f - \hat{f}(\psi)1 + \hat{f}(\psi)1$. Since $f - \hat{f}(\psi)1 \in N$, $d(f) = d(\hat{f}(\psi)1)$. But $1^2 = 1$ so $d(1) = 1(\psi)d(1) + 1(\psi)d(1) = 2d(1) = 0$. Thus, $d(f) = 0$ for all $f \in B$. That completes the proof of the Proposition. □

Proposition 8.5.1 implies that if ψ is a strong boundary point for either the Banach algebra B or for the closure of B in the spectral radius norm, then there is no nonzero point derivation at ψ that is continuous in the spectral radius topology.

Examples of point derivations are simple to produce for $M(G)$, as the next Proposition shows.

8.5.2. Proposition. *Let G be a locally compact abelian group. Let $\psi \in \Delta M(G)$ be such that $\psi \geq 0$ and $\psi \neq \psi^2$. For each $\mu \in M(G)$, we define*

$$d(\mu) = \frac{d}{dz} \int \psi^z \, d\mu \bigg|_{z=1}$$

Then d is a point derivation at ψ that is bounded in the spectral radius norm.

Proof. That d is a bounded linear functional follows from Cauchy's integral formula and the Fubini theorem:

$$d(\mu) = \frac{1}{2\pi i} \int_{|z-1|=1/2} (z-1)^{-2} \int \psi^z \, d\mu \, dz;$$

An obvious calculation shows that $|d(\mu)| \leq 2\|\hat{\mu}\|_\infty$. Now

$$\begin{aligned} d(\mu * \nu) &= \frac{d}{dz} [\hat{\mu}(\psi^z)\hat{\nu}(\psi^z)]|_{z=1} \\ &= \hat{\mu}(\psi) \frac{d}{dz} \hat{\nu}(\psi^z)|_{z=1} + \hat{\nu}(\psi) \frac{d}{dz} \hat{\mu}(\psi^z)|_{z=1} \\ &= \hat{\mu}(\psi)d(\nu) + \hat{\nu}(\psi)d(\mu). \end{aligned}$$

The proof of the Proposition is complete. □

A more general form of 8.5.2 is provided by the notion of analytic disc. An *analytic disc* in $M(G)$ is the image $\{\varphi(z): |z| < 1\}$ where φ is a non-constant analytic function with range in Δ; that is, each composition $\hat{\mu} \circ \varphi$ is analytic for $\mu \in M(G)$ and at least one such composition is not constant. Clearly if φ is analytic, then $\mu \mapsto [(d/dz)\hat{\mu} \circ \varphi(z)]|_{z=z_o}$ defines a point derivation on $M(G)$ for each z_o of modulus less than one. Also, that point derivation is continuous with respect to spectral radius norm.

Let H be a Borel subgroup of the locally compact abelian group G. Then there exists $\psi_H \in \Delta M(G)$ such that $(\psi_H)_\mu = 1$ a.e. $d\mu$ if μ is concentrated on a countable union of translates of H and $(\psi_H)_\nu = 0$ a.e. $d\nu$ if $|\nu|(x + H) = 0$ for all $x \in G$. The element ψ_H is that associated with the prime L-subalgebra of measures that are concentrated on sets in the Raĭkov system generated by H. See Section 5.2 for details.

8.5.3. Theorem. *Let G be a locally compact abelian group and let H be a σ-compact subgroup of G that has zero Haar measure.*

(i) *If $\rho \in \Delta M(G)$ and $|\rho| \leq \psi_H$, then there is an analytic disc in $\Delta M(G)$ that contains ρ.*
(ii) *There exists a point derivation on $M(G)$ at ψ_H that is continuous with respect to the total variation norm and not continuous with respect to the spectral radius norm.*

Proof. (i) Let N_0 be the L-subalgebra of $M(G)$ that is generated by $M(H) \cup M_d(G)$. Thus N_0 is the set of measures that are concentrated on countable unions of translates of H. Let E be a perfect compact perfect subset of G that is independent mod H. Such a set exists by 6.3.14(i). Let $\mathscr{A}(E)$ be the L-subalgebra of $M(G)$ that is generated by $M_c(E)$. For $j \geq 1$, let $N_j = N_0 * (\mathscr{A}(E) \cap M(jE))$. Then $N_j \perp N_k$ for $0 \leq j < k < \infty$ by 6.3.6. Therefore each $\omega \in M(G)$ has a unique representation

$$\omega = \omega_o + \sum_{j=1}^{\infty} (\mu_j * \omega_j) + \omega',$$

where $\mu_j \in N_0$ for $j \geq 0$, $\omega_j \in \mathscr{A}(E) \cap M(jE)$ for $j \geq 1$, and $\omega' \perp N_j$ for $j \geq 0$. By 6.3.11, the L-subalgebra $N = (\sum_{j=0}^{\infty} N_j)^-$ is prime, so the projection of $M(G)$ onto N is multiplicative. Hence each element of ΔN defines an element of $\Delta M(G)$: just compose with the projection onto N. Of course if $\rho \in \Delta M(G)$ and $|\rho| \leq \psi_H$, then ρ is already such a composition. Let $\rho \in \Delta M(G)$ with $|\rho| \leq \psi_H$. It will suffice to find an analytic disc in ΔN that contains ρ. For $|z| < 1$ and $\omega \in N$, let

$$\hat{\omega}(\varphi(z)) = \hat{\omega}_0(\rho) + \sum_{j=1}^{\infty} \hat{\mu}_j(\rho)\hat{\omega}_j(0)z^j.$$

By 6.3.7, $\|\omega_j * \mu_j\| = \|\omega_j\| \|\mu_j\|$ for all $j \geq 1$, so $|\hat{\omega}(\varphi(z))| \leq \|\omega\|$ for all ω and z. Because of the singularity of the N_j, $\hat{\omega}(\varphi(z))$ is well-defined. Because $N_j * N_k \subseteq N_{j+k}$ by 6.3.5, the map $\omega \mapsto \hat{\omega}(\varphi(z))$ is multiplicative for each z. Clearly φ is analytic and nonconstant. Of course $\varphi(0) = \rho$. Therefore ρ is contained in an analytic disc in ΔN. That ends the proof of 8.5.3(i). For the proof of (ii), see Saeki and Sato [1]. □

The proof of the following Corollary is left to the reader, who may apply 8.5.3, or argue directly, as in Brown and Moran [14].

8.5.4. Corollary. *Let G be a non-discrete locally compact abelian group. Let K be a perfect compact independent subset of G and let $\psi \in \Delta M(G)$ be such that*

$$\hat{\mu}(\psi) = \hat{\mu}_d(\psi) \quad \textit{for all } \mu \in M(G),$$

where $\mu = \mu_d + \mu_c$ is the decomposition of μ into its discrete and continuous parts. Then

$$d(\mu) = \sum_x \psi(\delta_x)\mu_c(x + K)$$

defines a point derivation at ψ that is continuous in the spectral radius norm.

The construction of an analytic disc in 8.5.3 depended on the existence of an idempotent $\psi \in \Delta M(G)$ that was strictly greater than ψ_H and was not the trivial idempotent (the identity of Γ). We define an order on $\Delta M(G)$ as follows. Let ψ, $\rho \in \Delta M(G)$. Then $\psi \gg \rho$ if $|\psi| \geq |\rho|$ and if $\psi/|\psi| = \rho/|\rho|$ wherever $|\rho| \neq 0$. We say that ψ is a *proper maximal element* if $|\psi| \neq 1$ and if ψ is maximal in $\Delta M(G)\backslash\Gamma$ with respect to the order $\gg$. Consideration of the discrete measures shows that for each $\psi \in \Delta$, there exists at most one $\gamma \in \Gamma$ such that $\gamma \gg \psi$. It is obvious that if ψ is proper maximal, then $|\psi|^2 = |\psi|$. A routine application of Zorn's Lemma shows that there exists, for each $\rho \in \Delta M(G)\backslash\Gamma$, a proper maximal element $\psi \gg \rho$. An application of Riesz products along the lines of 7.5.3 shows that there exist many proper maximal elements in $\Delta M(G)$ whenever G is a non-discrete locally compact abelian group.

8.5.5. Lemma. *Let G be a locally compact abelian group with dual group Γ. Let $\psi \in \Delta M(G)\backslash\Gamma$. If either ψ or $|\psi|$ is proper maximal, then $I = \{\rho\colon \rho \gg \psi, \rho \neq \psi\}$ is closed.*

Proof. Suppose that ψ is proper maximal. Clearly $\rho \gg \psi$ if and only if $\rho\bar{\psi} = |\psi|^2$. But multiplication in $\Delta M(G)$ is continuous in each variable separately, by 5.1.4(i). Therefore $\{\rho\colon \rho \gg \psi\}$ is closed. Since $I \cap \Gamma$ contains at most one element, there exists a neighborhood U of ψ such that $U \cap I \cap \Gamma = \varnothing$. It will suffice to show that $U \cap I = \varnothing$. But if $\rho \in U \cap I$, then $\rho \notin \Gamma$, so $|\rho| \neq 1$. Since $\rho \in I$, $\rho \gg \psi$. Since ψ is maximal, $\rho = \psi$. Hence $\rho \notin I$. That contradiction shows that $U \cap I = \varnothing$. Therefore I is closed. A similar argument applies in the case that ψ is proper maximal. $\square$

8.5.6. Theorem. *Let G be a locally compact abelian group with dual group Γ. Let $\psi \in \Delta M(G)\backslash\Gamma$. If either ψ or $|\psi|$ is proper maximal, then ψ is a strong boundary point for $M(G)$.*

Proof. Let Q be the set of probability measures $\mu \in M(G)$ such that $|\psi|\mu = \mu$. For each $\mu \in Q$, set

$$X_\mu = \{\rho\colon \rho\bar{\psi}\mu \geq \mu\}.$$

Then each set X_μ is compact in Δ and contains $I = \{\rho \in \Delta: \rho \gg \psi, \rho \neq \psi\}$. Since Q is convex, $\lambda = \frac{1}{2}(\mu + \nu) \in Q$ for all $\mu, \nu \in Q$. Of course, $X_\lambda \subseteq X_\mu \cap X_\nu$. It is clear that $\bigcap \{X_\mu : \mu \in Q\} = \{\psi\} \cup I$. Let U and V be neighborhoods of ψ and I such that $U \cap V = \varnothing$. Such neighborhoods exist by 8.5.5. Let $\mu \in Q$ be such that $X_\mu \subset U \cup V$. Let $\nu = \frac{1}{2}(\bar{\psi}\mu + (\bar{\psi}\mu)^2)$. Then $\|\hat{\nu}\| = \|\nu\| = \hat{\nu}(\psi) = 1$. Also, if $\hat{\nu}(\rho) = 1$, then $\rho\bar{\psi}\mu = \mu$, so $\rho \in X_\mu$. Therefore $X_\mu \cap U$ is a local peak set in $\Delta \backslash V$. By Rossi's Local Peak Set Theorem 8.4.4, there exists $\omega \in M(G)$ such that $\|\hat{\omega}\|_\Delta = 1$, $\hat{\omega}(\rho) = 1$ if and only if $\hat{\nu}(\rho) = 1$, and $|\hat{\omega}| < 1$ outside U. Since the above holds for all neighborhoods U of ψ with $U \cap V = \varnothing$, ψ is a strong boundary point. The Theorem is proved. □

Remarks and Credits. Proposition 8.5.1 is standard. Proposition 8.5.2 is from Taylor [1, 3]. Theorem 8.5.3(i) is an improvement of Brown and Moran [15, p. 47] and Saeki and Sato [1, Theorem 1]. (The argument that Saeki and Sato credit to an early version of this book is actually due to Brown and Moran, appearing in their [15].) Theorem 8.5.3(ii) is Saeki and Sato [1, Theorem 2]. Corollary 8.5.4 is from Brown and Moran [14], and was the surprising result that led to the other results to this Section. Theorem 8.5.6 is a slight improvement of Brown and Moran [16, Theorem 3] for the case of $M(G)$.

8.6. Gleason Parts for Convolution Measure Algebras

Let B be a commutative Banach algebra with identity. If $\psi, \rho \in \Delta B$, we say that $\psi \sim \rho$ if and only if

(1) $$\sup\{|\hat{\mu}(\psi) - \hat{\mu}(\rho)| : \mu \in B, \|\hat{\mu}\|_\Delta \leq 1\} < 2.$$

Using linear fractional transformations, it is not hard to show that (1) holds if and only if

(2) $$\sup\{|\hat{\mu}(\psi)| : \mu \in B, \|\hat{\mu}\|_\Delta \leq 1, \hat{\mu}(\rho) = 0\} < 1$$

and that $\sim$ is an equivalence relation on ΔB. The equivalence classes are called the *Gleason parts* for B. Any analytic disc must be contained entirely in one Gleason part. For proofs, see Gamelin [1, Chapter IV].

8.6.1. Theorem. *Let B be a commutative convolution measure algebra with an identity. Let S denote the structure semigroup of B and $\hat{S}$ the maximal ideal space of B. Let $\psi, \rho \in \hat{S}$. Then $\psi \sim \rho$ if and only if*

(3) $$\psi(s) = \rho(s) \text{ whenever either } |\psi(s)| = 1 \text{ or } |\rho(s)| = 1.$$

Proof. For simplicity we shall write $\psi \approx \rho$ in place of (3). The argument will follow these steps: We shall show that $\psi \sim \rho$ implies that $\psi \approx \rho$. Then we shall show that

$$(4) \qquad \psi \sim \rho \quad \text{and} \quad \lambda \in \hat{S} \quad \text{simply that } \lambda\psi \sim \lambda\rho$$

and that

$$(5) \qquad \psi \geq 0 \quad \text{and} \quad \psi \approx \rho \quad \text{imply that } \psi \sim \psi\rho.$$

Suppose that (4) and (5) have been established for all $\psi, \rho, \lambda \in \hat{S}$, and that $\psi \approx \rho$. Let $\psi = \psi_o|\psi|$ and $\rho = \rho_o|\rho|$ be the polar decompositions of ψ and ρ. Then $|\psi| \approx \bar{\psi}_o\rho_o|\psi|$, so $|\psi| \sim \bar{\psi}_o\rho_o|\psi|^2$ by (5). Since $z \mapsto \bar{\psi}_o\rho_o|\psi|^z$ is an analytic function into $\hat{S}, \bar{\psi}_o\rho_o|\psi|^2 \sim \bar{\psi}_o\rho_o|\psi|$. Thus $|\psi| \sim \bar{\psi}_o\rho_o|\psi|$. Multiply both sides by ψ_o and apply (4); then we obtain

$$\psi = \psi_o|\psi| \sim \rho_o|\psi|.$$

But $|\psi| \approx |\sigma|$, so (5) again implies that $|\psi| \sim |\psi||\rho|$. Similarly $|\rho| \sim |\psi||\rho|$, so $|\psi| \sim |\rho|$. We multiply this last by ρ_o and applying (4) we find that

$$\rho_o|\psi| \sim \rho_o|\rho| = \rho.$$

The last two displayed equations show that $\psi \sim \rho$.

It remains to be shown that $\psi \sim \rho$ implies that $\psi \approx \rho$ and that (4) and (5) hold.

$\psi \sim \rho$ implies that $\psi \approx \rho$. Suppose that $\psi \not\approx \rho$. Then there exists $s \in S$ such that $\psi(s) \neq \rho(s)$ and either $|\psi(s)| = 1$ or $|\rho(s)| = 1$. Assume that $|\psi(s)| = 1$. Let $\mu_\alpha \in B$ be probability measures that converge weak$*$ in $M(S)$ to δ_s. Then there exist choices of integers $n_\alpha \geq 1$ such that the measures

$$\nu_\alpha = \tfrac{1}{2}(\psi\mu_\alpha + (\psi\mu_\alpha)^2)^{n_\alpha}$$

have $\lim_\alpha \hat{\nu}_\alpha(\psi) = 1$ and $\lim \hat{\nu}_\alpha(\rho) = 0$. Since $\|\nu_\alpha\| \leq 1$, we obtain a contradiction of (2) unless $\psi \nsim \rho$. Therefore $\psi \nsim \rho$ implies $\psi \approx \rho$.

Proof of (4). Let (1) hold and let $\lambda \in \hat{S}$. For every $\mu \in B$ with $\|\hat{\mu}\|_\Delta \leq 1$,

$$|\hat{\mu}(\lambda\psi) - \hat{\mu}(\lambda\rho)| = |\hat{\nu}(\psi) - \hat{\nu}(\rho)|,$$

where $\nu = \bar{\lambda}\psi$. It follows that $\lambda\psi \nsim \lambda\rho$.

Proof of (5). Suppose that $\psi, \rho \in \hat{S}$, $\psi \geq 0$ and $\psi \approx \rho$. Let $\psi = |\psi|\psi_o$ and $\rho = |\rho|\rho_o$ be the polar decompositions of ψ and ρ. Let $\varphi \in A(T)$ be an odd function such that $\varphi(e^{it}) = t$ for $|t| < \delta$ where $t \in (-\pi, \pi)$, $\delta > 0$, and such that $\|\varphi\|_{A(T)} < \log(\frac{5}{4})$. Then $1 - e^{i\varphi(e^{it})} = \sum_{-\infty}^{\infty} a_n e^{int}$ where $\sum_{-\infty}^{\infty} |a_n| < e^{\log(5/4)} - 1 = \frac{1}{4}$. Define $\Phi \in C(S)$ by

$$\Phi(s) = \sum_{-\infty}^{\infty} a_n \rho_0(s)^n,$$

where $\rho_o(s)^{-n} = \overline{\rho_o(s)}^n$ for $n = 1, 2, \ldots$. Then $\Phi(s) = 1 - \rho_o(s)$ whenever $|t| = \arg \rho_o(s) < \delta$, where arg takes on values in $[-\pi, \pi]$. Because $\psi \approx \rho$ and $\psi \geq 0$, there exists an open set $U \supseteq X = \rho^{-1}(1)$ such that $u \in U$ implies $\arg \rho(u) = \arg \rho_o(u) \in (-\delta, \delta)$.

Similarly, $\exp(\varphi(e^{it})) = \sum_{-\infty}^{\infty} b_n e^{int}$, where $b_o = 1$ and $\sum_{n \neq 0} |b_n| < \frac{1}{4}$. Define $\Psi \in C(S)$ by

$$\Psi(s) = \sum b_n e^{in \log|\rho|} = \sum_{-\infty}^{\infty} b_n |\rho|^{in}.$$

Then $\Psi(s) = |\rho(s)|$ if $|\log|\rho(s)|| < \delta$. Let $V \subseteq U$ be a neighborhood of X such that $s \in V$ implies $|\log|\rho(s)|| < \delta$, and $\log|\psi(s)| > -1$. Define Λ by

$$\Lambda = (1 - \Psi) + \Psi\Phi.$$

Then

$$\Lambda(s) = 1 - |\rho(s)| + |\rho(s)|(1 - \rho_o(s)) = 1 - \rho(s), \ s \in V. \tag{6}$$

Let $\mu \in B$ and $\sup_S |\hat{\mu}| \leq 1$. We claim that then

$$\left| \int \Lambda \psi^z \, d\mu \right| \leq \tfrac{9}{16}. \tag{7}$$

Indeed,

$$\begin{aligned} \left| \int_S \Lambda \psi^z \, d\mu \right| &\leq \left| \int (1 - \Psi)\psi^z \, d\mu \right| + \left| \int \Psi\Phi\psi^z \, d\mu \right| \\ &\leq \sum_{n \neq 0} |b_n| |\hat{\mu}(\psi^z |\rho_o|^{-in})| + \sum_{m, n = -\infty}^{\infty} |b_n a_m| |\hat{\mu}(|\rho|^{-in} \rho_o^m \psi^z)| \\ &\leq \tfrac{1}{4} + (\tfrac{5}{4})(\tfrac{1}{4}) = \tfrac{9}{16}. \end{aligned}$$

Therefore

$$\left| \int_S (1 - \rho - \Lambda)\psi^z \, d\mu \right| \leq 2 + \tfrac{9}{16}. \tag{8}$$

Since $\psi > e^{-1}$ on V, (6) implies that

$$\text{(9)} \qquad \left|\int_S (1 - \rho - \Lambda)\psi^z \, d\mu\right| \le \left|\int_{\psi < e^{-1}} (1 - \rho - \Lambda)\psi^z \, d\mu\right| \le \sup_{s \in S} |1 - \rho(s) - \Lambda(s)| e^{-\operatorname{Re} z} \|\mu\|.$$

Let $\Gamma(z) = \int (1 - \rho - \Lambda)\psi^z \, d\mu$. Then Γ is analytic in the right-half plane. The inequalities (9) imply that $e^z\Gamma(z)$ is bounded. A straightforward computation using (8) shows that $|e^z\Gamma(z)| \le 3$ in the strip $0 < \operatorname{Re} z < \log 3 - \log(2 + \frac{9}{16})$. Therefore $|e^z\Gamma(z)| \le 3$ in the half-plane. Therefore

$$|\Gamma(z)| = \left|\int (1 - \rho - \Lambda)\psi^z \, d\mu\right| \le 3e^{-\operatorname{Re} z} \quad \text{for} \quad \operatorname{Re} z > 0.$$

But

$$|\hat{\mu}(\psi^z) - \hat{\mu}(\rho\psi^z)| = \left|\int (1 - \rho)\psi^z \, d\mu\right| \le \left|\int \Lambda\psi^z \, d\mu\right| + \left|\int (1 - \rho - \Lambda)\psi^z \, d\mu\right|.$$

Fix $n = z > 0$ such that $3e^{-n} < \frac{5}{16}$. Then (7) and (9) show that

$$|\hat{\mu}(\psi^n) - \hat{\mu}(\rho\psi^n)| \le \tfrac{9}{16} + \tfrac{5}{16} < 1.$$

Now n is independent of μ, so we see that $\psi^n \sim \rho\psi^n$. But $\psi \sim \psi^n$ and $\rho\psi \sim \rho\psi^n$. Thus $\psi \sim \psi\rho$ and (5) is proved. That completes the proof of 8.6.1. $\square$

Remarks and Credits. Theorem 8.6.1 is due to Brown and Moran [13]. That $\psi \sim \rho \Rightarrow \psi \approx \rho$ is due to R. R. Miller [1], who conjectured that (3) characterized Gleason parts and established that conjecture for non-negative ψ and ρ. Miller also established (4); Brown and Moran [13] give a simpler proof. The proof of (5) given here was suggested by Saeki.

Miller observed that (given his result) 8.6.3 would follow from "$\chi \approx |\chi|$ implies $\chi \sim |\chi|$." One way to attempt this would be to consider $\rho = \lim_{n\to\infty} |\chi|^n$ and work with ρ. One feels that ρ should be $\approx$ to $|\chi|$ so $\rho \sim |\chi|$ by Miller's result. Multiplying by χ_0, we see that $\chi_0\rho = \rho \sim \chi_0|\chi| = \chi$. Hence $|\chi| \sim \chi$.

Unfortunately, ρ may not be $\approx$ to $|\chi|$. In fact, ρ may be zero! That is because $\rho = \lim |\chi|^n$ is a weak $*$ limit. If $\{s \in S: |\rho(s)| = 1\}$ is not the closure of its interior, then $\rho \not\approx |\chi|$. Here is an example.

Let $E \subseteq T$ be a compact perfect independent set. Let B the L-subalgebra of $M(T)$ generated by $M_c(E)$ and δ_0. By 6.2.9, $\Delta B = \hat{S}$ may be identified with the set of elements $\psi \in M_c(E)^*$ with $\|\psi\| \le 1$.

Let $E_1, E_2, \ldots$ be a sequence of perfect pairwise disjoint subsets of E. Define $\psi, \rho, \lambda \in \hat{S}$ by

$$\psi(y) = \begin{cases} 1 - 2^{-n} & \text{if } y \in E_n, \\ 0 & \text{otherwise,} \end{cases} \qquad \rho(y) = \begin{cases} i(1 - 2^{-n}) & \text{if } y \in E_n, \\ 0 & \text{otherwise;} \end{cases}$$

and $\lambda(y) = 0$ for all $y \in E$. Then $\lambda = \lim \psi^n = \lim \rho^n$ weak$*$ in $\Delta B = \hat{S}$.

8.6.2. Proposition. *Under the preceding hypotheses, ψ, ρ, and λ are in three distinct Gleason parts of $\hat{S}$.*

Proof. For $n = 1, 2, \ldots$, let μ_n be a continuous probability measure concentrated on E_n. Then $\sup_{\Delta B}|\hat{\mu}_n| \leq 1$, $\hat{\mu}_n(\psi) = 1 - 2^{-n}$, $\hat{\mu}_n(\rho) = i - 2^{-n}i$ and $\hat{\mu}_n(\lambda) = 0$, for $n = 1, 2, \ldots$. Thus $\psi \not\sim \lambda$ and $\rho \not\sim \lambda$. Also,

$$\lim(\hat{\mu}_n^2(\psi) - \hat{\mu}_n^2(\rho)) = 2,$$

so $\psi \not\sim \rho$.

Chapter 9

The Wiener–Lévy Theorem and Some of Its Converses

9.1. Introduction

Let G and Γ denote locally compact abelian groups, each the dual of the other. Let $U \subseteq \mathbb{C}$ and let F be a complex-valued function defined on U. Suppose that $\hat{\mu}$ is a Fourier-Stieltjes transform on Γ with $\hat{\mu}(\Gamma) \subseteq U$. If $F \circ \hat{\mu}$ is also a Fourier-Stieltjes transform, we say F *operates on* μ, and we let $F \circ \mu$ denote the measure whose transform is $F \circ \hat{\mu}$. This chapter discusses necessary and sufficient conditions under which F operates on all μ that belong to varying classes of measures. These results have their origin in the theorem of Wiener and Lévy.

9.1.1. The Wiener–Lévy Theorem. *Let U be open in the plane and let the function F be analytic in U. If $\hat{f} \in A(T)$ and $\hat{f}(T) \subseteq U$, then $F \circ \hat{f} \in A(T)$.*

Wiener proved that for $F(z) = 1/z$ (Wiener [1, Lemma II_e]) and Lévy [1] gave the general case. A slightly more general form is available. A complex valued function F defined on an open set U is *real-analytic* on U if for each $x_0 + iy_0 \in U$, F has a power series expansion $\sum_{m,n\geq 0} b_{mn}(x - x_0)^m(y - y_0)^n$ which converges absolutely in a neighborhood of $x_0 + iy_0$. When U is the plane and one power series converges absolutely everywhere, F is *real-entire*. (The function $F(x + iy) = (1 + x^2 + y^2)^{-1}$ is real-analytic everywhere but not real-entire.) It is apparent from the definition that F is real-analytic if and only if $F(x + iy) = H(x, y)$, where $H(w, z)$ is analytic in both w and z.

9.1.2. Theorem. *Let Γ be an infinite abelian group. Let U be an open subset of the plane and F a real-analytic function defined on U. If Γ is not compact, let $0 \in U$ and $F(0) = 0$. Then $F \circ \hat{\mu} \in A(\Gamma)$ whenever $\hat{f} \in A(\Gamma)$ and $\hat{f}(\Gamma) \subseteq U$.*

Theorem 9.1.2 will be proved in Section 9.2. In Section 9.2 we shall state and prove also a stronger version of the Wiener-Lévy Theorem that is due to Marcinkiewicz. The remainder of the Chapter is devoted to various global converses to the Wiener-Lévy Theorem, which we now discuss.

Let $X \subseteq B(\Gamma)$ and $F\colon U \to \mathbb{C}$, where $U \subseteq \mathbb{C}$. We say that F *operates from* X *to* $B(\Gamma)$ if $\hat{\mu} \in X$ and $\hat{\mu}(\Gamma) \subseteq U$ imply that $F \circ \hat{\mu} \in B(\Gamma)$. We say that F

operates in X if $\hat{\mu} \in X$ and $\hat{\mu}(\Gamma) \subseteq U$ imply that $F \circ \hat{\mu} \in X$. The simplest way to describe the first converse to the Wiener-Lévy Theorem is this: real-analyticity is *necessary* as well as sufficient for F to operate in $A(\Gamma)$. That result and its proof appear in Section 9.3.

When Γ is not compact then a necessary and sufficient condition for F to operate in $B(\Gamma)$ is that F be (extendible to a function that is) real-entire. That (Theorem 9.4.1) is proved in Section 9.4.

In the remaining sections the functions operating in or on other classes of functions are discussed: $B_o(\Gamma) = \{\hat{\mu} \in B(\Gamma): \hat{\mu} \in C_o(\Gamma)\}$, in Section 9.5; $PD(\Gamma) = \{\hat{\mu} \in B(\Gamma), \mu \geq 0, \|\mu\| = 1\}$ in Section 9.6.

Each of sections 9.4 through 9.6 is essentially independent of the others. We have endeavored to illustrate as many different methods of proof as possible. The reader new to this subject might try using methods from one section to prove the theorems of another.

The reader who is familiar with the Gel'fand theory of Banach algebras will see that the Wiener–Lévy Theorem follows at once from that theory. In a number of places we do use Banach algebra theory in our proofs. We do not, however, use Banach algebra theory explicitly in proving the Wiener–Lévy Theorem, because we want to expose the particularities of the algebra at hand, namely $A(\Gamma)$.

The results of this Chapter are global results. For a discussion of local results, that is, of results concerning functions operating in $A(E)$ and $B(E)$ for various sets $E \subseteq \Gamma$, see Section 13.1.

9.2. Proof of the Wiener–Lévy Theorem and Marcinkiewicz's Theorem

A complex-valued function f on the locally compact abelian group Γ *belongs to* $A(\Gamma)$ *locally at* $\gamma \in \Gamma$ if there exists a neighborhood U of γ and $g \in A(\Gamma)$ such that $g = f$ on U. We say that f *belongs to* $A(\Gamma)$ *locally at infinity* if there exists a compact set $K \subseteq \Gamma$ and $g \in A(\Gamma)$ such that $f = g$ on $\Gamma \backslash K$. The following Lemma is easily proved by a partition of unity argument (for the details, see Rudin [1, p. 134]).

9.2.1. Lemma. *If the complex-valued function f on Γ belongs to $A(\Gamma)$ locally at γ for all $\gamma \in \Gamma$ and locally to $A(\Gamma)$ at infinity, then $f \in A(\Gamma)$.*

Proof of 9.1.2. Let $F(s + it)$ be real-analytic in a neighborhood U of $f(\Gamma)$. It will suffice to show that $F \circ f$ belongs to $A(\Gamma)$ locally at each $\gamma \in \Gamma$ and at infinity. Let us begin with $\gamma_0 \in \Gamma$ and let $s_0 + it_0 = f(\gamma_0)$. Then for some $\varepsilon > 0$,

$$F(s + it) = \sum b_{mn}(s - s_0)^m(t - t_0)^n$$

and

$$\sum |b_{mn}||s - s_0|^m |t - t_0|^n < \infty$$

if $|s - s_0| < \varepsilon$ and $|t - t_0| < \varepsilon$.

Since one-point sets obey spectral synthesis (A.3.1), there exists $g \in A(\Gamma)$ such that $\|g\| < \varepsilon/2$ and $f = f(\gamma_0) + g$ in a compact neighborhood U of γ_0. For $\gamma \in U, F \circ f(\gamma) = \sum b_{mn}(\text{Re } g)^m(\text{Im } g)^n$, and hence $F \circ f|_U$ has $A(U)$-norm bounded by $\sum |b_{mn}|\varepsilon^{m+n}$. Therefore $F \circ f \in A(\Gamma)$ locally at γ_0. It follows that $F \circ f \in A(\Gamma)$ if Γ is compact.

Now consider the case when Γ is not compact. Then $F(0) = 0$. Let $\varepsilon > 0$ be such that $F(s + it) = \sum b_{mn} s^m t^n$ and $\sum |b_{mn}||s|^m |t|^n < \infty$ if $|s| < \varepsilon$ and $|t| < \varepsilon$. Choose an element $g \in A(\Gamma)$ such that $\|g\| < 2$, $E = \text{supp } g$ is compact, $0 \le g \le 1$, and $\|f - gf\| < \varepsilon$. If $\gamma \in \Gamma \backslash E$, then

$$F \circ f(\gamma) = F \circ (f(\gamma) - g(\gamma)f(\gamma))$$

and the sum

$$h = \sum b_{mn}(\text{Re}(f - gf))^m(\text{Im}(f - gf))^n$$

converges in $A(\Gamma)$-norm. Hence $F \circ f$ agrees with the element h of $A(\Gamma)$ on $\Gamma \backslash E$. Thus $F \circ f$ belongs to $A(\Gamma)$ locally at infinity. That completes the proof of Theorem 9.1.2. □

9.2.2. Lemma. *(i) Let $0 < s \le 1$, and let $\{x_j\} \subseteq R$. If $\sum |x_j|^s < \infty$, then for each $\varepsilon > 0$, there exists $K = K(\varepsilon)$ such that*

$$\log \prod_{n=1}^{\infty} (1 + k|x_j|) \le \varepsilon k^s \quad \text{if } k \ge K. \tag{1}$$

(ii) Let F be a continuous periodic function on R with period $2\pi P$ and Fourier series expansion $F(x) = \sum_{-\infty}^{\infty} d_k \exp(ikx/P)$. Let $0 < s \le 1$. Suppose that F is infinitely often differentiable and that there exists $C > 0$ such that

$$|F^{(n)}(x)| \le C^n n^{n/s} \quad \textit{for } n \ge 1 \quad \textit{and all} \quad x \in [0, 2\pi P]. \tag{2}$$

Then there exist $C_1 > 0$, $\delta > 0$ such that

$$|d_k| \le C_1 \exp(-\delta |k|^s) \quad \textit{for } k \in Z. \tag{3}$$

Proof. (i) There exists $C > 0$ such that $1 + x \le \exp(Cx^s)$ for all $x \ge 0$. Then

$$\prod_{j=1}^{\infty} (1 + k|x_j|) \le \prod_{j=1}^{N} (1 + k|x_j|) \exp\left(C \sum_{j=N+1}^{\infty} |kx_j|\right)^s.$$

It is now easy to show that for each $\varepsilon > 0$, there exists K for which (1) holds.

(ii) Integration by parts shows that

$$|d_k| \leq |P/k|^n C^n n^{n/s} \quad \text{for } 1 \leq n, |k| < \infty.$$

There exists $K_1 > 0$ such that for each integer k with $|k| \geq K_1$, there exists $n = n(k)$ for which

$$\tfrac{1}{4} \leq |P/k| C n^{1/s} \leq \tfrac{1}{2}.$$

Therefore for $|k| \geq K_1$ and $n = n(k)$,

$$\begin{aligned} \log|d_k| &\leq n \log(|P/k| C n^{1/s}) \leq -n \log 2 \\ &\leq -(|k|/4PC)^s \log 2. \end{aligned}$$

Let $\delta = (\tfrac{1}{4}PC)^s \log 2$. Then

$$|d_k| \leq \exp(-\delta |k|^s) \quad \text{for } |k| \geq K_1,$$

and it follows that (ii) holds for some finite C_1. That ends the proof of 9.2.2. □

9.2.3. Marcinkiewicz's Theorem. *Let $f \in A(T)$ be a real-valued function with Fourier coefficients $\{c_n\}$. Let $0 < s \leq 1$. Suppose that $\sum |c_n|^s < \infty$. If F is an infinitely often differentiable periodic function and if for some $C > 0$,*

(4) $$|F^{(n)}(x)| \leq C^n n^{n/s} \quad \textit{for } n \geq 1 \quad \textit{and all} \quad x \in R,$$

then $F \circ f \in A(T)$.

Proof. Let $n, k \in Z$ and $P > 0$. Because $\|g\|_{A(T)} \leq \|g\|_\infty + 2\|g'\|_\infty$ for all $g \in C^1(T)$,

$$\begin{aligned} \|\exp(ik(c_n e^{inx} + c_{-n}e^{-inx})/P)\|_{A(T)} &= \|\exp(ik(c_n e^{ix} + c_{-n}e^{-ix})/P)\|_{A(T)} \\ &\leq (1 + 2|kc_n|/P)(1 + 2|kc_{-n}|/P). \end{aligned}$$

Therefore, for all $k \in Z$ and $P > 0$,

(5) $$\|\exp(ikf/P)\|_{A(T)} \leq \prod_{n=-\infty}^{\infty} (1 + 2|kc_n|/P).$$

Let $P \geq 1$ be a period for F such that $[-\pi P, \pi P]$ contains $f(T)$. Then 9.2.2(i) and (5) imply that for all $\varepsilon > 0$ and $K = K(\varepsilon)$,

$$\|\exp(ikf/P)\|_{A(T)} \leq \exp(\varepsilon |k|^s) \quad \text{for } k \geq K.$$

Let $F(x) = \sum_{k=-\infty}^{\infty} d_k \exp(ikx/P)$ be a Fourier series expansion for F. Let C_1 and $\delta > 0$ be given by 9.2.2(ii), and let $0 < \varepsilon < \delta$. Then for some $C_2 > 0$,

$$\|F \circ f\|_A \leq C_2 + \sum_{|k| \geq K(\varepsilon)} |d_k| \exp(\varepsilon |k|^s)$$

$$\leq C_2 + C_1 \sum \exp((\varepsilon - \delta)|k|^s) < \infty.$$

That proves 9.2.3. □

Remarks and Credits. The proof we have given of 9.1.2 is essentially that of Wiener and Lévy. For a simple proof, due to A. P. Calderòn, of 9.1.1, see Zygmund [1, Vol. 1, pp. 245–246]. Lemma 9.2.2(i) is adapted from Kahane [2, p. 78]; 9.2.2(ii) is from Marcinkiewicz [1]; the argument here was suggested by Saeki. Theorem 9.2.3 is due to Marcinkiewicz; for a further discussion, see Kahane [2, pp. 77–80]. For the converse to 9.2.3, see 9.3.9 and Rivière and Sagher [1].

9.3. Converses to the Wiener–Lévy Theorem

This section is devoted to the proof of the following theorems.

9.3.1. Theorem. *Let Γ be a non-discrete abelian group. Let U be an open connected subset of $\mathbb{C}$ and let $F: U \to \mathbb{C}$. If Γ is non-compact, suppose also that $0 \in U$. If $F \circ f \in A(\Gamma)$ whenever $f \in A(\Gamma)$ and $f(\Gamma) \subseteq U$, then F is real-analytic in U.*

9.3.2. Theorem. *Let Γ be an infinite discrete abelian group. Let U be an open neighborhood of 0 in the plane and let $F: U \to \mathbb{C}$. If $F \circ f \in A(\Gamma)$ whenever $f \in A(\Gamma)$ and $f(\Gamma) \subseteq U$, then F is real-analytic in a neighborhood of 0.*

9.3.3. Theorem. *Let Γ be a non-discrete abelian group. Let $F: [-1, 1] \to \mathbb{C}$. If $F \circ f \in A(\Gamma)$ whenever $f \in A(\Gamma)$ and $f(\Gamma) \subseteq [-1, 1]$, then F extends to a function analytic in a neighborhood of $[-1, 1]$.*

9.3.4. Theorem. *Let Γ be an infinite discrete abelian group. Let $F: [-1, 1] \to \mathbb{C}$. If $F \circ f \in A(\Gamma)$ whenever $f \in A(\Gamma)$ and $f(\Gamma) \subseteq [-1, 1]$, then F agrees with an analytic function in a neighborhood of 0.*

Remark. The hypotheses of 9.3.3 and 9.3.4 may be weakened slightly. Instead of requiring that $F \circ f \in A(\Gamma)$, it is enough to require only that $F \circ f \in B(\Gamma)$ as the reader will find easy to see.

The proofs of 9.3.3 and 9.3.4 will be given first, followed by an indication of the modifications necessary to obtain 9.3.1 and 9.3.2.

9.3.5. Lemma. *Let Γ be an infinite abelian group. Let $F:[-1, 1] \to \mathbb{C}$. If $F \circ f \in B(\Gamma)$ whenever $f \in A(\Gamma)$ and $f(\Gamma) \subseteq [-1, 1]$, then there exist $K > 0$ and $\delta > 0$ such that*

$$(1) \qquad \|F \circ f\| < K \quad \text{if} \quad f \in A(\Gamma), \|f\| < \delta \quad \text{and} \quad f(\Gamma) \subseteq [-1, 1].$$

Proof. We may assume that $F(0) = 0$. There are two cases.

Case I, Γ is not compact. Suppose that the conclusion is false. Then, for each $j \geq 1$, there exists $f_j \in A(\Gamma)$ such that

$$(2) \qquad f_j(\Gamma) \subseteq R, \|f_j\| < 2^{-j-1} \quad \text{and} \quad \|F \circ f_j\| > 2^j.$$

Then, by Theorem 1.8.1, there exists a compact set $E_j \subseteq \Gamma$ such that if $g \in B(\Gamma)$ and $g = F \circ f_j$ on E_j, then $\|g\| > 2^j$.

Let $k_j \in A(\Gamma)$ be chosen such that $\|k_j\| \leq 2$, k_j has compact support, $k_j = 1$ on E_j, and $0 \leq k_j \leq 1$. Let $h_j \in A(\Gamma)$ be such that $\|h_j\| < 2$, h_j has compact support D_j, and $h_j = 1$ on the support of k_j. Then

$$(3) \quad h_j F \circ (k_j f_j) = F \circ f_j \quad \text{on} \quad E_j \quad \text{and} \quad \|h_j F \circ (k_j f_j)\| > 2^j \quad \text{for } j \geq 1.$$

Since Γ is non-compact, there exist $\gamma_1, \gamma_2, \ldots, \in \Gamma$ such that

$$(4) \qquad (\gamma_n + D_n) \cap \bigcup_{j=1}^{n-1} (\gamma_j + D_j) = \varnothing.$$

Now for $g \in B(\Gamma)$, $\gamma \in \Gamma$ denote by g^γ the function $g^\gamma(p) = g(p - \gamma)$. Let $f = \sum k_j^{\gamma_j} f_j^{\gamma_j}$. Then $\|f\| < \sum_{j=1}^{\infty} 2 \cdot 2^{-j-1} < 1$. Since each f_j was real-valued, f is also real-valued. Hence $F \circ f \in B(\Gamma)$ by the hypothesis of the Lemma. Now, using (3) and (4), we see that for each j

$$(5) \qquad h_j^{\gamma_j} F \circ f = h_j^{\gamma_j} F \circ (k_j^{\gamma_j} f_j^{\gamma_j}) = F \circ (k_j^{\gamma_j} f_j^{\gamma_j}) = (F \circ k_j f_j)^{\gamma_j}.$$

By (3) and (5),

$$\|h_j^{\gamma_j}(F \circ f)\| = \|(F \circ k_j f_j)\| > 2^j$$

while

$$\|h_j^{\gamma_j} F \circ f\| \leq \|h_j^{\gamma_j}\| \, \|F \circ f\| \leq 2\|F \circ f\|.$$

That contradiction completes the proof for non-compact Γ.

Case II, Γ is compact. We will again suppose that the conclusion is false and reach a contradiction. Let $\{U_j\}$ be a sequence of non-empty open disjoint subsets of Γ. For each j we select functions k_j and f_j as follows. Let $k_j \in A(\Gamma)$ have norm at most 2, range contained in $[0, 1]$, and value 1 on a non-empty open set. To find f_j, we shall need several auxillary functions. Let g and h be elements of $A(\Gamma)$ such that each has norm at most 2, range contained in $[0, 1]$, and such that supp $h \subseteq k_j^{-1}(1)$, supp $g \subseteq h^{-1}(1)$, and $g^{-1}(1)$ has non-empty interior V. Let $\{\gamma_i + V: 1 \leq i \leq m\}$ be a finite set of translates of V whose union covers Γ. For each i, let $g_i = g^{\gamma_i}$. Then $u = \sum g_i > 0$ everywhere on Γ. By 9.1.2, $u^{-1} \in A(\Gamma)$. Therefore $\{u^{-1}g_i\}$ is a partition of unity. Let K be the supremum of the norms $\|u^{-1}g_i\|$, and let $f \in A(G)$ be such that $\|f\| < 2^{-j-1}$, and $\|F \circ f\| > 2^j mK$. Then for some i, $\|u^{-1}g_i F \circ f\| > 2^j$. For an appropriate translate h_j of h and an appropriate translate f_j of $h_j f$ the following hold: $\|f_j\| < 2^{-j}$, supp $f_j \subseteq k_j^{-1}(1)$, and $\|k_j F \circ f_j\| > 2^j$. To see the last, we merely apply the conditions on the supports of h, g and f; the details are left to the reader. That completes our selection of k_j and f_j.

We set $f = \sum_1^\infty f_j$ and suppose that $F \circ f \in A(\Gamma)$. Then $k_j F \circ f = k_j F \circ f_j$, so $\|F \circ f\| > 2^{j-1}$ for all j, a contradiction. The Lemma is proved. □

9.3.6. Lemma. *Let Γ be an infinite locally compact abelian group, let $\delta > 0$, and let r be a positive integer. Then*

$$(6) \qquad \sup\{\|e^{irf}\| : f \in A(\Gamma),\ f(\Gamma) \subseteq [-\delta, \delta],\ \|f\| \leq 1\} = e^r.$$

Proof. Obviously $\|e^{if}\| \leq e^{\|f\|}$, so we need to show that the value $e^r - \varepsilon$ is less than the supremum in (6) for all $\varepsilon > 0$.

Suppose that there exists a measure μ on G, the dual group of Γ, such that $\hat{\mu}(\Gamma) \subseteq [-\delta, \delta]$, $\|\mu\| \leq 1$, and $\|\exp(ir\mu)\| > e^r - \varepsilon$. Let $\{f_\alpha\} \subseteq L^1(G)$ be an approximate identity with $\|f_\alpha\| \leq 1$ and $\hat{f}_\alpha(\Gamma) \subseteq [-1, 1]$ for all α. Then for some α, $\|\exp(irf_\alpha * \mu)\| > e^r - \varepsilon$ and $f = f_\alpha * \mu$ will have $\|f\| \leq 1$, $\hat{f}(\Gamma) \subseteq [-\delta, \delta]$, and $\|\exp(ir\mu)\| > e^r - \varepsilon$. It remains to find μ, given ε.

First method. Suppose first that Γ is discrete, so its dual group G is compact. We may then suppose that Γ is countable and hence that G is metrizable. Then G contains a compact perfect independent M_o-set E. (See Theorem 4.7.1 for the proof when $G = T$; for the general case, the reader may adapt the argument of 4.7 or consult Lindahl and Poulsen [1, Section XIII.3]).

Let μ_1 and μ_2 be mutually singular probability measures in $M_o(E)$. Then the sets $E_j = \{\gamma \in \Gamma : |\hat{\mu}_j(\gamma)| \geq \frac{1}{2}\}$ are finite. Let $\gamma_1 \in \Gamma$ be such that $(E_1 + \gamma_1) \cap E_2 = \varnothing$. Then $\nu = \frac{1}{2}(\gamma_1\mu_1 + \mu_2)$ has $|\hat{\nu}| \leq \frac{3}{4}$ on Γ. Also, $\nu^r \perp \nu^s$ if $0 \leq r \neq s$, by 6.2.6. It is now clear that an appropriate power $\mu = \nu^k$ will do.

We now reduce to the case that Γ is discrete. In general, we can find $\mu \in M(bG)$ and a finite set $H \subseteq \Gamma_d$ such that $\|e^{ir\hat{\mu}}|_H\|_{A(H)} > e^r - \varepsilon$. We may find a discrete measure μ' on G such that $\hat{\mu}'(\Gamma) \subseteq [-\delta, \delta]$, $\|\mu'\| = 1$ and $|\hat{\mu}' - \hat{\mu}|$ is so small on H that $\|e^{ir\hat{\mu}'}|_H\|_{A(H)} > e^r - \varepsilon$. Then μ' will suffice.

Second method. As in the first method, we may suppose that G is compact. We let Θ be any infinite dissociate subset of Γ and let $a, b: \Theta \to [0, \delta/2]$ be such that

$$(7) \qquad \sum |a(\theta)^m b(\theta)^n - a(\theta)^p b(\theta)^q|^2 = \infty$$

if $1 \leq m, n, p, q < \infty$ and $(m, n) \neq (p, q)$. For example, if $a(\theta) = \delta/2$ for all θ and $b(\theta) = \delta/\pi$ for all θ, then (7) holds. Let μ_1 be the Riesz product based on Θ and a, and μ_2 the Riesz product based on Θ and b. Then 7.2.1(i) and (7) imply that

$$\mu_1^m * \mu_2^n \perp \mu_1^p * \mu_2^q \quad \text{if} \quad (m, n) \neq (p, q).$$

It is now easy to see that $\mu = \frac{1}{2}(\mu_1 - \mu_2)$ has the required properties. That ends the proof of the second method. □

9.3.7. Lemma. *Let Γ be an infinite abelian group, and suppose that $F:[-1, 1] \to \mathbb{C}$ operates from $A(\Gamma)$ to $B(\Gamma)$. Then F is continuous in $(-\delta, \delta)$ where δ is given by* 9.3.5.

Proof. We argue by contradiction, supposing that for some $|t| < \delta$ and sequence $\{t_j\} \subseteq (-\delta, \delta)$, $\lim t_j = t$ and $\lim F(t_j)$ exists but is not equal to $F(t)$. Let K be given by 9.3.5. By replacing $\{t_j\}$ by a subsequence and adding an absolutely convergent series to F, we may assume that $F(t) = 1$ and $F(t_j) = 0$ for all j. (That may increase K). Let E_1 and E_2 be disjoint compact subsets of Γ. Let $g, h \in A(\Gamma)$ be such that $g = t$ on $E_1 \cup E_2$, $\|g\| < \delta$, $h = 1$ on E_1, and $h = 0$ on E_2. Then for all j, $f = g + (t_j - t)h$ equals t_j on E_1 and t on E_2. Let j be so large that $\|g\| + |t_j - t|\,\|h\| < \delta$. Then $\|F \circ f\| < K$ by 9.3.5, while $F \circ f$ equals 0 on E_1 and 1 on E_2. It follows at once that every idempotent function belongs to $B(\Gamma_d)$, which contradicts 9.3.8 below. Lemma 9.3.7 is proved, subject to 9.3.8 being proved. □

9.3.8. Lemma. *Let Γ be an infinite discrete abelian group. Then there exists $E \subseteq \Gamma$ such that $\chi_E \notin B(\Gamma)$.*

Proof. An elementary induction argument shows that there exist $\gamma_1, \gamma_2, \ldots, \in \Gamma$ such that each $\gamma \in \Gamma$ may be written in at most one way as a sum

$$\gamma = \pm\varepsilon_i\gamma_i \pm \varepsilon_j\gamma_j, (i \neq j), \tag{8}$$

where ε_i may be $+1$ or -1 if the term γ_i has order at least three and must be $+1$ if γ_i has order 2. Let $E = \{\gamma_j: 1 \leq j < \infty\}$. Then $\chi_E \notin B(\Gamma)$.

Indeed, if $\chi_E \in B(\Gamma)$, then the translate $\chi_E^{\gamma_j}$ of χ_E by γ_j is also in $B(\Gamma)$ and $\|\chi_E^{\gamma_j}\| = \|\chi_E\|$.

It is immediate from (8) that $\lim_j \chi_E^{\gamma_j}(\gamma) = 0$ if $\gamma \neq 0$ and $\lim_j \chi_E^{\gamma_j}(0) = 1$. Hence $f = \lim_j \chi_E^{\gamma_j} \in A(\Gamma)$. That contradicts 1.2.2, and 9.3.8 is proved. □

Proof of Theorem 9.3.3 for compact Γ. Suppose that we can show that if H operates, then H is analytic in a neighborhood of 0. On setting $H(t) = F(t - t_0)$, we conclude that F is analytic in a neighborhood of t_0 for all

$t_0 \in (-1, 1)$. On setting $H(t) = F(1 - t^2)$ we conclude that $F(1 - t^2) = \sum a_n t^{2n}$ where $\sum a_n |t|^{2n} < \infty$ for $|t| < \varepsilon$ where $\varepsilon > 0$. Then $F(1 - t) = \sum a_n t^n$ converges absolutely in a neighborhood of 0, so F is analytic in a neighborhood of 1. A similar argument applied to $H(t) = F(t^2 - 1)$ shows that F is analytic in a neighborhood of -1.

It remains to show that F is analytic in a neighborhood of 0. Let $\delta > 0$ and $K > 0$ be given by 9.3.5, let $\alpha \in (0, \delta e^{-2\pi - 1})$, and let $F_1(t) = F(\alpha \sin t)$. Then F is analytic in a neighborhood of 0 if and only if F_1 is analytic in a neighborhood of 0. By 9.3.7, F_1 is continuous everywhere, and $\|F_1 \circ f\| \leq K$ for all $f \in A(\Gamma)$ such that $\|f\| \leq 2\pi + 1$ and $f(\Gamma) \subseteq R$.

Let $\hat{F}_1(k)$ denote the k-th Fourier coefficient of F_1. Let $\gamma \in \Gamma$ and $f \in A(\Gamma)$ with $\|f\| \leq 1$ and $f(\Gamma) \subseteq R$. Then

$$\hat{F}_1(k)\exp(ikf(\gamma)) = \frac{1}{2\pi}\int_0^{2\pi} F_1(f(\gamma) + x))e^{-ikx}\,dx.$$

But $\|F_1 \circ (f + x1)\| \leq K$ for all $x \in [0, 2\pi]$ and all such f. Therefore

(9) $\quad \|\hat{F}_1(k)\exp(ikf)\| \leq K, \quad$ for all $\quad k \in Z \quad$ and all $\quad f \in A(\Gamma) \quad$ with

$$\|f\| \leq 1 \quad \text{and} \quad f(\Gamma) \subseteq R.$$

Lemma 9.3.6 and (9) show that $|\hat{F}_1(k)|e^{|k|} \leq K$ for all k, which implies that the series $\sum_{-\infty}^{\infty} \hat{F}_1(k)e^{ikz}$ converges absolutely for $|\text{Im } z| < 1$. Therefore F_1 is analytic at 0. That completes the proof of 9.3.3 for compact Γ.

Proof of 9.3.3 for non-compact Γ. We reduce to the compact case. By the structure theorem, we may assume that Γ has the form $R^n \times H$ where H is compact.

If H is infinite, then F operates in $A(H)$, because real-valued elements of $A(H)$ extend to real-valued elements of $A(\Gamma)$ with no increase in either A-norm or in the supremum norm. Then F is analytic in a neighborhood of $[-1, 1]$. If H is finite, we may assume that $n = 1$ (by the same observation on extensions). We have reduced to the case $\Gamma = R$.

By 9.3.3 for compact Γ, it will suffice to show that F operates in $A(T)$. Let $f \in A(T)$ and let f_1 be f regarded as a 2π-periodic function on R. Then $\|f\|_{A(T)} = \|f_1\|_{B(R)}$, since $f(x) = \sum a_n e^{inx}$ and $f_1(x) = \sum a_n \hat{\delta}_n(x)$. Since F operates in $A(R)$, $F \circ f_1$ belongs to $A(R)$ locally in each compact set. Let E_1 be any interval of R of length $b < 2\pi$ and let E_2 be any interval of T of length b. Then $A(E_1, R)$ and $A(E_2, T)$ are isomorphic (Rudin [1, 2.7.6]). Therefore $F \circ f \in A(T)$ locally everywhere. Therefore F operates in $A(T)$. That completes the proof of 9.3.3. □

Proof of 9.3.4. Let $\delta > 0$ and $K > 0$ be given by 9.3.5. By 9.3.7, F is continuous in $(-\delta, \delta)$. We replace $F(t)$ by $F_1(t) = F(\alpha \sin t)$ where $0 < \alpha < \delta e^{-2\pi - 1}$.

Then $\|f\|_A \leq 2\pi + 1$ implies that $\|F_1 \circ f\|_A \leq K$ for real-valued $f \in A(\Gamma)$ and that F_1 is continuous and 2π periodic on R. Let $\hat{\mu} \in B(\Gamma)$ be real-valued with $\|\hat{\mu}\|_B \leq 2\pi + 1$, and let $\hat{\mu} = \lim f_\alpha$ where $\|f_\alpha\|_A \leq \|\hat{\mu}\|_B, f_\alpha(\Gamma) \subseteq R$, and $f_\alpha \to \hat{\mu}$ uniformly on compact sets. Then $F_1 \circ \hat{\mu} = \lim F_1 \circ f_\alpha$, so $F_1 \circ \hat{\mu} \in B(\Gamma)$ and $\|F_1 \circ \hat{\mu}\|_B \leq K$. In particular $F_1 \circ (f + x1) \in B(\Gamma)$ for $0 \leq x \leq 2\pi$ and all real-valued $f \in A(\Gamma)$ of norm one. For such f and x, $\|F_1 \circ (f + x1)\| \leq K$. The proof of 9.3.4 now proceeds exactly as in the case of 9.3.1 for compact Γ, obtaining the analyticity of F_1 in a neighborhood of 0. Therefore F is analytic in a neighborhood of 0. That ends the proof of 9.3.4. □

Proofs of 9.3.1 and 9.3.2. These proofs are almost identical with the proofs of 9.3.3 and 9.3.4. Here are the changes required in those latter proofs.

(a) 9.3.5 is stated and proved for $F: U \to \mathbb{C}$. No essential changes in the argument are required. (Recall that $0 \in U$ if Γ is non-compact).
(b) 9.3.6 can be used as given above.
(c) 9.3.7 is modified to conclude that F is continuous in $\{z: |z| < \delta\}$. No essential changes in the proof are required.
(d) F is replaced by $F_1(s, t) = F_1(\alpha \sin, \beta \sin t)$ where $\alpha, \beta \in (0, \delta e^{-2\pi - 1})$. F_1 is then shown to extend to an analytic function in two variables in a neighborhood of (0, 0). Here we use the estimates

$$|\hat{F}_1(j, k)| \leq K e^{-|j| - |k|} \quad \text{for } (j, k) \in Z \times Z.$$

The remaining details are left to the reader. □

Remarks and Credits. Marcinkiewicz [1] seems to be the first published recognition of the importance of the growth rate of $\|e^{ikf}\|_{A(T)}$ as $|k| \to \infty$. It also contains examples of $g, f \in A(T)$ such that $g \circ f$ is well-defined and *not* in $A(T)$. Unfortunately, this paper went virtually unnoticed from its publication in 1940 until its inclusion in 1964 by Zygmund in Marcinkiewicz's *Collected Works.*

The idea of using estimates of $\|e^{inf}\|$ was rediscovered by Leĭbenson [1] in 1954, and led eventually to the proofs of Section 9.3. In 1955, Kahane, citing Leĭbenson, showed that absolute value does not operate in $A(T)$ (Kahane [3, 4]), nor in $A(Z)$ (Kahane [5]). In that same year, Rudin [4] showed that $f^{1/2} \notin A(T)$ for some non-negative $f \in A(T)$. In 1956 Rudin [5] proved that operating functions obeyed a Lipschitz condition at zero. In 1958, Kahane [6] proved that if $\mathscr{F}$ was a quasi-analytic class of functions on $[-1, 1]$ that contained a non-analytic function, then $\mathscr{F}$ contained a function that did not operate in $A(T)$. That same paper contained a proof that if F operates boundedly, then F is analytic. Katznelson [2] then proved 9.3.1 for $A(T)$ while Helson and Kahane [1] gave 9.3.2 for $A(Z)$, and the most general cases came in Helson, Kahane, Katznelson and Rudin [1]. Proofs of 9.3.1–9.3.4 that do not depend on the structure theorem are indicated in Section 11.3.

Rudin [13] establishes the following theorem.

9.3.9. Theorem. *Let G be a compact abelian group with dual group Γ and F: $[-1, 1] \to \mathbb{C}$. Suppose that for every $f \in A(G)$ with $f(G) \subseteq [-1, 1]$, there exist $p \in [1, 2)$ and $g \in L^p(\Gamma)$ such that $F \circ f(x) = \hat{g}(x)$ a.e. on G. Then F extends to be analytic in a neighborhood of $[-1, 1]$.*

Theorem 9.3.9 has been extended by Rivière and Sagher [1] to yield a converse of Marcinkiewicz's Theorem 9.2.3. Rider [2, 6] considers functions operating from $L^p(T)^\wedge$ to $L^1(T)^\wedge = A(Z)$ when $1 < p \leq \infty$.

A multivariable version of 9.3.3 appears in Helson and Kahane [1]. For individual symbolic calculus results see, for example, Malliavin [4], Leblanc [1] and Kahane [2, Chapter VI].

For functions operating in $A(G)$ and $A_p(G)$ (see 10.1 for the definition) for not necessarily abelian groups G, see Rider [5] ($A(G)$ is analytic if G is compact and infinite), and de Michele and Soardi [1, 2, 3] ($A_p(G)$ is analytic for many G's).

There is a rich theory of functional calculus for Banach algebras and topological algebras. For the now classical results, the reader might consult Chapter III of Gamelin [1]. Generalizations to "non-Banach" algebras are surveyed by Waelbroeck [1].

We conclude with a simple proof, due to Kahane, of 9.3.3 for the special case $G = T^2$. Suppose that F is defined on R and F operates on the real-valued elements of $A(T^2)$. Let $g \in A(T)$ be real valued and such that $g(t) = t$ for $|t| \leq 2\pi/3$. Let $h \in A(T)$ be real valued and such that $h(t) = t$ for $|t| \geq \pi/3$. Then for all real-valued $f \in A(T)$, $F \circ (f(s) + g(t))$ and $F \circ (f(s) + h(t))$ belong to $A(T^2)$. Hence for each $t \in [-\pi, \pi]$, $F(f(s) + t) \in A(T)$ locally with a bound on the norms that is independent of t. Therefore

$$\left\| \int_{-\pi}^{\pi} F(f(s) + t)\exp(-int)dt \right\| \leq C.$$

We now argue as in the proof of 9.3.3.

9.4. Functions Operating in $B(\Gamma)$

9.4.1. Theorem. *Let Γ be a locally compact, non-compact abelian group. Let U be an open and connected subset of $\mathbb{C}$ and let $F: U \to \mathbb{C}$. If $F \circ f \in B(\Gamma)$ whenever $f \in B(\Gamma)$ and $f(\Gamma) \subseteq U$, then F agrees with a real-entire function on U.*

9.4.2. Theorem. *Let Γ be a locally compact, non-compact abelian group. Let $F: [-1, 1] \to \mathbb{C}$. If $F \circ f \in B(\Gamma)$, whenever $f \in B(\Gamma)$ and $f(\Gamma) \subseteq [-1, 1]$, then F extends to an entire function.*

This Section will show how the methods of Section 9.3 may be extended to prove 9.4.2. The proof of 9.4.1 will be left to the reader. The methods of each of Sections 9.5 and 9.6 may also be applied to prove 9.4.1 and 9.4.2.

9.4.3. Theorem. *Let Γ be an infinite discrete abelian group. Let $F: R \to \mathbb{C}$ be continuous and periodic. If $F \circ f \in B(\Gamma)$ whenever $f \in B(\Gamma)$ and $f(\Gamma) \subseteq R$, then F extends to an entire function.*

How 9.4.3 implies 9.4.2. Let Γ and F satisfy the hypotheses of 9.4.2. We first show that F is analytic in a neighborhood of $[-1, 1]$. For $|t| < 1$, let $F_t(s) = F(s + t)$. Let $f \in A(\Gamma)$ and suppose that $f(\Gamma) \subseteq [-1,+|t|, 1 - |t|]$. Then $F_t \circ f = F \circ (f + t1) \in B(\Gamma)$, that is, F_t operates from $A(\Gamma)$ to $B(\Gamma)$. That F extends to a function analytic in a neighborhood of $[-1, 1]$ follows as in the first paragraph of the proof of 9.3.3. In particular F is continuous on $[-1, 1]$.

We now show that we may assume that Γ is discrete. If Γ has a compact open subgroup Λ, then Γ/Λ is infinite and discrete. The elements of $B(\Gamma)$ that are constant on cosets of Λ define elements of $B(\Gamma/\Lambda)$ (Rudin [1, 2.7.1]); therefore F operates in $B(\Gamma/\Lambda)$. If Γ has no compact open subgroup, then the structure theorem implies that Γ contains a copy of R as a closed subgroup. Therefore Γ contains a copy of Z as a closed subgroup. It follows that $F \circ f \in B(Z)$ whenever $f \in B(Z)$ and $f(Z) \subseteq [-1, 1]$. That completes the reduction to the case that Γ is discrete.

We now show that we may assume that F is periodic. For $0 < r \leq 1$, let $F_r(t) = F(r \sin t)$. Then Γ and F_r satisfy the hypotheses of 9.4.3, so F_r is entire. Since $F(t) = F_r(\arcsin(t/r))$, F extends to a (possibly many-valued) analytic function in $\mathbb{C}\backslash\{r, -r\}$ with possible branch points at $+r, -r$. If $r_1 \neq r_2$, then the equation $F(t) = F_{r_1}(\arcsin(t/r_1)) = F_{r_2}(\arcsin(t/r_2))$ shows that the branch points are, in fact, non-existent. That implies that F is entire. Thus 9.4.2 follows from 9.4.3. □

The proof of 9.4.3 will use the following variant of Lemma 9.3.5.

9.4.4. Lemma. *Let Γ be an infinite discrete abelian group. Let $F: R \to \mathbb{C}$ be a continuous periodic function such that $F \circ f \in B(\Gamma)$ whenever $f \in B(\Gamma)$ and $f(\Gamma) \subseteq R$. If $g \in B(\Gamma)$ and $g(\Gamma) \subseteq R$, then there exist $K > 0$, $\delta > 0$ such that*

(1) $$\|F \circ (g + f)\| \leq K$$

if $f \in B(\Gamma)$, $f(\Gamma) \subseteq R$ and $\|f\| < \delta$.

Proof. Let $g \in B(\Gamma)$ with $g(\Gamma) \subseteq R$. Suppose that there exist $K > 0, \delta > 0$ such that (1) holds whenever $f \in B(\Gamma), f(\Gamma) \subseteq R$, $\|f\| < \delta$ and f has finite support. Then (1) holds for all real-valued $f \in B(\Gamma)$ with $\|f\| < \delta$. To see that, let us fix $f \in B(\Gamma)$ with $f(\Gamma) \subseteq R$ and $\|f\| < \delta$. Let $k \in A(\Gamma)$ have finite support and satisfy $0 \leq k \leq 1$ and $\|k\| \|f\| < \delta$. Then $\|F \circ (g + kf)\| \leq K$. But a net

$\{k_\alpha\}$ of such k may be chosen such that $\lim k_\alpha(\gamma) = 1$ for each $\gamma \in \Gamma$. By 1.8.1 and the continuity of F, $\|F \circ (g + f)\| \le K$.

Thus it is sufficient to find $K > 0$, $\delta > 0$ such that (1) holds for all finitely supported f with $\|f\| < \delta$ and $f(\Gamma) \subseteq R$. Assuming that to be false, we produce sequences f_j and k_j as follows. Let f_1 and k_1 be finitely supported real-valued functions in $A(\Gamma)$ such that $\|f_1\| \le \frac{1}{2}$, $\|k_1\| \le 3$, $0 \le k_1 \le 1$ and $\|k_1 F \circ (g + f_1)\| > 3$. Now suppose that we have found $f_1, \ldots, f_n, k_1, \ldots, k_n \in A(\Gamma)$ such that the f_j and k_j are real-valued, finitely supported, and such that for $1 \le j \le l \le n$,

$$\|f_j\| < 2^{-j}, \ \|k_j\| < 3, \tag{2}$$

$$\|k_j F \circ (g + f_j)\| > 3j, \tag{3}$$

and

$$0 \le k_j \le 1, \ k_j F \circ \left(g + \sum_{i=1}^{l} f_i\right) = k_j F \circ (g + f_j). \tag{4}$$

It is immediate from (2)–(4) that $\|F \circ (g + \sum_{i=1}^{l} f_i)\| > l$ for $1 \le l \le n$. To find f_{n+1} and k_{n+1}, several auxiliary functions are needed.

Choose α, $\beta \in A(\Gamma)$ with finite support such that $\|\alpha\| < \frac{3}{2}$, $\|\beta\| < \frac{3}{2}$, $0 \le \alpha \le 1$, $0 \le \beta \le 1$, $\alpha = 1$ on the union of the supports of the $2n$ functions $k_1, \ldots, k_n$ and $f_1, \ldots, f_n$, and $\beta = 1$ on the support of α. Choose $h \in A(\Gamma)$ such that $\|h\| < 2^{-n-3}$, $h(\Gamma) \subseteq R$, supp h finite, and

$$\|F \circ (g + h)\| > 3(n + 1) + \left(\sup_{x \in R} |F(x)|\right)(\# \text{ supp } \beta). \tag{5}$$

(Since F is periodic and continuous, F is bounded. Therefore (5) makes sense). Set $f_{n+1} = (1 - \alpha)h$. Then it is easy to see that

$$(1 - \beta)F \circ (g + (1 - \alpha)h) = (1 - \beta)F \circ (g + h) = (1 - \beta)F \circ (g + f_{n+1}).$$

Therefore $\|(1 - \beta)F(g + f_{n+1})\| > 3(n + 1)$.

Choose a finitely supported $k \in A(\Gamma)$ such that $0 \le k \le 1$, $\|k\| \, \|1 - \beta\| < 3$, and $\|k(1 - \beta)F \circ (g + f_{n+1})\| > 3(n + 1)$. Set $k_{n+1} = k(1 - \beta)$. Then $k_{n+1} = 0$ on the supports of $f_1, \ldots, f_n$, and that implies that

$$k_{n+1} F \circ \left(g + \sum_{i=1}^{n+1} f_i\right) = k_{n+1} F \circ (g + f_{n+1}). \tag{6}$$

That (2)–(4) hold is trivial. That completes the induction.

Set $f = \sum_i^\infty f_i$. Then (4) and (6) imply that for all j,

$$3\|F \circ (g + \sum f_i)\| \geq \|k_j F \circ (g + \sum f_i)\| = \|k_j F \circ (g + f_j)\| > 3j.$$

Therefore $\|F \circ (g + f)\| > j$ for $j = 1, 2, \ldots$. That contradiction completes the proof of Lemma 9.4.4. □

Proof of Theorem 9.4.3. Using Lemma 9.4.4 and the compactness of $[0, 2\pi]$, we see that if $g \in B(\Gamma)$, then $\sup_{0 \leq x \leq 2\pi} \|F \circ (g + x1)\| \leq K_g < \infty$, where K_g is a constant which depends only on g. Let $\hat{F}(k)$ denote the kth Fourier coefficient of F. Then for $g \in B(\Gamma)$ with $g(\Gamma) \subseteq R$, we find that

$$(7) \qquad |\hat{F}(k)| \, \|e^{ikg}\| \leq K_g, \; k \in Z.$$

All that is needed to prove 9.4.3 is to show that there exists $g \in B(\Gamma)$ such that $g(\Gamma) \subseteq R$ and $\|e^{ikg}\| = e^{|k| \, \|g\|}$. For (7) then implies that

$$|\hat{F}(k)| \leq K_g e^{-|k| \, \|g\|}.$$

On replacing g by rg (and K_g by K_{rg}), we have

$$|\hat{F}(k)| \leq K_{rg} e^{-|k|R \, \|g\|}.$$

Therefore F is analytic in a strip of width $r\|g\|$; r being arbitrary, F must be entire.

In Chapter 6 and 7, many examples of $g \in B(\Gamma)$ with $\|e^{ikg}\| = e^{|k| \, \|g\|}$ and $g(\Gamma) \subseteq R$ are given, and the proof of Theorem 9.4.3 is complete. □

Remarks and Credits. Theorem 9.4.2 is due to Kahane and Rudin [1] for Z and to Helson, Kahane, Katznelson and Rudin [1] for arbitrary Γ, as is 9.4.1.

For many groups Γ one can easily show the following. If $F:[-1, 1] \to \mathbb{C}$ operates in $B(\Gamma)$, then for each $\delta > 0$ there exists $K > 0$ such that $\|F \circ f\| \leq K$ if $f \in B(\Gamma)$ is real-valued with $\|f\| \leq \delta$. Here is a sketch of the argument for $\Gamma = Z$. See Katznelson [1, p. 248, exercise 7] for details.

We argue by contradiction and conclude that there exist trigonometric polynomials f_j such that $\|f_j\| \leq \delta$ while $\|F \circ \hat{f}_j\| \to \infty$. The problem is to sum the f_j. We may assume that $\operatorname{supp} \hat{f}_j \subseteq \{n_j k : |k| \leq m_j\}$ for all j, where n_j is much larger than

$$\sum_{k=1}^{j-1} n_k m_k \quad \text{for } 1 \leq j < \infty.$$

It is easy to see that then whenever g_j is a trigonometric polynomial with $\operatorname{supp} \hat{g}_j \subseteq \operatorname{supp} \hat{f}_j$ for each j, $\|\sum_1^n g_j\|$ is of the same order of magnitude as

$\sum_1^n \|g_j\|$ for all n. It follows that there exists $\mu \in M(T)$ such that $\mu = f_j$ on supp f_j, so $\|F \circ \hat{\mu}\| = \infty$.

9.5. Functions Operating in $B_o(\Gamma)$

Recall that $B_o(\Gamma) = \{f \in B(\Gamma): f \in C_o(\Gamma)\} = M_o(G)^\wedge$. In this section the following result of Varopoulos [3] is proved. For another proof, see 7.3.7.

9.5.1. Theorem. *Let Γ be an infinite abelian group. Let $F: [-1, 1] \to \mathbb{C}$ be such that $F \circ f \in B_o(\Gamma)$ whenever $f \in B_o(\Gamma)$ and $f(\Gamma) \subseteq R$. Then F agrees with an entire function in a neighborhood of 0.*

9.5.2. Lemma. *Let $\nu \in M(G)$ be a singular measure on the non-discrete group G. Let E_0 be a compact subset of the dual group Γ of G. Then for each $\varepsilon > 0$, there exists a compact subset $E \subseteq \Gamma \backslash E_0$ and $p \in C_E(G)$ such that $\|p\|_\infty \leq 1$ and*

$$\|\hat{\nu}|_E\|_{A(E)} \geq \left|\int p\, d\nu\right| \geq (1 - \varepsilon)\|\nu\|.$$

Proof. By 8.3.1, $\lim_{x \to 0} \sup\|\nu - \delta_x * \nu\| = 2\|\nu\|$. Let $k \in L^1(G)$ be such that $k^\wedge = 1$ on E_0 and $\hat{k}$ has compact support. Then there exists $x \in G$ such that $\|\delta_x * k - k\| < \frac{1}{2}\varepsilon$ and $\|\nu - \delta_{-x} * \nu\| > (2 - \varepsilon)\|\nu\|$. From the last inequality we conclude that there exists $q \in A(G) \cap L^1(G)$ such that q has compact support, $\|q\|_\infty \leq 1 - \frac{1}{2}\varepsilon$ and

$$\left|\int q\, d(\nu - \delta_x * \nu)\right| > (2 - \varepsilon)\|\nu\|(1 - \tfrac{1}{4}\varepsilon).$$

We may assume that $\varepsilon < \frac{1}{2}$. Let

$$p = \tfrac{1}{2}(q - k * q - \delta_x * (q - k * q)),$$

so that $p \in A(G) \cap L^1(G)$. Then $\hat{p}(\gamma) = 0$ if $\gamma \in E_0$ and

$$\begin{aligned}\|p\|_\infty &\leq \tfrac{1}{2}\|q - \delta_x * q\|_\infty + \tfrac{1}{2}\|q * (k - \delta_x * k)\|_\infty \\ &\leq (1 - \tfrac{1}{4}\varepsilon) + \tfrac{1}{2}(1 - \tfrac{1}{4}\varepsilon)(\varepsilon) \leq 1 - \tfrac{1}{4}\varepsilon + \tfrac{1}{4}\varepsilon = 1.\end{aligned}$$

Now

$$\begin{aligned}\int p\, d\nu &= \tfrac{1}{2}\int q\, d\nu - \tfrac{1}{2}\int k * q\, d\nu - \tfrac{1}{2}\int \delta_x * (q - k * q) d\nu \\ &= \tfrac{1}{2}\left(\int q\, d\nu - \int q\, d\delta_{-x} * \nu\right) + \tfrac{1}{2}\int (\delta_x * k - k) * q\, d\nu\end{aligned}$$

so

$$\left|\int p\,d\nu\right| \geq \tfrac{1}{2}(2 - \varepsilon)\|\nu\| - \tfrac{1}{4}\varepsilon(1 - \tfrac{1}{4}\varepsilon)\|\nu\| \geq (1 - \varepsilon)\|\nu\|.$$

Let E be the support of $\hat{p}$. Then the inequalities $\|p\|_\infty \leq 1$ and $|\int p\,d\nu| \geq (1 - \varepsilon)\|\nu\|$ show that $\|\hat{\nu}|_E\|_{A(E)} \geq (1 - \varepsilon)\|\nu\|$. That completes the proof of the Lemma. □

9.5.3. Lemma. *Let G be an infinite compact abelian group with dual group* Γ. *Then there exists a probability measure* $\mu \in M_0(G)$ *such that for all* $\varepsilon > 0$ *and* $N \geq 1$, *there exists a finite subset E of* Γ *such that*

(1) $$\left\|\sum_1^N c_j \hat{\mu}^j|_E\right\|_{A(E)} \geq (1 - \varepsilon)\sum_1^N |c_j| \quad \textit{whenever} \quad \{c_1, \ldots, c_N\} \subseteq \mathbb{C},$$

and

(2) $$\|\hat{\mu}^{N+1}|_E\|_{A(E)} < \varepsilon.$$

Proof. Let Θ be a fixed infinite dissociate subset of Γ. We shall construct a function $a: \Theta \to (0, \frac{1}{2})$ such that the Riesz product based on Θ and a will satisfy the conclusions of the Lemma. For more about Riesz products, see A.1 and Section 7.1.

Let $N_1, N_2, \ldots$ be any increasing sequence of integers with $\lim N_j = \infty$. Let $a_1: \Theta \to (0, \frac{1}{2})$ be such that

$$\sum_{\theta \in \Theta} |a_1(\theta)|^{2N_1} = \infty$$

and

$$\sum_{\theta \in \Theta} |a_1(\theta)|^{2N_1 + 2} < \infty.$$

Let μ_1 be the Riesz product based on Θ and a_1, and let ν be the Riesz product based on Θ and the function equal to $\frac{1}{2}$ identically on Θ. Then $\nu * \mu_1, \ldots, \nu * \mu_1^{N_1}$ are mutually singular and $\nu * \mu_1^{N_1 + 1} \in L^1(G)$ by 7.2.1.

Let E_0 be a finite set such that $\|\hat{\mu}^{N_1 + 1}|_{E_0}\|_{A(E_0)} \leq \frac{1}{2}$. By 9.5.3, there exists, for each finite set $\{c_1, \ldots, c_{N_1}\} \leq \mathbb{C}$, a polynomial p such that $\operatorname{supp} \hat{p} \subseteq E_0$, $\|p\|_\infty \leq 1$, and $|\int p\,d(\sum c_j \nu * \mu^j)| \leq \frac{1}{2}\sum|c_j|$. Of course, if $q = \nu * p$, then $\int p\,d(\sum c_j \nu * \mu^j) = \int q\,d(\sum c_j \mu^j)$, and $\operatorname{supp} \hat{q} \subseteq \Omega(\Theta)\backslash E_0$. By the compactness of $\{(c_1, \ldots, c_{N_1}): |c_j| \leq 1 \text{ for } 1 \leq j \leq N_1\}$, there exists a finite set $E_1 \subseteq \Omega(\Theta)\backslash E_0$ such that for all $c_1, \ldots, c_{N_1} \in \mathbb{C}$ there exists a polynomial q with $\operatorname{supp} \hat{q} \subseteq E_1$ such that $\|q\|_\infty \leq 1$ and $|\int q\,d(\sum c_j \mu^j)| \geq \frac{1}{2}\sum|c_j|$.

Let Φ_1 be a finite subset of Θ such that $E_1 \subseteq \Omega(\Phi_1)$. If a_1 is changed on $\Theta\backslash\Phi_1$, then the transform of μ on E_1 is not affected. That begins our induction.

Suppose that $n \geq 1$ and that disjoint finite subsets $\Phi_1, \ldots, \Phi_n$ of Θ, functions $a_j: \Phi_j \to [0, 2^{-j}]$ and subsets $E_j \subseteq \Omega(\Phi_j)$ have been found such that for $1 \leq j \leq n$, and $c_1, \ldots, c_j \in \mathbb{C}$,

$$(3) \qquad \left\| \sum_1^j c_k \hat{\mu}_j^k |_{E_j} \right\|_{A(E_j)} \geq (1 - 2^{-j}) \sum_1^j |c_k| \quad \text{and} \quad \|\hat{\mu}^{j+1}|_{E_j}\|_{A(E_j)} < 2^{-j}.$$

In (3), μ_j is the Riesz product based on a_j and Φ_j.

Let $a: \Theta \setminus (\bigcup_1^n \Phi_j) \to [0, 2^{-n-1}]$ be such that

$$\sum |a(\theta)|^{2N_{n+1}} = \infty \quad \text{and} \quad \sum |a(\theta)|^{2N_{n+1}+2} < \infty.$$

The argument used above shows that there exists a finite subset Φ_{n+1} of $\Theta \setminus \bigcup_1^n \Phi_j$ and a finite subset E_{n+1} of $\Omega(\Phi_{n+1})$ such that (3) holds for $j = n + 1$ and all $\{c_1, \ldots, c_{n+1}\} \subseteq \mathbb{C}$. We define a_{n+1} to equal a on Φ_{n+1}. That completes the induction. We now let b be the function on $\bigcup_1^\infty \Phi_j$ that agrees with a_j on each Φ_j, and let μ be the Riesz product based on $\bigcup \Phi_j$ and b. Then $\mu \in M_o(G)$. It is clear from (3) that (1) and (2) hold for μ. The Lemma is proved. □

Proof of 9.5.1. Suppose that 9.5.1 holds for infinite discrete Γ. If Γ is any general infinite locally compact abelian group, and if Γ has a compact open subgroup Λ, then F operates in $B_o(\Gamma/\Lambda)$, so F agrees with an entire function in a neighborhood of 0. We may thus assume that Γ has an open subgroup of the form $R^n \times \Lambda$, where $n \geq 1$, and Λ is compact. Then F operates in $B_o(R^n \times \Lambda)$. An application of A.7.1 shows that every real-valued function in $B_o(Z^n \times \Lambda)$ extends to a real-valued element of $B_o(R^n \times \Lambda)$ with no increase in the supremum of the Fourier-Stieltjes transform. Hence, F operates in $B_o(Z^n \times \Lambda)$. We have thus reduced to the case of discrete Γ.

By the Remark following 9.3.4, F is analytic in a neighborhood of 0: $F(t) = \sum a_n t^n$ with $\sum |a_n| |t|^n < \infty$ for $|t| < \delta$ where $\delta \in (0, 1)$. It will suffice to show that $\{|a_n| r^n\}$ is a bounded sequence for each $r > 0$. Let us note that we may extend F to all of R, in any way, and still have F operate in $B_o(\Lambda)$. (We are assuming that Γ is discrete.)

Let $\mu \in M_o(G)$ be given by 9.5.3, and let $r > 0$. Let $n \geq 1$ and $\varepsilon \in (0, \delta^{N+1}/r)$. Then

$$F(r\hat{\mu}(\gamma)) = \sum_1^N a_n \hat{\mu}(\gamma)^n r^n + \sum_{N+1}^\infty a_n \hat{\mu}(\gamma)^n r^n$$

whenever $\gamma \in E$, the set for which (1) and (2) hold, since for such γ, $|r\hat{\mu}(\gamma)| < \delta$. By (1)

$$\|F \circ (r\hat{\mu})|_E\|_{A(E)} \geq (1 - \varepsilon) \sum_1^N |a_n| r^n - \sum_{N+1}^\infty |a_n| r^n.$$

Applying (2), we have

$$\|F \circ (r\hat{\mu})|_E\|_{A(E)} \geq (1 - \varepsilon) \sum_1^N |a| r^n - \sum_{j=1}^N \sum_{k=1}^\infty |a_{Nk+j}| r^{Nk+j} \varepsilon^k.$$

For sufficiently small $\varepsilon > 0$,

$$\|F \circ (r\mu)\| \geq \|F \circ (r\hat{\mu})|_E\|_{A(E)} \geq \tfrac{1}{2} \sum_1^N |a_n| r^n - 1.$$

Since N is arbitrary, $\sum |a_n| r^n < \infty$. Therefore $\{|a_n| r^n\}$ is a bounded sequence. Theorem 9.5.1 is proved. □

Remarks and Credits. Theorem 9.5.1 is due to Varopoulos [3]; see Varopoulos [2] for an exposition of the basic idea of his argument. Lemma 9.5.2 is from Doss [1]. The proof of 9.5.1 that is here is from Graham [11], where other measures satisfying the conclusion of 9.5.3 are exhibited. A version of 9.5.1 for functions defined on a neighborhood of 0 in $\mathbb{C}$ appears in Saeki [3].

9.6. Functions Operating on Norm One Positive-Definite Functions

Let Γ be an infinite abelian group, and let $PD(\Gamma)$ denote the set of functions on Γ which are the Fourier–Stieltjes transforms of probability measures on the dual group G of Γ. By Bochner's Theorem (Rudin [1, 1.4.3]), $PD(\Gamma)$ is exactly the set of continuous positive-definite functions on Γ that equal one at the identity. In this section we characterize the functions that operate from $PD(\Gamma)$ to $PD(\Gamma)$ and from $PD(\Gamma)$ to $B(\Gamma)$ for most groups Γ.

It is obvious that $PD(\Gamma)$ is convex and closed under multiplication, and that if $\hat{\mu} \in PD(\Gamma)$, then $|\hat{\mu}| \leq 1$. The functions that operate on $PD(\Gamma)$ will have to have domains containing the set

$$D(\Gamma) = \bigcup \{\hat{\mu}(\Gamma) : \hat{\mu} \in PD(\Gamma)\}$$

and, of course, we shall only be able to discuss the behavior of F on $D(\Gamma)$. The most interesting results concern the behavior of F on the open set Int $D(\Gamma)$, and we shall be concerned almost exclusively with this open set. (For the behavior on the boundary, see Graham [11].) Note that if Γ contains elements of infinite order, or of arbitrarily high finite order, then Int $D(\Gamma) = \{z : |z| < 1\}$, and if the orders of the elements of Γ are at most p and p *is* the order of one element of Γ, then Int $D(\Gamma)$ is the interior of the regular convex p-gon with vertices at the pth roots of unity. In that case, we say that Γ has *exponent p*.

9.6.1. Theorem. *Let Γ be an infinite abelian group that is not the product of a finite group with a group of exponent two. Let $F: D(\Gamma) \to \mathbb{C}$ be such that $F \circ f \in PD(\Gamma)$ if $f \in PD(\Gamma)$. Then there exist numbers $a_{mn} \geq 0$ ($m, n = 0, 1, 2, \ldots$) such that*

(1) $$F(z) = \sum a_{mn} z^m \bar{z}^n, \quad \textit{for } z \in \text{Int } D(\Gamma)$$

and

(2) $$\sum a_{mn} < \infty.$$

9.6.2. Theorem. *Let Γ be an infinite abelian group such that either (i) the dual group G of Γ does not contain an infinite open-compact subgroup of exponent two, or (ii) for some open compact subgroup H of G, G/H has an element of infinite order. Let $F: D(\Gamma) \to \mathbb{C}$ be such that $F \circ f \in B(\Gamma)$ whenever $f \in PD(\Gamma)$. Then there exist numbers $a_{mn} \in \mathbb{C}$ such that*

(3) $$F(z) = \sum a_{mn} z^m \bar{z}^n, \quad \textit{for } z \in \text{Int } D(\Gamma)$$

and

(4) $$\sum |a_{mn}| < \infty.$$

Rather than using the methods of Section 9.5 to prove Theorems 9.6.1 and 9.6.2 (the method in Graham [11]), we use the method of Moran [3]. The idea is to exploit analytic behavior in the maximal ideal space of $B(\Gamma)$, namely in the weak$*$ closure Γ^- of Γ. The reader who is not familiar with generalized characters should consult Section 5.1. We shall use measures concentrated on Kronecker and K_p-sets and remind the reader that a *Kronecker set* K (K_p-*set*) in a locally compact abelian group G is a set K (totally disconnected if K_p) such that every continuous function from K to T (K to the pth roots of unity) is the uniform limit on K of continuous characters (is the restriction to K of a continuous character). Such sets are independent. See Section 6.2 for properties of measures on independent sets.

9.6.3. Proposition. *Let G be a non-discrete abelian group with dual group Γ. Let μ be a continuous probability measure concentrated on a Kronecker or K_p-set $K \subseteq G$ where $p > 3$. If $F: \hat{\mu}(\Gamma)^- \to \mathbb{C}$ is continuous on* Int $\hat{\mu}(\Gamma)^-$ *and $F \circ \hat{\mu} \in B(\Gamma)$, then there exist $a_{mn} \in \mathbb{C}$ with*

(5) $$\sum |a_{mn}| < \infty \quad \textit{and} \quad F(z) = \sum a_{mn} z^m \bar{z}^n \quad \textit{for all } z \in \text{Int } \hat{\mu}(\Gamma)^-.$$

The progression of the proofs is this: some lemmas, culminating in the proof of 9.6.3; the proof of 9.6.2, whose difficult part lies in showing that F is continuous; then the proof of 9.6.1. Some examples are given that show that 9.6.1 and 9.6.2 are sharp.

The basic idea of this section is illustrated in the simple proof of the following Proposition. We let $c(\mu)$ denote the set of elements of Γ^- that are constant a.e. $d\mu$.

9.6.4. Proposition. *Let* $F: [0, 1] \to \mathbb{C}$ *be continuous on* $(0, 1)$. *Let* μ *be a probability measure with* $\hat{\mu}(\Gamma) \subseteq [0, 1]$ *and* $c(\mu) = [0, 1]$. *If* $F \circ \mu \in B(\Gamma)$, *then* F *extends to a function holomorphic and bounded on* $\{z: |z| < 1, z \notin (-1, 0]\}$.

Proof. For $m = 1, 2, \ldots$ let $\xi_m \in \Gamma^-$ be such that $(\xi_m)_\mu = \exp(-2^{-m})$ a.e. $d\mu$. By replacing ξ_m by $\xi_{m+1}\bar{\xi}_{m+1}$, we may assume that $\xi_m \geq 0$, that is, that $(\xi_m)_\omega \geq 0$ a.e. $d\omega$ for all $\omega \in M(G)$. (See Section 5.1 for material that will clarify that.) If $m = 1, 2, \ldots$ and $|z| < 1$, then $\xi_m^{-2^m \log z} \in \Delta M(G)$ since Re log $z < 0$. (We choose the branch of log z defined on $X = \mathbb{C}\backslash(-\infty, 0]$ and zero at 1.) Therefore for each m,

$$G_m(z) = (F \circ \mu)\hat{\ }(\xi_m^{-2^m \log z})$$

is a well-defined analytic function for those $z \in X$ with $|z| < 1$. Also, $|G_n(z)| \leq \|F \circ \mu\|$ for all n.

We claim that lim G_m exists. Indeed, let $x = \exp(-p^{2^{-q}})$ where p and q are positive integers. If $m \geq q$, then

$$\xi_m^{p2^{m-q}} \in \Gamma^-$$

because Γ^- is a semigroup under pointwise multiplication. Furthermore,

$$G_m(x) = F(\hat{\mu}(\xi_m^{p2^{m-q}})) = F(x),$$

since

$$(\xi_m^{+p2^{m-q}})_\mu = (e^{-2^{-m}})^{p2^{m-q}} = e^{-p2^{-q}} = x, \text{ a.e. } d\mu.$$

Thus $\{G_m\}$ converges on a dense subset of $(0, 1)$. By Vitali's theorem (Hille [1, vol. II, p. 251]), lim $G_m = G$ exists uniformly on compact subsets of $\{z: |z| < 1, z \notin (-1, 0]\}$. By the continuity of F and G, $F = G$ on $(0, 1)$, and that completes the proof of 9.6.4. $\square$

The next result will be used below. Its proof is a complicated version of the proof of 9.6.4.

9.6.5. Lemma. *Let* μ *and* F *satisfy the hypotheses of Proposition* 9.6.3. *Then* F *is real-analytic in a neighborhood of* 0.

Proof. We may write the continuous measure μ as a sum

$$\mu = \tfrac{1}{5}(\lambda + \nu + 2\sigma + \tau)$$

where λ, ν, σ, τ are pairwise mutually singular probability measures on K.

Because K is a Kronecker (or K_p-)set, there exists $\eta \in \Gamma^-$ and there exists for all $m \geq 1$, α_m, and $\beta_m \in \Gamma^-$ such that

$$\begin{aligned}
(\alpha_m)_\lambda &= e^{-2^{-m}}, & (\beta_n)_\lambda &= 1, & \eta_\lambda &= 1,\\
(\alpha_m)_\nu &= 1, & (\beta_n)_\nu &= 1, & \eta_\nu &= -\tfrac{1}{2},\\
(\alpha_m)_\sigma &= 1, & (\beta_n)_\sigma &= e^{-2^{-n}}, & \eta_\sigma &= \tfrac{1}{2}i,\\
(\alpha_m)_\tau &= 1, & (\beta_n)_\tau &= 1, & \eta_\tau &= -\tfrac{1}{2}i,
\end{aligned}$$

and such that $\alpha_m \geq 0$, $\beta_n \geq 0$.

Let $F \circ \mu = \rho \in M(G)$. We now define a sequence of functions of two complex variables. The domain common of these functions will be the set $X = \{(w, z) \in \mathbb{C}^2 : |w| < \frac{1}{10}, |z| < \frac{1}{10}\}$. For $(w, z) \in X$ and $m, n \geq 1$, define $H_{m,n}(w, z)$ by

$$H_{m,n}(w, z) = \hat{\rho}(\eta\alpha_m^{-2^m \log(5w + 1/2)}\beta_n^{-2^n \log(5z + 1/2)}).$$

Then the $H_{m,n}$ are holomorphic in X and bounded by $\|\rho\| = \|F \circ \mu\|$. We shall show that the $H_{m,n}$ converge to a function H holomorphic in X and that $H(s, t) = F(s + it)$ near zero. Thus F will be shown to be real-analytic in a neighborhood of zero. Let $w_0 = \frac{1}{5}(\exp(-p2^{-q}) - \frac{1}{2})$ and $z_0 = \frac{1}{5}(\exp(-r2^{-s}) - \frac{1}{2})$, where p, q, r, s are positive integers. Then, if $m \geq q$, $n \geq s$ we find, as in 9.6.4, that

$$\begin{aligned}
H_{m,n}(w_0, z_0) &= \hat{\rho}(\eta\alpha_m^{p2^{m-q}}\beta_n^{r2^{n-s}})\\
&= F \circ (\tfrac{1}{5}(\hat{\lambda}(\alpha_m^{p2^{m-q}}) + \hat{\nu}(\eta) + 2\hat{\sigma}(\eta\beta_n^{r2^{n-s}}) + \hat{\tau}(\eta))\\
&= F(\tfrac{1}{5}(\exp(-p2^{-q}) - \tfrac{1}{2}) + \tfrac{1}{5}i(\exp(-r2^{-s}) - \tfrac{1}{2})) = F(w_0 + iz_0),
\end{aligned}$$

where we have used the continuity of F and the fact that $\alpha_m^{p2^{m-q}}$ and $\beta_n^{r2^{n-s}}$ belong to Γ^-.

The set $\{(w_0, z_0) \in X : w_0 = \frac{1}{5}(\exp(-p2^{-q} - \frac{1}{2}), z_0 = \frac{1}{5}(\exp(-r2^{-s}) - \frac{1}{2}), p, q, r, s \geq 1\}$ is a set of uniqueness for the analytic functions in X and $\lim H_{m,n}$ exists there. By the two-dimensional version of Vitali's theorem (Gunning and Rossi [1, p. 11]), $H(w, z) = \lim H_{m,n}(w, z)$ exists uniformly on compact subsets of X, and therefore $F(s + it) = H(s, t)$ if $|s| < \frac{1}{10}$, $|t| < \frac{1}{10}$. That proves the Lemma. □

Remark. A more complicated version of the preceding argument can be used to show that F is real-analytic at each $z_0 \in \operatorname{Int} \hat{\mu}(\Gamma)^-$. That proof involves adding another term ω to the sum $\mu = \frac{1}{5}(\lambda + \nu + 2\sigma + \tau)$ and carefully estimating how much mass ω needs to have for $\hat{\omega}(\eta)$ to equal z_0.

Proof of 9.6.3. We now write $F(u + iv) = \sum b_{mn} u^m v^n$ where the convergence is absolute when $|u| < \delta$ and $|v| < \delta$ for some $\delta > 0$. We set $p_k(u + iv) = \sum_{m+n=k} b_{mn} u^m v^n$ and observe that, on replacing u by $\frac{1}{2}(z + \bar{z})$ and v by $(z - \bar{z})/2i$, we uniquely determine numbers a_{mn} such that

$$p_k(z) = \sum_{m+n=k} a_{mn} z^m \bar{z}^n \quad \text{for } k \geq 0, 1, 2, \ldots .$$

We claim that $F(z) = \sum_{m,n \geq 0} a_{mn} z^m \bar{z}^n$ with absolute convergence if $|z| < \delta$. To establish the claim, note that

$$p_k(z) = \sum_{m+n=k} a_{m,n} z^m \bar{z}^n = \sum_{m+n=k} 2^{-k} b_{m,n} \sum_{r=0}^{n} \sum_{s=0}^{n} \binom{m}{r}\binom{n}{s} i^{-n}(-1)^s z^{r+s} \bar{z}^{k-r-s}.$$

Therefore

$$\sum_{m+n=k} |a_{m,n}||z|^k \leq \sum_{m+n=k} 2^{-k} |b_{m,n}| \sum_{r=0}^{m} \sum_{s=0}^{n} \binom{m}{r}\binom{n}{s} |z|^k$$

$$= 2^{-k} \sum_{m+n=k} |b_{m,n}| 2^m 2^n |z|^k = \sum_{m+n=k} |b_{m,n}||z|^k.$$

Since $|z| < \delta$, $\sum_{k=0} \sum_{m+n=k} |b_{m,n}||z|^k < \infty$. That establishes the claim. Note that we have lost something, since $\sum a_{mn} z^m \bar{z}^n$ converges only in the disc $|z| < \delta$ while $\sum b_{mn} s^m t^n$ converges in the square $|s| = |\text{Re } z| < \delta$, $|t| = |\text{Im } z| < \delta$.

Since μ is concentrated on a Kronecker set or K_p-set ($p \geq 3$) E, $E \cup -E$ is algebraically scattered and $E \cap -E = \varnothing$. Therefore 6.2.6 implies that $\mu^m * \tilde{\mu}^n \perp \mu^p * \tilde{\mu}^q$ if $(m, n) \neq (p, q)$. We may therefore write

$$F \circ \mu = \sum d_{mn} \mu^m * \tilde{\mu}^n + \lambda$$

where $\lambda \perp \mu^m * \tilde{\mu}^n$ for all $m, n = 0, 1, \ldots$. Let $\beta \in \Gamma^-$ be such that $\beta \geq 0$ and $\beta_\mu = \delta/2$ a.e. $d\mu$. Let $\gamma_\alpha \in \Gamma$ have $\lim \gamma_\alpha = \beta$. Then for each $\gamma \in \Gamma$

$$\begin{aligned}(\beta F \circ \mu)\hat{}(\gamma) &= \sum b_{mn}(\delta/2)^{m+n}(\mu^m * \tilde{\mu}^n)\hat{}(\gamma) + (\beta\lambda)\hat{}(\gamma)\\ &= \lim F(\hat{\mu}(\gamma_\alpha \gamma)) + (\beta\lambda)\hat{}(\gamma)\\ &= \sum a_{mn}(\delta/2)^{m+n} \hat{\mu}(\gamma)^m (\hat{\mu}(\gamma)^-)^n + (\beta\lambda)\hat{}(\gamma).\end{aligned}$$

Therefore $\beta F \circ \mu = \sum a_{mn}(\delta/2)^{m+n} \mu^m * \tilde{\mu}^n + \beta\lambda$ and $a_{mn}(\delta/2)^{m+n} = d_{mn}(\delta/2)^{m+n}$ for all m, n. Therefore $\sum |a_{mn}| = \sum |d_{mn}| < \infty$.

It remains to show that $\sum a_{mn} z^m \bar{z}^n = F(z)$ for $z \in \text{Int}\, \hat{\mu}(\Gamma)^-$. Let $H(z) = \sum a_{mn} z^m \bar{z}^n$. Let $z \in \text{Int}\, \hat{\mu}(\Gamma)^-$. Let $\eta \in \Gamma^-$ and $\beta_m \in \Gamma^-$ for $m \geq 1$ be such that $\eta_\mu = z$ a.e. $d\mu$, $\beta_m \geq 0$, and $(\beta_m)_\mu = \exp(-2^{-m})$ a.e. $d\mu$. For $w \in X_z = \text{Int}\, \hat{\mu}(\Gamma)^- \cap \{w : zw \notin (-1, 0]\}$ and $m = 1, 2, \ldots$, let

$$G_m(w) = (F \circ \mu)^\wedge(\eta \beta_m^{-2^m \log w}).$$

(Here we use the branch of log defined on X_z, with $\log 1 = 0$ or $\log i = \pi i/2$.) If $w = \exp(-p2^{-q})$ and $m \geq q$, then $\eta \beta_m^{-2^m \log w} \in \Gamma^-$ and $G_m(w) = F(wz)$. By Vitali's Theorem (Hille [1, Volume II, p. 251], $\lim G_m = G$ exists uniformly on compact subsets of X_z, $G(t) = F(tz)$ for $0 < t < 1$, and $F(tz) = \sum a_{mn} z^m \bar{z}^n t^{m+n}$ for $0 < t < \delta$. Since $H(tz)$ extends to a function analytic in t for $|t| < 1$, we see that $G(t) = H(tz) = F(tz)$ for $0 < t < 1$. Thus $F(w) = H(w)$ if $w = tz$, $0 < t < 1$ and $z \in \text{Int}\, \hat{\mu}(\Gamma)^-$. It is now immediate that $F = H$ on $\text{Int}\, \hat{\mu}(\Gamma)^-$. That completes the proof of 9.6.3. □

9.6.6. Lemma. *Let F, G, and Γ satisfy the hypotheses of* 9.6.2. *Then F is continuous in* $\text{Int}\, D(\Gamma)$.

Proof. Let $z_0 \in \text{Int}\, D(\Gamma)$. Then there exist $x_1, x_2 \in G$, $\alpha_1, \alpha_2 \in [0, 1]$, and $\gamma_0 \in \Gamma$ such that

$$z_0 = \alpha_1 \langle x_1, -\gamma_0 \rangle + \alpha_2 \langle x_2, -\gamma_0 \rangle, \quad \text{and} \quad \alpha_1 + \alpha_2 < 1.$$

Without loss of generality, we may assume that for some $x_0 \in G$, $\{x_1, x_2\} \subseteq Gp\{x_0\}$.

Case I, x_0 has finite order $\neq 2$. Let $\Lambda = \{\gamma \in \Gamma : \langle x_0, \gamma \rangle = 1\}$. Then Λ is an open infinite subgroup of Γ. We claim that there exists $\delta > 0$ such that $F(z_0 + \hat{\mu}) \in B(\Lambda)$ for all $\mu \in B(\Lambda)$ with $\|\mu\| < \delta$.

Since x_0 has order $\neq 2$, there exist $y_1, y_2 \in K$ and $\beta_1, \beta_2 \geq 0$ such that

$$\beta_1 \langle y_1, -\gamma_0 \rangle + \beta_2 \langle y_2, -\gamma_0 \rangle = i \quad \text{and} \quad \beta_1 + \beta_2 \leq 2.$$

Let $\omega = \alpha_1 \delta_{x_1} + \alpha_2 \delta_{x_2}$ and $\tau = \beta_1 \delta_{y_1} + \beta_2 \delta_{y_2}$. Then $\|\omega\| < 1$ and $\|\tau\| \leq 2$. Let $0 < \delta < (\frac{1}{16})(1 - \|\omega\|)$, so $\alpha_1 + \alpha_2 + 16\delta < 1$. Suppose that $\hat{\mu} \in B(\Lambda)$ and $\|\mu\| < \delta$. Let $\hat{\mu}' \in B(\Gamma)$ have $\hat{\mu}'|_\Lambda = \hat{\mu}$ and $\|\mu'\| < \delta$. Then $\gamma_0 \mu' = \mu_1 - \mu_2 + i\mu_3 - i\mu_4$, where $\mu_1 \geq 0, \ldots, \mu_4 \geq 0$ and $\sum \|\mu_j\| \leq 2\|\mu'\| < 2\delta$. Let

$$\nu = \omega + \mu_1 + \tau^2 * \mu_2 + \tau * \mu_3 + \tau^3 * \mu_4 + \beta m_K, \tag{6}$$

where $\beta \geq 0$ is such that $\|\nu\| = \hat{\nu}(0) = 1$. Then

$$\hat{\nu}(\gamma_0 + \lambda) = z_0 + (\gamma_0 \mu')^\wedge(\gamma_0 + \lambda) + 0 = z_0 + \hat{\mu}(\gamma)$$

for $\lambda \in \Lambda$, $\lambda \neq 0$. Therefore $F \circ v^\wedge = F(z_0 + \hat{\mu})$ on $\Lambda \backslash \{0\}$. It follows at once that $F(z_0 + \hat{\mu}) \in B(\Lambda)$ for $\hat{\mu} \in B(\Lambda)$, $\|\mu\| < \delta$. By a simple variant of the proof of 9.3.7, F is continuous at z_0.

Case II, x_0 *has infinite order*. Then either $\mathrm{Gp}(\{x_0\}) = Zx_0$ is closed, in which case 9.6.7 below applies, or $\mathrm{Gp}(\{x_0\})^-$ is compact, in which case 9.6.8 applies.

Case III, x_0 *has order* 2. It is easy to see that we may replace x_1, x_2 by other elements x'_1, x'_2 such that x'_1, x'_2 belong to $\mathrm{Gp}(\{x'_0\})$ where $2x'_0 \neq 0$. That ends the proof of 9.6.6, subject to 9.6.7 and 9.6.8. □

9.6.7. Lemma. *Let F and Γ satisfy the hypotheses of Theorem* 9.6.2. *Suppose also that Γ contains a closed subgroup isomorphic to Z. Then F is continuous in* Int $D(\Gamma)$.

Proof. Since $\Gamma \supseteq Z$, $D(\Gamma) = \{z : |z| \leq 1\}$. Since each element of $PD(Z)$ has an extension to an element of $PD(\Gamma)$, F operates from $PD(Z)$ to $B(Z)$, and therefore we may assume that $\Gamma = Z$.

Choose $z_0 \in \mathrm{Int}\, D(\Gamma) = \{z : |z| < 1\}$. Let $0 < \varepsilon < 1 - |z_0|$. For each $n > 1$, there is a discrete probability measure v such that $\hat{v}(k) = e^{2\pi i/n}$ for k in the coset $nZ + 1$.

By choosing an appropriate n and appropriate powers of the resulting measure v, we see that there exist non-negative integers p, q and numbers $\alpha_1, \alpha_2 \geq 0$ such that $\alpha_1 + \alpha_2 \leq |z_0| + \varepsilon/2$ and if $v' = \alpha_1 v^p + \alpha_2 v^q$, then

$$\hat{v}'(k) = \alpha_1 \hat{v}^p(k) + \alpha_2 \hat{v}^q(k) = z_0 \quad \text{for all } k \in nZ + 1.$$

We may also assume that n is a multiple of 4. Then there exists μ' with $\hat{\mu}'(k) = i$ on $nZ + 1$, where μ' is a discrete probability measure. We now use the argument that gave us (6) to show that if $f \in B(nZ + 1)$ and $\|f\| < \varepsilon/8$, then $F(z_0 + f) \in B(nZ + 1)$. Since $B(Z)$ is isomorphic to $B(nZ + 1)$, we see that $F_1(z) = F(z_0 + z)$ operates on $\{v : v \in M(T), \|v\| < \varepsilon/8\}$. By 9.3.7, F_1 is continuous at 0. Therefore F is continuous at z_0. That proves Lemma 9.6.7. □

9.6.8. Lemma. *Let F and Γ satisfy the hypotheses of Theorem* 9.6.2. *Suppose also that Γ is compact. Then F is continuous in* Int $D(\Gamma)$.

Proof. We shall consider two separate cases: (1) G contains an element of infinite order; and (2) G is a torsion group. By passing to subgroups of G, we see that we may assume that either (1′) $G = Z$ and $\Gamma = T$; or (2′) $G = \sum Z_{p_j}, \Gamma = \prod Z_{p_j}$ is totally disconnected and $p_j > 2$ for at least one value of j. In both (1′) and (2′), if U is any open subset of Γ, then $A(\overline{U})$ contains a subalgebra B isomorphic to $A(T)$ in case (1′) or to $A(\prod_{j \geq n} Z_{p_j})$ in case (2′). Thus if $f \in B$ and $F \circ f \in A(\overline{U})$, then $F \circ f \in B$. (That follows from the fact that every neighborhood contains a translate of $\prod_{j \geq n} Z_{p_j} \times \{(0, 0, \ldots, 0)\}$ in case (2′), and the use of $2\pi/k$ periodic functions in case (1′).) We may therefore apply 9.3.4 as soon as we know that, given $z_0 \in \operatorname{Int} D(\Gamma)$, there exists an open set $U \subseteq \Gamma$ and $\delta > 0$ such that $F(z_0 + f) \in A(\bar{U})$ whenever $f \in A(\bar{U})$ and $\|f\| < \delta$.

We shall use a convergent series argument. Since Γ does not have exponent two (by hypothesis), there exist $x \in G$ and $\gamma_0 \in \Gamma$ such that z_0 is in the interior of the convex hull of $\{(kx, \gamma_0): k = 1, 2, \ldots\}$ and $\langle x, \gamma_0 \rangle \neq \pm 1$.

Let $\alpha_1, \alpha_2 \geq 0$, and $x_1, x_2 \in Z^+x$ be such that

$$\alpha_1 \langle x_1, -\gamma_0 \rangle + \alpha_2 \langle x_2, -\gamma_0 \rangle = z_0$$

and

$$\alpha_1 + \alpha_2 < 1.$$

Let $\omega = \alpha_1 \delta_{x_1} + \alpha_2 \delta_{x_2}$. It is obvious that there exist $\beta_1, \beta_2 > 0$, $y_1, y_2 \in Z^+x$ such that $\beta_1 + \beta_2 \leq 2$ and

$$\beta_1 \langle y_1, -\gamma_0 \rangle + \beta_2 \langle y_2, -\gamma_0 \rangle = i.$$

Let $\mu = \beta_1 \delta_{y_1} + \beta_2 \delta_{y_2}$.

Let $\delta > 0$ be such that $6\delta^2 \|\mu\| < \delta/4$ and $\|\omega\| + \|\mu\|^3 \delta < 1$. Let U be a neighborhood of γ_0 such that $0 \notin \bar{U}$ and these estimates hold:

$$\|i - \hat{\mu}|_{\bar{U}}\|_{A(\bar{U})} < \delta \quad \text{and} \quad \|z_0 - \hat{\omega}|_U\|_{A(U)} < \delta/4.$$

Let $\hat{f} \in A(\bar{U})$ have $\|\hat{f}\|_{A(\bar{U})} < \delta$, and let $f_1^{(1)}, \ldots, f_1^{(1)}$ be non-negative elements of $L^1(G)$ such that $\sum_1^4 \|f_j^{(1)}\| < 2\|\hat{f}\|_{A(\bar{U})}$ and

$$\hat{f} = \hat{f}_1^{(1)} - \hat{f}_2^{(1)} + i\hat{f}_3^{(1)} - i\hat{f}_4^{(1)}$$

on $\bar{U}$. Let $v_1 = \omega + f_1^{(1)} + \mu^2 * f_2^{(1)} + \mu * f_3^{(1)} + \mu^3 * f_4^{(1)}$. Then $\|v_1\| \leq \|\omega\| + 4\|\mu\|^3 < 1$. Also

$$\begin{aligned} \|z_0 + \hat{f} - \hat{v}_1|_{\bar{U}}\|_{A(\bar{U})} &\leq \|z_0 - \hat{\omega}|_{\bar{U}}\|_{A(\bar{U})} + \|f - \hat{v}|_{\bar{U}}\|_{A(\bar{U})} \\ &\leq \delta/4 + 6\delta^2\|\mu\|^2 \\ &\leq \delta/2. \end{aligned}$$

Let $\hat{f}_2 = z_0 + \hat{f} - \hat{v}_1|_{\bar{U}} \in A(\bar{U})$, so $\|\hat{f}_2\| \leq \delta/2$. Then there exists $f_1^{(2)}, \ldots, f_4^{(2)} \in L^1(G)$ with $f_1^{(2)}, \ldots, f_4^{(2)} \geq 0$ such that

$$\hat{f}_2 = \hat{f}_1^{(2)} - \hat{f}_2^{(2)} + i\hat{f}_3^{(2)} - i\hat{f}_4^{(2)}$$

on U and

$$\sum \|f_j^{(2)}\| \leq 2\|\hat{f}_2\|_{A(\bar{U})} < \delta.$$

Let

$$\nu_2 = f_1^{(2)} + \mu^2 * f_2^{(2)} + \mu * f_3^{(2)} + \mu^3 * f_4^{(2)}.$$

Then

$$\|\hat{f}_2 - \hat{\nu}_2|_U\| \leq 4\|\mu\|^3\delta^2/4,$$

and

$$\|\nu_2\| < 2\delta\|\mu\|^3.$$

Proceeding in this manner, we find $\nu_3, \nu_4, \ldots$ such that for all $k \geq 1$,

$$\|\nu_k\| < 2^{3-k}\delta\|\mu\|^3, \quad \nu_k \geq 0$$

and

$$\left\| z_0 + \hat{f} - \sum_{j=1}^{k} \nu_j|_{\bar{U}} \right\|_{a(\bar{U})} < 4\|\mu^3\|\delta^2 2^{-k}.$$

Let $\nu = \sum_1^\infty \nu_j$. Then $\nu \geq 0$ and

$$\|\nu\| < \|\omega\| + 4\|\mu\|^3\delta + \sum_{k=2}^{\infty} \delta\|\mu\|^3 2^{3-k} = \|\omega\| + 8\delta\|\mu\|^3 < 1.$$

Furthermore, $\hat{\nu} = z_0 + \hat{f}$ on $\bar{U}$. Thus if $\hat{f} \in A(\bar{U})$ and $\|\hat{f}\| < \delta$, then there exists $\nu \in L^1(G)$, $\nu \geq 0$, $\|\nu\| < 1$ such that $\hat{\nu} = z_0 + \hat{f}$ on $\bar{U}$. To complete the proof of the Lemma, we need only to augment ν so that $\|\nu\| = 1$. And that is easily done, as follows. Choose a symmetric neighborhood $V \subseteq \Gamma$ of zero such that $(V - V) \cap \bar{U} = \varnothing$. (That is possible since $0 \notin \bar{U}$.) Let $\chi = \chi_V$ be the indicator function of V. Then $\chi * \chi \in A(G)$ and $(\chi * \chi)^\wedge \geq 0$. Since $\chi * \chi = 0$ outside $V - V = V + V$, there is an appropriate multiple $t\chi * \chi$ of $\chi * \chi$ such that $\|\hat{\nu} + t\chi * \chi\| = 1$. That ends the proof of 9.6.8. □

Proof of Theorem 9.6.2. If G contains a perfect Kronecker set or a perfect K_p-set for some $p \geq 3$, then 9.6.2 follows at once from 9.6.3 and 9.6.6. Thus we may assume that G contains no perfect Kronecker or K_p-sets.

By the structure theorem, G must contain an open compact subgroup H, since R contains perfect Kronecker sets. That subgroup H must be of exponent two, since otherwise it would contain either a Kronecker or K_p-set with $p > 2$ (Rudin [1, 5.2] or Hewitt and Ross [1, 41.5]). That excludes (i) of the hypotheses of 9.6.2, so by (ii) of 9.6.2, G/H must contain an element of infinite order. It is now easy to see that we may assume that $H = D_2$ and $G = Z \times D_2$.

Let μ be a continuous probability measure on H that is concentrated on a K_2-set. Let $x \in Z \times \{0\}$ be such that $x \neq 0$. We claim that

$$(\delta_x * \mu)^m * (\delta_x * \mu)^{\sim n} \perp (\delta_x * \mu)^p * (\delta_x * \mu)^{\sim q}$$

if $(m, n) \neq (p, q)$. Indeed

$$(\delta_x * \mu)^m * (\delta_x * \tilde{\mu})^m = \delta_{(m-n)x} * \mu^{n+m}.$$

Thus, if $(m, n) \neq (p, q)$, then either $m - n \neq p - q$, in which case $\delta_{(m-n)x} * \mu^{m+n} \perp \delta_{(p-q)x} * \mu^{p+q}$ since $[(m - n)x + H] \cap [(p - q)x + H] = \varnothing$, or $r = m - n = p - q$ and $m + n \neq p + q$, so (by 6.2.6)

$$\mu^{m+n} \perp \mu^{p+q}, \quad \text{and} \quad \delta_{rx} * \mu^{m+n} \perp \delta_{rx} * \mu^{p+q}.$$

Therefore $F \circ \mu = \sum b_{mn}(\delta_x * \mu)^m * (\delta_x * \mu)^{\sim n} + \lambda$ where $\lambda \perp (\delta_x * \mu)^m * (\delta_x * \mu)^{\sim n}$ for all $m, n = 0, 1, 2, \ldots$, and $\sum |b_{mn}| < \infty$.

As in 9.6.4, F is real-analytic in a neighborhood of 0. The remainder of the proof follows the lines of the proof of 9.6.3. The details are left to the reader. That ends the proof of 9.6.2. □

Proof of 9.6.1. Suppose that G contains a perfect Kronecker set or set of type K_p for some $p \geq 3$. Then 9.6.2 shows that $F(z) = \sum a_{mn} z^m \bar{z}^n$ for $z \in \text{Int } D(\Gamma)$ and $\sum |a_{mn}| < \infty$. Of course, if μ is a continuous probability measure on a Kronecker set or K_p set for some $p \geq 3$, then $F \circ \mu = \sum a_{mn} \mu^m \tilde{\mu}^n \geq 0$. By the mutual singularity of the products $\mu^m * \mu^n$ (see 6.2.6), $a_{mn} \geq 0$ for all m, n.

We now reduce to the preceding case. If G contains neither a perfect Kronecker set nor a perfect K_p-set for some $p \geq 3$, then G has an open compact subgroup H of exponent two. Either G/H is not a torsion group, or G contains an infinite independent set $\{x_j\}$ such that $2x_j \neq 0$ for all j. In both cases the Bohr compactification bG of G contains either a perfect Kronecker set or a K_p-set for some $p \geq 3$. Let ρ be a probability measure on bG. Then $\rho = \lim \rho_\alpha$ where the ρ_α are probability measures on G. Since 9.6.6 can be adapted to show that F is continuous on $\text{Int } D(\Gamma)$, $F \circ \hat{\rho}(\gamma) = \lim F \circ \hat{\rho}_\alpha(\gamma)$ if $\hat{\rho}(\gamma) \in \text{Int } D(\Gamma)$. Since $\|F \circ \rho_\alpha\| = F \circ \hat{\rho}_\alpha(1) \leq 1$, we may assume that for each $\hat{\rho} \in PD(\Gamma_d)$, there exists $\lambda \in PD(\Gamma_d)$ such that $\hat{\lambda}(\gamma) = F \circ \hat{\rho}(\gamma)$ whenever $\hat{\rho}(\gamma) \in \text{Int } D(\Gamma)$. Therefore

$$F(z) = \sum a_{mn} z^m \bar{z}^n \quad \text{for all } z \in \text{Int } \hat{\mu}(\Gamma_d), \quad \text{where} \quad a_{mn} \geq 0, \sum a_{mn} < \infty,$$

and μ is any continuous probability measure concentrated on a perfect Kronecker or K_p-set ($p \geq 3$).

Now fix $z_0 \in \operatorname{Int} D(\Gamma)$, and let $v \geq 0$ be such that $\hat{v}(\gamma_0) = z_0$ for some $\gamma_0 \in \Gamma$, $\|v\| < 1$, and v is finitely supported. Let μ be a continuous probability measure supported on a Kronecker or K_p-set ($p \geq 3$), and $r = 1 - \|v\|$. Then for each $|w| < r$, there exists $\eta \in \Gamma^-$ such that $\eta_v \equiv 1$ a.e. dv and $\eta_v = w$ a.e. $d\mu$. It is now easy to show that F is real-analytic in a neighborhood of z_0. Therefore F is real-analytic in $\operatorname{Int} D(\Gamma)$. It is also easily shown that $H(z) = \sum a_{mn} z^m \bar{z}^n$ is real-analytic in $|z| < 1$ when $\sum |a_{mn}| < \infty$. Since $H = F$ in a neighborhood of $z = 0$, it follows that $H = F$ in $\operatorname{Int} D(\Gamma)$. That ends the proof of 9.6.1. □

EXAMPLES (*9.6.1 is sharp*). Suppose that $\Gamma = \Gamma_0 \times \Gamma_1$ where Γ_0 is finite and Γ_1 has exponent two. Let q be the order of Γ_0, and let

$$F(z) = \left(q! \sum_{j=0}^{q-1} z^j \bar{z}^{q-j}\right) - z^q$$

We claim that F maps $PD(\Gamma)$ into $PD(\Gamma)$.

If $\mu \geq 0$ is a measure on G, we let $v(E) = \mu(-E) = \tilde{\mu}(E)$. Then

$$F \circ \mu = q! \sum_{j=0}^{q-1} \mu^j v^{q-j} - \mu^q.$$

Let $E \subseteq G$ be a Borel set and $\sigma \in S$, the set of permutations of $\{1, \ldots, q\}$. Define sets E' and $E_\sigma^i (1 \leq i \leq q)$ by

$$E' = \{(x_1, \ldots, x_q) \in G^q : \textstyle\sum x_j \in E\}$$

and

$$E_\sigma^i = \{(x_1, \ldots, x_q) \in G^q : x_{\sigma(1)} + \cdots + x_{\sigma(i)} - x_{\sigma(i+1)} - \cdots - x_{\sigma(q)} \in E\}.$$

Then

$$\mu^q(E) = (\mu \times \cdots \times \mu)(E')$$

and

$$\mu^i * v^{q-i}(E) = (\mu \times \cdots \times \mu)(E_\rho^i) \quad \text{for all } \sigma \in S.$$

Therefore, since S has $q!$ elements,

$$\begin{aligned} F \circ \mu(E) &= \sum_{\sigma \in S} \sum_{i=0}^{q-1} (\mu \times \cdots \times \mu)(E_\sigma^i) - (\mu \times \cdots \times \mu)(E') \\ &\geq (\mu \times \cdots \times \mu)(A) - (\mu \times \cdots \times \mu)(E'), \end{aligned}$$

where $A = \bigcup_{\sigma \in S} E^i$.

Thus to show $F \circ \mu \geq 0$, we need only show that $A \supseteq E'$. Let $(x_1, \ldots, x_q) \in E'$; then $x_1 + \cdots + x_q \in E$. Each $x_i = y_i + z_i$, where $y_i \in \Gamma_0 \times \{0\}$ and $z_i \in \{0\} \times \Gamma_1$, so $z_i = -z_i$. Consider the q elements $y_1, y_1 + y_2, \ldots, y_1 + \cdots + y_q$. Since Γ_0 has order q, either two are the same or one is zero. In either case, there are integers $1 \leq s \leq t \leq q$ such that

$$y_s + \cdots + y_t = 0.$$

Therefore

$$x_s + \cdots + x_t = z_s + \cdots + z_t = -z_s - \cdots - z_t = -x_s - \cdots - x_t.$$

Hence $(x_1, \ldots, x_q) \in \bigcup_{\sigma \in S} E_\sigma^{q-t+s-1} \subseteq A$. That completes the verification.

(*9.6.2 is sharp*). Let $\beta \in T$ be an element of infinite order and let $F(z) = (2 - \beta z - \bar{z})^{-1}$. It is easy to see that F is the restriction to $\{(z, w); z = \bar{w}\}$ of $H(w, z) = (2 - \beta w - z)^{-1}$ which is analytic in $\{(w, z)\colon \beta w + z \neq 2\}$. Hence F is real-analytic in the plane, but the power series of F in z and $\bar{z}$ cannot be of the form $\sum a_{mn} z^m \bar{z}^n$ where $\sum |a_{mn}| < \infty$, since that would imply that

$$H(w, z) = \sum a_{mn} w^m z^n$$

would have no pole at $w = \bar{\beta}, z = 1$. We show that F operates from $PD(\Gamma)$ to $B(\Gamma)$ whenever the dual group G of Γ has an open subgroup H with exponent two and G/H is a torsion group.

Let $\mu \in M(G)$ be a probability measure; then $\mu = \sum a_j \delta(x_j) * v_i$ where $v_j \in M(H)$ are probability measures, the $x_j + H$ are distinct cosets of H and $a_j \geq 0$ with $\sum a_j = 1$. Then (formally)

$$F \circ \mu = \sum_0^\infty 2^{-n}(\beta\mu + \tilde{\mu})^n. \tag{7}$$

We may assume that $a_1 \neq 0$. Let k be the order of x_1, and choose n so that $|\beta^{kn} + 1| < \frac{1}{10}$. Then

$$(\beta\mu + \tilde{\mu})^{nk} = \beta^{kn} a_1^{kn} \delta(knx_1) v_1^{kn} + a_1^{kn} \delta(-knx_1) v_1^{kn} + \text{other terms}.$$

Thus,

$$\|(\beta\mu + \tilde{\mu})^{kn}\| \leq 2 - 2a_1^{kn}\|v_1^{kn}\| + a_1^{kn}\|v_1\|^{kn}|\beta^{kn} + 1| < \infty,$$

and the sum in (7) converges in norm. Hence $F \circ \hat{\mu} \in B(\Gamma)$.

Remarks and Credits. Theorem 9.6.1 for general groups was first proved by Rider [4], a special case having been established by Herz [6]. Theorem 9.6.2

and Proposition 9.6.3 are due to Moran [3]. An incomplete proof of 9.6.2 appeared in Graham [3]. The first example is due to Rider [4]; the second to Graham [11].

For functions defined on $[-1, 1]$ and operating from $PD(\Gamma)$ to $PD(\Gamma)$, analogous results were obtained by Rudin [7] before those of Herz and Rider; the reader is encouraged to formulate and prove those results.

Mason [1] shows that if X is an L-subspace of $M(T)$ such that $\hat{\mu} \in X$ whenever $\mu \in X$, and if $F: [-1, 1] \to \mathbb{C}$ has $F \circ \hat{\mu} \in B(Z)$ for all $\mu \in X$, then F is analytic in a neighborhood of 0. The proof does not appear to extend to all groups.

For extensions of 9.6.1 to matrices, see Christensen and Ressel [1] and their bibliography.

Chapter 10

The Multiplier Algebras $M_p(\Gamma)$, and the Theorem of Zafran

10.1. Introduction

For $1 \le p < \infty$, consider the Banach space of functions on the integer group Z,

$$L^p(T)^\wedge = \{\hat{f} : f \in L^p(T)\},$$

with norm

$$\|\hat{f}\|_{L^p(T)^\wedge} = \|f\|_{L^p(T)}.$$

Let $M_p(Z)$ be the algebra, under pointwise multiplication, of functions f defined on Z that multiply $L^p(T)^\wedge$ into itself, in the sense that $fg \in L^p(T)^\wedge$ whenever $g \in L^p(T)^\wedge$. If $f \in M_p(Z)$, then the operator on $L^p(T)^\wedge$ that maps g to fg is easily proved continuous, by using the closed graph theorem, and the norm of f in $M_p(Z)$ is defined to be the norm of this operator:

$$\|f\|_{M_p(Z)} = \sup\{\|fg\|_{L^p(T)^\wedge} : \|g\|_{L^p(T)^\wedge} \le 1\}.$$

With that norm, $M_p(Z)$ is a Banach algebra. The purpose of this Chapter is to present a deep theorem (10.3.1) about the Banach subalgebra $M_{po}(Z) = M_p(Z) \cap C_o(Z)$, and to make it accessible even to a reader who is unfamiliar with multipliers. Such a reader will have to take on faith two results from the Littlewood-Paley theory, Theorems 10.3.5 and 10.3.7. For an exposition of that subject, we recommend the book of Edwards and Gaudry [1]. The reader will need also to be conversant with the basic theory of commutative Banach algebras, for which a good reference is Rudin [3, Chapters 10 and 11].

The Littlewood–Paley theory provides an ample supply of functions (including many idempotents) that belong to $\bigcap_{1<p\le 2} M_p(Z)$ but not to $M_1(Z)$. The algebras $M_p(Z)$ are known to grow steadily as p increases from 1 to 2, and techniques are known for constructing, for $p > 1$, functions f that are in $M_p(Z)$ but not in $\bigcup_{r<p} M_r(Z)$. One strength of Zafran's work is that it produces many such functions, with striking properties.

Zafran's Theorem belongs to the family of converses of the Wiener-Lévy Theorem, introduced in Chapter 9. Its presentation here will allow the reader to compare the structural theory of multiplier algebras with that of measure algebras treated at length in earlier chapters. Nevertheless, except for a modest number of specific references, this Chapter may be read independently.

In Section 10.2 we shall define the algebras $M_p(\Gamma)$ for arbitrary locally compact abelian groups Γ, and treat some of the softer aspects of the theory.

The goals of this Chapter are quite limited. Multiplier algebras are part of a large and active area of harmonic analysis of which we do not even attempt an adequate survey. A survey is provided by Coifman and Weiss [2], and an extensive introduction by the books of Stein [1, 2] and Stein and Weiss [1].

To place Zafran's work in perspective, we offer some elementary remarks about Banach algebras. Consider a Banach algebra B, consisting of a class of complex-valued functions defined on a set X. Let the algebra structure of B be provided by the usual pointwise operations. Let Δ denote the maximal ideal space of B. Suppose that whenever x and y are distinct points of X, there exists $f \in B$ such that $f(x) = 0$ and $f(y) = 1$. Then X may be regarded as a subset of Δ, since for each $x \in X$ the mapping $f \to f(x)$ is a distinct homomorphism of X onto the complex field. The Gelfand transform $\hat{f}$ of an element $f \in B$ is an extension of f to all of Δ.

When studying a particular algebra of this type, it is natural to ask whether $X = \Delta$, and if not, how important a part of Δ is X. Equality holds, for example, if $B = C(T)$ or $A(Z)$ or $C_o(Z)$. Sometimes Δ is a compactification of X, as when $B = C_o(Z) \oplus \mathbb{C}$ or $AP(Z)$ or $l^\infty(Z)$. Some other possible conditions, in order of nondecreasing strength, are as follows.

(1) X is not dense in Δ.

(2) $\sup\left\{\dfrac{\sup\{|\hat{f}(h)|: h \in \Delta\}}{\sup\{|f(x)|: x \in X\}}: f \in B\right\} = \infty.$

(3) There exists $g \in B$ such that g is real-valued (on X) but $\hat{g}$ is not everywhere real-valued.

(4) B is an entire algebra.

We shall define statement (4) presently. The point we are making is that when B is entire, its maximal ideal space is especially large and interesting. Zafran's result is that for $1 < p < 2$ or $2 < p < \infty$, $M_{po}(Z)$ is entire. (It was known earlier that $M_{1o}(Z)$ is entire (Theorem 9.5.1), and that $M_p(Z)$ is entire for $1 \leq p < 2$ or $2 < p < \infty$).

A function F defined on $[-1, 1]$ *operates in* B (considered as an algebra of functions on X) if $F \circ f \in B$ whenever $f \in B$ and $f(X) \subseteq [-1, 1]$. We say that B is *analytic* if for every such F there exists $\eta > 0$ such that $F(x) = \sum_{j=0}^{\infty} c_j x^j$, a convergent power series, for $|x| < \eta$. We say that B is *entire* if for every F defined on $[-1, 1]$ that operates in B, there exists $\eta > 0$ such that F agrees on

$[-\eta, \eta]$ with some entire function. (If B has an identity, we may as well say that F agrees on $[-1, 1]$ with some entire function.) Of course, the general theory of Banach algebras tells us that if $f \in B$ and F is analytic on (the closure of) $\hat{f}(\Delta)$, (and if $F(0) = 0$), then $F \circ \hat{f} = \hat{g}$ for some $g \in B$ and thus $F \circ f \in B$ (the words in parentheses may be omitted if B contains an identity). Thus (4) implies (3), since if (3) failed then every F analytic in $\{x + iy: |y| < \varepsilon\}$ for some $\varepsilon > 0$ would operate in B.

Condition (2) obviously implies (1). To see that (3) implies (2), note that if g is real on X but $i \in \hat{g}(\Delta)$, then the functions $f_n = \sin ng$ are in B and establish (2).

If for every $g \in B$ there exists $\tilde{g} \in B$ such that $\hat{\tilde{g}}(h) = \overline{\hat{g}(h)}$ for every $h \in \Delta$, then B is *symmetric*, or *self-adjoint*. Evidently condition (3) implies that B is not symmetric.

The reverse implications. Condition (1) does not imply (2), as a familiar example shows. Let B be the algebra of functions f defined on $T = \{z: |z| = 1\}$ such that $f(z) = \sum_{n=0}^{\infty} c_n z^n$ where $\|f\|_B = \sum |c_n| < \infty$. Then $\Delta = \{z: |z| \le 1\}$.

Condition (2) does not imply (3). There exist algebras B of quasi-analytic functions on T,

$$B = B_{\{w_n\}} = \left\{ f(z) = \sum_{n=-\infty}^{\infty} a_n z^n: \|f\|_B = \sum |a_n| w_{|n|} < \infty \right\}$$

for an appropriate choice of $\{w_n\}_{n=0}^{\infty}$, such that $\Delta = T$ and such that an element $f \in B$ is determined by its values on any infinite set $X \subseteq T$. Choose an infinite proper subset $X \subseteq T$ and regard B as an algebra of functions on X. See Katznelson [1, Section V.2].

We do not know whether (3) implies (4). A different but related question is considered in deLeeuw and Katznelson [1]. See also Katznelson and McGehee [2, Section 3].

10.2. The Basic Theory of the Algebras $M_p(\Gamma)$

For $1 \le p \le \infty$, $f \in L^p(G)$, and $x \in G$, let $\tau_x f$ be the element of $L^p(G)$ determined by the condition

$$\tau_x f(y) = f(y - x) \quad \text{for } y \in G.$$

Let $L_p^p(G)$ denote the Banach algebra of bounded operators U on $L^p(G)$ that commute with τ_x for every $x \in G$, with the operator norm, denoted by $\|U\|_{p,p}$ or $\|U\|_{L_p^p(G)}$. It turns out that if $1 \le p < \infty$ and $U \in L_p^p(G)$, then there is a unique element $\hat{U}$ of $L^\infty(\Gamma)$ such that

$$(1) \qquad (Uf)^\wedge = \hat{U}\hat{f} \quad \text{for } f \in L^p(G).$$

(Of course, when G is not compact and $p > 1$, $\hat{f}$ requires care to define and if $p > 2$ it must be considered as a distribution; see Katznelson [1, Chapter VI] for a discussion in the setting $G = R$. But the reader of this Chapter will not need to be familiar with those matters.) The Fourier representation of $L_p^p(G)$ is the algebra $M_p(\Gamma) = \{\hat{U}: U \in L_p^p(G)\}$, with pointwise multiplication and with the norm of an element $\hat{U}$ denoted by $\|\hat{U}\|_{M_p(\Gamma)}$ or $\|\hat{U}\|_{p,p}$ and defined to equal $\|U\|_{p,p}$.

10.2.1. Note. Let $1 \leq p \leq \infty$. Every measure $\mu \in M(G)$ determines an operator U on $L^p(G)$: $Uf = \mu * f$. We shall write $\|\mu\|_{p,p}$ for $\|U\|_{p,p}$. The function $\hat{U}$ determined by (1) is evidently $\hat{\mu}$. Since $\|\mu * f\|_p \leq \|\mu\|_M \|f\|_p$, it follows that the natural inclusion is norm-decreasing: $M(G) \subseteq L_p^p(G)$ or, equivalently, $B(\Gamma) \subseteq M_p(\Gamma)$. In particular, $L^1(G) \subseteq L_p^p(G)$ or $A(\Gamma) \subseteq M_p(\Gamma)$.

10.2.2. Proposition. *If $1 \leq p < \infty$, $U \in L_p^p(G)$, $\mu \in M(G)$, and $f \in L^p(G)$, then $U(f * \mu) = (Uf) * \mu$.*

Proof. For $g \in C_c(G)$,

$$\begin{aligned}\langle U(f * \mu), g\rangle &= \langle f * g, U * g\rangle \\ &= \iint \tau_y f(x)d\mu(y)\, U * g(x)dx = \int \langle \tau_y f, U * g\rangle d\mu(y) \\ &= \int \langle \tau_y Uf, g\rangle d\mu(y) = \langle Uf * g, \mu\rangle.\end{aligned}$$

Since $C_c(G)$ is dense in $L^p(G)$, the result follows. □

10.2.3. Proposition. *For $1 < p < \infty$ and $p + q = pq$, $L_p^p(G)$ is naturally isometrically isomorphic to $L_q^q(G)$.*

Proof. Since the relation between p and q is symmetric, it suffices to describe a suitable norm-decreasing map, $U \to U'$, from $L_p^p(G)$ into $L_q^q(G)$. For $U \in L_p^p(G)$, $f \in C_c(G)$, and $g \in C_c(G)$, using 10.2.2 we find that

$$(2) \qquad \begin{aligned}\left|\int_G f(-x)Ug(x)dx\right| &= |(f * Ug)(0)| = |(Uf * g)(0)| \\ &\leq \|Uf\|_p\|g\|_q \leq \|U\|_{p,p}\|g\|_q\|f\|_p.\end{aligned}$$

It follows that for every $g \in C_c(G)$, $Ug \in L^q(G)$ and $\|Ug\|_q \leq \|U\|_{p,p}\|g\|_q$. Since $C_c(G)$ is dense in $L^q(G)$, U extends uniquely to an operator $U' \in L_q^q(G)$, with $\|U'\|_{q,q} \leq \|U\|_{p,p}$. The Proposition is proved. □

10.2.4. Lemma. *Let U' be a bounded operator from $C_o(G)$ into $L^\infty(G)$ that commutes with all translations; for example, $U' \in L_\infty^\infty(G)$. If $f \in C_o(G)$, then $U'f$ is uniformly continuous; that is, the equivalence class $U'f$ contains a uniformly continuous function.*

Proof. $\|\tau_x U'f - U'f\|_\infty = \|U'(\tau_x f - f)\|_\infty \le \|U'\| \|\tau_x f - f\|_\infty \to 0$ as $x \to 0$. Let g be a representative of $U'f$ as an element of $L^\infty(G)$, and let $\{k_V\}$ be an approximate identity indexed by the class of open neighborhoods V of 0 in G: $\int k_V = 1$, $k_V \ge 0$, $k_V = 0$ off V. Then

$$\|k_v * g(x + \cdot) - k_v * g(\cdot)\|_\infty \le \|\tau_x g - g\|_\infty,$$

so $\{k_V * g\}$ is an equicontinuous family. Therefore a subnet converges, uniformly on each compact subset, to some uniformly continuous function h. This subnet also converges weak* to g. Therefore $g = h$ locally a.e. □

10.2.5. Wendel's Theorem. *If $U \in L_1^1(G)$, there exists $\mu \in M(G)$ such that $Uf = \mu * f$ for every $f \in L^1(G)$ and $\|U\|_{1,1} = \|\mu\|_M$. Thus $L_1^1(G)$ is isometrically isomorphic to $M(G)$.*

Proof. With $p = 1$ and $q = \infty$, line (2) in the proof of 10.2.3 is valid for all $f, g \in C_c(G)$. It follows that for every $g \in C_c(G)$, $Ug \in L^\infty(G)$ and $\|Ug\|_\infty \le \|U\|_{1,1}\|g\|_\infty$. Therefore U extends uniquely to an operator U': $C_o(G) \to L^\infty(G)$, with norm bounded by $\|U\|_{1,1}$, that commutes with all translations. By 10.2.4, the mapping $f \to U'f(0)$ is well-defined on $C_o(G)$, and its norm is bounded by $\|U\|_{1,1}$. By the Riesz Representation Theorem, there exists $\mu \in M(G)$ such that $Uf(0) = \int f(-y)d\mu(y)$ for $f \in C_o(G)$, and $\|\mu\|_M \le \|U\|_{1,1}$. Then $Uf(x) = \tau_x Uf(0) = U(\tau_x f)(0) = \int f(x - y)d\mu(y)$ for $f \in C_o(G)$. It follows that $Uf = f * \mu$ for all $f \in L^1(G)$, and $\|\mu\|_M \le \|U\|_{1,1}$. The reverse inequality also holds, by 10.2.1. The Theorem is proved. □

10.2.6. The Riesz–Thorin Theorem. *Let X and Y be measure spaces. Let $B = L^{p_o} \cap L^{p_1}(X)$, $B' = L^{p'_o} \cap L^{p'_1}(Y)$, where the p's are in $[1, \infty]$. Let S be a linear mapping from B into B' and suppose that*

$$\|S\|_j = \sup\{\|Sf\|_{p'_j}: f \in B, \|f\|_{p_j} \le 1\}$$

is finite for $j = 0$ and for $j = 1$. For $0 < a < 1$, let

$$\frac{1}{p_a} = \frac{a}{p_o} + \frac{1-a}{p_1}, \qquad \frac{1}{p'_a} = \frac{a}{p'_o} + \frac{1-a}{p'_1}.$$

Then $\|S\|_a \le \|S\|_o^a \|S\|_1^{1-a}$.

For two different proofs, see Zygmund [1, Section XII.1] and Katznelson [1, Chapter IV].

10.2.7. The Hausdorff–Young Theorem. *If* $1 \le p \le 2$ *and* $(1/p) + (1/q) = 1$, *the Fourier transform is a norm-decreasing map from* $L^p(G)$ *into* $L^q(\Gamma)$.

Proof. The Fourier transform has norm one as a map from $L^1(G)$ into $L^\infty(\Gamma)$ and as a map from $L^2(G)$ onto $L^2(\Gamma)$. The Theorem is therefore a consequence of 10.2.6. $\square$

10.2.8. Proposition. *For* $U \in L_2^2(G)$, *there exists* $\hat{U} \in L^\infty(\Gamma)$ *such that* $(Uf)^\wedge = \hat{U}\hat{f}$ *for all* $f \in L^2(G)$. *The mapping* $U \to \hat{U}$ *is an isometric isomorphism of* $L_2^2(G)$ *onto* $L^\infty(\Gamma)$.

Proof. Let $U \in L_2^2(G)$. For $\gamma \in \Gamma$, let $\varphi, \psi \in C_c(G)$ such that $\hat{\varphi}$ and $\hat{\psi}$ are nonzero near γ. Since $U\varphi * \psi = U(\varphi * \psi) = U\psi * \varphi$ by 10.2.2, it is true near γ that

$$\frac{(U\varphi)^\wedge}{\hat{\varphi}} = \frac{(U\psi)^\wedge}{\hat{\psi}}.$$

It follows that there is a unique locally measurable function $\hat{U}$ such that $(U\varphi)^\wedge = \hat{U}\hat{\varphi}$ for $\varphi \in C_c(G)$. If $f \in L^2(G)$, there exists a sequence $\{\varphi_n\} \subseteq C_c(G)$ such that $\varphi_n \to f$ in $L^2(G)$. Then $(U\varphi_n)^\wedge \to (Uf)^\wedge$ in $L^2(\Gamma)$. Replacing $\{\varphi_n\}$ by a subsequence if necessary, we may suppose that $\hat{\varphi}_n \to f$ a.e. and $(U\varphi_n)^\wedge \to (Uf)^\wedge$ a.e., and hence that $(Uf)^\wedge = \hat{U}\hat{f}$ a.e. If $|\hat{U}| > a$ on a compact set V of positive measure, let $\hat{g} = \chi_V \operatorname{sgn} \hat{U}$ and note that $\|U_g\|_2 = \|\hat{U}\hat{g}\|_2 > a\|\hat{g}\|_2 = a\|g\|_2$. It follows that $\hat{U} \in L^\infty(\Gamma)$ and $\|\hat{U}\|_\infty \le \|U\|_{2,2}$.

It remains to show that the map $U \to \hat{U}$ is onto and norm-preserving. For $h \in L^\infty(\Gamma)$ and $f \in L^2(G)$, $\|h\hat{f}\|_2 \le \|h\|_\infty \|f\|_2$. Then $h = \hat{U}$, where U is the element of $L_2^2(G)$ such that $(Uf)^\wedge = h\hat{f}$ for $f \in L^2(G)$; and $\|\hat{U}\|_\infty \ge \|U\|_{2,2}$. The Theorem is proved. $\square$

10.2.9. Remark. Let $PM(G)$ be the space of distributions S (called *pseudomeasures*) on G whose transforms belong to $L^\infty(\Gamma)$; by definition, $\|S\|_{PM} = \|\hat{S}\|_\infty$ and $S * f$ is determined by: $(S * f)^\wedge = \hat{S}\hat{f}$. An operator belonging to $L_2^2(G)$ is convolution with a pseudomeasure, and vice versa. Thus $PM(G)$ is identified with $L_2^2(G)$, and $L^\infty(\Gamma) = M_2(\Gamma)$.

10.2.10. Summary. If $1 < p < \infty$ and $p + q = pq$, then $L_p^p(G)$ and $L_q^q(G)$ are isometrically isomorphic. By 10.2.1 and 10.2.5, $L_1^1(G)$ is identified with $M(G)$ and enjoys a norm-decreasing inclusion in $L_\infty^\infty(G)$. If $1 \le p < r \le 2$, there is a norm-decreasing inclusion of $L_p^p(G)$ in $L_r^r(G)$, by 10.2.6, since $p < r < q$ and since every U in L_p^p belongs also to L_q^q. For $1 \le p \le 2$, by 10.2.8 and the fact that $L_p^p \subseteq L_2^2$, every operator $U \in L_p^p(G)$ determines an element $\hat{U}$ of $L^\infty(\Gamma)$

such that $(Uf)^\wedge = \hat{U}\hat{f}$ for every $f \in L^p(G)$. Let $M_p(\Gamma) = \{\hat{U}: U \in L_p^p(G)\}$, which forms a Banach algebra under pointwise multiplication, with norm $\|\hat{U}\|_{p,p} = \|U\|_{p,p}$. The correspondence $U \leftrightarrow \hat{U}$ is then an isometry of $L_p^p(G)$ and $M_p(\Gamma)$.

We have proved that for $1 \leq p < r \leq 2$,

$$M(G) \cong L_1^1(G) \subseteq L_p^p(G) \subseteq L_r^r(G) \subseteq L_2^2(G) \cong PM(G)$$

or, in other terms,

$$B(\Gamma) = M_1(\Gamma) \subseteq M_p(\Gamma) \subseteq M_r(\Gamma) \subseteq M_2(\Gamma) = L^\infty(\Gamma).$$

The inclusion $L_p^p \subseteq L_r^r$ is proper if G is infinite; for if it were not, then by the open mapping theorem the L_p^p and L_r^r norms would be equivalent; and there are measures μ of Rudin-Shapiro type for which the ratio of $\|\mu\|_{p,p}$ to $\|\mu\|_{r,r}$ is arbitrarily large, as we are about to explain.

10.2.11. Remark. Consider a discrete measure $\mu = \sum_{j=1}^J \alpha_j \delta_{t_j}$, where the points t_j are distinct. We claim that for $1 \leq p < \infty$,

$$\|\mu\|_{p,p} \geq \left(\sum |\alpha_j|^p\right)^{1/p}. \tag{3}$$

(Note that by 10.2.5, equality always holds if $p = 1$). The proof consists in convolving μ with χ_V, where V is a neighborhood of 0 with finite Haar measure d. Then $\|\chi_V\|_p = d^{1/p}$ and $\mu * \chi_V = \sum \alpha_j \chi_{V+t_j}$. If V is small enough, the sets $V + t_j$ are disjoint, so that $\|\mu * \chi_V\|_p = (d \sum |\alpha_j|^p)^{1/p}$. The inequality (3) follows. We are about to see that for $1 < p \leq 2$, the estimate (3) is about the best possible. Consider a Rudin-Shapiro measure $\mu = 2^{-n} \sum_{j=1}^{2^n} \pm \delta_{t_j}$ as in Section 1.6. Then $\|\mu\|_M = \|\mu\|_{1,1} = 1$ and $\|\mu\|_{PM} = \|\mu\|_{2,2} \leq 2^{(1-n)/2}$. By (3) and the Riesz-Thorin Theorem 10.2.6, if $p + q = pq$ then

$$2^{-n/q} \leq \|\mu\|_{p,p} \leq \|\mu\|_{1,1}^{1-(2/q)} \|\mu\|_{2,2}^{2/q} \leq 2^{1/q} 2^{-n/q}.$$

On any infinite G, we may find such a measure for each $n = 1, 2, \ldots$. It follows that even in $M_d(G)$, $\|\mu\|_{p,p}$ and $\|\mu\|_{r,r}$ are not equivalent norms.

The following Lemma will be of use in Section 10.3.

10.2.12. Lemma. *For $1 \leq p < 2$, there is a constant $c_p > 1$ such that for every positive integer n,*

$$\sup\{\|e^{inf}\|_{M_p(Z)} : f \in A(Z),\ f \text{ real-valued},\ \|f\|_{A(Z)} \leq 1\} > c_p^n.$$

Proof. Let πx be irrational, and let $\mu = (\tfrac{1}{2})(\delta_x + \delta_{-x})$. Then $e^{i\mu}$ is a discrete measure, writable as $\sum_{j=1}^\infty \alpha_j \delta_{t_j}$, and it follows from 10.2.11 that $\|e^{i\mu}\|_{p,p} \geq c_p$, where $c_p = (\sum |\alpha_j|^p)^{1/p}$. It is easy to see that c_p is independent of x, and that $c_p > 1$. Now fix n, let $\{x(1), \ldots, x(n)\}$ be an independent set, and let $\mu_j = (\frac{1}{2})(\delta_{x(j)} + \delta_{-x(j)})$. Then no point of T can be written in more than one way as $\sum_{j=1}^n u_j$, where $u_j \in \operatorname{supp} e^{i\mu_j}$. Therefore, if $v_n = (1/n)\sum_{i=1}^n \mu_j$, then $\|e^{inv_n}\|_{p,p} \geq c_p^n$. That proves the Lemma with "$B(Z)$" instead of "$A(Z)$." The rest of the proof consists in replacing v_n by $v_n * \Delta_\varepsilon$ for arbitrarily small $\varepsilon > 0$, where $\|\Delta_\varepsilon\|_1 = 1$, $\Delta_\varepsilon \geq 0$, and $\Delta_\varepsilon = 0$ off $[-\varepsilon, \varepsilon]$. We leave the details to the reader (see 10.2.17 and its proof). □

The next two results concern the closure, denoted by $m_p(\Gamma)$, of $A(\Gamma)$ in $M_p(\Gamma)$. They are formulated here only for the setting of discrete Γ, but in fact they generalize; see Hörmander [1, Section 1.4].

10.2.13. Lemma. *Let Γ be discrete and let $1 \leq r \leq 2$. Let f be a complex-valued function on Γ. Then these two conditions are equivalent:*

(a) $f \in m_r(\Gamma)$.
(b) *For every $\varepsilon > 0$, there exists a finite set $K \subseteq \Gamma$ such that $\|f\|_{M_r(\Gamma\setminus K)} < \varepsilon$.*

Proof. (a) ⇒ (b). If $f \in m_r(\Gamma)$ and $\varepsilon > 0$, then there exists $g \in A(\Gamma)$ such that $\|f - g\|_{M_r(\Gamma)} < \varepsilon/2$. Since $g \in A(\Gamma)$, there is a finite set K such that $\|g\|_{A(\Gamma\setminus K)} < \varepsilon/2$. Then

$$\begin{aligned} \|f\|_{M_r(\Gamma\setminus K)} &\leq \|f - g\|_{M_r(\Gamma\setminus K)} + \|g\|_{M_r(\Gamma\setminus K)} \\ &\leq \|f - g\|_{M_r(\Gamma)} + \|g\|_{A(\Gamma\setminus K)} < \varepsilon. \end{aligned}$$

(b) ⇒ (a). For each $\varepsilon > 0$, according to (b), there exists $f_1 \in M_r(\Gamma)$ such that $f_1 = f$ off K and $\|f_1\|_{M_r(\Gamma)} < \varepsilon$. Then $f - f_1$ has compact support and is therefore in $A(\Gamma)$. Its distance from f in $M_r(\Gamma)$ is less than ε. It follows that $f \in m_r(\Gamma)$. The Lemma is proved. (Note: 10.2.13 is a variant of 9.5.2.) □

10.2.14. Proposition. *If Γ is discrete and $1 \leq p < r \leq 2$, then $M_{po}(\Gamma) \subseteq m_r(\Gamma)$.*

Proof. Let $f \in M_{po}(\Gamma)$ and let $\varepsilon > 0$. It suffices to show that for some finite set $K \subseteq \Gamma$, $\|f\|_{M_r(\Gamma\setminus K)} < \varepsilon$. Note that $f \in M_r(\Gamma)$. Choose $a \in (0, 1)$ such that

$$\frac{1}{r} = \frac{a}{p} + \frac{(1 - a)}{2}.$$

Choose ε' sufficiently small so that the inequality (4), below, holds true. Let $C = \{\gamma : |f(\gamma)| \geq \varepsilon'\}$, a finite subset of Γ. There exists $\mu \in M(G)$ such that $\hat{\mu} = 1$ on C, $0 \leq \hat{\mu} \leq 1$ on Γ, $\|\mu\|_M \leq 2$, and $K = \operatorname{supp} \hat{\mu}$ is finite. Let

$g = (1 - \hat{\mu})f$. Then $\|g\|_{M_p(\Gamma)} \leq 3\|f\|_{M_p(\Gamma)}$, $g = f$ off K, and $|g| < \varepsilon'$ everywhere on Γ. With the aid of the Riesz-Thorin Theorem we find that

$$\begin{aligned} \|f\|_{M_r(\Gamma\setminus K)} &\leq \|g\|_{M_r(\Gamma)} \leq \|g\|^a_{M_p(\Gamma)}\|g\|^{1-a}_{M_2(\Gamma)} \\ &\leq (3\|f\|_{M_p(\Gamma)})^a(\varepsilon')^{1-a} < \varepsilon. \end{aligned} \tag{4}$$

The Proposition is proved. □

10.2.15. Remark. It is well known (see Chapter 4, 6, or 7) that if Γ is not compact, then there are singular measures on G whose transforms vanish at infinity. Thus there exist functions $f \in M_{1o}(\Gamma)\setminus m_1(\Gamma)$. As we shall show, it is also the case that for $1 < p < 2$, there exist functions $f \in M_{po}(\Gamma)\setminus m_p(\Gamma)$. Proposition 10.2.14 implies that for such an f, $f \notin M_s(\Gamma)$ if $1 \leq s < p$ and $f \in m_r(\Gamma)$ if $p < r \leq 2$. In other words, p is the smallest index value for which f is a multiplier function, and the largest index value for which f is outside the closure of $A(\Gamma)$. For $\Gamma = Z$, we shall find such an f in the next section. In fact, the f that we shall describe will have the stronger property that for every $N > 0$,

$$\liminf_{j\to\infty}(\|f^j\|_{M_p(\{n\in Z:\, |n|\geq N\})})^{1/j} > 0.$$

Now, although we have no use for this fact in the Chapter, we shall prove that for $1 < p < 2$, L^p_p is identifiable with the Banach space dual of a certain algebra, denoted by A_p, of functions on G. For $p = 1$ and $p = 2$, of course, such a result is already familiar: $A_1 = C_o$ and $A_2 = A$. A hint about the nature of A_p for other values of p comes from the fact that $A = \{f * g : f, g \in L^2\}$ (see Rudin [1, 1.6.3]).

Let $1 < p < 2$, $p + q = pq$. Note that if $f \in L^p(G)$ and $g \in L^q(G)$, then $f * g \in C_o(G)$. Let $A_p(G)$ consist of all the functions $h \in C_o(G)$ that can be written in the form

$$h = \sum_{j=1}^{\infty} f_j * g_j, \quad \text{where } \{f_j\} \subseteq L^p(G),\ \{g_j\} \subseteq L^q(G), \tag{5}$$

and where the sum

$$\sum_{j=1}^{\infty} \|f_j\|_p \|g_j\|_q \tag{6}$$

is finite. If the norm of $h \in A_p(G)$ is defined to be the infimum of the sums (6) such that (5) holds, then $A_p(G)$ becomes a Banach space. It may be defined equivalently as $(L^p \hat{\otimes} L^q)/N$, where $L^p \hat{\otimes} L^q$ is the complete tensor product with greatest cross norm, and N is the subspace of elements f such that $f(x) = 0$ for all $x \in G$.

A norm-decreasing linear map $\alpha\colon A_p(G)^* \to L_p^p(G)$ may be defined as follows. Fix $\varphi \in A_p(G)^*$. Each $f \in L^p(G)$ determines a linear functional on $L^q(G)$ by the rule:

$$g \to \varphi(f * g).$$

Its norm is evidently bounded by $\|\varphi\|\,\|f\|_p$. Therefore there exists an element $Uf \in L^p(G)$ such that $\|Uf\|_p \le \|\varphi\|\,\|f\|_p$ and

$$\varphi(f * g) = \int_G Uf(-x)g(x)dx = Uf * g(0) \quad \text{for } g \in L^q(G).$$

The mapping U is evidently an operator on $L^p(G)$ with norm at most $\|\varphi\|$. In order to prove that $U \in L_p^p(G)$, it suffices to point out that for $y \in G$, $f \in L^p$, and $g \in L^q$,

$$\int_G U\tau_y f(-x)g(x)dx = \varphi(\tau_y f * g) = \varphi(f * \tau_{-y} g)$$

$$= \int_G Uf(x - t)g(t + y)dt = \int_G \tau_y Uf(x - t)g(t)dt.$$

Thus $U\tau_y f = \tau_y Uf$ for all $f \in L^p$, so that $U \in L_p^p(G)$. Therefore the mapping $\alpha\colon \varphi \to U$ is a norm-decreasing linear map from $A_p(G)^*$ into $L_p^p(G)$.

10.2.16. Theorem. *The mapping α is an isometric isomorphism from $A_p(G)^*$ onto $L_p^p(G)$ $(1 < p < 2)$.*

Proof. For $F \in L^1$, let V_F denote the operator in L_p^p such that $V_F f = F * f$. We shall use the fact, to be proved below as a lemma, that $\{V_F\colon F \in L^1\}$ is dense in L_p^p in the weak operator topology.

Let $U \in L_p^p$. It suffices to find $\varphi \in A_p^*$ such that $\|\varphi\| \le \|U\|_{p,p}$ and $\alpha\varphi = U$. Let $h \in A_p$. Choose a representation of h as in (5). Let

$$\varphi(h) = \sum_{j=1}^{\infty} (Uf_j * g_j)(0).$$

To show that this formula defines an element of A_p^* it suffices to show that if $\sum f_j * g_j = 0$, then $\varphi(h) = 0$.

Let $\varepsilon > 0$. There exists N such that $\sum_{j>N} \|f_j\|_p \|g_j\|_q < \varepsilon$. By 10.2.17 (below), there exists $F \in L^1$ such that

$$\sum_{j=1}^{N} |((Uf_j - V_F f_j) * g_j)(0)| < \varepsilon$$

and $\|V_F\|_{p,p} \le \|U\|_{p,p}$. Then

$$\begin{aligned} |\varphi(h)| &= \left|\sum_j (Uf_j * g_j)(0)\right| \\ &\le \left|\sum_j (F * f_j * g_j)(0)\right| \\ &\quad + \sum_{j=1}^{N} |((Uf_j - V_F f_j) * g_j)(0)| \\ &\quad + \sum_{j>N} |((Uf_j - V_F f_j) * g_j)(0)| \\ &\le 0 + \varepsilon + 2\|U\|_{p,p}\varepsilon. \end{aligned}$$

It follows that $\varphi(h) = 0$. It is an easy exercise to prove that $U = \alpha\varphi$ and that $\|\varphi\|_{A_p^*} \le \|U\|_{p,p}$. The Theorem is proved. □

10.2.17. Lemma. *For every $U \in L_p^p(G)$, there is a net $\{F_n\} \subseteq L^1(G)$ such that $\|V_{F_n}\|_{p,p} \le \|U\|_{p,p}$ and $V_{F_n} \to U$ in the weak operator topology of $L_p^p(G)$ $(1 < p < 2)$.*

Proof. Let $\{u_n\} \subseteq L^1 \cap L^p$ be an L^1-bounded approximate identity such that $\hat{u}_n$ has compact support for each n. For $U \in L^p$ and $n \ge 1$, let $U_n f = U(u_n * f)$ for $f \in L^p$. Then $U_n \to U$ in the *strong* operator topology of L_p^p. Since $\hat{u}_n$ has compact support so does $\hat{U}_n = (Uu_n)^\wedge = \hat{U}\hat{u}_n$. Since $\hat{U}_n$ is also bounded, it follows that $\hat{U}_n \in L^2$. Thus in proving the Lemma, we may suppose that $\hat{U} \in L^2$.

For $\gamma \in \Gamma$, let

$$U_\gamma f = \gamma U(\bar{\gamma} f) \quad \text{for } f \in L^p.$$

Then $\hat{U}_\gamma(\gamma) = \hat{U}(\lambda - \gamma)$. Using the fact that the mapping: $\gamma \to \hat{U}_\gamma$ is continuous from Γ into $L^2(\Gamma)$, we shall now prove that the mapping: $\gamma \to U_\gamma$ is continuous from Γ into $L_p^p(G)$, when the latter space is equipped with the weak operator topology. It suffices to prove continuity at 0.

Let $\varepsilon > 0$. For $f \in L^p$ and $g \in L^q$, choose f_0 and g_0 in C_c such that $\|f - f_0\|_p < \varepsilon$, $\|g - g_0\|_q < \varepsilon$, and (for convenience) $\|f_0\|_p \le \|f\|_p$ and $\|g_0\|_q \le \|g\|_q$. Then

$$\begin{aligned} |\langle U_\gamma f, g\rangle - \langle Uf, g\rangle| &\le |\langle U_\gamma f, g - g_0\rangle| + |\langle U_\gamma(f - f_0), g_0\rangle| \\ &\quad + |\langle (U_\gamma - U)f_0, g_0\rangle| + |\langle Uf_0, g_0 - g\rangle| \\ &\quad + |\langle U(f_0 - f), g\rangle| \\ &\le \|U\|_{p,p}\|f\|_p\varepsilon + \|U\|_{p,p}\varepsilon\|g\|_q \\ &\quad + \|\hat{U}_\gamma - \hat{U}\|_2\|\hat{f}_0\hat{g}_0\|_2 + \|U\|_{p,p}\|f\|_p\varepsilon \\ &\quad + \|U\|_{p,p}\varepsilon\|g\|_q. \end{aligned}$$

It follows that if $\varepsilon' > 0$; $f_1, \ldots, f_k \in L^p$; and $g_1, \ldots, g_k \in L^q$, then there exists a neighborhood N of 0 in Γ such that

$$\gamma \in N \Rightarrow |\langle U_\gamma f_j, g_j\rangle - \langle Uf_j, g_j\rangle| < \varepsilon' \quad \text{for } 1 \le j \le k.$$

The mapping $\gamma \to U_\gamma$ is thus continuous as claimed.

Let $\{h_n\} \subseteq L^1 \cap L^2(\Gamma)$ be an approximate identity, with $\|h_n\|_1 = 1$. Let

$$U_n = \int_\Gamma U_\gamma h_n(\gamma) d\gamma.$$

The integrand has values in the topological vector space $L_p^p(G)$, equipped with the weak operator topology. (See Hille and Phillips [1, pp. 58–92] for a general treatment of vector-valued integrals.) Since the ball of radius $\|U\|_{p,p}$ in L_p^p is compact in that topology (Dunford and Schwartz [1, Section VI.8]), we know that $U_n \in L_p^p$ and $\|U_n\|_{p,p} \le \|U\|_{p,p}$. For f and g in C_c,

$$\begin{aligned}\langle U_n f, g\rangle &= \int_\Gamma \langle U_\gamma f, g\rangle h_n(\gamma) d\gamma \\ &= \int_\Gamma \int_\Gamma \hat{U}(\lambda - \gamma) \hat{f}(\lambda) \hat{g}(\lambda) h_n(\gamma) d\gamma \, d\lambda \\ &= \int_\Gamma (\hat{U} * h_n)(\lambda) \hat{f}(\lambda) \hat{g}(\lambda) d\lambda.\end{aligned}$$

Therefore $\hat{U}_n = \hat{U} * h_n$. Since $\hat{U}$ and h_n are in $L^2(\Gamma)$, $\hat{U}_n \in L^2 * L^2 = A(\Gamma) = L^1(G)^\wedge$. Therefore $U_n = V_{F_n}$ for some $F_n \in L^1(G)$. Clearly $\langle U_n f, g\rangle \to \langle Uf, g\rangle$ for all $f \in L^p$ and $g \in L^q$. The Lemma is proved. □

10.2.18. Theorem. *$A_p(G)$ is a Banach algebra under pointwise multiplication.*

Proof. It suffices to show that if $f_j, g_j \in C_c(G)$ for $j = 1$ and 2, then

$$\|(f_1 * f_2)(g_1 * g_2)\|_{A_p} \le \|f_1\|_p \|g_1\|_p \|f_2\|_q \|g_2\|_q.$$

Let $h(x) = (f_1 * f_2)(x)(g_1 * g_2)(x)$. Then

$$\begin{aligned}h(x) &= \iint g_1(x - z - y) f_1(x - y) f_2(y) g_2(y + z) dz\, dy \\ &= \int ((\tau_z g_1) \cdot f_1) * (f_2 \cdot \tau_{-z} g_2) dz.\end{aligned}$$

Therefore

$$\|h\|_{A_p} \leq \int \|\tau_z g_1 \cdot f_1\|_p \|f_2 \cdot \tau_{-z} g_2\|_q \, dz$$

$$= \int \left(\int |g_1(x - z) f_1(x)|^p \, dx \right)^{1/p} \left(\int | f_2(y) g_2(y + z)|^q \, dy \right)^{1/q} dz$$

$$\leq \left(\iint |g_1(x - z) f_1(x)|^p \, dx \, dz \right)^{1/p} \left(\iint | f_2(y) g_2(y + z)|^q \, dy \, dz \right)^{1/q}$$

$$\leq \|g_1\|_p \|f_1\|_p \|f_2\|_q \|g_2\|_q.$$

The theorem is proved. □

Credits and Remarks. Ludvik [1, p. 30] shows that when G is compact, there exist discontinuous operators on $L^1(G)$ that commute with translation. That is not true when $G = R$; see Johnson [2]. Stafney [1] shows that not every operator in L_∞^∞ commutes with all measures, so that 10.2.2 does not extend to the case $p = \infty$. Oberlin [1] shows that 10.2.3 does not extend to all non-abelian G. Theorem 10.2.5 is from Wendel [2]. The fact that the inclusion $L_p^p(G) \subseteq L_r^r(G)$ is proper whenever $1 \leq p < r \leq 2$ and G is an infinite abelian group appears in Figà-Talamanca and Gaudry [1] and in Price [1]; Gaudry [1] deals with inclusion results involving other multiplier algebras as well. Cowling and Fournier [1] give an extensive treatment of inclusion questions, dealing with both abelian and non-abelian G. Lemma 10.2.12 is due to Igari [1, p. 309]. Theorem 10.2.16 is due to Figà-Talamanca [1]; 10.2.18 is due to Herz [7]. The well known theorem of Bernstein was generalized by Lohoué [2] to say that for every $p \in (1, 2]$, $\Lambda_\alpha(T) \subseteq A_p(T)$ if and only if $\alpha > 1/q$; see also Uno [1].

Hörmander [1] asked whether there exists $f \in M_{po}(R^n) \backslash m_p(R^n)$; affirmative answers (for R^n and Z^n) were given by Figà-Talamanca and Gaudry [2]; an exposition for the case of Z (from which the other cases follow easily) appears in Edwards and Gaudry [1, Section 9.3]. But the f constructed in that paper has its square in m_p, unlike the f in condition (5) above.

Lohoué [1] shows that a distribution in L_p^p can be approximated weak$*$ by measures with the same L_p^p norm.

Let $n > 1$ and let B be the unit ball in R^n, in the sense of the usual Euclidean norm. Charles Fefferman [1] proved that $\chi_B \in M_p(R^n)$ only for $p = 2$. His proof uses Besicovitch's solution to the Kakeya needle problem. Carleson and Sjölin [1] have given a different proof.

H. S. Shapiro [2] considers $f \in M_p(Z)$ for which the norm is an attained value. One result obtained is that for $p \neq 2$, if $f(0) = f(-1) = \|f\|_{M_p} = 1$, then f is identically one. Several interesting questions are raised. In a related paper, Fefferman and Shapiro [1] have shown that for $1 < p < \infty$, there exists $\alpha(p) > 0$ such that if $f \in M_p(Z)$, $f(0) = 1$, and $\|f - \chi_{\{0\}}\|_{M_p(Z)} < \alpha(p)$, then $\|f\|_{M_p(Z)} = 1$; in other words, the unit sphere in $M_p(Z)$ has a planar face.

It turns out that $B_p(G)$, the space of functions that multiply $A_p(G)$ into itself, is the dual space of $m_p(\Gamma)$. For studies of A_p, B_p, and M_p extending even into the setting of non-abelian groups, see Eymard [1], Lohoué [5, 6], and Herz [10, 11], in that order. Herz's papers feature the language of category theory. Drury [2] proved that $A_p(T)$ is an analytic algebra. A proof for all infinite abelian G, and for more general settings as well, may be found in de Michele and Soardi [1, 2].

A characterization of functions in $M_p(\Gamma)$ analogous to Theorem 1.8.1 was proved by de Leeuw [1], and the analogue of 1.8.4 is due to Cowling [1].

Some interesting variants on the question of identifying multiplier functions of $L^1(R)^\wedge$ are studied in Glicksberg and Wik [1] and in Friedberg and Spence [1].

We have neglected a number of basic results about the restrictions of multiplier functions to (and their extensions from) subgroups (and homomorphically imbedded groups). For such matters, see the early paper of deLeeuw [1]; the thesis of Lohoué [4]; the papers by Jodeit [1, 2], Saeki [5], Lohoué [1], and Gaudry [2]; and the treatment by Stein and Weiss [1, Section VII.3].

Hörmander's 1960 paper [1] provides an extensive survey of the theory of multipliers from $L^p(R^n)^\wedge$ into $L^{p'}(R^n)^\wedge$, where p and p' range from 1 to ∞, as of that date. The 1971 thesis of Lohoué [4] treats $A_p(G)$ and $B_p(G)$ systematically. The recent book by Edwards and Gaudry [1] provides a useful introduction to $L_p^p(G)$ as well as to the Littlewood-Paley theory.

10.3. Zafran's Theorem about the Algebra $M_{po}(Z)$

Recall that for $1 \leq p < \infty$, $M_p = M_p(Z)$ is a Banach algebra of functions $\hat{U}$ defined on Z. It is the Fourier representation of the algebra of operators $U \in L_p^p(T)$. Zafran's theorem is concerned with the subalgebra $M_{po} = M_p(Z) \cap C_o(Z)$ for $p \neq 2$. The set Z may be regarded as a subset of the maximal ideal space Δ of M_{po}, and the Gel'fand transform of $\hat{U} \in M_{po}$ (or of the corresponding element $U \in L_p^p$) as an extension of $\hat{U}$ to all of Δ. We shall sometimes use symbols like φ, instead of $\hat{U}$, to stand for elements of M_p. Recall that if $1 < p < 2$ and $p + q = pq$, then $M_p = M_q$. In this section we shall most often suppose, for convenience, that $1 < p < 2$.

10.3.1. Zafran's Theorem. *Let $1 < p < 2$. Let $F: [-1, 1] \to \mathbb{C}$, and suppose that $F \circ \varphi \in M_{po}$ whenever $\varphi \in M_{po}$ and $\varphi(Z) \subseteq [-1, 1]$. Then for some $\eta > 0$, F agrees on $[-\eta, \eta]$ with an entire function.*

The Section is devoted largely to preliminaries; the proof proper has two main parts, formulated as Propositions 10.3.9 and 10.3.10.

10.3.2. Lemma. *Let $1 < p < 2$. Let $F: [-1, 1] \to \mathbb{C}$, and suppose that $F \circ \varphi \in M_p$ whenever $\varphi \in A(Z)$ and $\varphi(Z) \subseteq [-1, 1]$. Then for some $\eta > 0$, F is given by a convergent power series, $F(x) = \sum_{j=0}^{\infty} c_j x^j$, for $|x| \leq \eta$.*

Proof. An argument like the proof of 9.3.5, Case 1, shows that there exist $K < \infty$ and $\delta > 0$ such that

$$\|F \circ \varphi\|_{M_p} < K \quad \text{if} \quad \|\varphi\|_{A(Z)} < \delta \quad \text{and} \quad \varphi(Z) \subseteq [-1, 1].$$

The method used in the proof of 9.3.8 allows one to show that if F is discontinuous at 0, then every idempotent element of l^∞ belongs to M_p with norm less than $K + 1$; it follows that $l^\infty = M_p$, which is false (see 10.2.10). Therefore F is continuous at 0, and the proof of 9.3.7 also shows that F is continuous on $[-a, a]$ for some $a > 0$. Let $F_1(x) = F(r \sin x)$ for $x \in R$, where $r < \min(\delta, a)/10e$. As in the proof of 9.3.4, one may show by using 10.2.12 that F_1 extends to a function that is analytic on a horizontal strip $\{z: |\text{Im } z| < b\}$ for some $b > 0$. It follows that F extends to a function analytic near 0. The Lemma is proved. □

10.3.3. Proposition. *Let $1 < p < 2$. Let $\Phi(x) = \sum_{j=1}^{\infty} a_j x^j$ for $|x| \leq \varepsilon$, where $\varepsilon > 0$. Suppose that $\Phi \circ \varphi \in M_{po}$ whenever $\varphi \in M_{po}$ and $\varphi(Z) \subseteq [-\varepsilon, \varepsilon]$. Then $a_j = 0(6^j)$ as $j \to \infty$.*

How 10.3.2 and 10.3.3 imply 10.3.1. A function F that satisfies the hypothesis of 10.3.1 also satisfies that of 10.3.2. Also, $F(0) = 0$. By 10.3.2, then, we may write

$$F(x) = \sum_{j=1}^{\infty} c_j x^j \quad \text{for } |x| \leq \eta.$$

For every $m > 0$, we may apply 10.3.3 with $\Phi(x) = F(mx) = \sum c_j m^j x^j$ and $\varepsilon = \eta/m$. Therefore $c_j m^j = 0(6^j)$ as $j \to \infty$, for every $m > 0$; $\limsup |c_j|^{1/j} = 0$; and the function $z \mapsto \sum c_j z^j$ is entire. Thus to prove 10.3.1, it suffices to prove 10.3.3.

As explained in Section 10.1, a consequence of Zafran's Theorem is that for every $C > 0$, there exists $\varphi \in M_{po}$ such that

(1) $$\sup\{|\hat{\varphi}(h)|: h \in \Delta\} > C \sup\{|\varphi(n)|: n \in Z\}.$$

To prove that weaker result, it would suffice to construct an element φ such that for some $c > 0$,

(2) $$\liminf_{j \to \infty} (\|\varphi^j\|_{M_p(\{n: |n| \geq N\})})^{1/j} \geq c \quad \text{for every } N > 0.$$

It follows from (2) that

$$(3) \qquad \sup_{h\in\Delta}|\hat{\varphi}(h)| = \lim_{j\to\infty}\|\varphi^j\|_{M_p}^{1/j} \geq c.$$

Choose N so that $|\varphi(n)| < c/C$ for $|n| \geq N$. If φ is modified by making $\varphi(n) = 0$ whenever $|n| < N$, then the lim infs in (2) are unchanged, (3) still holds, φ still belongs to M_{po}, and (1) is satisfied.

An element φ satisfying (2) may be constructed as follows. Suppose that $\{F_j\}$ is a sequence of disjoint finite subsets of Z, and that $\{\varphi_j\}$ is a sequence of functions on Z such that

(i) $\varphi_j(n) = 0 \quad \text{for } n \notin F_j$,
(ii) $\varphi \in M_p$, where $\varphi = \sum_{j=1}^{\infty} \varphi_j$,
(iii) $\max_{n\in F_j}|\varphi_j(n)| \to 0$ as $j \to \infty$, and
(iv) $\|\varphi_j^j\|_{M_p(F_j)} > e^{-j}$.

Evidently $\varphi \in M_{po}$. For each N, $(-N, N) \cap F_j = \emptyset$ and hence

$$\|\varphi^j\|_{M_p(\{n:|n|\geq N\})} \leq \|\varphi_j^j\|_{M_p(F_j)}$$

for all sufficiently large j. Condition (2) follows, with $c = 1/e$. Such a construction was achieved in Zafran [2]. In Zafran [3] the procedure was refined and the following result obtained. Here we return to using $\hat{U}$ instead of φ for elements of M_p.

10.3.4. Proposition. *Let $1 < p < 2$. Let $\varepsilon > 0$, and let $\{a_j\}_{j=1}^{\infty}$ be a sequence of complex numbers. Then there exist a sequence of disjoint sets $F_j \subseteq Z$, a sequence of real-valued functions $\hat{U}_j$ on Z, and a sequence of trigonometric polynomials ψ_j with $\|\psi_j\|_p = 1$, such that*

(i) *$\hat{U}_j(n) = 0$ and $\psi_j(n) = 0$ for $n \notin F_j$,*
(ii) *the sum $\sum_{j=1}^{\infty} \hat{U}_j$ belongs to M_p,*
(iii) *$\sup_{n\in F_j}|\hat{U}_j(n)| \to 0$ as $j \to \infty$,*
(iv) *$|\sum_{k=1}^{j} a_k U_j^k * \psi_j\|_p \geq 6^{-j}|a_j|$, and $\|U_j^k * \psi_j\|_p < (\varepsilon/2)^k$ for $k > j$.*

How 10.3.4 implies 10.3.3. Let Φ, ε, and $\{a_j\}$ be as in the hypothesis of 10.3.3, and apply 10.3.4. Choose j_ε so that $\max_{n\in F_i}|\hat{U}_j(n)| \leq \varepsilon$ for $j \geq j_\varepsilon$. Let $\hat{U} = \sum_{j\geq j_\varepsilon} \hat{U}_j$. Then $\hat{U} \in M_{po}$ and $\Phi \circ \hat{U} \in M_{po}$. For $j \geq j_\varepsilon$, the restriction of $\Phi \circ \hat{U}$ to F_j equals $\Phi \circ \hat{U}_j$, and therefore $\|(\Phi \circ \hat{U})\hat{\psi}_j\|_{L^p(T)^\wedge} = \|\sum_{k=1}^{\infty} a_k U_j^k * \psi_j\|_p \geq 6^{-j}|a_j| - \sum_{k>j}|a_k|(\varepsilon/2)^k$. Hence for $j \geq j_\varepsilon$, $|a_j| \leq 6^j(\|\Phi \circ \hat{U}\|_{M_p} + \sum_{k=1}^{\infty}|a_k|(\varepsilon/2)^k)$; and 10.3.3 is proved.

We have shown that to prove Zafran's Theorem, if suffices to prove 10.3.4. The "piecing together" of suitable "pieces" $\hat{U}_j$ so that their sum belongs to M_p

is a subtle matter. The solution is provided in part by the Littlewood-Paley theory. Let

$$(4)\qquad E_j = \begin{cases} \{n: 2^{j-1} \le n < 2^j\} & \text{if } j > 0, \\ \{0\} \quad \text{if } j = 0, \quad \text{and} \\ \{n: -2^j < n \le -2^{j-1}\} & \text{if } j < 0. \end{cases}$$

For $f \in L^1(T)$, let $S_j f(x) = \sum_{n \in E_j} \hat{f}(n) e^{inx}$. Then of course, $S_j \in L_p^p(T)$ and $\hat{S}_j = \chi_{E_j} \in M_p$ for every $p \ge 1$.

10.3.5. The Littlewood-Paley Theorem for the circle group. *For each $p \in (1, \infty)$, there exist positive constants A_p and B_p such that*

$$(5)\qquad A_p \|f\|_p \le \left\| \left(\sum_{j=-\infty}^{\infty} |S_j f|^2 \right)^{1/2} \right\|_p \le B_p \|f\|_p \quad \textit{for all } f \in L^p(T).$$

We omit the proof and recommend to the reader the thorough development of the Theorem in the book of Edwards and Gaudry [1]. Other treatments with further information appear in Zygmund [1, XV.2], J. Schwartz [1], and Stein [1, Chapter IV].

Let $L^p(l^2)$ denote the Banach space of all strongly-measurable, p-integrable, l^2-valued (equivalence classes of) functions defined on T. If an element of the space is denoted by $\{g_j\}_{j=-\infty}^{\infty}$, its value at $x \in T$ is $\{g_j(x)\}_{j=-\infty}^{\infty} \in l^2$, and its norm is

$$\|\{g_j\}\|_{L^p(l^2)} = \left(\frac{1}{2\pi} \int_{-\pi}^{\pi} \left(\sum_{j=-\infty}^{\infty} |g_j(x)|^2 \right)^{p/2} dx \right)^{1/p}$$

The inequality (5) may be rewritten:

$$(5)\qquad A_p \|f\|_p \le \|\{S_j f\}\|_{L^p(l^2)} \le B_p \|f\|_p \quad \text{for} \quad f \in L^p(T).$$

When X is a Banach space, let $O(X)$ denote the space of bounded linear operators on X. The next three results establish a connection between M_p and the space of operators $O(L^p(l^2))$. Proposition 10.3.8 will explain how to construct an element of M_p from a certain kind of operator in $O(L^p(l^2))$. It thus provides a device for "piecing together."

10.3.6. Lemma. *For $U \in L_p^p(T)$, let $\tilde{U}\{g_j\} = \{Ug_j\}$. Then $\tilde{U} \in O(L^p(l^2))$ and $\|\tilde{U}\| \le 2\|U\|_{p,p}$.*

Proof. Let $Vf = \operatorname{Re}(Uf)$ for all $f \in L^p$, or $Vf = \operatorname{Im}(Uf)$ for all $f \in L^p$. Then V is a bounded mapping from L^p into L^p that is linear over the reals and maps real-valued functions to real valued functions; and $\|V\| \le \|U\|_{p,p}$. Let

$g_1, \ldots, g_n$ be real-valued elements of L^p. Let $G = (\sum |g_j|^2)^{1/2}$, $H = (\sum |Vg_j|^2)^{1/2}$. Let $u = (u_1, \ldots, u_n)$ be a point on the unit sphere Σ in R^n. If $g = \sum u_j g_j$, then $\sum u_j Vg_j = Vg$. Therefore

$$(6) \qquad \frac{1}{2\pi} \int_{-\pi}^{\pi} |\sum u_j Vg_j(t)|^p \, dt \le \frac{\|U\|_{p,p}^p}{2\pi} \int_{-\pi}^{\pi} |\sum u_j g_j(t)|^p \, dt.$$

Let $\delta(t, u)$ be the angle between the vectors $(g_1(t), \ldots, g_n(t))$ and $(u_1, \ldots, u_n)$. Let $\delta'(t, u)$ be the angle between $(Vg_1(t), \ldots, Vg_n(t))$ and $(u_1, \ldots, u_n)$. Then (6) may be rewritten:

$$\frac{1}{2\pi} \int_{-\pi}^{\pi} |H(t) \cos \delta'(t, u)|^p \, dt \le \frac{\|U\|_{p,p}^p}{2\pi} \int_{-\pi}^{\pi} |G(t) \cos \delta(t, u)|^p \, dt.$$

Integrating both sides over Σ, with respect to the $(n - 1)$-dimensional surface area form, and reversing the order of integration, we obtain the inequality

$$\frac{1}{2\pi} \int_{-\pi}^{\pi} |H(t)|^p \int_{\Sigma} |\cos \delta'(t, u)|^p \, du \, dt$$

$$\le \frac{\|U\|_{p,p}^p}{2\pi} \int_{-\pi}^{\pi} |G(t)|^p \int_{\Sigma} |\cos \delta(t, u)|^p \, du \, dt.$$

The two integrals over Σ are independent of t and are equal. Therefore

$$\frac{1}{2\pi} \int_{-\pi}^{\pi} (\sum |Vg_j(t)|^2)^{p/2} \, dt \le \frac{\|U\|_{p,p}^p}{2\pi} \int_{-\pi}^{\pi} (\sum |g_j(t)|^2)^{p/2} \, dt.$$

This inequality still holds if we consider complex-valued g_j's in L^p, because $|g|^2 = |\mathrm{Re}\, g|^2 + |\mathrm{Im}\, g|^2$ and $|Vg|^2 = |V(\mathrm{Re}\, g)|^2 + |V(\mathrm{Im}\, g)|^2$, and we can apply the same argument with $2n$ in the role of n. It follows that

$$\left(\frac{1}{2\pi} \int_{-\pi}^{\pi} (\sum |Ug_j(t)|^2)^{p/2} \, dt\right)^{1/p} \le 2\|U\|_{p,p} \left(\frac{1}{2\pi} \int_{-\pi}^{\pi} (\sum |g_j(t)|^2)^{p/2} \, dt\right)^{1/p}$$

It follows that $\|\tilde{U}\|_{O(L^p(l^2))} \le 2\|U\|_{p,p}$. □

10.3.7. Theorem. *Let $1 < p < \infty$. There exists a constant C_p such that if for each $j \in Z$, $\hat{R}_j$ is the characteristic function of an interval $\{n \in Z : n_j \le n \le n_j'\}$ and if $R\{g_j\} = \{R_j * g_j\}$, then $R \in O(L^p(l^2))$ with norm less than C_p.*

For a proof, we refer the reader to Edwards and Gaudry [1, 6.1.1].

10.3.8. Proposition. *Let $\{P_j\}$ be a sequence of trigonometric polynomials such that $\hat{P}_j(n) = 0$ for $n \notin E_j$, where E_j is as defined by (4). Let $\hat{U} = \sum_{j=-\infty}^{\infty} \hat{P}_j$. For*

$\{g_j\} \in L^p(l^2)$, *let* $V\{g_j\} = \{P_j * g_j\}$. *Then* $V \in O(L^p(l^2))$ *if and only if* $U \in L_p^p$ *(that is,* $\hat{U} \in M_p$*).*

Proof. Suppose that $V \in O(L^p(l^2))$. Applying (5), we find that for every trigonometric polynomial f,

$$A_p\|Uf\|_p \leq \|\{S_j Uf\}\|_{L^p(l^2)} = \|\{P_j * S_j f\}\|_{L^p(l^2)}$$

$$\leq \|V\| \, \|\{S_j f\}\|_{L^p(l^2)} \leq \|V\| B_p \|f\|_p.$$

Therefore $\hat{U} \in M_p$ (or, $U \in L_p^p$) with norm bounded by $\|V\| B_p A_p^{-1}$.

Suppose now that $U \in L_p^p$. Then

$$\begin{aligned} \|V\{g_j\}\|_{L^p(l^2)} &= \|\{S_j U g_j\}\|_{L^p(l^2)} \\ &\leq C_p \|\{U g_j\}\|_{L^p(l^2)} \quad \text{(by 10.3.7)} \\ &\leq 2C_p \|U\|_{p,p} \|\{g_j\}\|_{L^p(l^2)} \quad \text{(by 10.3.6).} \end{aligned}$$

Therefore $V \in O(L^p(l^2))$, and the proposition is proved. □

The next two results represent the two major steps of the proof: the piecing-together, and the design of the pieces. They make it possible to prove 10.3.4 and thus complete the proof of Zafran's Theorem.

If $\mu, \nu \in M(T)$, we define $\mu \leq \nu$ to mean that $\mu(E) \leq \nu(E)$ for every Borel set E.

10.3.9. Proposition. *Let* $1 < p < 2$ *and* $p + q = pq$. *Let* $\{\lambda_j\}$ *be a sequence of measures, and* λ *a positive measure, in* $M(T)$, *such that* $|\lambda_j| \leq \lambda$ *for all* j. *Let* $K > 0$ *and* $M_j > 0$ *be such that* $M_j \|\hat{\lambda}_j\|_\infty \leq K$ *for all* j. *Let* $Y\{g_j\} = \{M_j^{2/q} \lambda_j * g_j\}$. *Then* $Y \in O(L^p(l^2))$.

10.3.10. Proposition. *Let* $1 < p < 2$ *and* $p + q = pq$. *Let* $\varepsilon > 0$, *and let* $\{a_j\}_{j=1}^\infty$ *be a sequence of complex numbers. Then there exist a positive measure* $\lambda \in M(T)$; *a sequence of other measures* $\{\lambda_j\} \subseteq M(T)$ *such that* $\hat{\lambda}_j$ *is real-valued; a sequence of positive integers,* $n(j) \to \infty$; *and a sequence of trigonometric polynomials* $\{w_j\}$ *with* $\|w_j\|_p = 1$, *such that for every* j,

(a) $|\lambda_j| \leq \lambda$;
(b) $\|\hat{\lambda}_j\|_\infty \leq 2^{(1-n(j))/2}$,
(c) $\|\sum_{k=1}^j a_k (2^{n(j)/q}\lambda_j)^k * w_j\|_p \geq 6^{-j}|a_j|$, and
$\|(2^{n(j)/q}\lambda_j)^k * w_j\|_p \leq (\varepsilon/2)^k$ for $k > j$.

Before proving the two propositions, we shall explain how to use them to obtain 10.3.4. We will need the following easy lemma.

10.3.11. Lemma. *Let $1 \leq p < \infty$. Let $Y \in O(L^p(l^2))$ be given by*

$$Y\{g_j\} = \{Y_j g_j\},$$

where $Y_j \in L_p^p$ for each j. Let $\hat{V}_j(n) = \hat{Y}_j(n - m_j)$, where $\{m_j\}$ is an arbitrary sequence of integers, and let $V\{g_j\} = \{V_j g_j\}$. Then $V \in O(L^p(l^2))$.

Proof. Let $A\{g_j\} = \{e^{im_j x} g_j(x)\}$. Evidently $\|A\|_{O(L^p(l^2))} = 1$ and $V = A \circ Y \circ A^{-1}$. □

Proof of 10.3.4, using 10.3.9 and 10.3.10. Let p, ε, and $\{a_j\}$ be as in the hypothesis of 10.3.4; and let λ, $\{\lambda_j\}$, $\{n(j)\}$, and $\{w_j\}$ be the items provided by 10.3.10. By 10.3.9, the operator Y such that

$$Y\{g_j\} = \{2^{n(j)/q} \lambda_j * g_j\}$$

belongs to $O(L^p(l^2))$. Let $\{m_j\}$ be a sequence of integers. Let $\{\xi_j\}$ be a new sequence of measures defined by letting

$$d\xi_j(x) = 2^{n(j)/q} e^{im_j x}\, d\lambda_j(x).$$

Then

$$\hat{\xi}_j(n) = 2^{n(j)/q} \hat{\lambda}_j(n - m_j). \tag{7}$$

If $V\{g_j\} = \{\xi_j * g_j\}$, then $V \in O(L^p(l^2))$, by Lemma 10.3.11. The proof of 10.3.4 depends on making a suitable choice of the integers m_j.

We choose $\{m_j\}$ inductively so that the functions

$$\hat{\psi}_j(n) = \hat{w}_j(n - m_j)$$

have disjoint supports and, more precisely,

$$\operatorname{supp} \hat{\psi}_j \subseteq \{n\colon 2^{k(j)-1} \leq n < 2^{k(j)}\} = E_{k(j)}$$

for an increasing sequence $\{k(j)\}$. Let $\hat{R}_j$ be the characteristic function of $E_{k(j)}$. By Theorem 10.3.7, the operator $R\{g_j\} = \{R_j * g_j\}$ belongs to $O(L^p(l^2))$. Therefore so does the composition $R \circ V$, which maps $\{g_j\}$ to $\{R_j * \xi_j * g_j\}$. Define a sequence $\{P_k\}$ by letting $P_k = 0$ if $k \notin \{k(j)\}$, $\hat{P}_k = \hat{R}_j \hat{\xi}_j$ if $k = k(j)$. By Proposition 10.3.8, the sum $\hat{U} = \sum \hat{P}_k$ belongs to M_p.

Altering the notation, we let $F_j = E_{k(j)}$ and

$$\hat{U}_j = \hat{P}_{k(j)} = \chi_{F_j} \hat{\xi}_j. \tag{8}$$

Properties (i) and (ii) in the conclusion of 10.3.4 are now clear. Property (iii) follows from 10.3.10 (b), (7), (8), and the fact that $q > 2$. Property (iv) follows from 10.3.10 (c) because for every j and k,

$$U_j^k * \psi_j(x) = e^{im_jx}(2^{n(j)/q}\lambda_j)^k * w_j(x).$$

Proposition 10.3.4 is proved, using 10.3.9 and 10.3.10.

Proof of 10.3.9. For the sake of somewhat greater simplicity in the argument, we pass to the finite-dimensional space $l_N^2 = \{\{a_j\} \in l^2 : a_j = 0 \text{ for } |j| > N\}$. It suffices to show that for arbitrary N,

(9) $$\|Y\|_{O(L^p(l_N^2))} \leq K^{2/q}(2\|\lambda\|)^{(q-2)/q}.$$

We claim that if $|c| = 1$ and $V\{f_j\} = \{c\lambda_j * f_j\}$, then

(10) $$\|V\|_{O(L^1(l_N^2))} \leq 2\|\lambda\|.$$

To prove this, we note that convolution with λ is an operator in L_1^1 with norm $\|\lambda\|$; by 10.3.6, then, the mapping $\{f_j\} \to \{\lambda * f_j\}$ is in $O(L^1(l_N^2))$ with norm no greater than $2\|\lambda\|$. Therefore since $|\lambda_j| \leq \lambda$ for all j,

$$\|V\{f_j\}\|_{L^1(l_N^2)} = \int_T \left(\sum_j |\lambda_j * f_j(x)|^2\right)^{1/2} dx$$
$$\leq \int_T \left(\sum_j |(\lambda * |f_j|)(x)|^2\right)^{1/2} dx \leq 2\|\lambda\| \, \|\{f_j\}\|_{L^1(l_N^2)},$$

and (10) is proved. (The factor 2 in (10) can be eliminated; see Zafran [2, pp. 358–359]).

The proof of (9) relies on a complex interpolation method. Consider the operator-valued function R, defined on the strip $S = \{x + iy : 0 \leq x \leq 1\}$ by letting

$$R(z)\{f_j\} = \{M_j^z \lambda_j * f_j\}.$$

For each real y, $R(iy)$ is like V above, so that

(11) $$\|R(iy)\|_{O(L^1(l_N^2))} \leq 2\|\lambda\|.$$

Also for each real y,

(12) $$\|R(1 + iy)\|_{O(L^2(l_N^2))} \leq K,$$

because

$$\begin{aligned}\|R(1+iy)\{f_j\}\|^2_{L^2(l^2_N)} &= \int_T \sum |M_j^{1+iy}\lambda_j * f_j(x)|^2\,dx \\ &\le \sum M_j^2 \int_T |\lambda_j * f_j(x)|^2\,dx \\ &\le \sum M_j^2 \|\hat{\lambda}_j\|_\infty \int_T |f_j(x)|^2\,dx \le K^2 \|\{f_j\}\|^2_{L^2(l^2_N)}.\end{aligned}$$

If $s = 2/q$, then

$$\frac{1-s}{1} + \frac{s}{2} = \frac{1}{p},$$

so that (9) is equivalent to

$$\|R(s)\|_{L^p(l^2_N)} \le (2\|\lambda\|)^{1-s} K^s. \tag{13}$$

The fact that (11) and (12) imply (13) is a special case of a vector-valued version of a theorem found in Stein and Weiss [1, Section V.4], but we shall present an *ad hoc* argument, making use of the following Phrägmen-Lindelöf ("three-lines") theorem.

10.3.12. Theorem. *Let F be a bounded continuous complex-valued function on the strip S, analytic on the interior of S. If $|F(iy)| \le m_o$ and $|F(1+iy)| \le m_1$ for all real y, then $|F(x+iy)| \le m_o^{1-x}$ for all $x + iy \in S$.*

Proof. See Stein and Weiss [1, p. 180] or Rudin [3, 12.8].

Let $\{f_j\}$ and $\{g_j\}$ be simple l^2_N-valued functions whose norms, in $L^p(l^2_N)$ and $L^q(l^2_N)$ respectively, both equal one. Define $e_j(x)$ and $h_j(x)$ by requiring:

$$f_j(x) = \|\{f_j(x)\}\|_{l^2_N} e_j(x),$$
$$g_j(x) = \|\{g_j(x)\}\|_{l^2_N} h_j(x).$$

For $z \in S$, let

$$f_j^{(z)}(x) = (\|\{f_j(x)\}\|_{l^2_N})^{p(1-z/2)} e_j(z),$$
$$g_j^{(z)}(x) = (\|\{g_j(x)\}\|_{l^2_N})^{qz/2} h_j(z),$$

and let

$$F(z) = \int_T \langle R(z)\{f_j^{(z)}\}(x), \{g_j^{(z)}(x)\}\rangle dx.$$

Since for each real y,

$$\sup_x \|\{g_j^{(iy)}(x)\}\|_{l_N^2} = 1$$

and

$$\|\{f_j^{(iy)}\}\|_{L^1(l_N^2)} = \int_T \|\{f_j(x)\}\|_{l_N^2}^p \, dx = 1,$$

it follows from (11) that $|F(iy)| \le 2\|\lambda\|$. Since for each real y,

$$\|\{f_j^{(1+iy)}\}\|_{L_2(l_N^2)}^2 = \int_T \|\{f_j(x)\}\|_{l_N^2}^p \, dx = 1$$

and

$$\|\{g_j^{(1+iy)}\}\|_{L^2(l_N^2)}^2 = \int_T \|\{g_j(x)\}\|_{l_N^2}^q \, dx = 1,$$

it follows from (12) that $|F(1 + iy)| \le K$. Therefore by 10.3.12, in particular,

(14) $$|F(s)| \le (2\|\lambda\|)^{1-s}K^s.$$

Note that

$$\|\{f_j^{(s)}\}\|_{L^p(l_N^2)} = \left(\int_T \|\{f_j(x)\}\|_{l_N^2}^p \, dx\right)^{1/p} = \|\{f_j\}\|_{L^p(l_N^2)},$$

and

$$\|\{g_j^{(s)}\}\|_{L^q(l_N^2)} = \left(\int_T \|\{g_j(x)\}\|_{l_N^2} \, dx\right)^{1/q} = \|\{g_j\}\|_{L^q(l_N^2)}.$$

Since (14) holds for every choice of $\{f_j\}$ and $\{g_j\}$, (13) follows. Proposition 10.3.9 is proved. □

Proof of 10.3.10. Let j and n be positive integers. Let $\{t_1, \ldots, t_j\}$ be an independent subset of T, and let $1 \le s \le j$. As explained in Section 1.6, there is a measure $\rho_{n,s}$ that assigns mass 2^{-n} or -2^{-n} to each point of the set $E_{n,s} = \{at_s: a = 1, 2, \ldots, 2^n\}$ such that

(16) $$\|\rho_{n,s}\|_M = 1 \quad \text{and} \quad \|\hat{\rho}_{n,s}\|_\infty \le 2^{1/2}2^{-n/2}.$$

Since the support of $\rho_{n,s}^2 = \rho_{n,s} * \rho_{n,s}$ is contained in the set $\{at_s: a = 2, \ldots, 2^{n+1}\}$, it follows from 1.6.1 that the total-mass norm of $\rho_{n,s}^2$ falls significantly below 1:

(17) $$\|\rho_{n,s}^2\|_M \le 2^{(n+1)/2}\|(\rho_{n,k}^2)^\wedge\|_\infty \le 2^{3/2}2^{-n/2}.$$

The conjugate measure $\tilde{\rho}_{n,s}$ also satisfies (16), and its support is $-E_{n,s}$. The estimate (17) holds for each of the measures $\rho_{n,s} * \tilde{\rho}_{n,s}$ and $\tilde{\rho}^2_{n,s}$, because the support of each contains at most 2^{n+1} points. Consider now the symmetric measure $\mu_{n,s} = (\frac{1}{2})(\rho_{n,s} + \tilde{\rho}_{n,s})$. It assigns a mass of absolute value $2^{-(n+1)}$ to each of the 2^{n+1} points in its support $E_{n,s} \cup (-E_{n,s})$. Evidently

$$\|\mu_{n,s}\|_M = 1 \quad \text{and} \quad \|\hat{\mu}_{n,s}\|_\infty \leq 2^{1/2}2^{-n/2}. \tag{18}$$

The methods of 10.2.11 show that

$$2^{-n/q} \leq \|\mu_{n,s}\|_{p,p} \leq 2^{1/q}2^{-n/q}. \tag{19}$$

Since $\mu^2_{n,s} = (\frac{1}{4})(\rho^2_{n,s} + 2\rho_{n,s} * \tilde{\rho}_{n,s} + \tilde{\rho}^2_{n,s})$, evidently $\|\mu^2_{n,s}\|_M \leq 2^{3/2}2^{-n/2}$. If $s = 2/q$ then

$$\|\mu^2_{n,s}\|_{p,p} \leq \|\mu^2_{n,s}\|_M^{1-s}\|\hat{\mu}^2_{n,s}\|_\infty^s$$

and hence

$$\|\mu^2_{n,s}\| < 3\alpha^n 2^{-2n/q} \quad \text{where} \quad \alpha = 2^{(1/q)-(1/2)} < 1. \tag{20}$$

Note that in particular,

$$\|\mu^2_{n,s}\|_{p,p} \leq 3\alpha^n\|\mu_{n,s}\|^2_{p,p}.$$

Thus the powers of $\mu_{n,s}$ decline rapidly in L^p_p-norm. Our objective requires us to seek a measure whose powers stand up better. Such a measure is the average

$$\nu_{n,j} = \frac{1}{j}\sum_{s=1}^{j} \mu_{n,s}.$$

We shall show that

$$\left\|\sum_{k=1}^{j} a_k 2^{nk/q}\nu^k_{n,j}\right\|_{p,p} > 6^{-j}|a_j|; \tag{21}$$

and that there exists $n(j) > 0$ such that

$$2^{nk/q}\|\nu^k_{n,j}\|_{p,p} < (\varepsilon/2)^k \quad \text{whenever} \quad k > j \quad \text{and} \quad n \geq n(j). \tag{22}$$

Note that according to these inequalities, polynomials w_j could be chosen so that the measure $\nu_{n(j),j}$ would play the role of λ_j in conditions (b) and (c) of the Proposition. However, another major step will be required to obtain the measures λ_j that will satisfy (a) as well.

To prove (21) and (22) we study the expansion

$$(23) \qquad v_{n,j}^k = \frac{1}{j^k} \sum_{s_1=1}^{j} \cdots \sum_{s_k=1}^{j} \mu_{n,s_1} * \cdots * \mu_{n,s_k}.$$

Each summand has the form

$$(24) \qquad \mu_{n,1}^{m_1} * \cdots * \mu_{n,j}^{m_j} \quad \text{where} \quad \sum_{i=1}^{j} m_i = k, \quad \text{each} \quad m_i \geq 0.$$

Consider (21) first. When $k = j$, the measure $\mu_{n,1} * \cdots * \mu_{n,j}$ occurs $j!$ times as a summand in (23). It assigns a mass of absolute value $2^{-(n+1)j}$ to each of the $2^{(n+1)j}$ points of the set

$$F = \left\{ \sum_{s=1}^{j} x_s : x_s \in E_{n,s} \cup (-E_{n,s}) \right\}.$$

Therefore $v_{n,j}^j$ assigns a mass of absolute value $j!\,j^{-j}2^{-(n+1)j}$ to each of those points—as well as other masses to other points. For $k < j$, the measure $v_{n,j}^k$ assigns no mass to F. Therefore the measure $\sum_{k=1}^{j} a_k 2^{nk/q} v_{n,j}^k$ assigns a mass of absolute value $|\alpha_j| 2^{-nj/q} j!\,j^{-j} 2^{-(n+1)j}$ to each point of F (as well as other masses to other points), and therefore by 10.2.11 (3), its L_p^p-norm is at least

$$(|a_j| 2^{nj/q} j!\,j^{-j} 2^{-(n+1)j}) 2^{nj/p} \geq |a_j| 2^{-j} j!\,j^{-j} \geq (2e)^{-j} |a_j| > 6^{-j} |a_j|,$$

so that (21) is proved.

Considering (22), let $k > j$ and fix attention on an arbitrary summand (24). Then $k = j - 1 + 2r$ or $j + 2r$ for some positive integer r, and each m_k equals $2\mu_i + v_i$, where u_i and v_i are nonnegative integers, $v_i = 0$ or 1, $\sum u_i = r$, and $\sum v_i = k - 2r$. Therefore by (19) and (20), the L_p^p-norm of the measure (24) is bounded by

$$(3\alpha^n 2^{-2n/q})^r (2^{1/q} 2^{-n/q})^{k-2r} = \alpha^{nr} 2^{-nk/q} 3^r 2^{(k-2r)/q}$$
$$\leq \alpha^{n(k-j)/2} 2^{-nk/q} 3^{(k+1-j)/2} 2^{j/q}$$

Therefore

$$2^{nk/q} \| v_{n,j}^k \|_p \leq \alpha^{n(k-j)/2} 3^{(k+1-j)/2} 2^{j/q}.$$

Evidently (22) is true if $n(j)$ is sufficiently large.

We may suppose that $n(j) \to \infty$. We shall now complete the proof by producing the measures $\lambda \geq 0$ and $\{\lambda_j\}$. It follows from (21) that there is a trigonometric polynomial α_j such that $\|\alpha_j\|_p = 1$ and

$$\left\| \sum_{k=1}^{j} a_k 2^{n(j)k/q} v_{n(j),j}^k * \alpha_j \right\|_p > 6^{-j} |a_j|.$$

Let $\{K_r\}_{r=1}^{\infty}$ denote the Fejér kernel. Choose $r(j)$ so large that if $f_j = K_{r(j)} * v_{n(j),j}$, then

(25) $$\left\| \sum_{k=1}^{j} a_k 2^{n(j)k/q} f_j^k * \alpha_j \right\|_p > 6^{-j}|a_j|;$$

and also large enough so that supp $\hat{\alpha}_j \subseteq \{n : |n| \leq r(j)\}$. Let $g_j = K_{r(j)} * |v_{n(j),j}|$. Let $\{m_k\}$ be a sequence of positive integers such that $m_{k+1} > 2\sum_{j=1}^{k} m_j$ for every k. That lacunarity condition makes it easy to evaluate the transforms of the products

$$h_N(x) = \prod_{k=1}^{N} g_k(m_k x) \quad \text{for } N \geq 1,$$

$$h_{N,j}(x) = f_j(m_j x) \prod_{\substack{k=1 \\ k \neq j}}^{N} g_k(m_k x) \quad \text{for } N \geq j \geq 1.$$

Then $h_N \geq 0$ and $\hat{h}_N(0) = 1 = \|h_N\|_1$, $|h_{N,j}| \leq h_N$ so that $\|h_{N,j}\|_1 \leq 1$, and the transform converges pointwise for each of the sequences $\{h_N\}_{N=1}^{\infty}$ and $\{h_{N,j}\}_{N=1}^{\infty}$. Let λ, λ_j be their respective weak$*$ limits in $M(T)$. Then $\lambda \geq 0$, $\|\lambda\| = 1$, and

(a) $$|\lambda_j| \leq \lambda \quad \text{for all } j.$$

Note that

(26) $$\hat{\lambda}_j\left(\sum_{k=1}^{N} u_k m_k\right) = \hat{f}_j(u_j) \prod_{\substack{k=1 \\ k \neq j}}^{N} \hat{g}_k(u_k)$$

whenever N is finite, $u_k \in Z$ and $|u_k| \leq r(k)$ for $1 \leq k \leq N$. In particular, $\hat{\lambda}_j(u_j m_j) = \hat{f}_j(u_j)$ for $|u_j| \leq r(j)$. Also, $\hat{\lambda}_j(n) = 0$ for all n that are not of the form accounted for by (26). Thus

(b) $$\|\hat{\lambda}_j\|_\infty \leq \|\hat{f}_j\|_\infty \leq 2^{(1-n(j))/2}.$$

Let $w_j(x) = \alpha_j(m_j x)$. Since $\|F(mx)\|_p = \|F(x)\|_p$ whenever $F \in L^p$ and $m \in Z$, and since $\lambda_j^k * w_j(x) = f_j^k * \alpha_j(m_j x)$, it follows that

(c) $$\left\| \sum_{k=1}^{j} a_k 2^{n(j)k/q} \lambda_j^k * w_j \right\|_p > 6^{-j}|a_j|, \qquad \text{by (25), and}$$

$$\|2^{n(j)k/q} \lambda_j^k * w_j\|_p \leq 2^{n(j)k/q} \|f_j^k\|_{L_p^p} \|w_j\|_p \leq 2^{n(j)k/q} \|v_{n(j),j}^k\|_{L_p^p} < (\varepsilon/2)^k.$$

Proposition 10.3.10 is proved. $\square$

Note that we cannot claim that $\|\lambda_j^k\|_{L_p^p} \leq \|f_j\|_{L_p^p}$. Therefore (c) includes only a "weak" reflection of condition (23).

Remarks. The fact that for an arbitrary non-compact locally compact abelian group Γ, the algebra $M_{1o}(\Gamma)$ (which is identical to $B_o(\Gamma)$) is entire, has been known for some time; see Section 9.5. Igari [1, 2] proved (among other results) that for $p \in (1, 2) \cup (2, \infty)$, $M_p(\Gamma)$ is an entire algebra. The results of Zafran [2, 3], though foreshadowed by Igari's work, are more delicate and difficult. As Zafran [3] points out, it is not difficult to prove from Theorem 10.3.1 that $M_{po}(\Gamma)$ is entire whenever Γ is a discrete group containing an element of infinite order. The latter half of Zafran [3] is devoted to the highly non-trivial task of proving that $M_{po}(T^n) = M_p(T^n) \cap C(T^n)$ is entire. The methods above, techniques from Igari [2], and a vector-valued analogue of the theory of the space BMO are involved.

Lemma 10.3.6 and its proof are adapted from Zygmund [1, Section XV.2].

Chapter 11

Tensor Algebras and Harmonic Analysis

11.1. Introduction and Initial Results

Let U and W be Banach spaces. The *complete tensor product with greatest cross norm* $U \hat{\otimes} W$ is the space obtained as follows. The usual tensor product $U \otimes W$ is given the norm

$$\left\| \sum_{j=1}^{m} g_j \otimes h_j \right\| = \inf\left\{ \sum_{k=1}^{p} \|r_j\|_U \|s_j\|_W : \sum_{1}^{p} r_j \otimes s_j = \sum_{1}^{m} g_j \otimes h_j,\ 1 \leq p < \infty \right\},$$

and $U \hat{\otimes} W$ is the completion of $U \otimes W$ with respect to that norm. If U and W are also commutative Banach algebras, then $U \hat{\otimes} W$ is a Banach algebra under the multiplication induced by $(u \otimes w)(x \otimes y) = (ux) \otimes (wy)$ where u and x range over U and w and y range over W. The maximal ideal space of $U \hat{\otimes} W$ is easily seen to be identifiable with $\Delta U \times \Delta W$. When $U = C(X)$ and $W = C(Y)$ where X and Y are compact Hausdorff spaces, we write $V = V(X, Y)$ in place of $C(X) \hat{\otimes} C(Y)$. Then the elements of V are the functions $f: X \times Y \to \mathbb{C}$ that can be represented as sums

$$f(x, y) = \sum_{j=1}^{\infty} g_j(x) h_j(y), \tag{1}$$

where $g_j \in C(X)$ and $h_j \in C(Y)$ for all j and

$$\sum_{j=1}^{\infty} \|g_j\|_\infty \|h_j\|_\infty < \infty. \tag{2}$$

The norm on V is equal to the infimum of the numbers (2) subject to (1). Of course, if $f \in C(X)$ and $g \in C(Y)$, then $f \otimes g$ is the function $f \otimes g\,(x, y) = f(x)g(y)$ on $X \times Y$.

This Chapter is about the algebra V, its variants, and their relationship to Fourier algebras. Tensor algebras first became important in harmonic analysis through their applications to harmonic synthesis and functional calculus. The reader who wants to learn about those applications should first read Theorems 11.1.1 and 11.1.2, and then Sections 11.2 and 11.3.

In Section 11.4 infinite tensor products and their relationship to restriction algebras are discussed. In Section 11.5 conditions relating smoothness to membership in $V(T, T)$ are discussed. In the first five sections the relationship with group algebras is stressed. The V-Sidon constants for finite products of finite sets are defined and estimated in Section 11.6, where applications to Helson constants are also given, including the "Kns" Theorem 11.6.5. In Section 11.7 the automorphisms of $V(X, Y)$ are identified for connected X and Y. In Section 11.8 the theories of V-Sidon sets and V-interpolation sets are surveyed and "tilde" extensions of tensor algebras are covered in Section 11.9.

11.1.1. Theorem. *Let G be a compact abelian group. Let $M: C(G) \to C(G \times G)$ be defined by $Mf(x, y) = f(x + y)$. Let $P: C(G \times G) \to C(G)$ be defined by $Pf(z) = \int_G f(z - x, x)dx$. Then the following hold.*

(i) *$P(gMf) = (Pg)f$ for all $f \in C(G)$ and $g \in C(G \times G)$.*
(ii) *M maps $A(G)$ isometrically into $V(G, G)$.*
(iii) *P maps $V(G, G)$ onto $A(G)$ and the norm of this mapping is one.*

Proof. (i) For $z \in G$,

$$P(gMf)(z) = \int_G g(z - x, x)Mf(z - x, x)dx$$

$$= \int_G g(z - x, x)f(z - x + x)dx = (Pg(z))f(z),$$

which establishes (i).

(ii)–(iii) Let $f \in A(G)$. Then for all $x \in G, f(x) = \sum_{k=1}^{\infty} a_k\langle\gamma_k, x\rangle$, where the γ_k are continuous characters on G and $\sum_{k=1}^{\infty} |a_k| = \|f\|_A$. Then $Mf(x, y) = \sum_{k=1}^{\infty} a_k\langle\gamma_k, x + y\rangle = \sum_{k=1}^{\infty} a_k\langle\gamma_k, x\rangle\langle\gamma_k, y\rangle$. Thus $Mf \in V(G, G)$ and $\|Mf\|_V \le \sum |a_k| = \|f\|_A$. Now suppose that $f(x, y) \in V(G, G)$. Let $\varepsilon > 0$, and suppose that $g_k, h_k \in C(G)$ are such that $f(x, y) = \sum_{k=1}^{\infty} g_k(x)h_k(y)$ and $\sum_{k=1}^{\infty} \|g_k\|_\infty \|h_k\|_\infty \le \|f\|_V + \varepsilon$. Then $Pf(z) = \sum_{k=1}^{\infty} g_k * h_k(z)$. But if $g, h \in C(G)$, then $g * h \in A(G)$ and $\|g * h\|_A \le \|g\|_2 \|h\|_2 \le \|g\|_\infty \|h\|_\infty$. Therefore $Pf \in A(G)$ and $\|Pf\|_A \le \|f\|_V + \varepsilon$. Since $\varepsilon > 0$ was arbitrary, $\|Pf\|_A \le \|f\|_V$ for all $f \in V(G, G)$.

We have shown that as operators between $A(G)$ and $V(G, G)$, P and M both have norm at most one. Since $P \circ M$ is the identity, M must be an isometry. We have proved Theorem 11.1.1. □

11.1.2. Theorem. *Let G be a locally compact abelian group and let X and Y be compact disjoint subsets of G. Suppose that $X \cup Y$ is either a Kronecker set or a set of type K_p for some $p > 1$. Then $A(X + Y)$ is isomorphic to $V(X, Y)$, where the isomorphism φ is given by $(\varphi f)(x, y) = f(x + y)$. Furthermore, φ is an isometry if $X \cup Y$ is a totally disconnected Kronecker set.*

Proof. The hypotheses imply that the map from $X \times Y$ to G given by $(x, y) \to x + y$ is a homeomorphism. (That is immediate from the independence of $X \cup Y$.) If $g(x + y) = \sum a_k \langle \gamma_k, x + y \rangle$ is an element of $A(X + Y)$ then $\varphi g = \sum a_k \langle \gamma_k, x \rangle \langle \gamma_k, y \rangle$. Thus, $\varphi g \in V(X, Y)$ and $\|\varphi g\|_V \leq \|g\|_A$. Clearly, φ is one to one.

To complete the proof, we show that φ is onto. For that it will suffice to show that if

(3) $$f_1 = f \otimes g \in V, \ \|f_1\|_\infty = 1, \text{ and } \varepsilon > 0,$$

then there exists $h \in A(X + Y)$ such that $\|\varphi h - f_1\|_V < \varepsilon$ and $\|h\|_A \leq C\|f_1\|_V$, where C depends only on $X \cup Y$. Note that for a product such as f_1, V-norm and supremum norm are equal.

Case I, when $X \cup Y$ is a totally disconnected Kronecker set. Let f_1 be as in (3). We may assume that $|f_1| > 0$ on $X \times Y$. Then f and g are each the average of two continuous unimodular functions on X and Y respectively. Thus, on expanding the product (3), we see that we may assume that f and g are unimodular. Since $X \cup Y$ is a Kronecker set, there exists, for each $\varepsilon > 0$, a continuous character γ on G such that $|\langle \gamma, x \rangle - f(x)| < \varepsilon/2$ for $x \in X$ and $|\langle \gamma, y \rangle - g(y)| < \varepsilon/2$ for $y \in Y$. It now follows at once that $\|\varphi\gamma - f_1\|_V < \varepsilon$. Since $\|\gamma\|_A = 1$, we have established that φ is onto and that φ is an isometry in this case.

Case II, when $X \cup Y$ is a Kronecker set. Let $f_1 \in V(X, Y)$ have the form (3). By replacing f and g with their real parts, we see that we may assume that f and g are real. That change will introduce a factor of at most 4 into the calculation of C.

Since f and g are real, each is the average of two unimodular continuous functions. The argument proceeds as in Case I, and $\|\varphi\| \leq 4$.

Case III, when $X \cup Y$ is a K_p-set, $p \geq 3$. If f_1 has the form (3) then f_1 has the form

$$f_1 = \tfrac{1}{4} \sec^2(\pi/p)(g_1 + h_1) \otimes (g_2 + h_2),$$

where g_1, h_1 are continuous functions from X to the pth roots of unity and g_2, h_2 are continuous functions from Y to the pth roots of unity. The proof proceeds as in Case I. In this case, $C = \sec^2 \pi/p$.

Case IV, where $X \cup Y$ is a K_2-set. We write $f_1 = (g_1 + ih_1) \otimes (g_2 + ih_2)$ where g_1, h_1 are continuous real-valued functions on X and g_2, h_2 are continuous real-valued functions on Y. We then approximate g_1, h_1, g_2, h_2 by convex combinations of functions taking on only the values ± 1. In this case, $C = 4$. We leave the details to the reader. That ends the proof of 11.1.2. □

11.1.3. Corollary. *Let X and Y be compact Hausdorff spaces. Then $V(X, Y)$ is a regular commutative Banach algebra with maximal ideal space $X \times Y$ and such that every singleton in $X \times Y$ is a set of synthesis for $V(X, Y)$.*

Proof. We leave to the reader the simple but instructive task of providing a direct proof. Here is one based on 11.1.2.

Let Γ be the set of all continuous unimodular functions on the disjoint union of X and Y. Then Γ forms a discrete abelian group, and $X \cup Y$ is a Kronecker set in $\Gamma^\wedge$. Therefore $V(X, Y)$ is isomorphic to $A(X + Y)$, and the Corollary follows. □

We now turn to the useful "block" method. Let X and Y be compact Hausdorff spaces, and let π_1, π_2 denote the projections of $X \times Y$ on X and on Y. We say that two subsets E and F of $X \times Y$ are *bidisjoint* if $\pi_j E \cap \pi_j F = \varnothing$ for $j = 1, 2$.

11.1.4. Theorem. *Let X and Y be compact Hausdorff spaces. Let f and g be functions in $V(X, Y)$ whose supports E and F are bidisjoint. Then $\|f + g\|_V = \max(\|f\|_V, \|g\|_V)$.*

Proof. Let $\varepsilon > 0$. Let $f = \sum_{m=1}^{\infty} f_{1m} \otimes f_{2m}$ and $g = \sum_{j=1}^{\infty} g_{1m} \otimes g_{2m}$, where $\sum \|f_{1m}\|_\infty \|f_{2m}\|_\infty \le \|f\|_V + \varepsilon$ and $\sum \|g_{1m}\|_\infty \|g_{2m}\|_\infty \le \|g\|_V + \varepsilon$. We may assume that there are bidisjoint neighborhoods U and W of E and F, respectively, such that for each m, $f_{1,m} \otimes f_{2,m}$ is supported in U and $g_{1,m} \otimes g_{2,m}$ is supported in W. We may assume also that $\sum \|f_{1m}\|_\infty \|f_{2m}\|_\infty > \sum \|g_{1m}\|_\infty \|g_{2m}\|_\infty$. We claim that we may further assume that

$$\|f_{1m}\|_\infty \|f_{2m}\|_\infty \ge \|g_{1m}\|_\infty \|g_{2m}\|_\infty, \text{ for } m = 1, 2, \ldots. \tag{4}$$

For simplicity in notation, let the left hand side of (4) be denoted by a_m and the right hand side by b_m. Then it will suffice to show that whenever $\{a_m\}$ and $\{b_m\}$ are sequences of non-negative numbers and $\sum a_m > \sum b_m$, one may choose by an inductive procedure sequences $\{s_{m,n}\}_{n=1}^{\infty}$ and $\{t_{m,n}\}_{n=1}^{\infty}$ of non-negative numbers for each m such that $\sum_{n=1}^{\infty} s_{m,n} = \sum_{n=1}^{\infty} t_{m,n} = 1$ for all m and such that if the sets

$$\{s_{m,n} a_m : 1 \le m, n < \infty\}, \{t_{m,n} b_m : 1 \le m, n < \infty\}$$

are arranged in monotonic order, then the lth term of the first sequence is at least as large as the lth term of the second. We leave that exercise to the reader. The claim will follow.

For $j = 1, 2$ and each $k \ge 1$, there exist $h_{j,k}$ such that $h_{j,k} = f_{j,k}$ on $\pi_j U$ and $h_{j,k} = g_{j,k}$ on $\pi_j W$ and $h_{j,k}$ is supported on $\pi_j U \cup \pi_j W$. Let $h' = \sum_{k=1}^{\infty} h_{1,k} \otimes h_{2,k}$. Then $\|h'\|_V \le \sum \|f_{1,k}\|_\infty \|f_{2,k}\|_\infty$ by (4). Unfortunately, the support of h' may be too large, though $h' = f$ on U and $h' = g$ on V. For $j = 1, 2$, let u_j equal one on $\pi_j E$ and be supported in $\pi_j U$, and let v_j equal one on $\pi_j F$ and be supported in $\pi_j V$ and $\|u_j\|_\infty = \|v_j\|_\infty = 1$. Let h'' be defined by

$$h'' = \tfrac{1}{2}(u_1 + v_1) \otimes (u_2 + v_2) + \tfrac{1}{2}(u_1 - v_1) \otimes (u_2 - v_2).$$

Then $\|h''\|_V = 1$ and h'' equals one on $E \cup F$ and zero outside $U \cup V$. Then $h'h'' = f + g$, and $\|h'h''\|_V < \max(\|f\|_V, \|g\|_V) + \varepsilon$. Since $\varepsilon > 0$ is arbitrary, we have proved 11.1.4. □

The dual space of $V(X, Y)$ is denoted by $BM(X, Y)$ and its elements are called *bimeasures* and the dual space norm of $\mu \in BM(X, Y)$ is denoted by $\|\mu\|_{BM}$.

11.1.5. Theorem. *Let X and Y be compact Hausdorff spaces, and $f \in V(X, Y)$. Then*

$$\|f\|_V = \sup\left\{\left|\int f\, d\mu\right| : \mu \text{ is a finitely supported measure, } \|\mu\|_{BM} \leq 1\right\}.$$

It is obvious that 11.1.5 follows at once from the next lemma, whose proof is easy, depending as it does on the construction of partitions of unity for X and Y.

11.1.6. Lemma. *For $j = 1, 2$, let E_j be a dense subset of the compact Hausdorff space X_j. Let $E = E_1 \times E_2$, and let $M_f(E)$ denote the set of finitely supported measures on E. Then there exists a directed family of linear operators $\{L_w: w \in I\}$ from $BM(X_1, X_2)$ to $M_f(E)$ such that the following hold for all $S \in BM(X_1, X_2)$.*

(i) *The range of each L_w is finite dimensional.*
(ii) $\|L_w S\|_{BM} \leq \|S\|_{BM}$.
(iii) *$L_w S \to S$ weak*.*
(iv) *support $L_w S \to$ support S.*

Remarks and Credits. The product $L \hat{\otimes} M$ is also called the *projective tensor product* and its norm the *projective norm* or *greatest cross-norm*. $L \hat{\otimes} M$ has the functorial property that any multilinear bounded map from $L \times M$ to a Banach space factors through $L \hat{\otimes} M$ continuously and linearly. See Shatten [1] for general results about tensor products of Banach spaces.

The use of the maps M and P is due to Herz [8], but the embedding of $A(G)$ in $V(G, G)$ and Theorem 11.1.1 and 11.1.2 and Corollary 11.1.3 are due to Varopoulos [5, 6, 7]. Theorem 11.1.4 is stated in Varopoulos [11], and a proof is sketched there. Complete details of the proof appeared in Saeki [8] and Stegeman [4].

S. Kaijser [1, 2] has shown that if $K_1 \cup K_2$ has Helson constant $1/(1 - \beta)$ where $\beta(2 + K_c) < 1$ (K_c is the complex Grothendieck constant), then $A(K_1 + K_2)$ is isomorphic to $V(K_1, K_2)$.

Theorem 11.1.5 was proved by Varopoulos [11] for a special case. The proof here is from Saeki [8, Theorem 4.5].

Saeki [6] shows that a continuous function f on $X \times Y$ belongs to $V(X, Y)$ if and only if $f \in U \hat{\otimes} W$, where U is the Banach space of bounded Borel functions on X and W is the space of bounded Borel functions on Y.

Doss [4] has shown that there exist two sequences of continuous functions $\{f_n\}$, $\{g_n\}$ on $[0, 1]$ such that for every continuous function f on $[0, 1] \times [0, 1]$, there exists a sequence $\{a_j\} \subseteq \mathbb{C}$ such that $\sum |a_k| < \infty$ and $f(x, y) = \sum a_k \exp(2\pi i f_k(x) g_k(y))$ for all $x, y \in [0, 1]$. Because $V(X, Y) \neq C(X \times Y)$, it follows that $\lim \sup \|f_k\|_\infty \|g_k\|_\infty = \infty$.

Ito and Schreiber [1] have shown that if U and V are unitary operators on the Hilbert space H and if A is any bounded operator on H, then the double sequence $a_{mn} = \langle AU^m x, V^n y\rangle$ represents the Fourier transform of a bimeasure on T^2, for all $x, y \in H$. Whether that characterizes the Fourier transforms of the bimeasures appears to be unknown.

11.2. Transfer Methods: Harmonic Synthesis and Non-finitely Generated Ideals in $L^1(G)$

In this section we show how to transfer results for harmonic synthesis from one group to another and to $V(X, Y)$, and how to construct ideals in group algebras that are not finitely generated. For terminology and preliminary results about harmonic synthesis, see Section 3.1.

The transfer procedure is summed up in (i)–(iii) of the following theorem. Part (iv) is a famous result of L. Schwartz [1], and (v) is Malliavin's theorem [3].

11.2.1. The Transfer Theorem. (i) *If synthesis fails for* $V(T^\infty, T^\infty)$, *then synthesis fails for* $V(D, D)$.

(ii) *If synthesis fails for* $V(D, D)$, *then synthesis fails for* $A(G)$ *where G is any non-discrete locally compact abelian group.*

(iii) *If synthesis fails for* $A(G)$ *for some non-discrete locally compact abelian group G, then synthesis fails for* $A(T^\infty)$ *and for* $V(T^\infty, T^\infty)$.

(iv) *Synthesis fails for* $A(R^3)$.

(v) *Synthesis fails for* $A(G)$ *for all non-discrete locally compact abelian groups G.*

Proof. (i) We shall write elements of D as doubly indexed sequences $\{x_{jk}\}_{j,k=1}^\infty$ of zeroes and ones. For each k, D_k is the subset of sequences $\{x_{j,k}\}_{j=1}^\infty$ and $u_k: D_k \to T_k$ (the kth copy of T) is the map given by $u_k(\{x_{j,k}\}_{j=1}^\infty) = 2\pi \sum_{j=1}^\infty x_{j,k}/2^j$. It is clear that except for countable subsets of D_k and T_k, the map u_k is surjective and one-to-one. Also u_k is continuous and preserves Haar measure.

Let $u: D \to T^\infty$ be given by $u(\{x_{j,x}\}_{j,k=1}^\infty) = \{u_k(\{x_{j,k}\}_{k=1}^\infty)\}_{k=1}^\infty$. Then u is continuous and preserves Haar measure. Also, there exist null Borel sets $X \subseteq D$ and $Y \subseteq T^\infty$ such that $u: D\backslash X \to T^\infty \backslash Y$ is one-to-one and onto. Let $U: T^\infty \to D$ be a Borel mapping such that $u \circ U$ is the identity on $D\backslash X$. We define $\varphi: V(T^\infty, T^\infty) \to V(D, D)$ and $\Phi: V(D, D) \to L^\infty(T^\infty) \hat{\otimes} L^\infty(T^\infty)$ by $\varphi f(x, y) = f(u(x), u(y))$ and $\Phi g(w, z) = g(U(w), U(z))$ for all $f \in V(T^\infty, T^\infty)$

and $g \in V(D, D)$. Then φ and Φ are both norm decreasing. We claim that φ and Φ are isometries. To see that, let $\{h_j\}$ be an approximate identity for $L^1(T^\infty)$ such that $\|h_j\| \leq 1$ for all j. For each j, and $f \in L^\infty(T^\infty) \hat{\otimes} L^\infty(T^\infty)$, let

$$H_j f(x, y) = \iint_{T^\infty \times T^\infty} f(x - x', y - y')h_j(x')h_j(y')dx'\,dy'.$$

Then $H_j f \in V(T^\infty, T^\infty)$ and $\|H_j f\|_V \leq \|f\|_{L^\infty \hat{\otimes} L^\infty}$. Also if $f \in V(T^\infty, T^\infty)$, then $\|f - H_j f\|_V \to 0$ as $j \to \infty$.

Let $E \subseteq T^\infty \times T^\infty$ be a closed set of non-synthesis for $V(T^\infty, T^\infty)$. Let $f \in V(T^\infty, T^\infty)$ be zero on E and not be approximable by elements of $V(T^\infty, T^\infty)$ that vanish in a neighborhood of E. Suppose that φf could be approximated in norm by elements of $V(D, D)$ that vanish in a neighborhood of $(u^{-1} \times u^{-1})(E)$. Then $\Phi\varphi f = f$ could be approximated in $L^\infty(T^\infty) \hat{\otimes} L^\infty(T^\infty)$-norm by functions f_k that vanish in a neighborhood of E. On replacing the f_k by $H_j f_k$ we obtain elements of $V(T^\infty, T^\infty)$ that approximate f and that vanish in neighborhoods of E. (We will need to choose k and then $j = j(k)$.) That contradiction proves (i).

(ii). Let $K \subseteq G$ be a perfect compact metrizable set that is either a Kronecker set or a set of type K_p for some $p \geq 2$. Such a set exists by Rudin [1, 5.2.2]. Let E_1 and E_2 be disjoint perfect subsets of K.

Then E_1 and E_2 are homeomorphic to D, and therefore, by 11.1.2, $A(E_1 + E_2)$ is isomorphic to $V(D, D)$.

Let $E \subseteq D \times D$ be closed and let $f \in V(D, D)$ be zero on E and such that f cannot be approximated in V-norm by functions vanishing in a neighborhood of E. Let $g \in A(G)$ be such that the restriction of g to $E_1 + E_2$ is carried, by the isomorphism of $A(E_1 + E_2)$ to $V(D, D)$, to f. It is clear that if $\{g_n\}$ is a set of functions in $A(G)$ that vanish in a neighborhood of $g^{-1}(0)$, then $\{g_n\}$ does not converge to g in A-norm. That proves (ii).

(iii) We reduce to the case that G is compact. Let $E \subseteq G$ be a compact set for which synthesis fails. The group G has an open subgroup of the form $R^n \times K$, where K is compact. We may assume that $E \subseteq R^n \times K$. Since E is compact, there is a number $\alpha > 0$ such that $(\alpha Z^n \times \{0\}) \cap E$ is at most one point. Then $G' = R^n \times K/(\alpha\mathbb{Z}^n \times \{0\})$ is compact and the image of E in G' also disobeys synthesis. We may thus suppose that G is compact.

We now reduce the general compact case to the case of compact metrizable G. Let G be compact and $f \in A(G)$ be such that synthesis fails for f. Let Λ be the subgroup of the dual group Γ of G that is generated by supp $\hat{f}$. Let H be the annihilator of Λ in G, that is $H = \{x \in G: \langle x, \lambda\rangle = 1 \text{ for all } \lambda \in \Lambda\}$. Let $G'' = G/H$. Then $f \in A(G'')$ in the obvious abuse of notation, so $A(G'')$ contains a function, namely f, for which synthesis fails. Since Λ is discrete and countable, G'' is compact and metrizable. We may thus assume that G is compact and metrizable. But every compact metrizable abelian group G is a closed subgroup of T^∞ (Rudin [1, 2.2.6]). Therefore synthesis fails for $A(T^\infty)$.

Let $E \subseteq T^\infty$ be a closed subset for which synthesis fails, and let $f \in A(T^\infty)$ be such that $f = 0$ on E and such that f is not the limit in A-norm of functions $g \in A(T^\infty)$ that vanish in a neighborhood of E. Let

$$E^* = \{(x, y) \in T^\infty \times T^\infty : x + y \in E\}.$$

Then $Mf = 0$ on E^*. Suppose that $g \in V(T^\infty, T^\infty)$ vanishes in a neighborhood of E^*. Then Pg vanishes in a neighborhood of E. By 11.1.1, $\|f - Pg\| \leq \|Mf - g\|$. Therefore Mf cannot be approximated in V-norm by functions $g \in V(T^\infty, T^\infty)$ that vanish in a neighborhood of E^*. That completes the proof of (iii).

(iv) Let $S = \{(x, y, z) \in R^3 : x^2 + y^2 + z^2 = 1\}$. We shall show that S is not a set of synthesis. Let μ denote normalized surface area measure on S. Then

$$\hat{\mu}(x, y, z) = \int \exp(i(xx' + yy' + zz')d\mu(x', y', z')$$

is rotationally symmetric, so $\hat{\mu}(x, y, z) = g(r)$ where $r = (x^2 + y^2 + z^2)^{1/2}$. But then

$$g(r) = \hat{\mu}(0, 0, r) = \frac{1}{4\pi} \int_0^{2\pi} \int_0^{\pi} \exp(ir \cos\theta)\sin\theta \, d\theta \, d\psi = \frac{\sin r}{r},$$

by integration in polar coordinates. Let Q denote the distribution derivative $\partial\mu/\partial x$. Then $\hat{Q} = ix\hat{\mu}$, so $Q \in PM(R^3)$. We claim that Q is supported on S and that $\langle Q, f\rangle \neq 0$ for some $f \in A(R^3)$ that vanishes on S. For the first claim, note that if $f \in A(R^3)$ vanishes in a neighborhood of S, then there exist functions $f_n \in A(R^3) \cap C^\infty(R^3)$ vanishing in neighborhoods of S with $\lim f_n = f$ in A-norm. But $\langle Q, f_n\rangle = \langle \partial u/\partial x, f_n\rangle = -\langle \mu, \partial f_n/\partial x\rangle = 0$ for all such f_n. Hence $\langle Q, f\rangle = 0$ for all f zero in a neighborhood of S, so Q is supported on S.

Let $f(x, y, z) = x[\exp(-r^2 + 1) - \exp(-2r^2 + 2)]$ where $r^2 = x^2 + y^2 + z^2$. Then f is zero on S. Because f and all its derivatives are in $C_0(R^3)$, $f \in A(R^3)$. Finally $\langle Q, f\rangle = -\int 2x^2 \, d\mu \neq 0$. That completes the proof of (iv).

(v) is obvious from (i)–(iv). That ends the proof of 11.2.1. □

Remark. The proof of (iii) shows that if E is not of synthesis for the compact abelian group G, then E^* is not of synthesis for $V(G, G)$.

A closed ideal I in a commutative Banach algebra B *has exactly k generators* if there exists k elements $f_1, \ldots, f_k$ of I such that $Bf_1 + \cdots + Bf_k$ is dense in I and such that whenever $\{g_1, \ldots, g_{k-1}\} \subseteq I$, then $Bg_1 + \cdots + Bg_{k-1}$ is not dense in I. We shall show below that for each $k \geq 2$, there exist ideals in

$A(G)$ that have exactly k generators, and that there are ideals that are not finitely generated. The following proposition illustrates the connection of that with synthesis.

11.2.2. Proposition. *Let G be a metrizable σ-compact abelian group and I a closed ideal in $A(G)$. If I is not singly generated, then $E = \{x \in G: f(x) = 0$ for all $f \in I\}$ is not a set of synthesis.*

Proof. Since G is metrizable and σ-compact, there exists $f \in I$ such that $f \geq 0$ on G and $f(x) = 0$ if and only if $x \in E$. If E were of synthesis, then the products gf, for $g \in A(G)$, would generate I and I would be singly generated. The Proposition is proved. □

Remark. The converse of 11.2.2 is false. If $G = R^3$ and S is the unit sphere, then $I(S) = \{f \in A(R^3): f = 0 \text{ on } S\}$ is singly generated, as Lemma 11.2.10 below shows.

11.2.3. Theorem. *Let G be a non-discrete locally compact abelian group. Then the following hold.*

(i) *For each $k = 2, 3, \ldots, A(G)$ contains a closed ideal that has exactly k generators.*
(ii) *$A(G)$ contains a closed ideal that is not finitely generated.*

The basic idea of the proof of 11.2.3 can be illustrated by consideration of $C^1(R^k)$: the ideal generated by functions vanishing at the origin has exactly k generators, as can be seen by applying Lemma 11.2.4 below to the functional $\Psi(f_1, \ldots, f_k) = \mathrm{Det}(\partial f_i/\partial x_j)|_0$. As the reader will see, there are some subtleties involved in applying the idea to $A(R^n)$ and then to $V(D, D)$. The map P, for example, is used to transfer the proof, rather than the conclusion as in 11.2.1. We now give the proof of 11.2.3, subject to Lemma 11.2.11, whose proof incorporates the subtleties to which we have referred.

Proof of 11.2.3. In Lemma 11.2.11 below, we show that for each $k = 2, 3, \ldots$ $V(D, D)$ contains a closed ideal that has exactly k generators. Since $A(G)$ has a quotient algebra isomorphic to $V(D, D)$, (i) follows.

For (ii) we let $E_1, E_2, \ldots$ be a sequence of pairwise disjoint compact subsets of G whose union has compact closure such that for each $j \geq 1$, no element of E_j is a cluster point of $\bigcup_{i \neq j} E_i$, and such that $A(E_j)$ is isomorphic to $V(D, D)$ for $1 \leq j \leq \infty$. For each $1 \leq j \leq \infty$, we choose j functions $f_{j_1}, \ldots, f_{jj}$ such that $f_{j,k} = 0$ on $\bigcup_1^\infty E_m$ for $1 \leq k \leq j$, and such that the ideal I_j in $A(E_j)$ generated by $\{f_{j,m|E_j}\}_{m=1}^j$ has exactly j generators. We let I be the ideal in $A(G)$ generated by $\{f_{j,m}: 1 \leq j < \infty, 1 \leq m \leq j\}$. Note that if $f \in I$, then $f \in I_j$ for all j. Thus, if $f_1, \ldots, f_k$ generated I, the restrictions $\{f_j|_{E_{k+1}}\}_{j=1}^k$ would generate I_{k+1}, a contradiction. Therefore I is not finitely generated. That proves 11.2.3, subject to proving 11.2.11. □

11.2.4. Lemma. *Let B be a Banach algebra and I a closed left ideal in B. Let $k \geq 2$ be an integer and Ψ a bounded k-linear functional on I that is not identically zero and is such that if $g_1, \ldots, g_k \in I$ are not distinct, then*

$$\Psi(h_1 g_1, \ldots, h_k g_k) = 0 \text{ for all } h_1, \ldots, h_k \in B. \tag{1}$$

Then I is not generated (even topologically) by $k - 1$ elements.

Proof. Since $\Psi \not\equiv 0$, there exist $f_1, \ldots, f_k \in I$ such that $\Psi(f_1, \ldots, f_k) = 1$. Let $g_1, \ldots, g_{k-1} \in I$, and suppose that, for each $j = 1, \ldots, k$ and each $\varepsilon_j > 0$, there exist $h_{j,m} \in B$ such that

$$\left\| f_j - \sum_{m=1}^{k-1} h_{j,m} g_m \right\| < \varepsilon_j . \tag{2}$$

The boundedness of Ψ implies that for some small $\varepsilon > 0$, if $0 < \varepsilon_j < \varepsilon$ for all j, then

$$|\Psi(f_1, \ldots, f_k) - \Psi(\sum h_{1,m} g_m, \ldots, \sum h_{k,m} g_m)| < 1/2.$$

In particular,

$$\Psi(\sum h_{1,m} g_m, \ldots, \sum h_{k,m} g_m) \neq 0. \tag{3}$$

By the multilinearity of Ψ, the left hand side of (3) equals

$$\sum_{j(1)=1}^{k-1} \cdots \sum_{j(k)=1}^{k-1} \Psi(h_{1,j(1)} g_{j(1)}, \ldots, h_{k,j(k)} g_{j(k)}). \tag{4}$$

Since there are k variables in Ψ and $k - 1$ functions $g_1, \ldots, g_{k-1}$, each summand in (4) is zero, by (1). That contradicts (3). The Lemma is proved. □

If n and k are positive integers, we identify elements x of R^{nk} with k-tuples of n-tuples: $x = (x_1, \ldots, x_k)$. A function f on R^{nk} is *k-radial* if $f(x_1, \ldots, x_k)$ depends only on $\|x_1\|, \ldots, \|x_k\|$. A function f on R^{nk} is *harmonic* if $f \in C^2(R^{nk})$ and $\Delta f = 0$, where Δf is the sum of all second partial derivatives of f. The function f is *homogeneous of degree m* if $f(tx) = t^m f(x)$ for all $x \in R^{nk}$ and all $t > 0$. For each k-tuple $m = (m_1, \ldots, m_k)$ of non-negative integers, we let H_m be the set of products $f_1(x_1) \cdots f_k(x_k)$, where each f_j is harmonic and homogeneous of degree m_j. For $r, s \in (0, \infty)^k$, and such a k-tuple m, let

$$G(m; r) = \prod_{j=1}^{k} [2\pi i^{-m_j} (r_j)^{-(n+2m_j-2)/2}]$$

and

$$J(m; r, s) = \prod_{j=1}^{k} [J_{(n+2m_j-2)/2}(2\pi r_j s_j)(s_j)^{(n+2m_j)/2}].$$

Here J_p denotes the Bessel function of order p of the first kind. The following lemma can be proved by adapting the proof of Stein and Weiss [1, IV.3.10]. The reader will note that the formula given in the next lemma is quite manageable when $k = 1$ (the case in Stein and Weiss), and that spherical harmonics play an important, though implicit, role in what follows.

11.2.5. Lemma. *Let n and k be positive integers and let $m = (m_1, \dots, m_k)$ be a k-tuple of non-negative integers. Let $f \in H_m$ and let g be a k-radial function on R^{nk}, with $g(x_1, \dots, x_k) = g_0(\|x_1\|, \dots, \|x_k\|)$. If $fg \in L^1(R^{nk})$, then*

$$(5) \quad (fg)\hat{}(x) = f(x)G(m; \|x_1\|, \dots, \|x_k\|) \times \int_0^\infty \cdots \int_0^\infty g_0(s)J(m; (\|x_1\|, \dots, \|x_k\|), s)\, ds_1 \cdots ds_k.$$

Let n and k be integers not less than 2. Let $S^{(n-1)} = \{x \in R^n : \|x\| = 1\}$. Let $E = E_{n,k} = S^{(n-1)} \times \cdots \times S^{(n-1)}$ (k times).

Let $I = I_{n,k} = \{f \in L^1(R^{nk}) : \hat{f} = 0 \text{ on } E\}$. Let ν denote normalized $(n-1)$-volume measure on $S^{(n-1)}$, and let $\mu = \nu \times \cdots \times \nu$ (k times). For $1 \le j \le k$, we let $r_j(x_1, \dots, x_k) = x_j$. Then

$$(6) \qquad r_j^m \frac{\partial^m \hat{\mu}}{\partial r_j^m} \varepsilon \cdot L^\infty(R^{nk}) \quad \text{if } 0 \le m \le (n-1)/2.$$

A proof of (6) is easily extracted from 11.2.5, where f is taken to be the identity function and g tends weak$*$ to μ.

11.2.6. Lemma. *Let $n \ge 3$ and $k \ge 1$. Let $f \in L^1(R^{nk})$ and $\varepsilon > 0$. If $\hat{f} = 0$ on $E_{n,k}$, then there exists $g \in L^1(R^{nk})$ such that $\|f - g\| < \varepsilon$, $\hat{g} = 0$ on $E_{n,k}$ and g has compact support.*

Proof. For $1 \le j \le k$, let $S(j) = \{(x_1, \dots, x_k) \in R^{nk} : x_i = 0 \text{ if } i \ne j \text{ and } \|x_j\| = 1\}$. For non-negative integers j and m, we let $H_{j,m}$ be the set H_p where $p = (0, \dots, 0, m, 0, \dots, 0)$ and m appears in the jth place. It is well-known that $\bigcup_{m=0}^\infty H_{j,m}|_{S(j)}$ spans a dense subspace of $C(S(j))$ and that $\int_{S(j)} YZ\, d\nu = 0$ if $Y \in H_{j,m}$ and $Z \in H_{j,q}$ with $m \ne q$. For proofs of those assertions, see Stein and Weiss [1, IV.2.3]. For functions f on R^{nk} and $g \in C(E_{n,k})$ we define

$$(7) \quad f \# g(x) = \int_{SO(n)} \cdots \int_{SO(n)} f(\sigma_1 x_1, \dots, \sigma_k x_k) g(\sigma_1 \xi, \dots, \sigma_k \xi)\, d\sigma_1 \cdots d\sigma_k$$

where $d\sigma_j$ denotes normalized Haar measure on the jth copy of $SO(n)$, the group of all rotations σ of R^n, and $\xi = (1, 0, \ldots, 0)$ is the North Pole of $S^{(n-1)}$. It is easy to see that when f is locally in $L^1(R^{nk})$, then $f \# g$ is defined for almost all x, that $f \# g$ is continuous if f is continuous, and that $\|f \# g\|_1 \leq \|f\|_1 \|g\|_1$, where $\|g\|_1 = \int |g| \, d\mu$.

Let H denote the subspace of $C(E_{n,k})$ that is generated by products of the form $g_1 \cdots g_k$, where $g_j \in H_{j,m(j)}$ for all j and some choice of non-negative integers $m(j)$. It is easy to see that there exists a sequence $g_j \in H$ such that $\|g_j\|_1 \leq 1$ for all j, and such that $\|f \# g_j - f\|_1 \to 0$ for all $f \in L^1(R^{nk})$.

Let H' denote the space of functions in $L^1(R^{nk})$ that is spanned by functions of the form fg where g is k-radial and $f \in \bigcup H_m$. It is easy to see that H' is a dense subspace of $L^1(R^{nk})$. We claim that if $f \in I_{n,k}$, then there exists a sequence of functions $f_s \in H'$ such that $\lim f_s = f$ and $f_s \in I_{n,k}$ for all s. To see that, let $\varepsilon > 0$, and let $g \in H$ be such that $\|f \# g - f\| < \varepsilon$. Let $\{h_s\} \subseteq H'$ be such that $\lim h_s = f$. Then $\lim h_s \# g = f \# g$. For each pair of integers j and m, we choose a basis $\{Y_{j,m,p}\}$ of $H_{j,m}$ such that $\int_{S(j)} Y_{j,m,p} \overline{Y_{j,m,q}} \, dv$ is zero if $p \neq q$ and one if $p = q$. It then follows that the set of products of the form $Y = Y_{1,m(1),p(1)} \cdots Y_{k,m(k),p(k)}$ forms an orthonormal basis for $L^2(\mu)$. Fixing such a Y, we see that $\int h_s \# g(r_1 x_1, \ldots, r_k x_k) \overline{Y(x_1, \ldots, x_k)} \, d\mu = \rho_Y^{(s)}$ is a function of the k non-negative numbers $r_1, \ldots, r_k$. Identifying k-radial functions with the corresponding function of $r_1, \ldots, r_k$, we conclude that $\rho_Y^{(s)}$ converges in $L^1(R^{nk})$-norm to the k-radial function $\rho_Y = \int f \# g(r_1 x_1, \ldots, r_k x_k) \overline{Y(x_1, \ldots, x_k)} \, d\mu$. It is now easy to see that $h_s \# g = \sum \{Y \rho_Y^{(s)} \colon Y$ is a term of $g\}$. Therefore $f \# g = \sum \{Y \rho_Y \colon Y$ is a term of $g\}$. Since g is a finite linear combination of the Y's, we see that $(f \# g)$ is a finite sum of terms of the form (5). Because $(f \# g)^\wedge = (\hat{f}) \# g$, it follows that each of the terms of the form (5) for $(f \# g)^\wedge$ vanishes on $E_{n,k}$, that is, $\rho_Y^\wedge$ vanishes on $E_{n,k}$ for each Y. We leave to the reader the task of showing that the functions $f_0(s)$ appearing in (5) may be replaced with functions that have compact support, are arbitrarily close to f_0 in L^1-norm, and yield a transform vanishing on $E_{n,k}$. That ends the proof of Lemma 11.2.6. □

For $x = (x_1, \ldots, x_k) \in R^{nk}$, $1 \leq j \leq k$, and $\rho > 0$, we define $\tau_{\rho,j} x = (x_1, \ldots, x_{j-1}, \rho x_j, x_{j+1}, \ldots, x_k)$.

11.2.7. Lemma. *Let k be an even positive integer, and let $n = 2k + 1$. Then there exists $C_1 > 0$ such that for every f in the ideal generated by the set of all products of k elements of $I_{n,k}$,*

(8) $$\left| \int_{E_{n,k}} \hat{f}(\tau_{\rho,j} x) d\mu \right| \leq C_1 (1 - \rho)^k \|f\|$$

for all $1 \leq j \leq k$ and $\rho \in [\frac{1}{2}, 1)$.

Proof. We first show that if $0 \le m < p \le k$ and $f_1, \ldots, f_p \in I_{n,k}$, then

$$\int_{R^{nk}} (f_1 * \cdots * f_p)(r_j)^m \frac{\partial^m \hat{\mu}}{\partial r_j^m}\, dy = 0. \tag{9}$$

Indeed, since $r_j^m\, \partial^m \hat{\mu}/\partial r_j^m \in L^\infty(R^{nk})$, we may apply Lemma 11.2.5 and conclude that there is no loss of generality in assuming that the $f_j \in C^\infty(R^n)$. But then for some constant C that is independent of $f_1, \ldots, f_p$,

$$\int (f_1 * \cdots * f_p) r_j^m \frac{\partial^m \hat{\mu}}{\partial r_j^m}\, dy = C \int \frac{\partial^m}{\partial r_j^m} (\hat{f}_1 \cdots \hat{f}_p) d\mu,$$

which is of course zero, since each $\hat{f}_j$ vanishes on the support of μ and $m < p$. Therefore (9) holds.

By Taylor's Theorem,

$$\begin{aligned} \hat{\mu}(\tau_{\rho,j} y) = {} & \sum_{l=0}^{m-1} \frac{1}{l!} r_j^l \frac{\partial^l \hat{\mu}}{\partial r_j^l}(y)(\rho - 1)^l \\ & + \frac{1}{m!} \lambda^{-m} r_j^m \frac{\partial^m \hat{\mu}}{\partial r_j^m}(\tau_{\lambda,j} y)(\rho - 1)^m, \end{aligned} \tag{10}$$

for $0 \le m \le k$, where $\rho < \lambda < 1$. We may assume that f is a finite sum of products of k elements of $I_{n,k}$. Applying (9) to each term, we conclude that

$$\int \hat{f}(\tau_{\rho,j} x) d\mu = \int f(y)(k!)^{-1} \lambda^{-k} r_j^k \frac{\partial^k \hat{\mu}}{\partial r_j^k}(\tau_{\lambda,j} y)(\rho - 1)^k\, dy.$$

Therefore (8) holds with $C_1 = (k!\, 2^k) \| r_j^k (\partial^k \hat{\mu}/\partial r_j^k) \|_\infty$. The Lemma is proved. □

11.2.8. Lemma. *Let k be a positive even integer and $n = 2k + 1$. Then there exists a constant $C > 0$ such that for all $f_1, \ldots, f_k \in I_{n,k}$,*

$$\left| (\rho - 1)^{-k} \int \operatorname{Det}(\hat{f}_m(\tau_{\rho,j} x)) d\mu \right| \le C \|f_1\| \cdots \|f_k\| \quad \textit{for } \rho \in [\tfrac{1}{2}, 1).$$

Proof. The terms in $\operatorname{Det}(\hat{f}_m(\tau_{\rho,j} x))$ have the form $\pm \prod_{j=1}^k f_{m(j)}(\tau_{\rho,j} x)$ where $j \mapsto m(j)$ is a permutation of $\{1, \ldots, k\}$. By the generalized Hölder inequality (Hewitt and Stromberg [1, p. 200])

$$\left| (\rho - 1)^{-k} \int \prod_{j=1}^k f_{m(j)}(\tau_{\rho,j} x) d\mu \right| \le \prod_{j=1}^k \left(\int (\rho - 1)^{-k} | f_{m(j)}(\tau_{\rho,j} x)|^k\, d\mu \right)^{1/k}.$$

Since k is even $|\hat{f}_{m(j)}|^k = \hat{g}_1 \cdots \hat{g}_k$, where $g_i \in I_{n,k}$. An application of 11.2.7 shows that $C = C_1 k!$ will do, where C_1 is the constant given by 11.2.7. That proves 11.2.8. □

Now suppose that k is a positive even integer and that $n = 2k + 1$. If $f_1, \ldots, f_k \in I_{n,k} \cap \{f: f \text{ has compact support in } R^n\}$, then the numbers

$$\Psi_\rho(f_1, \ldots, f_k) = \int \operatorname{Det} \hat{f}_m(\tau_{\rho,j} x) d\mu(x)/(\rho - 1)^k \tag{11}$$

converge to

$$\Psi(f_1, \ldots, f_k) = \int \operatorname{Det}(\partial \hat{f}_m/\partial r_j) d\mu \tag{12}$$

as $\rho \to 1$. Since such f's are dense in $I_{n,k}$ by 11.2.6, and since $\{\Psi_\rho\}$ is a uniformly bounded set of k-linear functionals on $I_{n,k}$, Ψ extends to a bounded k-linear functional on $I_{n,k}$.

11.2.9. Lemma. *Let k be a positive even integer and $n = 2k + 1$. Then Ψ, as defined above, satisfies the hypotheses of Lemma* 11.2.4.

Proof. First suppose that for $1 \leq j \leq k$, $f_j = \|x_j\| - 1$ in a neighborhood of $E_{n,k}$. Then $f_j \in I_{n,k}$ and $\Psi(f_1, \ldots, f_k) = 1$, so Ψ is not the zero functional.

Now suppose that $g_1, \ldots, g_k \in I_{n,k}$ are not distinct and that $h_1, \ldots, h_k \in L^1(R^{nk})$. By the boundedness of Ψ, and 11.2.6, we may assume that the functions g_j and h_j have compact support. Then

$$\Psi(h_1 g_1, \ldots, h_k g_k) = \int \operatorname{Det}\left[\hat{h}_m \frac{\partial \hat{g}_m}{\partial r_j} + \hat{g}_m \frac{\partial \hat{h}_m}{\partial r_j}\right] d\mu.$$

Since $\{g_j\} \subseteq I_{n,k}$, the second terms $\hat{g}_m(\partial \hat{h}_m/\partial r_j)$ vanish on support of μ. Since $g_l = g_m$ for some $l \neq m$, two rows of the remaining determinant are multiples of one another. Therefore, for each $x \in E_{n,k}$, $\operatorname{Det}[\partial(\hat{h}_m \hat{g}_m)/\partial r_j(x)] = 0$. Therefore (1) holds, and the Lemma is proved. □

11.2.10. Lemma. *Let n and k be positive integers. Then there are k functions that generate $I_{n,k}$.*

Proof. For $1 \leq j \leq k$, let $\hat{f}_j$ be a function in $\hat{I}_{n,k}$ that agrees with $r_j - 1$ in a neighborhood of $E_{n,k}$. It will suffice to show that the k functions f_j generate $I_{n,k}$. The reader will see from the proof of 11.2.6 that it will suffice to show that each k-radial function in $I_{n,k}$ belongs to the ideal generated by the f_j.

By 11.2.6, it will suffice to show that each k-radial function $f \in I_{n,k}$ of compact support belongs to the ideal generated by the f_j. We claim that $\hat{f}$, considered as a function of the positive variables $r_1, \ldots, r_k$, is an analytic function in those variables. Indeed, by 11.2.5, $\hat{f}$ is the uniform limit of (Riemann) sums of analytic functions, since the Bessel functions, and hence $J(m; r, s)$, are analytic in r and s. The claim follows.

Since $\hat{f}(1, \ldots, 1) = 0$, there exists a linear combination $\hat{g} = \sum c_j \hat{f}_j$ such that $\hat{f}/\hat{g}$ is analytic in a neighborhood of $(1, \ldots, 1)$. For a proof see, for example, Gunning and Rossi [1, I.B.2]. Therefore $\hat{f}/\hat{g} = \hat{h}$ in a neighborhood of $(1, \ldots, 1)$ where $\hat{h}$ is a k-radial function in $A(R^{nk})$. Hence $\hat{f} = \hat{g}\hat{h}$ belongs to $I_{n,k}$ locally in a neighborhood of $E_{n,k}$ and therefore $\hat{f} \in I_{n,k}$. That ends the proof of the Lemma. □

11.2.11. Lemma. *Let $k \geq 2$. Then $V(D, D)$ contains a closed ideal that has exactly k generators.*

Proof. It will suffice to prove the Lemma when k is an even integer. Let $n = 2k + 1$, and $G = T^{kn}$. We set $E = \{(x_1, \ldots, x_k) \in (T^n)^k : \|x_j\| = 1$ for $1 \leq j \leq k\}$. Here $\|x_j\| = (\sum x_{jp}^2)^{1/2}$ and each factor T is identified with $[-\pi, \pi)$. Only behavior of functions in a neighborhood of E will be relevant. The reader will see that we can transfer to the present setting the functions r_j, $\tau_{\rho,j}$ and the functionals Ψ_ρ and Ψ. We leave those details to the reader.

We define a function $u: D \to G$ as follows (compare the proof of 11.2.1). We write D as $D = D \times \cdots \times D$ (k times) and let $u = u_1 \times \cdots \times u_k$, where each u_j is a continuous mapping of D onto T^n that is onto and one-to-one except for a countable subset of D. Let U_j be a Borel inverse to u_j, and let $U = U_1 \times \cdots \times U_k$. Then $U(u(x)) = x$ and $u(U(y)) = y$ except for a null set of x's in D and a null set of y's in G. Let $\varphi: V(G, G) \to V(D, D)$ and $\Phi: V(D, D) \to L^\infty(G^2)$ be given by $\varphi f(x, y) = f(u(x), u(y))$ and $\Phi f(x, y) = f(U(x), U(y))$. Then φ and Φ are both norm decreasing. Let $M: A(G) \to L^\infty(G) \hat{\otimes} L^\infty(G)$ be given by $Mf(x, y) = f(x + y)$ and $P: L^\infty(G) \hat{\otimes} L^\infty(G) \to A(G)$ by $Pf(x) = \int f(x - y, y)dy$. Then $P(gMf) = fPg$ for all $g \in L^\infty(G) \hat{\otimes} L^\infty(G)$ and $f \in A(G)$. (It is trivial that P maps $L^\infty \hat{\otimes} L^\infty$ to A.) Let $f_1, \ldots, f_k \in L^1(Z^{nk})$ have transforms that agree with $r_1 - 1, \ldots, r_k - 1$ in a neighborhood of E. Let I' be the ideal in $V(D, D)$ that is generated by the k functions φMf_j. It will suffice to show that I' is not generated by $k - 1$ functions. We shall "lift" the functionals Ψ_ρ and Ψ from I to I' as follows. For $\rho \in [\frac{1}{2}, 1)$, and $g_1, \ldots, g_k \in I'$, let

$$\Psi'_\rho(g_1, \ldots, g_k) = (\rho - 1)^{-k} \int_E \int_G \mathrm{Det}(\Phi g_m(\tau_{\rho,j}x - y, y))dy\, d\mu(x).$$

That $\{\Psi'_\rho\}$ forms a set of uniformly bounded k-linear functionals is easily proved by mimicking the proof for $\{\Psi_\rho\}$. Here are the details.

Let $g = g_1 \cdots g_k$, where the $g_j \in I'$. We claim that

$$(13) \qquad \left| \int\int \Phi g(\tau_{\rho,j}x - y, y)dy\, d\mu \right| \leq C_1(\rho - 1)^k \|g\|_V,$$

where C_1 is independent of g and ρ. Indeed

$$\int\int \Phi g(\tau_{\rho,j}x - y, y)dy\, d\mu = \int P\Phi g(\tau_{\rho,j}x)d\mu.$$

We may assume that the g_j are finite sums of the form $\sum h_l M f_l$. But then it is obvious that $P\Phi g$ belongs to the ideal generated by the set of all products of k elements of I. By 11.2.7,

$$\left| \int P\Phi g(\tau_{\rho,j}x)d\mu(x) \right| \leq C_1 \|P\Phi g\|_A (1 - \rho)^k$$

for $1 \leq j \leq k$ and $\rho \in [\frac{1}{2}, 1)$. Since $\|P\| = \|\Phi\| = 1$, (13) follows. The boundedness of the Ψ'_ρ follows now exactly as in 11.2.8.

To show that the Ψ'_ρ converge to a functional Ψ', it will suffice to show that $\Psi'_\rho(h_1 \varphi M f_{j(1)}, \ldots, h_k \varphi M f_{j(k)})$ converges for all $h_1, \ldots, h_k \in V(D, D)$ and all choices $j(1), \ldots, j(k)$. But $\varphi M f_{j(l)}(\tau_{\rho,j}x - y, y)$ equals $\rho - 1$ if $j(l) = j$ and 0 otherwise. Therefore $\Psi'_\rho(h_1 M f_{j(1)}, \ldots, h_k M f_{j(k)})$ converges either to $\pm\int (\Pi_1^k P h_m)d\mu$ or to zero, the sign in the first case being the sign of the permutation $(1, \ldots, k) \mapsto (j(1), \ldots, j(k))$.

Suppose that $\{h_1, \ldots, h_k\} \subseteq V(D, D)$ and that $\{g_1, \ldots, g_k\}$ is a set of not distinct elements of I. We may assume that $g_1 = g_2$ and that $g_m = \sum_1^p a_i^{(m)} M f_i$, where $a_i^{(m)} \in V(D, D)$ for all i and m. (That uses the continuity of Ψ'.) Then

$$\begin{aligned} &\Psi'_\rho(h_1 g_1, \ldots, h_k g_k) \\ &\quad = (\rho - 1)^{-k} \int_E \int_G \operatorname*{Det}_{m,j} \left[\Phi\left(\sum_i a_i^{(m)} M f_i h_m \right)(\tau_{\rho,j}x - y, y) \right] dy\, d\mu \\ &\quad = \int_E \int_G \operatorname*{Det}_{m,j} [\Phi a_j^{(m)}(\tau_{\rho,j}x - y, y) h_m(\tau_{\rho,j}x - y, y)] dy\, d\mu. \end{aligned}$$

Therefore

$$(14) \quad \Psi'(h_1 g_1, \ldots, h_k g_k) = \int_E \int_G \operatorname*{Det}_{m,j} [\Phi a_j^{(m)}(x - y, y) h_m(x - y, y)] dy\, d\mu.$$

Since $a_j^{(1)} = a_j^{(2)}$ for all j, two rows of the determinant in (14) are multiples of each other. Therefore $\Psi'(g_1 h_1, \ldots, g_k h_k) = 0$ and (1) holds. An application of 11.2.4 completes the proof of 11.2.11. □

Remarks and Credits. In proving 11.2.1 we have followed the exposition of Kahane [7]; the transfer technique is due to Varopoulos [5, 6, 7, 10]. Theorem 11.2.1 can be proved by showing that synthesis fails for $A(D)$ (instead of $A(R^3)$); see Katznelson [1, Chapter 9]. Chapter 3 gives other proofs of the failure of synthesis.

Theorem 11.2.3 is due to Atzmon [1, 4], and we have followed him closely. Lust-Piquard [2] shows that the set $\{(x_1, \ldots, x_{2n}): \sum x_j^2 = 1\}$ is a set of synthesis and a strong Ditkin set for $V_0(R^n, R^n)$. Varopoulos [8] shows that the ideals $I = I(S^{(n-1)}), I^2, \ldots, I^k$ are all distinct for $k \leq \frac{1}{2}(n-1) + 1$; see also Schoenberg [2]. Gatesoupe [1] has shown that in the complement of any neighborhood of the origin, the following two algebras are isomorphic.

$$\{\hat{f}: f \in L^1(R^n), f \text{ radial}\}$$

and

$$\left\{\hat{f}: f \in L^1(R^n), \int |f(x)|(1 + \|x\|)^{(n-1)/2} < \infty\right\};$$

for related results, see Lust-Piquard [2].

A subset $E \subseteq X \times Y$ is *nontriangular* if $x_1, x_2 \in X$ and $y_1, y_2 \in Y$ imply $\#[E \cap (\{x_1, x_2\} \times \{y_1, y_2\})] \neq 3$. Drury [3] shows that if E is a nontriangular closed subset of $X \times Y$, then $I(E)$ has a bounded approximate identity $\{u_\alpha\}$ where each u_α is zero in a neighborhood of E. (If E is an infinite subset of an abelian group G, then $E^* = \{(x, y): x + y \in E\}$ is not nontriangular.)

11.3. Sets of Analyticity and Tensor Algebras

Let B be an algebra of functions on a set X. If $U \subseteq \mathbb{C}$, and if $F: U \to \mathbb{C}$ is such that $F \circ f \in B$ for all $f \in B$ such that $f(X) \subseteq U$, then *F operates in B.* If B is symmetric on X and the only functions $F: [-1, 1] \to \mathbb{C}$ that operate in B are those that have extensions to $\mathbb{C}$ that agree with an analytic function in a neighborhood of 0, then B is *analytic.*

In this section we show that $V(D, D)$ is analytic, and then prove via our transfer methods that $A(G)$ is analytic for any infinite abelian group G. The proof in this section, unlike the one in Section 9.3, does not use the structure theorem. We do use 9.3.6 and the fact that if F operates in $A(E)$, then there exist

(1) $$\varepsilon > 0, K > 0 \text{ such that } \|F \circ f\| < K \text{ if } \|f\| < \varepsilon.$$

That follows from the proof of 9.3.5.

The last two results of the Section, 11.3.5 and 11.3.6, deal with a more special problem: if $F \circ \hat{\mu} \in B(\Gamma)$ for one special μ, what can we conclude about F? For those proofs, the reader will need a few facts established in Section 9.6.

For a further discussion of functions that operate, see Chapter 9, Chapter 10, and Section 13.1.

11.3.1. Lemma. (i) *Let G be a compact abelian group and $F\colon [-1, 1] \to \mathbb{C}$. Suppose that for some $\varepsilon > 0$ and $K > 0$, $\|F \circ f\| \leq K$ whenever $f \in A(G)$ is real-valued and $\|f\| < \varepsilon$. Then F agrees with an analytic function in a neighborhood of zero.*

(ii) *Suppose that $F\colon[-1, 1] \to \mathbb{C}$ is such that for every neighborhood U of 0 in $\mathbb{C}$ and all functions g that are analytic in U, F does not agree with g in $U \cap [-1, 1]$. Then for all $\varepsilon > 0$, $K > 0$, there exists an integer $m = m(\varepsilon, K, F)$ such that whenever X and Y are discrete spaces of cardinality at least m, then there exists $f \in V(X, Y)$ such that f is real-valued, $\|f\|_V < \varepsilon$, and $\|F \circ f\|_V \geq K$.*

Proof. (i) Let $0 < \alpha < \varepsilon e^{-2\pi-1}$. Then for all real-valued $f \in A(G)$ of norm at most one and all $t \in [0, 2\pi)$, $\|F(\alpha \sin(f(\cdot) + t))\| \leq K$. Let $F_1(t) = F(\alpha \sin t)$. Then for all $y \in G$, $n \in Z$, and all such $f \in A(G)$

$$\hat{F}_1(n)\exp(\mathrm{i}nf(y)) = \int_{-\pi}^{\pi} F_1(f(y) + t)\exp(-\mathrm{i}nt)dt/2\pi.$$

Therefore $|\hat{F}_1(n)|\,\|\exp(\mathrm{i}nf)\| \leq K$ whenever f is real-valued and $\|f\| \leq 1$. By 9.3.6, $\sup\{\|\exp(\mathrm{i}nf)\| : f$ is real-valued, $\|f\| \leq 1\} = e^{|n|}$, and therefore $|F_1(n)| \leq Ke^{-|n|}$. Therefore F_1 is analytic and F is analytic in a neighborhood of 0.

(ii) Let $\varepsilon > 0$ and $K > 0$. By part (i) applied to $G = D$, there exists $f_1 \in A(D)$ such that f_1 is real-valued, $\|f_1\|_A < \varepsilon$, and $\|F \circ f_1\|_A > K$. Therefore $Mf_1 \in V(D, D)$ is real-valued, $\|Mf_1\|_V < \varepsilon$, and $\|F \circ Mf_1\|_V = \|M(F \circ f_1)\|_V > K$. By 11.1.5, there exist finite sets $X_1 \subseteq D$, $X_2 \subseteq D$ such that $\|F \circ Mf_1|_{X_1 \times X_2}\|_{V(X_1, X_2)} > K$. It is now clear that the integer $m = \max(\# X_1, \# X_2)$ has the required property. That ends the proof of the Lemma. □

11.3.2. Lemma. (i) *Let X and Y be subsets of the abelian group G and let $m \leq \min(\# X, \# Y)$. Then there exist subsets $X' \subseteq X$, $Y' \subseteq Y$ and elements $x_0 \in X \cup \{0\}$, $y_0 \in Y \cup \{0\}$, such that $(x_0 + X') \cup (y_0 + Y')$ is dissociate and*

$$\# X' = \# Y' \geq ((\log m) - 6)/4 \log 5. \tag{2}$$

(ii) *Let X and Y be disjoint finite subsets of the abelian group G such that $X \cup Y$ is dissociate. Then $A(X + Y)$ is isomorphic to $V(X, Y)$, the norm of the* *natural mapping from $V(X, Y)$ to $A(X + Y)$ is at most four, and the norm of the natural mapping from $A(X + Y)$ to $V(X, Y)$ is one.*

Proof. (i) If $G = Z$, (i) is easily proved. The difficulty arises (for example) when $G = D$. For the general case we proceed as follows. If $m \le 64$, then (2) holds with $\#X = \#Y = 0$. We may thus assume that $m \ge 64$ and $m/2 \ge m^{1/2}$.

If $\#\{2x: x \in X\} \ge m^{1/2}$, we let $x_0 = 0$ and let $X_0 \subseteq X$ be such that $\#X_0 = \#\{2x: x \in X_0\} \ge m^{1/2}$. If $\#\{2x: x \in X\} < m^{1/2}$, then there exists $x_0 \in X$ such that $X_0 = \{x \in X: 2x = 2x_0\}$ has cardinality at least $m^{1/2}$. Y_0 and y_0 are chosen similarly. We set $X_1 = X_0 - x_0$ and $Y_1 = Y_0 - y_0$. Then either $\#\{2x: x \in X_1\} = \#X_1$ or $2x = 0$ for all $x \in X_1$, and similarly for Y_1.

Let $x_1 \in X_1\backslash\{0\}$ be arbitrary. If $2x_1 = 0$, then there exists $y_1 \in Y_1\backslash\{0\}$ such that $\pm y_1 \ne x_1$ and $\pm 2y_1 \ne x_1$ unless Card $Y_1 \le 4$. If $2x_1 \ne 0$, there exists $y_1 \in Y_1\backslash\{0\}$ such that

$$\{y_1, 2y_1\} \cap \{x_1, 2x_1, -x_1, -2x_1\} = \varnothing$$

unless $\#Y_1 \le 8$. In either case, $m^{1/2} > 8$ implies there exist $x_1 \in X_1$, $y_1 \in Y_1$ such that $\{x_1, y_1\}$ is dissociate. That begins the induction.

We now suppose that $k \ge 1$, $x_1, \ldots, x_k \in X_1$ and $y_1, \ldots, y_k \in Y_1$ have been found such that $\{x_j\} \cup \{y_j\}$ is dissociate. Suppose that

$$k + 1 \le ((\log m) - 6)/4 \log 5.$$

Then

$$e^2 5^{2k+3} \le m^{1/2} \le m = \min(\#X_1, \#X_2).$$

Let

$$F = \left\{\sum_{j=1}^{k} \varepsilon_j x_j + \delta_j y_j: \varepsilon_j = \pm 2, \pm 1, 0; \delta_j = \pm 2, \pm 1, 0\right\}.$$

Then $\#F \le 5^{2k}$.

Suppose that $2X_1 = \{0\}$. Let $x_{k+1} \in X_1\backslash F$. Such an x_{k+1} exists since $\#X_1 > \#F$. Then $\{x_j\}_1^{k+1} \cup \{y_j\}_1^k$ is dissociate. Suppose that $2X_1 \ne \{0\}$. Because $\#X_1 > 2\#F$, there exists $x_{k+1} \in X_1$ such that $\{x_{k+1}, 2x_{k+1}\} \cap F = \varnothing$. Again $\{x_j\}_1^{k+1} \cup \{y_j\}_1^k$ is dissociate. Let

$$F' = F \cup (F + x_{k+1}) \cup (F - x_{k+1}) \cup (F + 2x_{k+1}) \cup (F - 2x_{k+1}).$$

Then $\#F' \le 5^{2k+3}$. Similarly, there exists $y_{k+1} \in Y_1$ such that $\{x_j\}_1^{k+1} \cup \{y_j\}_1^{k+1}$ is dissociate. That ends the induction. Let $X' = \{x_j\} + x_0$ and $Y' = \{y_j\} + y_0$. Then X' and Y' have the required properties.

(ii) We may assume that G is discrete and generated by $X \cup Y$. Let X' and Y' be disjoint subsets of the discrete abelian group G' such that $X' \cup Y'$ is Kronecker, $\# X' = X$, $\# Y' = Y$, and $X' \cup Y'$ generates G'. Then any (fixed) map $u: X' \cup Y' \to X \cup Y$ that is one-to-one with $u(X') = X$ induces a group homeomorphism from G' to G. Therefore $A(X + Y) \to A(X' + Y')$ is norm reducing. By 11.1.2, $A(X' + Y')$ is isometrically isomorphic to $V(X', Y')$, which is of course isometrically isomorphic to $V(X, Y)$. To obtain the mapping the other way, we use Riesz products, as follows. Let $g \in C(X)$, $h \in C(Y)$. Since $X \cup Y$ is dissociate, there exists a measure μ on the (compact) dual group of G such that $\mu \geq 0$, $\|\mu\| \leq 4\|g\|_\infty \|h\|_\infty$, and $\hat{\mu}(x + y) = g(x)h(y)$ for all $x \in X$, $y \in Y$. It follows at once that if $f \in V(X, Y)$, then $f'(x + y) = f(x, y)$ has $\|f'\|_{A(X+Y)} \leq 4\|f\|_V$. That ends the proof of 11.3.2.

□

11.3.3. Theorem. (i) *Let X and Y be infinite locally compact Hausdorff spaces. Then $V_0(X, Y) = C_0(X) \hat{\otimes} C_0(Y)$ is analytic.*

(ii) *Let E_1 and E_2 be infinite subsets of the locally compact abelian group G, and $E = (E_1 + E_2)^-$. Then $A(E)$ is analytic.*

Proof. (i) Suppose that for all functions g analytic in a neighborhood U of 0, F and g do not agree on $U \cap [-1, 1]$. For each $k = 1, 2, \ldots$ let $m_k = m(2^{-k}, k, F)$ be given by 11.3.1 (ii), and let X_k and Y_k be open subsets of X and Y such that $\# X_k$, $\# Y_k \geq m_k$ and such that the X_k are disjoint and the Y_k are disjoint. Let $h_k \in V(X, Y)$ be a real-valued function such that $\|h_k\|_V < 2^{-k}$, $\|F \circ h_k\|_V > k$, and $\operatorname{supp} h_k \subseteq X_k \times Y_k$. Such an h_k is easily obtained by extension from an $m_k \times m_k$ "square." Then $h_k \in V_0(X, Y)$ and $\sum_{k=1}^\infty h_k \in V_0(X, Y)$. Of course, since the h_k have disjoint supports $X_k \times Y_k$,

$$F \circ (\textstyle\sum h_k)|_{X_j \times Y_j} = F \circ h_j.$$

Since

$$\|F \circ (\textstyle\sum h_k)\|_{V_0(X,Y)} \geq \|F \circ (\sum h_k)_{|X_j \times Y_j}\|_{V(X_j, Y_j)}$$

by 11.1.4, we see that $\|F \circ (\sum h_k)\|_{V_0(X,Y)} \geq k$ for all k. Therefore F does not operate in $V_0(X, Y)$. That contradiction proves (i).

(ii) Let $F: [-1, 1] \to \mathbb{C}$ operate in $A(E)$. If F did not agree with an analytic function in a neighborhood of 0, then for each $\varepsilon > 0$, $K > 0$, there would exist m such that $\|F \circ f\|_V > K$ for some real-valued $f \in V(X_1, X_2)$ with $\|f\|_V \leq \varepsilon$ whenever Card X_1 = Card $X_2 \geq m$. By 11.3.2(i), there exist finite subsets $X_j \subseteq E_j$ and $x_j \in E_j$ for $j = 1, 2$, such that Card $X_j \geq m$ for $j = 1, 2$ and $(X_1 + x_1) \cup (X_2 + x_2)$ is dissociate. By 11.3.3(ii), $A(X_1 + X_2 + x_1 + x_2)$ is isomorphic to $V(X_1, X_2)$. It follows that for each $\varepsilon > 0$, $K > 0$, there exists a real-valued $f \in A(E)$ such that $\|f\| < \varepsilon$ and $\|F \circ f\| > K$. Now (1) shows that F cannot operate. That ends the proof of 11.3.3. □

The proof of the next Corollary is left to the reader.

11.3.4. Corollary. *Let G be a locally compact abelian group and let $E \subseteq G$ be a closed subset such that for all n, E contains a subset of the form $X + Y$ where $\# X = \# Y \geq n$. Then $A(E)$ is analytic.*

Remarks. (i) If X and Y in 11.3.3(i) are non-discrete, then $F: [-1, 1] \to \mathbb{C}$ operates in $V(X, Y)$ if and only if F extends to a function analytic in a neighborhood of $[-1, 1]$. The proof uses 11.3.3(i) combined with the argument given in 9.3 for the corresponding assertion for $A(G)$.

(ii) If X and Y are discrete and infinite, then $F: [-1, 1] \to \mathbb{C}$ operates in $V(X, Y)$ if and only if $F(0) = 0$ and F is analytic in a neighborhood of 0.

(iii) 11.3.1(ii) has a version for $A(E)$: given F, ε, K as in the statement of 11.3.1(ii), there exists m such that if E contains a coset of a finite subgroup, or an arithmetic progression of cardinality at least m, then there exists $f \in A(E)$ with $\|f\| < \varepsilon$ and $\|F \circ f\|_{A(E)} > K$. The proof is similar to the proof of 11.3.1(ii).

Let $U = \{z \in \mathbb{C}: |z| \leq 1\}$, and let U_p be the closed convex hull of the pth roots of unity, for $2 \leq p < \infty$.

11.3.5. Theorem. *Let G be a locally compact abelian group with dual group Γ. Let μ be a continuous probability measure on a Kronecker subset E (or K_p-set, $p > 2$). Let F be a continuous function on the closed unit disc U^- (or U_p^-). If $F \circ \hat{\mu} \in B(\Gamma)$, then*

(3) $$F(z) = \sum_{n, m = 0}^{\infty} a_{n,m} z^n \bar{z}^m, \quad \text{for all } z \in U \text{ (or } z \in U_p\text{)},$$

where

$$\sum |a_{m,n}| < \infty.$$

Proof. In view of the method of proof of Theorem 9.6.1, and the continuity of F, it is sufficient to show that F agrees with a real-analytic function in a neighborhood of every z in the open unit disc (or Int U_p).

Let $|z| < 1$ (or $z \in \text{Int } U_p$). Then there exists a Borel subset E_0 of E such that $|z| = \mu(E_0)$. Let $\varepsilon = 1 - \mu(E_0)$.

We now assume that F does not agree with any real-analytic function in any neighborhood of z. Let $C > 0$. By the obvious extension of Lemma

11.3.1(ii), there exists an integer $m \geq 1$ such that if X, Y are discrete finite sets with m elements each, then there exists $h \in V = V(X, Y)$ such that

$$\|h\|_V < \varepsilon/4 \text{ and } \|F \circ h\|_V > C. \tag{4}$$

The function h has a representation $h = \sum_1^\infty g_j \otimes h_j$ with $\sum \|g_j\|_\infty \|h_j\|_\infty < \varepsilon/4$. From the continuity of F and the finite dimensionality of V, we see that we may assume that

$$h = \sum_1^m g_j \otimes h_j, \sum_1^m \|g_j\|_\infty \|h_j\|_\infty < \varepsilon/4.$$

It is a straightforward computation to show that h has another representation

$$h = \sum_1^{m'} r_j g'_j \otimes h'_j, \quad \text{with } r_j \geq 0 \ (1 \leq j \leq m') \text{ and } \sum r_j < \varepsilon, \tag{5}$$

where g'_j, h'_j have ranges in T (or Z_p). We will now suppress the primes in (5).

We divide $E \backslash E_0$ into m Borel sets $E_1, \ldots, E_m$ such that

$$\mu(E_j) = r_j \quad \text{and} \quad E_i \cap E_j = 0, 1 \leq i \neq j \leq m.$$

Now we enumerate the elements of $X(Y)$ as $x_1, \ldots, x_m$ $(y_1, \ldots, y_m)$. Let $\delta > 0$. Since K is a Kronecker set (or K_p-set) and since the g_j, h_j take values in T (or Z_p), there exists $\gamma_1, \ldots, \gamma_n, \rho_1, \ldots, \rho_n \in \Gamma$ such that for $1 \leq i, k \leq n$, $1 \leq j \leq m$,

$$\mu\{w: |(\gamma_i, w) - g_j(x_i)| \geq \delta, w \in E_j\}) < \delta, \tag{6}$$

$$\mu(\{w: |(\rho_k, w) - h_j(y_k)| \geq \delta, w \in E_j\}) < \delta, \tag{7}$$

$$\mu(w: |(\gamma_i, w) - 1| \geq \delta, w \in E_0\}) < \delta, \tag{8}$$

$$\mu(w: |(\rho_k, w) - z| \geq \delta, w \in E_0\} < \delta, \tag{9}$$

and

$$\left| \int_{K \backslash \cup_0^m E_j} (z, \gamma_i + \rho_k) d\mu \right| < \delta. \tag{10}$$

Conditions (6)–(9) are easily obtained because they are independent. To obtain (10), choose the γ's first and then "adjust" the ρ's on $K \backslash \bigcup_0^m E_j$ so

(10) holds. (For example, we may assume the γ's are almost one on $K\setminus\bigcup_0^m E_j$.) Also, we may assume that

$$\#\{\gamma_i + \rho_k: 1 \le i, k \le m\} = m^2.$$

Set $X' = \{\gamma_i: 1 \le i \le n\}$ and $Y' = \{\rho_k: 1 \le k \le n\}$. Let φ be the map from $X \times Y$ to $X' + Y'$ given by $\varphi(x_i, y_k) = \gamma_i + \rho_k$ for $1 \le i, k \le m$, and consider

$$\check{\varphi}: B(X' + Y', \Gamma) \to V(X, Y),$$

given by $\check{\varphi}f = f \circ \varphi$.

The estimates (6)–(10) imply that

$$\|\check{\varphi}\hat{\mu}|_{X'+Y'} - (z1 + h)\|_V = o(\delta) \text{ as } \delta \to 0. \tag{11}$$

Since $F \circ (\check{\varphi}\hat{\mu}|_{X'+Y'}) = \check{\varphi}((F \circ \hat{\mu})|_{X'+Y'})$, the continuity of F, (4) and (11) imply (for small $\delta > 0$) that

$$\|\check{\varphi}(F \circ \hat{\mu})|_{X'+Y'}\|_V > C.$$

As in the proof of 11.3.2 (ii), $\check{\varphi}$ reduces norm, so

$$\|F \circ \hat{\mu}|_{X'+Y'}\|_{A(X'+Y')} \ge \|\check{\varphi}(F \circ \hat{\mu})|_{X'+Y'}\|_V.$$

Thus, $\|F \circ \hat{\mu}|_{X'+Y'}\|_{A(X'+Y')} \ge C$. Since C was arbitrary, F cannot operate. That contradiction proves the Theorem. □

We close this section by pointing out that a modification of these techniques yields the following result. The proof is left to the reader.

11.3.6. Theorem. *Let μ be a continuous Hermitian probability measure on $E \cup -E$ where E is a Kronecker set (or K_p-set, $p \ge 2$). Let F be a continuous function on $[-1, 1]$ (or $D_p \cap [-1, 1]$) and suppose that $F \circ \mu \in B(\Gamma)$. Then F extends to a function G on U (or U_p) where*

$$G(z) = \sum_{n=0}^{\infty} a_n z^n$$

with $\sum_{n=0}^{\infty} |a_n| < \infty$.

Remarks and Credits. Theorems 11.3.1–11.3.4 are essentially due to Varopoulos [6]. The proof of 11.3.4 is a modification of Salinger and Varopoulos [1], who also show that if G is a metrizable abelian group, $\mu, \nu \in M_c(G)$, and $\mu * \nu(E) \ne 0$, then E satisfies the hypothesis of 11.3.4. The metrizability requirement for that last result is not necessary, as Saeki [16] shows.

The log m in 11.3.2(i) is to be expected; compare the "Kns" Theorem 11.6.5.

For a tensor algebra version of Katznelson and Malliavin's [1, 2] statistical verification of the dichotomy conjecture, see Salinger [1].

Theorems 11.3.5 and 11.3.6 are due to Moran [3]; the proof here is new; for Moran's proof, see Section 9.6.

11.4. Infinite Tensor Products and the Saucer Principle

Let $X_1, X_2, \ldots$ be compact Hausdorff spaces, and let $X = \prod_{j=1}^{\infty} X_j$. The *infinite tensor algebra* $V = V(X_1, X_2, \ldots)$ *on* X is the set of all functions of the form

(1) $$f = \sum_{j=1}^{\infty} f_{j,1} \otimes \cdots \otimes f_{j,n(j)},$$

where $f_{j,k} \in C(X_k)$ $1 \leq n(j) < \infty$, for all j and k, and

(2) $$\sum_{j=1}^{\infty} \|f_{j,1}\|_\infty \cdots \|f_{j,n(j)}\|_\infty < \infty.$$

The norm of f is, of course, the infimum of the numbers (2) subject to (1). The case that each X_j has cardinality two is most important, and we shall denote the resulting V by "V_2." Before stating our main theorem, we establish the "saucer principle."

11.4.1. The Saucer Principle. *Let $\Lambda \subseteq R$ be a finite arithmetic progression of step t and let $0 < \delta < t/\pi$. Let $\varphi\colon [-\delta, \delta] \times \Lambda \to [-\delta, \delta] + \Lambda$ be defined by $\varphi(x, \lambda) = x + \lambda$ for $|x| \leq \delta$ and $\lambda \in \Lambda$. Then for all $f \in A([-\delta, \delta] + \Lambda)$, $f \circ \varphi \in A([-\delta, \delta] \times \Lambda)$ and*

(3) $$\begin{aligned} \|f \circ \varphi\|_{A([-\delta,\delta]\times\Lambda)} &\leq \|f\|_{A([-\delta,\delta]+\Lambda)} \\ &\leq (1 - \delta^2\pi^2/2t^2)^{-1}\|f \circ \varphi\|_{A([-\delta,\delta]\times\Lambda)}. \end{aligned}$$

Proof. It is obvious that $f \circ \varphi \in A([-\delta, \delta] + \Lambda)$. Now let T be a pseudomeasure supported on $[-\delta, \delta] + \Lambda$. Then

$$\hat{T}(y) = \sum_{\lambda \in \Lambda} \hat{\psi}_\lambda(y) e^{i\lambda y},$$

where the ψ_λ are pseudomeasures supported on $[-\delta, \delta]$. Let

$$\hat{S}(y, z) = \sum_{\lambda \in \Lambda} \hat{\psi}_\lambda(y) e^{i\lambda z}.$$

Then $\hat{S}$ is the transform of a pseudomeasure on $[-\delta, \delta] \times \Lambda$, and $\langle S, f \circ \varphi \rangle = \langle T, f \rangle$ for all $f \in A(\Lambda + [-\delta, \delta])$. Obviously $\|\hat{T}\|_\infty \leq \|S\|_\infty$. That proves the left-hand inequality in (3).

To establish the right-hand inequality in (3), it will suffice to work with pseudofunctions T. Then all the ψ_λ's are pseudofunctions also. Let (y_0, z_0) be a point at which $|\hat{S}|$ attains its supremum. Such a point exists since $\psi_\lambda(y) \to 0$ as $|y| \to \infty$ and $\hat{S}$ is periodic with period t in z. We may assume that $|y_0 - z_0| \leq \pi/t$ (because $\hat{S}$ has period $2\pi/t$ in the z variable). By Bernstein's inequality A.4.1

$$|\partial\hat{S}/\partial y| \leq \delta\|\hat{S}\|_\infty, \quad \text{and} \quad |\partial^2\hat{S}/\partial y^2| \leq \delta^2\|\hat{S}\|_\infty.$$

Therefore

$$\hat{S}(y, z_0) = \hat{S}(y_0, z_0) + A(y - y_0) + B(y - y_0)^2/2, \tag{4}$$

where $|B| \leq \delta^2\|\hat{S}\|_\infty$. ($B$ may vary with y, but that is of no moment here.)

Since $|\hat{S}(y_0, z_0)| = \|\hat{S}\|_\infty$, A must be of the form $A = i\alpha\hat{S}(y_0, z_0)$ where α is real. Therefore

$$|\hat{S}(y, z_0)| \geq |\hat{S}(y_0, z_0)| - \delta^2\|\hat{S}\|_\infty |y - y_0|^2/2.$$

Since $|y_0 - z_0| \leq \pi/t$,

$$|\hat{S}(y_0, y_0)| \geq |\hat{S}(y_0, z_0)| - (\delta^2\pi^2/2t^2)\|\hat{S}\|_\infty = (1 - \delta^2\pi^2/2t^2)\|\hat{S}\|_\infty.$$

Since $\|\hat{S}\|_\infty \geq |\hat{S}(y_0, z_0)| = |\hat{T}(y_0)|$, we have the right-hand inequality of (3). That proves 11.4.1. □

Remark. It is easy to establish the following generalization of 11.4.1. Let $m \geq 1$, $X \subseteq R^m$ be compact, and Λ, δ, t be as in the statement of 11.4.1 and $E \subseteq [-\delta, \delta]$. Then $A(X \times \Lambda \times E)$ and $A(X \times (\Lambda + E))$ are isomorphic, and the following holds, where $\varphi: X \times \Lambda \times E \to X \times (\Lambda + E)$ is given by $\varphi(x, \lambda, \theta) = (x, \lambda + e)$ for all x, λ, and e.

$$\|f \circ \varphi\|_{A(X \times \Lambda \times E)} \leq \|f\|_{A(X \times (\Lambda + E))} \leq (1 - \delta^2\pi^2/2t^2)^{-1}\|f \circ \varphi\|_{A(X \times \Lambda \times E)}, \tag{5}$$

for all $f \in A(X \times (\Lambda + E))$.

11.4.2. Theorem. *Let $\{t_j\}_{j=1}^\infty$ be a decreasing sequence of positive real numbers such that $\sum (t_{j+1}/t_j)^2 < \infty$ and $t_j > \sum_{k=j+1}^\infty t_k$ for all j. Let $E = \{\sum_1^\infty \varepsilon_j t_j : \varepsilon_j = 0$ or 1 for $1 \leq j < \infty\}$. Then $A(E)$ is isomorphic to V_2.*

Proof. Let $\{t'_j\}$ be a decreasing sequence of positive numbers such that $t'_j = t_j$ for j sufficiently large and $t'_j \geq t_j$ for all j. Let $E' = \{\sum \varepsilon_j t'_j : \varepsilon_j = 0$ or 1, for $1 \le j < \infty\}$. It is easy to see that $A(E') = A(E)$. We may therefore assume that $t_{j+1}/t_j \le 1/2$ for all j. Then for all j, $\sum_{k=j+1}^{\infty} t_k/t_j \le t_1/2^{-j}$. Let $b_j = \pi \sum_{k=j+1}^{\infty} t_k$, so that $\sum b_j^2/t_j^2 < \infty$. We may assume that $b_j^2/t_j^2 < 2$ for all j. (As in the assumption that $t_{j+1}/t_j \le 1/2$, that requires merely increasing the first few t_j's.)

We shall define sets and functions as follows. Here m may be $1, 2, \ldots$. $E_m = \{\sum_{j=m+1}^{\infty} \varepsilon_j t_j : \varepsilon_j = 0, 1\}$; $X_{m+1} = \prod_{j=1}^{m} \{0, t_j\} \subseteq R^m$; $X_1 = \{0\}$; $\Lambda_m = \{0, t_m\}$; if $f \in A(E)$, $f_0 = f$ and $f_m: X_m \times \Lambda_m \times E_m \to \mathbb{C}$ is defined inductively by $f_m(\varepsilon_1, \ldots, \varepsilon_{m-1}, \varepsilon_m, x) = f_{m-1}(\varepsilon_1, \ldots, \varepsilon_{m-1}, \varepsilon_m + x)$. Then (5) may be applied to yield $\|f_{m+1}\| \le \|f_m\| \le (1 - b_m^2/2t_m^2)^{-1}\|f_{m+1}\|$. Let $\Phi: V_2 \to A(E)$ be defined by $\Phi f(\sum \varepsilon_j t_j) = f(\varepsilon_1, \varepsilon_2, \ldots)$. If $f \in V_2$ depends only on $\varepsilon_1, \ldots, \varepsilon_n$, then $f(\varepsilon_1, \varepsilon_2, \ldots) = f_n(\varepsilon_1, \varepsilon_2, \ldots, \varepsilon_n, x)$ for all $\varepsilon_1, \ldots, \varepsilon_n \in X_n \times \Lambda_n$ and all $x \in E_n$. Therefore $\|f\|_{V_2} = \|(\Phi f)_n\|_{V_2}$ and $\|f\|_{V_2} \le \|\Phi f\|_{A(E)} = \|(\Phi f)_n\| \le \prod_{j=1}^{n} (1 - b_m^2/2t_m^2)^{-1} \|f\|_{V_2}$. Of course, $f \in V_2$ depends only on $\varepsilon_1, \ldots, \varepsilon_n$ if and only if Φf is locally constant on E. Since those sets of functions are dense, respectively, in V_2 and $A(E)$, we see that

$$\|f\|_{V_2} \le \|\Phi f\|_{A(E)} \le \prod_{j=1}^{\infty} (1 - b_m^2/2t_m^2)^{-1} \|f\|_{V_2}$$

for all $f \in V_2$. Since $\sum b_m^2/t_m^2 < \infty$ and $b_m^2/t_m^2 < 2$ for all m, we see that Φ is indeed an isomorphism. That completes the proof of 11.4.2. □

Let $\{t_j\} \subseteq (0, \infty)$ be such that $t_n > \sum_{j=n+1}^{\infty} t_j$ for all $n = 1, 2, \ldots$. Let $E = \{\sum \varepsilon_j t_j : \varepsilon_j = 0, 1\}$. The set E is *ultrathin* if $\sum t_{j+1}^2/t_j^2 < \infty$.

11.4.3. Corollary. *Let E and F be two ultrathin sets. Then $A(E)$ and $A(F)$ are isomorphic.*

Proof. $A(E)$ and $A(F)$ are each isomorphic to V_2. □

Remarks and Credits. For more about isomorphisms between algebras, see Section 13.3. The presentation here is an adaptation of Kahane [2, Chapter IX]. Corollary 11.4.3 is due to Schneider [1], and is more-or-less sharp: if $t_{2j-1}/t_{2j} = 2j$ for all j, then the conclusion of 11.4.2 fails (Schneider [2]). A stronger result has been proved by Y. Meyer [6], as follows.

Let $t_j = \varepsilon_1 \cdots \varepsilon_{2j-1}(1 - \varepsilon_j)$ for all j where $\varepsilon_j > 0$ and $\sum \varepsilon_j^2 < \infty$. Suppose that $\{a_n\}$, $\{b_n\}$ are sequences of positive numbers such that for some $\delta > 0$,

$$n^{2\delta - 1/2} \le a_n \quad \text{and} \quad a_n + (n!)^{-\delta} \le b_n \le 1/2.$$

Let $\{\varepsilon_j^{(0)}\}$ be a sequence as above and E_0 the set $\{\sum \varepsilon_j t_j^{(0)} : \varepsilon_j = 0, 1\}$. Suppose that the ε_j are chosen between a_n and b_n for all n. Call the resulting set $E = E(\{\varepsilon_n\})$. Let $\prod_1^{\infty}[a_n, b_n]$ have the probability measure that is the product of normalized Lebesgue measure on the factors. Then with probability one

$A(E)$ is not isomorphic to $A(E_0)$. See Kahane [2; p. 119ff] for proofs of those and related results.

Meyer has also shown that if E is the set of 11.4.2, then E is a set of uniqueness and of synthesis. See Meyer [1, Chapter VIII] for the proof.

That infinite tensor algebras are nearly as ubiquitous as finite tensor algebras is shown by the following result of Saeki [9, 14].

11.4.3. Theorem. *Let $\{F_j\}$ be a sequence of non-empty finite discrete spaces. Let G be a locally compact abelian I-group. Then there exist finite subsets $E_1, E_2, \ldots$ of G such that:*

(i) *$x_j \in E_j (j = 1, 2, \ldots)$ implies $\sum x_j$ converges.*
(ii) *$\#E_j = \#F_j$ for all j.*
(iii) *$A(\sum_1^\infty E_j)$ is isometrically isomorphic to $V(F_1, \ldots)$.*
(iv) *$\sum_1^\infty E_j$ is a Dirichlet set and a set of synthesis.*

11.5. Continuity Conditions for Membership in $V(T, T)$

Let $0 < \alpha < 1$. We say that a function f on a space X with metric d belongs to $\Lambda_\alpha(X)$ if there is a constant C such that $|f(x) - f(y)| \leq C\, d(x, y)^\alpha$ for all x, y in X. If $0 < \alpha < \beta < 1$, then $\Lambda_\beta(X) \subseteq \Lambda_\alpha(X)$. It is well known that $\Lambda_\alpha(T) \subseteq A(T)$ if and only if $1 \geq \alpha \geq \frac{1}{2}$. For a proof, see Kahane [2, p. 14], Kahane and Salem [1, page 129] or Zygmund [1, Volume I, p. 197 ff]. The "same" result holds for V, as we shall now prove.

11.5.1. Theorem. $\Lambda_\alpha(T^2) \subseteq V(T, T)$ *if and only if* $\alpha > \frac{1}{2}$.

11.5.2. Lemma. *Let d denote the metric $d((x_j), (y_j)) = \sum 2^{-j}(x_j - y_j)$ on $D = \{(x_j): x_j = 0, 1\}$. If $1 \geq \alpha > \frac{1}{2}$ and $f \in \Lambda_\alpha(D \times D)$, then $f \in V(D, D)$.*

Proof. Let $1/2 < \beta < \alpha \leq 1$, and let $f \in \Lambda_\alpha(D \times D)$. Let μ_k denote Haar measure on $H_k = (\{0\}^k \times D) \times (\{0\}^k \times D)$. Then $f \to f_k = \mu_k * f$ defines a map from $\Lambda_\alpha(D \times D)$ to $C(D \times D)$ and the following three estimates hold.

(1) $\|f - f_k\|_\infty \leq 2^{-\beta k}\|f\|_{\Lambda_\beta}$, $\|f_k\|_\infty \leq \|f\|_\infty$, and $\|f_k\|_{\Lambda_\beta} \leq \|f\|_{\Lambda_\beta}$.

Furthermore, since f_k is effectively a function on $(Z_2)^k \times (Z_2)^k$, 11.6.8 below implies that, for some $C > 0$ and all k,

(2) $$\|f_k\|_V \leq C 2^{k/2}\|f_k\|_\infty.$$

Of course, $f = f_1 + \sum_{j=1}^\infty (f_{j+1} - f_j)$, where the convergence is uniform by (1). Since $(f_{j+1} - f_j)_j = (f_{j+1} - f_j)$, formula (2) implies that

$$\|f_{j+1} - f_j\|_V \leq C 2^{(j+1)/2}\|f_{j+1} - f_j\|_\infty.$$

Now (1) implies also that

$$\|f_{j+1} - f_j\|_\infty \leq 2(2^{-\beta j})\|f\|_{\Lambda_\beta} \quad \text{for all } j.$$

Therefore $\|f_{j+1} - f_j\|_V \leq C2^{3/2}2^{j(1/2-\beta)}$. Since $\beta > 1/2$, $\sum \|f_{j+1} - f_j\|_V < \infty$, so $f \in V(D, D)$ if $f \in \Lambda_\alpha(D \times D)$ and $1/2 < \beta < \alpha \leq 1$. The Lemma is proved. □

Proof of Theorem 11.5.1. Let $1/2 < \alpha \leq 1$ and $f \in \Lambda_\alpha(T^2)$. Let φ denote the map from $D \times D$ to T^2 given by $\varphi(\{\varepsilon_j\}, \{\varepsilon_j'\}) = (\sum \varepsilon_j/2^j, \sum \varepsilon_j'/2^j)$. Then $f \circ \varphi \in \Lambda_\alpha(D \times D)$. Therefore $f \circ \varphi \in V(D, D)$, by Lemma 11.5.2. The method of 11.2.1 easily shows that $g \in V(T, T)$ if and only if $g \circ \varphi \in V(D, D)$. Therefore $f \in V(T, T)$. If $\alpha \leq \frac{1}{2}$, let $f \in \Lambda_{1/2}(T)\backslash A(T)$. Then $Mf \in \Lambda_{1/2}(T^2)$ and $Mf \notin V(T, T)$. That ends the proof of the Theorem. □

11.5.3. Corollary. *Let $0 < \alpha \leq 1/2$. Let $\tilde{V}(T, T)$ be as defined in Section 11.9.*

(i) *There exists $f \in \Lambda_\alpha(T^2)$ such that $f \notin \tilde{V}(T, T)$.*
(ii) *There exists $f \in \tilde{V}(T, T) \cap \Lambda_\alpha(T^2)$ such that $f \notin V(T, T)$.*

Proof. (i) It is clear that if $f \in \tilde{V}(T, T)$, then $Pf \in A(T)\tilde{} = A(T)$; in particular, if $g \in \Lambda_\alpha(T)$, $g \notin A(T)$, then $Mg \notin \tilde{V}(T, T)$. Now (i) follows.

(ii). Since $\Lambda_\alpha(T^2) \not\subseteq V(T, T)$ by 11.5.1, there exists (by smoothing elements of $\Lambda_\alpha(T^2)\backslash V(T, T)$) a sequence of functions $\{f_n\}$ such that

(3) $$\text{support } f_n \subseteq (1/n, 1/n - 1) \times (1/n, 1/n - 1),$$

(4) $$\|f_n\|_{\Lambda_\alpha} = 1,$$

(5) $$\|f_n\|_V \geq 2^n,$$

and

(6) $$f_n \in V(T, T).$$

Let $a_n = \|f_n\|_{\tilde{V}}^{-1}$, and let $g = \sum a_n f_n$. Then by 11.1.4, $g \in \tilde{V}(T, T) \cap \Lambda_\alpha(T \times T)$, since the f_n have pairwise bidisjoint supports. Of course $g \notin V(T, T)$, because in every neighborhood of $(0, 0)$ g has $\tilde{V}$-restriction norm at least one while $g(0, 0) = 0$. □

Credit. Theorem 11.5.1 is due to Varopolous [10], and the method provides an alternative to the usual proof that $\Lambda_\beta(T) \subseteq A(T)$ if $\beta > 1/2$. The Corollary is new.

11.6. Sidon Constants of Finite Sets for Tensor Algebras and Group Algebras

Let B be a commutative Banach algebra with maximal ideal space ΔB. Let $E \subseteq \Delta B$ be a closed subset that is either countable or discrete. The *Sidon* (or *B-Sidon*) *constant* of E is the infimum $\alpha = \alpha_B = \alpha_B(E)$ of numbers c such that for all $f \in C_o(E)$, there exists $g \in B$ such that $\hat{g}|_E = f$ and $\|g\|_B \le c\|f\|_\infty$. If there is no such finite α, we take $\alpha = \infty$.

In this Section we estimate the Sidon constants of $X_1 \times \cdots \times X_n$ for the algebra $V(X_1, \ldots, X_n)$ (defined below), where the X_j are finite. We also estimate Sidon constants for certain other quotients of group algebras. The main results of this Section are the "Kns" Theorem 11.6.5 and Theorem 11.6.8, which estimates the V-Sidon constants for $X_1 \times \cdots \times X_n$.

Let n be a positive integer, and let $X_1, \ldots, X_n$ be compact Hausdorff spaces. The *n-dimensional tensor algebra* (*on* $X_1, \ldots, X_n$) is the set of all functions f on $X_1 \times \cdots \times X_n$ that have the form

(1)
$$f(x_1, \ldots, x_n) = \sum_{j=1}^{\infty} f_{1,j}(x_1) \cdots f_{n,j}(x_n),$$

where $f_{i,j} \in C(X_i)$ for all i, j and

(2)
$$\sum_{j=1}^{\infty} \|f_{1,j}\|_\infty \cdots \|f_{n,j}\|_\infty < \infty.$$

The norm of f is the infimum of the numbers (2) subject to (1). We leave to the reader the routine tasks of formulating and proving the n-variable forms of Theorem 11.1.2 and Corollary 11.2.3, and of showing that $V(X_1, \ldots, X_n)$ is isomorphic to $C(X_1) \hat{\otimes} \cdots \hat{\otimes} C(X_n)$, whose definition we leave to be formulated by the reader.

We begin with some elementary results.

11.6.1. Proposition. *Let G be a finite abelian group of cardinality k. Then the Sidon constant of G is $k^{1/2}$.*

Proof. For Haar measure on G, we use counting measure. For Haar measure on the dual group Γ of G we use $1/k$ times counting measure. By 1.6.1 it will suffice to show that $\alpha(G) = k^{1/2}$.

For that, we must exhibit a measure μ on G such that $\|\hat{\mu}\|_\infty = 1$ and $\|\mu\| = k^{1/2}$. We claim that we may assume that G is cyclic. Indeed, if $G = G_1 \oplus \cdots \oplus G_m$ is a decomposition of G into a finite sum of cyclic groups, and $\mu_j \in M(G_j)$ is such that $\|\mu_j\| = (\# G_j)^{1/2}$ and $\|\hat{\mu}_j\|_\infty = 1$ for $1 \le j \le m$, then $\mu = \mu_1 \times \cdots \times \mu_m$ has $\|\mu\| = k^{1/2}$ and $\|\hat{\mu}\|_\infty = 1$. We may thus assume that G is cyclic of order k.

To verify that a measure μ has $\|\hat{\mu}\|_\infty = 1$, it suffices to show that $\mu * \tilde{\mu} = \delta$, since that last holds if and only if $|\hat{\mu}| \equiv 1$. We have two cases.

Case I, $k = \#G$ is odd. We identify G with $\{0, \ldots, k-1\}$, using addition mod k. Let $\mu(j) = \exp(2\pi i j^2/k)/k^{1/2}$ for $1 \leq j \leq k$. Then for all $0 \leq j \leq k-1$, since addition is mod k,

$$\mu(-j) = \exp(2\pi i(k-j)^2/k)/k^{1/2} = \mu(j),$$

and

$$\tilde{\mu}(-j) = \overline{\mu(j)} = \exp(-2\pi i j^2/k)/k^{1/2}.$$

Then

$$\begin{aligned}\mu * \tilde{\mu}(j) &= \sum_{m=0}^{k-1} \mu(m)\tilde{\mu}(j-m) = \sum_{m=0}^{k-1} \mu(m)\overline{\mu(j-m)} \\ &= k^{-1} \sum_{l=0}^{k-1} \exp(2\pi i(l^2 - (j-l)^2)/k) \\ &= k^{-1} \exp(-2\pi i j^2/k) \sum_{l=0}^{k-1} \exp(2\pi i(2j\pi)/k).\end{aligned}$$

Since k is odd, $l \to \exp(2\pi i(2jl)/k)$ equals the identity character on G if and only if $j = k$. Therefore $\mu * \tilde{\mu}(j) = 0$ if $j \neq 0$, and $\mu * \tilde{\mu}(0) = 1$; that is, $\mu * \tilde{\mu} = \delta_0$, so $|\hat{\mu}|^2 \equiv 1$. Since $\|\mu\| = k^{1/2}$, μ has the required properties.

Case II, $k = \#G$ is even. We begin identifying G with $\{0, \ldots, k-1\}$ and set $\mu(j) = \exp(2\pi i j^2/2k)/k^{1/2} = \exp(\pi i j^2/k)/k^{1/2}$ for $0 \leq j \leq k-1$. Then

$$\mu(-j) = \exp(2\pi i(k^2 - 2jk + j^2)/2k)/k^{1/2}.$$

Since k is even, $2k$ divides k^2, so $\mu(-j) = \mu(j)$. We conclude that $\mu * \tilde{\mu} = \delta_0$ exactly as in Case I, using the equation

$$\mu * \tilde{\mu}(j) = [k^{-1} \exp(2\pi i j^2/2k)] \sum_{l=0}^{k-1} \exp(2\pi i jl/k).$$

The Proposition is proved. □

For cyclic G of order k, the measure μ defined above is also such that $\|\mu^m\| = k^{1/2}$ for $1 < m < k$, while $\mu^k = \delta_0$. That the results of the next Corollary are close to being best possible will be proved in 11.6.8.

11.6.2. Corollary. *Let X_1, X_2, and X_3 be finite spaces all having the same cardinality k. Then*

(i) *the $V(X_1, X_2)$-Sidon constant of $X_1 \times X_2$ is at least $k^{1/2}$, and*
(ii) *the $V(X_1, X_2, X_3)$-Sidon constant of $X_1 \times X_2 \times X_3$ is at least k.*

Proof. (i) We identify X_1 and X_2 with the finite cyclic group G of cardinality k. Let $f \in A(G)$ be such that $\|f\|_A = k^{1/2}$ and $\|f\|_\infty = 1$. Such an f exists by 11.6.1. Then $\|Mf\|_\infty = 1$ and $\|Mf\|_V = k^{1/2}$, since M is an isometry both from $C(G)$ to $C(G \times G)$ and from $A(G)$ to $V(G, G)$.

(ii) Let $\mu(j, l) = \exp(2\pi ijl/k)$ be defined on $G \times G$, where we identify G with $\{0, \ldots, k-1\}$ (addition modulo k). Then it is easy to see that

$$\|\mu\|_{[A(G)\hat{\otimes}C(G)]^*} = \sup_{1 \le r \le k} \sum_{l=0}^{k-1} \left| \sum_{j=0}^{k-1} \exp(2\pi ijl/k)\exp(2\pi ijr/k) \right| = k.$$

Thus, the $A(G) \hat{\otimes} C(G)$-Sidon constant of $G \times G$ is at least k.

Let $f \in A(G) \hat{\otimes} C(G)$ with $\|f\|_\infty = 1$ and $\|f\|_{A\hat{\otimes}C} = k$. Then $\|(M \otimes 1)f\|_\infty = 1$ and $\|(M \otimes 1)f\|_V = k$. Corollary 11.6.2 is proved. □

11.6.3. Lemma. *Let n and N be positive integers. Let $P(x_1, \ldots, x_n) = \sum r_t \cos(\sum t_j x_j + \theta_j)$ be a real-valued trigonometric polynomial on T^n, where $t = (t_1, \ldots, t_n)$ ranges over elements of Z^n with $\sum |t_j| \le N$. Then there exists a hypercube H of side $1/N$ such that $|P(x_1, \ldots, x_n)| \ge \|P\|_\infty/2$ on H.*

Proof. We may assume that $\|P\|_\infty = P(0, \ldots, 0)$. Then for all $x = (x_1, \ldots, x_n) \in T^n$,

$$P(x_1, \ldots, x_n) - P(0, \ldots, 0) = \sum x_j \frac{\partial P}{\partial x_j}(\xi),$$

where $\xi = (\xi_1, \ldots, \xi_n)$ is on the line segment $\{\delta x : 0 \le \delta \le 1\}$ between 0 and x. Suppose that $|x_j| \le 1/2N$ for all j. Let $a_j = \partial P/\partial x_j(\xi)/|\partial P/\partial x_j(\xi)|$ if $\partial P/\partial x_j(\xi) \ne 0$, and $a_j = 0$ otherwise. Then

$$|P(x) - P(0)| \le \sum \left| \frac{\partial P}{\partial x_j}(\xi) \right| \Big/ 2N = \sum a_j \frac{\partial P}{\partial x_j}(\xi)/2N.$$

Let $Q(y) = P(\xi_1 + ya_1, \ldots, \xi_n + ya_n)$. Then the degree of Q is at most N, and $Q'(0) = \sum a_j(\partial P/\partial x_j)(\xi)$. By Bernstein's inequality A.4.1, $|Q'(0)| \le N\|Q\|_\infty \le N\|P\|_\infty$. Therefore $|P(0) - P(x)| \le |Q'(0)|/2N = \|P\|_\infty/2$. Thus $|P| \ge \|P\|_\infty/2$ on $\{x : |x_j| \le 1/2N\}$, which is the required hypercube. The Lemma is proved. □

11.6.4. Lemma. *Let n and N be positive integers and let*

$$I \subseteq \{t \in Z^n : \sum |t_j| \le N\}.$$

For each $t \in I$, let $r_t \in R$. For each choice $\omega = \{\varepsilon_t : t \in I\}$ of signs $\varepsilon_t = \pm 1$, let

$$P_\omega(x_1, \ldots, x_n) = \sum_{t \in I} \varepsilon_t r_t \exp(2\pi i \sum t_j x_j).$$

Then there exists at least one ω such that $\|P_\omega\|_\infty < C[(n \log N) \sum r_t^2]^{1/2}$, where C is an absolute constant.

Proof. For each ω and $\theta \in T$, let $Q_\omega(x, \theta) = \sum \varepsilon_t r_t \cos(\sum t_j x_j + \theta)$. Then

$$\sup\{|Q_\omega(x, \theta)| : x \in T^k, \theta \in T\} = \|P_\omega\|_\infty.$$

We shall use a probabilistic argument. Each choice of sign $\varepsilon_t r_t = \pm r_t$ will occur with probability one-half, and $\mathscr{E}$ will denote mean value with respect to the resulting probability space. Then for each $\lambda > 0$ and $(x, \theta) \in T^{n+1}$,

$$\begin{aligned}\mathscr{E}(\exp(2\lambda Q_\omega)) &= \prod_{t \in I} \mathscr{E}(\exp(2\lambda_t r_t \cos(\textstyle\sum t_j x_j + \theta))) \\ &= \prod \operatorname{Cosh}(2\lambda r_t \cos(\textstyle\sum t_j x_j + \theta)) \leq \exp(2\lambda^2 \textstyle\sum r_t^2),\end{aligned}$$

since $\operatorname{Cosh} y \leq e^{y^2/2}$ for all y. Similarly $\mathscr{E}(\exp(-2\lambda Q_\omega)) \leq \exp(2\lambda^2 \sum r_t^2)$. Therefore $\mathscr{E}(\exp(2\lambda |Q_\omega|)) \leq 2 \exp(2\lambda^2 \sum r_t^2)$. Note that the last inequality holds independently of x and θ.

For each ω let $M_\omega = \|Q_\omega\|_\infty$. Since Q_ω has degree $N + 1$, 11.6.3 shows that there exists a hypercube H_ω of side $1/(N + 1)$ such that $|Q_\omega| \geq M_\omega/2$ on H_ω. Therefore

$$\begin{aligned}(N + 1)^{-n}\mathscr{E}(\exp(2\lambda M_\omega)) &\leq \mathscr{E}\left(\int_{H_\omega} \exp(2\lambda |Q_\omega|)\, dx_1 \cdots dx_n\, d\theta\right) \\ &\leq \mathscr{E}\left(\int_{T^{n+1}} \exp(2\lambda |Q_\omega|) dx_1 \cdots dx_1\, d\theta\right) \\ &\leq 2(2\pi)^{n+1} \exp(2\lambda^2 \textstyle\sum r_t^2).\end{aligned}$$

Therefore for each λ, there exists ω such that

$$\exp(2\lambda M_\omega) \leq 2(2\pi)^{n+1}(N + 1)^n \exp(2\lambda^2 \textstyle\sum r_t^2),$$

or

$$2\lambda M_\omega \leq (n + 1)\log 2\pi + \log 2 + n \log(N + 1) + 2\lambda^2 \textstyle\sum r_t^2$$

or

$$\lambda M_\omega \leq C_1 n \log N + \lambda^2 \textstyle\sum r_t^2.$$

Let $\lambda = (n \log N/\sum r_t^2)^{1/2}$. Then

$$\begin{aligned} M_\omega &\leq C_1(n \log N)^{1/2}(\sum r_t^2)^{1/2} + (n \log N)^{1/2}(\sum r_t^2)^{1/2} \\ &\leq C[(n \log N) \sum r_t^2]^{1/2}. \end{aligned}$$

Since $M_\omega = \|P_\omega\|_\infty$, the Lemma is proved. □

11.6.5. The "Kns" Theorem. *Let E be a Helson subset of the circle group T. Then there exists an integer $K \geq 1$ such that if $y_1, \ldots, y_n \in T$ and $s \geq 2$, then E contains at most Kns elements of the form $y_t = \sum t_j y_j$, where $t = (t_1, \ldots, t_n)$ ranges over all n-tuples of integers such that $\sum |t_j| \leq 2^s$.*

Proof. For each t, let $r_t = 1$ if $\sum t_j y_j \in E$, and $r_t = 0$ otherwise. (If some element of E has more than one representation of the form $\sum t_j y_j$, we let $r_t = 0$ for all but one such representation.) We must show that $\sum r_t^2 = \sum r_t \leq Kns$ where K is independent of n and s.

By 11.6.4, there exists a choice $\omega = \{\varepsilon_t\}$ of signs such that $\|P_\omega\|_\infty \leq C(ns \sum r_t)^{1/2}$, where $P_\omega(x) = \sum \varepsilon_t r_t \exp(2\pi i \sum t_j x_j)$. Let μ be the measure on E given by $\mu = \sum \varepsilon_t r_t \delta(\sum_{j=1}^n t_j y_j)$. Then $\hat{\mu}(m) = \sum \varepsilon_t r_t \exp(-2\pi i m \sum t_j y_j)$, so $\|\hat{\mu}\|_\infty \leq C(ns \sum r_t)^{1/2}$, while $\|\mu\| = \sum r_t$. Since E is a Helson set $\|\hat{\mu}\|_\infty \geq \beta\|\mu\|$ where $\beta > 0$ depends only on E. Thus

$$\beta(\sum r_t)^{1/2} \leq C(ns)^{1/2},$$

or

$$\sum r_t \leq (C^2/\beta^2)ns.$$

Therefore any integer $K \geq C^2/\beta^2$ will do. The Theorem is proved. □

The next result is an immediate consequence of 11.6.5. See Section 1.6 for the proof that $\alpha(\{1, \ldots, n\}) \geq cn^{1/2}$.

11.6.6. Theorem. *Let $E \subseteq T$ be a Helson set. Then there exists $C > 0$ such that if P is any arithmetic progression in T, then $\#(E \cap P) \leq C \log \#P$.*

11.6.7. Lemma. *Let Γ be an infinite discrete abelian group all of whose elements have order two. Let G be the dual group of Γ. Let $k \geq 1$ and let $X_1, \ldots, X_k$ be finite pairwise disjoint subsets of Γ whose union is independent. Let $f: X_1 \times \cdots \times X_k \to \mathbb{C}$ and let $F(x) = \sum f(\gamma)\langle x, \gamma\rangle$. Then*

$$2^{-k}(q)^{-k/2}\|F\|_2 \leq \|F\|_p \leq 2^k p^{k/2}\|F\|_2. \tag{3}$$

(Here $1/q + 1/p = 1$ and $1 \leq p < \infty$.)

Proof. We shall establish first the right hand inequality in (3). We begin by considering first the case that f is real, $k = 1$, and $p = 2n$ is an even integer. Then

$$\int |F(x)|^p \, dx = \tag{4}$$

$$\int \sum A(\alpha) f(\gamma_1)^{\alpha_1} \cdots f(\gamma_j)^{\alpha_j} \langle x, \gamma \rangle^{\alpha_1} \cdots \langle x, \gamma_j \rangle^{\alpha_j} \, dx,$$

where the sum is over all $2n$-tuples $\alpha = (\alpha_1, \ldots, \alpha_{2n})$ of non-negative integers such that $\sum \alpha_j = 2n$ and the $A(\alpha)$ are the multinomial coefficients $(2n)!/(\alpha_1! \ldots \alpha_j!)$. It is clear that the integral of a term on the right side of (4) is zero unless the α_j are all even. Thus

$$\int |F(x)|^{2n} \, dx = \sum \{A(2\alpha) f(\gamma_1)^{2\alpha_1} \cdots f(\gamma_j)^{2\alpha_j} : \sum \alpha_i = n\}.$$

Let

$$B^{2n} = \sup\{A(\alpha)/A(\beta) : \sum \alpha_j = 2n, \sum \beta_j = n\}.$$

Then

$$\sum \{A(\beta) f(\gamma_1)^{2\beta_1} \cdots f(\gamma_j)^{2\beta_j} : \sum \beta_i = 2n\} = \|F\|_2^{2n},$$

so $\int |F(x)|^{2n} \, dx \leq B^{2n} \|F\|_2^{2n}$. Straightforward calculations show that $B^{2n} \leq n^n$ (see Zygmund [1, Vol. 1, p. 213]). Therefore when $k = 1$ and f is real, $\|F\|_{2n} \leq n^{1/2} \|F\|_2$. Since $\|F\|_p$ increases with p, $\|F\|_p \leq n^{1/2} \|F\|_2$ for all $2n - 2 \leq p \leq 2n$. Since $\sqrt{2} p^{1/2} \geq n^{1/2}$ for such p, we have $\|F\|_p \leq \sqrt{2} p^{1/2} \|F\|_2$ whenever f is real-valued and $k = 1$. For general $f = f_1 + if_2$ where the f_j are real-valued, we let F_j correspond to f_j and then note that

$$\begin{aligned} \|F\|_p &\leq \|F_1\|_p + \|F_2\|_p \leq \sqrt{2} p^{1/2} (\|F_1\|_p + \|F_2\|_p) \\ &= \sqrt{2} p^{1/2} (\|f_1\|_2 + \|f_2\|_2) \leq \sqrt{2} \|F\|_2 = 2p^{1/2} \|F\|_2. \end{aligned}$$

That proves the right hand side of (3) in case $k = 1$.

We shall argue by induction when $k \geq 2$. We assume that the right-hand side of (3) holds for $k = 1$ and $k = n - 1$ where $n \geq 1$. Note first that since $\|F\|_p$ increases with p, $\|F\|_p \leq \|F\|_2$ when $1 \leq p \leq 2$, which yields the right hand inequality in (3) for those values of p. Therefore we may assume that $p > 2$. Let $X = X_1 \times \cdots \times X_n$ and $X' = X_2 \times \cdots \times X_n$. Because $X_1 \cup \cdots \cup X_n$ is independent, we may write $G = G_1 \times G'$, where X_1 generates the dual group of $G_1 \times \{0\}$ and $\bigcup_{j=2}^n X_j$ is contained in the annihilator of

$G_1 \times \{0\}$. Let $X_1 = \{\gamma_j: 1 \le j \le l\}$ be an enumeration of X_1. Then $F(x) = F(x_1, x') = \sum_j \sum_{\gamma' \in X'} f(\gamma_j, \gamma')\langle x_1, \gamma_j\rangle\langle x', \gamma'\rangle$. For $1 \le j \le l$ let $F_j(x') = \sum_{\gamma' \in X'} f(\gamma_j, \gamma')\langle x', \gamma'\rangle$. By (3) for $k = 1$,

$$\int |F(x_1, x')|^p \, dx_1 \le 2p^{1/2}\left(\sum_j |F_j(x')|^2\right)^{p/2} \quad \text{for all } x' \in G'. \tag{5}$$

Since $p \ge 2$, we may apply Minkowski's inequality for $p/2$ and obtain

$$\int \left(\sum_j |F_j(x')|^2\right)^{p/2} dx' \le \left[\sum_j \left(\int |F_j(x')|^p \, dx'\right)^{2/p}\right]^{p/2}.$$

By (3) for $k = n - 1$ and each $1 \le j \le l$,

$$\left(\int |F_j(x')|^p \, dx'\right)^{2/p} \le [2^{n-1}p^{(n-1)/2}]^2\left[\sum_{\gamma' \in X'} |f(\gamma_j, \gamma')|^2\right]. \tag{6}$$

Now (5) and (6) show that

$$\begin{aligned}\|F\|_p^p = \int |F(x_1, x')|^p \, dx_1 \, dx' &\le 2^p p^{p/2} 2^{p(n-1)} p^{p(n-1)/2}\left[\sum_j \sum_{\gamma'} |f(\gamma_j, \gamma')|^2\right]^{p/2} \\ &= 2^{pn}p^{pn/2}\|F\|_2^p,\end{aligned}$$

which establishes the right-hand side of (3) for $k = n$. That completes the induction.

For the left hand inequality we argue by duality and Hölder's inequality. Since $\|F\|_p$ increases with p, $\|F\|_p \ge \|F\|_2$ for $p \ge 2$. Therefore we may assume that $1 \le p < 2$. Then Hölder's inequality implies that

$$\|F\|_2^2 = \int |F(x)|^2 \, dx \le \|F\|_q \|F\|_p.$$

Since $\|F\|_q \le 2^k q^{k/2}\|F\|_2$, we have $\|F\|_2 \le 2^k q^{k/2}\|F\|_p$. That completes the proof. □

11.6.8. Theorem. *Let $n \ge 2$. Then there exist constants $C_n, D_n > 0$ such that if $X_1, \ldots, X_n$ are finite spaces, each with the discrete topology, and if $\# X_1 \ge \# X_j$ for $2 \le j \le n$, then the V-Sidon constant $\alpha = \alpha(X_1 \times \cdots \times X_n)$ satisfies*

$$C_n \le \alpha \Big/ \left(\prod_{j=2}^{n} \# X_j\right)^{1/2} \le D_n,$$

where $C_n = O(1/(n \log n)^{1/2})$ and $D_n = O(2^n)$.

Proof. The constants C_n exist. By 11.1.2, we may assume that $X_1, \ldots, X_n$ are disjoint subsets of generators of Z^∞, so that the union of the X_j is independent. For $1 \le j \le n$ let $X_j = \{x_{j,k}: 1 \le k \le m(j)\}$. For each choice of signs $\omega = \{\varepsilon_{j,k}: 1 \le k \le m(j), 1 \le j \le n\}$, let

$$P_\omega(t) = \sum_{j=1}^{n} \sum_{k(j)=1}^{m(j)} \varepsilon_{j,k(j)} \exp(-2\pi i(x_{1,k(1)} + \cdots + x_{n,k(n)})).$$

Then each P_ω is a polynomial of degree n and $m = \sum_1^n m(j)$ variables. By 11.6.4, there exists an ω such that

$$\|P_\omega\|_\infty \le C\left(m \log n \prod_{j=1}^{n} \# X_j\right)^{1/2}.$$

Let μ denote the measure $\sum_j \sum_{k(j)} \varepsilon_{j,k(j)} \delta(x_{1,k(1)} + \cdots + x_{n,k(n)})$. Then $\hat{\mu} = P_\omega$, and

$$\|\mu\|_{BM} = \|\hat{\mu}\|_\infty \le C\left(m \log n \prod_{j=1}^{n} \# X_j\right)^{1/2}.$$

Since $\# X_1 \ge \# X_j$ for $2 \le j \le n$, we see that $m \le n \# X_1$. Therefore

$$\|\hat{\mu}\|_\infty \le C(n \log n)^{1/2} \# X_1 \left(\prod_{j=2}^{n} \# X_j\right)^{1/2}.$$

Since $\|\mu\|_M = \prod_{j=1}^n \# X_j$, we see that the Helson constant of $X_1 + \cdots + X_n$ is at least

$$a = \|\mu\| / \|\hat{\mu}\|_\infty = \left(\prod_{j=2}^{n} \# X_j\right)^{1/2} \Big/ C(n \log n)^{1/2}.$$

Since $V(X_1, \ldots, X_n)$ is isometrically isomorphic to $A(X_1 + \cdots + X_n)$ by an easy generalization of 11.1.2, the V-Sidon constant of $X_1 \times \cdots \times X_n$ is also at least a. Thus $C_n = 1/(C(n \log n)^{1/2})$ will do.

The constants D_n exist. We apply 11.6.7 with $k = n - 1$, as follows. Let μ be a measure on $X_1 \times \cdots \times X_n$. We retain the identification above of $X_1 \times \cdots \times X_n$ with $X_1 + \cdots + X_n \subseteq Z^\infty$. Then the norm of as an element of V^* is given by

$$(7) \quad \|\mu\|_{BM} = \sup \sum_{z \in X_1} \left| \sum_{x_2 \in X_2} \cdots \sum_{x_n \in X_n} a_2(x_2) \cdots a_n(x_n) \mu(z, x_2, \ldots, x_n) \right|$$

where the supremum in (7) is taken over all functions $a_j: X_j \to \{z: |z| \le 1\}$ for $2 \le j \le n$. (By the maximum modulus principle, we may restrict the a_i to have ranges in the unit circle, which shows that $\|\mu\|_{BM} = \|\hat{\mu}\|_\infty$.) Of course, if we restrict (as we shall) the functions a_j to take on only the values ± 1, we obtain a possible smaller value for the supremum.

Let G denote the subset of T^∞ of elements of order two. Then

$$\text{(8)} \qquad \|\hat{\mu}\|_\infty \ge \sum_{z \in X_1} \int \left| \sum_{\gamma \in X_2 \times \cdots \times X_n} \langle y, \gamma \rangle \mu(z, \gamma) \right| dy.$$

We apply 11.6.7 with $p = 1$ to (8) to obtain

$$\text{(9)} \qquad \|\hat{\mu}\|_\infty \ge \sum_{z \in X_1} 2^{-n-1} \left(\sum_\gamma |\mu(z, \gamma)|^2 \right)^{1/2}$$

$$\ge 2^{-n-1} \|\mu\| \Big/ \left(\prod_{j=2}^n \# X_j \right)^{1/2},$$

where we have used the Cauchy-Schwarz inequality in the last step. Therefore $D_n = 2^{n+1}$ will do. That ends the proof of 11.6.8. $\square$

The proofs of the next two results are immediate. These results will be used to give an alternative method for estimating V-Sidon constants.

11.6.9. Lemma. *Let X be a finite set and μ a measure on X. Then $\|\mu\| \ge 2^{-3/2} \sup\{|\int f \, d\mu| : f: X \to \{0, 1\}\}$.*

11.6.10. Corollary. *Let $X_1, \ldots, X_n$ be finite discrete spaces and let $f \in V(X_1, \ldots, X_n)$. Then there exist positive integers $m(1), \ldots, m(j)$, idempotent functions $\varphi_{j,1}, \ldots, \varphi_{j,m(j)} \in C(X_j)$, and complex numbers $c(k(1), \ldots, k(n))$ for $1 \le k(j) \le m(j)$ and $1 \le j \le n$ such that*

$$f = \sum_{j=1}^{n} \sum_{k(j)=1}^{m(j)} c(k(1), \ldots, k(n)) \varphi_{1,k(1)} \otimes \cdots \otimes \varphi_{n,k(n)}$$

and

$$\sum_j \sum_{k(j)} |c(k(1), \ldots, k(n))| \le 2^{3n/2} \|f\|_V.$$

If $Y \subseteq R^m$ and $L > 0$, the *L-span* of Y is the set

$$\left\{ \sum_{j=1}^k \lambda_j y_j : y_j \in Y, \lambda_j \in R, \sum |\lambda_j| \le L, 1 \le k < \infty \right\}.$$

11.6.11. Lemma. *Let $m \geq 1$ and let $U = \{(x_1, \ldots, x_m) \in R^m : |x_j| \leq 1, 1 \leq j \leq m\}$. If $Y \subseteq U$ and $L > 0$, and the L-span of Y contains U, then $L \geq 1$ and $\# Y \geq \exp(m/2L^2)/2$.*

Proof. That $L \geq 1$ is obvious. Let V denote the set of vertices of U. We claim that

(10) $$\sup\{|y \cdot v| : y \in Y\} \geq m/L \quad \text{for all } v \in V.$$

(Here $x \cdot v$ denotes the usual dot product.) Indeed, if $v = \sum \lambda_j y_j$ with $\sum |\lambda_j| \leq L$ and $y_j \in Y$ for all j, then

$$\begin{aligned} v \cdot v = \sum \lambda_j y_j \cdot v &\leq \sum |\lambda_j| \, |v \cdot y_j| \\ &\leq L \sup\{|y \cdot v| : y \in Y\}. \end{aligned}$$

That establishes (10). Let $u = (u_1, \ldots, u_m) \in U$ and let $\lambda > 0$. Since $V = -V$,

(11) $$\begin{aligned} 2^{-m} \sum_{v \in V} \exp(\lambda |u \cdot v|) &\leq 2 \cdot 2^{-m} \sum_{v \in V} e^{\lambda u \cdot v} \\ &\leq 2 \prod_{j=1}^{m} \cosh \lambda u_j \leq 2(\cosh \lambda)^m \\ &\leq 2 \exp(\lambda^2 m/2). \end{aligned}$$

We now apply (11) to elements $y \in Y \subseteq U$ and use the inequality

$$\exp(\lambda \sup\{|y \cdot v| : y \in Y\}) \leq \sum_{y \in Y} \exp(\lambda |y \cdot v|) \quad \text{for all } \lambda > 0$$

to obtain

(12) $$2^{-m} \sum_{v \in V} \exp(\lambda \sup\{|y \cdot v| : y \in Y\}) \leq 2 \exp(\lambda^2 m/2) \# Y.$$

Then (10) and (11) imply that

(13) $$\exp(\lambda m/L) \leq 2^{-m} \sum_{v \in V} \exp\left(\lambda \sup_{y \in Y} |y \cdot v|\right) \quad \text{for all } \lambda > 0.$$

Note that (11) and (12) hold for $\lambda \leq 0$. By (12) and (13), $\exp(\lambda m/L) \leq 2 \exp(\lambda^2 m/2) \# Y$ for all $\lambda \in (-\infty, \infty)$. Therefore

(14) $$0 \leq \log(2 \# Y) + \lambda^2 m/2 - mL\lambda \quad \text{for all } \lambda \in (-\infty, \infty).$$

Therefore the discriminant of (14) must be non-positive:

$$m^2/L^2 - 2m\log(2\#Y) < 0.$$

The conclusion follows at once. □

The following Theorem provides a method of estimating V-Sidon constants alternative to that of 11.6.8.

11.6.12. Theorem. *Let $X_1, \ldots, X_n$ be finite discrete spaces, let $E \subseteq X_1 \times \cdots \times X_n$, and let $l = \sum_1^n \#X_j$. Then the V-Sidon constant $\alpha = \alpha(E)$ of E satisfies the following estimates:*

$$C_n(\#E/l)^{1/2} \leq \alpha(E) \leq D_n \sum_{z \in X_1} (\#[E \cap (\{z\} \times X_2 \times \cdots \times X_n)])^{1/2},$$

where $C_n = O(2^{-3n/2})$ and $D_n = O(2^n)$.

Proof. The constants C_n exist. We set $N = \prod_{j=1}^n \#X_j$. If $f \in C(E)$, then $\alpha(E)\|f\|_\infty \geq \|f\|_{V(E)}$, where $V(E)$ denotes the restriction of $V(X_1, \ldots, X_n)$ to E. By 11.6.10, each $f \in V(E)$ has the form

$$f = \sum c(k(1), \ldots, k(n))\varphi_{1,k(1)} \otimes \cdots \otimes \varphi_{n,k(n)}|_E$$

where $\sum |c(k(1), \ldots, k(n))| \leq 2^{3n/2}\|f\|_{V(E)}$ and the $\varphi_{j,k(i)}$ are idempotent elements of $C(X_j)$ for all $k(i)$ and $1 \leq j \leq n$.

Let $m = \#E$. We identify $U = \{f : f : E \to [-1, 1]\}$ with the m-cube in R^m and see that U is in the $\alpha(E) \cdot 2^{3n/2}$-span of the set Y of idempotents $\varphi_{1k(1)} \otimes \cdots \otimes \varphi_{nk(n)}|_E$. Let J denote the number of such idempotents and $L = \alpha(E)2^{3n/2}$. Then $J \geq \frac{1}{2}\exp(m/2L^2)$ by 11.6.11. Therefore $\#E = m \leq 2L^2(\log J + \log 2)$. Since $J \leq \prod_{j=1}^n 2^{\#X_j} = 2^l$,

$$\#E \leq 2L^2(l + 1)\log 2 \leq 2^{3n+1}(l + 1)(\log 2)\alpha(E)^2.$$

Thus $\alpha(E) \geq (2^{3n+2} l \log 2)^{-1/2}(\#E)^{1/2}$, so $C_n = (2^{3n+2}\log 2)^{-1/2}$ will do.

The constants D_n exist. We apply the method used in the corresponding part of the proof of 11.6.8.

It is easy to see that $\alpha(E)$ is at most $(\#E)/\beta$, where $\beta = 2^{-1/2}\inf\{\|\mu\|_{BM} : \mu \in M(E), \mu(\{x\}) = \pm 1 \text{ for all } x \in E\}$. By the first inequality of (9), if $\mu \in M(E)$ and if $\mu(\{x\}) = \pm 1$ for all $x \in E$, then

$$\begin{aligned}\|\mu\|_{BM} = \|\hat{\mu}\|_\infty &\geq \sum_{z \in X_1} 2^{-n-1}\left(\sum_\gamma \{1 : (z, \gamma) \in E\}\right)^{1/2} \\ &\geq 2^{-n-1} \sum_{z \in X_1} (\#\{\gamma : (z, \gamma) \in E\})^{1/2}.\end{aligned}$$

Thus, $D_n = 2^{n+1}$ will do. That ends the proof of 11.6.12. □

Remarks and Credits. 11.6.1 is from Graham [12]. It can be used to prove the following version of a result of Körner [6, Theorem 2.3A]. Though the result here is stronger than Körner's, there is no hope of obtaining his Theorem 2.3B from it.

11.6.13. Corollary. *Let G be a finite abelian group with cardinality n. Then for all $L > 1$, there exists $\mu \in M(G)$ with*

(i) $\|\mu\| = 1$,
(ii) $\#\ \mathrm{supp}\ \mu \le n/L$, *and*

(iii) $|\hat{\mu}(\gamma)| \le \dfrac{2\sqrt{2}}{\sqrt{2}-1}(n/L)^{-1/2}$ *for all* $\gamma \in \hat{G}$.

Proof. Let $G = \sum_1^m Z_{n(j)}$ be a decomposition of G into a direct sum of cyclic groups. Let $n(0) = n(m+1) = 1$, and let $0 \le r \le m$ be such that

$$n(0) \cdots n(r) \le n/L < n(0) \cdots n(r+1).$$

If $r = m$, then $L = 1$ and 11.6.1 implies that there exists $\mu \in M(G)$ such that $\#\mathrm{supp}\ \mu = n$, $\|\hat{\mu}\|_\infty \le n^{-1/2}$ and $\|\mu\| = 1$.

If $r < m$, let $1 \le p \le n(r+1)$ be such that

$$n(1) \cdots n(r)p \le n/L < n(1) \cdots n(r)(p+1). \tag{15}$$

Since $p + 1 \le 2p$, the inequalities (15) imply that

$$[n(1) \cdots n(r)p]^{-1/2} \le (n/L)^{-1/2}\sqrt{2}. \tag{16}$$

By 11.6.1, there exists $\nu \in M(Z_{n(1)} \times \cdots \times Z_{n(r)})$ such that $\|\nu\| = 1$ and $\|\nu\|_\infty = [n(1) \cdots n(r)]^{-1/2}$. By the Rudin-Shapiro construction (1.6.3), there exists $\omega \in M(Z_{n(r+1)})$ such that $\|\omega\| = 1$, $\#\mathrm{supp}\ \omega \le p$, and $\|\hat{\omega}\|_\infty \le 2p^{1/2}/(\sqrt{2}-1)$. Therefore the measure $\mu = \nu \times \omega$ has $\|\mu\| = 1$, $\#\mathrm{supp}\ \mu \le n(1) \cdots n(r)p \le n/L$, and

$$\|\hat{\mu}\|_\infty \le (2/(\sqrt{2}-1))[n(1) \cdots n(r)p]^{-1/2} \le (2\sqrt{2}/(\sqrt{2}-1))(n/L)^{-1/2},$$

by (16). The Corollary is proved. □

11.6.2 is new. Whether the estimates given there are the best possible seems to be unknown. (They have, of course, the right order of magnitude, as 11.6.8 shows.) The "Kns" Theorem 11.6.5, Corollary 11.6.6, and the Lemmas preceding 11.6.5 are from Kahane and Salem [1, pp. 146–148]. Lemma 11.6.7 is well-known; our proof is based on Zygmund [1, Vol. I, p. 213] and Stein [1, pp. 276–278]. It shows that if $X_1, \ldots, X_k$ are disjoint subsets of an independent set in D, then $X_1 + \cdots + X_k$ is a $\Lambda(p)$-set for all $p > 2$. Bonami [1] has shown that if X is dissociate and $k \geq 1$ then $X + \cdots + X$ (k times) is a $\Lambda(p)$-set. Sidon sets E have the property that $\|F\|_p \leq Cp^{1/2}\|F\|_2$ for all E polynomials F (Rudin [10]) and all $p > 2$, and Pisier [1, 2] has shown that that property characterizes Sidon sets.

Theorem 11.6.8 is due to Varopoulos [10]; the proof here is partly from Varopoulos and partly from Johnson and Woodward [1]. Theorem 11.6.12 is also from Varopoulos [10]; its proof is taken from Stegeman [1]. For another proof of the existence of the D_n in 11.6.12, see Kaijser [3].

11.7. Automorphisms of Tensor Algebras

Let B be a commutative semisimple Banach algebra with maximal ideal space ΔB. If Φ is an automorphism of B, then there exists a homeomorphism $\varphi: \Delta B \to \Delta B$ such that $(\Phi f)\hat{} = \hat{f} \circ \varphi$ for all $f \in B$. A problem for each commutative Banach algebra B is to determine which homeomorphisms φ of ΔB are induced by automorphisms Φ of B. In this section we give the solution for that problem when $B = V(X, Y)$ and X and Y are connected compact Hausdorff spaces. In that case it is obvious that if $\varphi_1: X \to X$ and $\varphi_2: Y \to Y$ are homeomorphisms, then the map $f(x, y) \mapsto f(\varphi_1(x), \varphi_2(y))$ defines an automorphism of $V(X, Y)$. The map $\varphi(x, y) = (\varphi_1(x), \varphi_2(y))$ is called a *bihomeomorphism*. Similarly, if $\varphi_3: X \to Y$ and $\varphi_4: Y \to X$ are homeomorphisms then the map $f(x, y) \to f(\varphi_4(y), \varphi_3(x))$ also defines an automorphism of $V(X, Y)$. The map $\varphi'(x, y) = (\varphi_4(y), \varphi_3(x))$ is called a *cross-homeomorphism*.

11.7.1. Theorem. *Let X and Y be compact connected Hausdorff spaces. Let $\Phi: V(X, Y) \to V(X, Y)$ be an automorphism. Then the homeomorphism $\varphi: X \times Y \to X \times Y$ induced by Φ is either a bihomeomorphism or a cross-homeomorphism. In particular, Φ is an isometry.*

We shall prove several lemmas before proving 11.7.1. In all of them we shall assume that $\varphi: X \times Y \to X \times Y$ is a homeomorphism induced by the automorphism Φ of $V(X, Y)$, where X and Y are compact and connected. We shall write "$x \times Y$" and "$X \times y$" for "$\{x\} \times Y$" and "$X \times \{y\}$".

A subset E of $X \times Y$ is *diagonal* if whenever $x, y \in E$ are distinct, then $\{x\}$ and $\{y\}$ are bidisjoint. For $E \subseteq X \times Y$, let $D(E) = \sup\{\#F: F \subseteq E, F$ is diagonal$\}$. A subset $F \subseteq X \times Y$ is *parallel to* X (or *parallel to* Y) if F is contained in a set of the form $X \times y$ (or $x \times Y$).

11.7.2. Lemma. *If $D(\varphi(x \times Y)) = 1$ and $D(\varphi(X \times y)) = 1$ for all $x \in X$ and all $y \in Y$, then either φ is a cross-homeomorphism or a bihomeomorphism.*

Proof. If $D(\varphi(x \times Y)) = 1$ for all x, then for all x, $\varphi(x \times Y)$ is either parallel to X or to Y. Let $U = \{x: \varphi(x \times Y)$ is not parallel to $X\}$. It is easily seen that

U is open in X. But $V = \{x : \varphi(x \times Y) \text{ is not parallel to } Y\}$ is also open, and $U \cup V = X$. Since X is connected and $U \cap V = \varnothing$, either $U = \varnothing$ or $V = \varnothing$. We conclude that either $\varphi(x \times Y)$ is parallel to X for all x or that $\varphi(x \times Y)$ is parallel to Y for all x.

Suppose that $\varphi(x \times Y)$ is parallel to Y, for all x. We claim that $\varphi(X \times y)$ is parallel to X for all y. For otherwise $\varphi(X \times y)$ would be parallel to Y for all y. Then for $y \in Y$ and $x \in X$, we let b_y and a_x be such that $\varphi(X \times y) \subseteq b_y \times Y$ and $\varphi(x \times Y) \subseteq a_x \times Y$. Fix x. Then for all y, $\pi_1\varphi(x, y) = a_x = b_y$; that is, $\varphi(X \times y) \subseteq a_x \times Y$ for all $y \in Y$. Thus $\varphi(X \times Y) \subseteq a_x \times Y$, so $\#X = 1$ and Theorem 11.7.1 follows. Thus the claim follows.

Let $\varphi_1(x) = \pi_1\varphi(x \times Y)$ and $\varphi_2(y) = \pi_2(X \times y)$. Then $\varphi(x, y) = (\varphi_1(x), \varphi_2(y))$ so φ is a bihomeomorphism. When $\varphi(x \times Y)$ is parallel to X for all x, a similar argument shows that φ is a cross-homeomorphism. That ends the proof of the Lemma. □

11.7.3. Lemma. *Let E be a finite subset of $X \times Y$, and let $p \in E$. Then there exists a neighborhood U of p such that if $q \in U$, then the V-Sidon constant of $E \cup \{q\}$ equals the V-Sidon constant of E.*

Proof. Let U be so small that $\pi_1 U \cap \pi_1 E = \{\pi_1 p\}$ and $\pi_2 U \cap \pi_2 E = \{\pi_2 p\}$, and let $q \in U$. If $\pi_1 p = \pi_1 q$ and $\pi_2 p = \pi_2 q$, then $p = q$. We may thus assume that either $\pi_1 q \notin \pi_1 E$ or $\pi_2 q \notin \pi_2 E$. Note that in either case, for any complex number z there exists $f \in V(X, Y)$ such that $\|f\|_V = 1$, $f = 1$ on E, and $f(q) = z$.

Let μ be any measure on $E \cup \{q\}$, and let $g \otimes h$ be any product function on $X \times Y$ such that $\|g \otimes h\|_\infty \le 1$. We must estimate $|\int g \otimes h \, d\mu|$. Of course, because $\pi_1 E$ is finite, $g|_{\pi_1 E}$ is the average of two unimodular functions. Thus, we may assume that $|g| \equiv 1$. Similarly, we may assume that $|h| \equiv 1$. Subject to the conditions that $|g| \equiv 1$ on X and that $|h| \equiv 1$ on Y, choose g and h such that $\int_E g \otimes h \, d\mu$ is non-negative and maximal. Its value is then $\alpha(E)^{-1}|\mu|(E)$. We may thus assume that $\int g \otimes h \, d\mu|_E = \alpha(E)^{-1}|\mu|(E)$. Let f be such that $\|f\|_V = 1$, $f|_E \equiv 1$ and $f(q)\mu(\{q\}) = |\mu(q)|$. Then

$$\int fg \otimes h \, d\mu = \alpha(E)^{-1}|\mu|(E) + |\mu(q)|.$$

Therefore $\|\mu\|_{BM} = \alpha(E)^{-1}|\mu|(E) + |\mu(\{q\})|$. That equality implies that the largest ratio $\|\mu\|_M/\|\mu\|_{BM}$ occurs when $|\mu(\{q\})| = 0$. But that shows that the Sidon constant of $E \cup \{q\}$ equals $\alpha(E)$. That proves 11.7.3. □

11.7.4. Lemma. *$D(x \times Y) \le \|\Phi^{-1}\|^2$ for all $x \in X$ and $D(X \times y) \le \|\Phi^{-1}\|^2$ for all $y \in Y$.*

Proof. Let $n > 1$, and let $x \in X$. Suppose that $y_1, \ldots, y_n \in Y$ are such that $\{\varphi(x, y_j) : j = 1, \ldots, n\}$ is diagonal. Let $U \times V_j$ be a compact neighborhood of (x, y_j), for $j = 1, \ldots, n$, such that the sets $\varphi(U \times V_1), \ldots, \varphi(U \times V_n)$ are bidisjoint. An induction argument, using 11.7.3 n times, shows that there exist distinct points $x_1, \ldots, x_n \in U$ such that the V-Sidon constant of the set $\{\varphi(x_k, y_j): j, k = 1, \ldots, n\}$ is one.

Let $f \in V(\{x_j\}_{j=1}^n, \{y_j\}_{j=1}^n)$ be such that $\|f\|_V \geq n^{1/2}$ and $\|f\|_\infty = 1$. Such a function exists by 11.6.2. Let F denote the function $f \circ \varphi^{-1}$, which is defined on $\varphi(\{x_j\} \times \{y_k\})$. Since $\{\varphi(x_j, y_k): 1 \leq j, k \leq n\}$ has V-Sidon constant one, for each $\varepsilon > 0$ there exists a function $g \in V(X, Y)$ such that $\|g\|_V \leq 1 + \varepsilon$ and $g = F$ on $\{\varphi(\{x_j\} \times \{y_k\})\}$.

But then $\Phi^{-1}g = f$ on the set $\{\{x_j\} \times \{y_j\}\}$, so $\|\Phi^{-1}g\| \geq n^{1/2}$. Therefore $\|\Phi^{-1}\| \geq n^{1/2}(1 + \varepsilon)$, and $D(x \times Y) \leq \|\Phi^{-1}\|^2$. A similar argument shows that $D(X \times y) \leq \|\Phi^{-1}\|^2$ for all $y \in Y$. That ends the proof of Lemma 11.7.4. □

We say that y is a *corner* of $\varphi(x \times Y)$ if every neighborhood of y contains points $r, s \neq y$ such that $\{\varphi(x, r), \varphi(x, s)\}$ is diagonal. We define corners of $\varphi(X \times y)$ similarly.

11.7.5. Lemma. *For all $x \in X$ and all $y \in Y$, $x \times Y$ contains at most $\|\Phi^{-1}\|^2 - 1$ corners and $X \times y$ contains at most $\|\Phi^{-1}\|^2 - 1$ corners.*

Proof. We argue by contradiction, and suppose that $\{y_1, y_2, \ldots\}$ is a set of corners of $x \times X$. Let $r_1, r_2 \in Y$ be near y_1 and such that $\{\varphi(x, r_1), \varphi(x, r_2)\}$ is diagonal.

Suppose that $k \geq 2$ and that $r_1, \ldots, r_k \in Y$ are such that $\{\varphi(x, r_j): 1 \leq j \leq k\} = W$ is diagonal. Let $E = \pi_1 W \times \pi_2 W$. By perturbing the $r_1, \ldots, r_k$ by arbitrarily small amounts, we may assume that $\varphi(x, y_{k+1}) \notin E$. Let $j = k + 1$. We may assume that $\pi_1 \varphi(x, y_j) \notin \pi_1 W$. Let V be any neighborhood of y_j such that $\pi_1 \varphi(x \times V) \cap \pi_1 W = \varnothing$ and $\pi_2 \varphi(x \times V) \cap \pi_2 W = \{\pi_2 \varphi(x, y_j)\}$. Let $r, s \in V$ be such that $\{\varphi(x \times r), \varphi(x, s)\}$ is diagonal. (Since y_j is a corner, such a choice is possible.) Then either $\pi_2 \varphi(x, r) \neq \pi_2 \varphi(x, y_j)$ or $\pi_2 \varphi(x, s) \neq \pi_2 \varphi(x, y_j)$. The element ($r$ or s) for which the *in*equality holds will be our r_{k+1}. It is clear that $\{\varphi(x, r_1)\}_{i=1}^{k+1}$ is diagonal. If $\pi_2 \varphi(x \times V) \cap \pi_2 W = \varnothing$, we argue similarly.

Thus, if $x \times Y$ contained more than $\|\Phi^{-1}\|^2 - 1$ corners, then it would contain a diagonal set of more than $\|\Phi^{-1}\|^2$ elements, which contradicts 11.7.4. That ends the proof of 11.7.5. □

11.7.6. Lemma. *Let y be a corner of $x \times Y$. Then y has a neighborhood V such that*

$$\varphi(x \times V) \subseteq (X \times \pi_2 \varphi(x, y)) \cup (\pi_1 \varphi(x, y) \times Y).$$

Proof. Otherwise, $\varphi(x \times V)$ would contain an infinite diagonal subset, as one may show easily after drawing a picture. □

11.7.7. Lemma. *Let $(x, y) \in X \times Y$, and let $U \times V$ be a neighborhood of (x, y). Then $\pi_1\varphi((x \times V) \cup (U \times y))$ is a neighborhood of $\pi_1\varphi(x, y)$, and $\pi_2\varphi((x \times V) \cup (U \times y))$ is a neighborhood of $\pi_2\varphi(x, y)$.*

Proof. By 11.7.5 we may assume that $x \times V$ and $U \times y$ have at most one corner each. Let $R \times S$ be a neighborhood of $(p, q) = \varphi(x, y)$ with $R \times S \subseteq \varphi(U \times V)$. Then

$$\varphi^{-1}((p \times S) \cup (R \times q)) \subseteq (x \times V) \cup (U \times y),$$

by 11.7.6 applied to φ^{-1}, provided $R \times S$ is sufficiently small. But that proves the Lemma. □

11.7.8. Lemma. *Let y be a corner of $x \times Y$ and V an open neighborhood of y. For $j = 1$ and 2, the connected component of*

$$\{v \in V: \pi_j\varphi(x, v) = \pi_j\varphi(x, y)\}$$

that contains y is non-trivial.

Proof. For $j = 1$ and 2, let $W_j = \{v \in V: \pi_j\varphi(x, v) = \pi_j\varphi(x, y)\}$. We may assume that $\varphi(x \times V) \subseteq (X \times \pi_2\varphi(x, y)) \cup (\pi_1\varphi(x, y) \times Y)$. Then $W_1\backslash\{y\}$ is an open subset of V, since

$$W_1\backslash\{y\} = \pi_2\varphi^{-1}((\pi_1\varphi(x, y) \times Y)\backslash\varphi(x, y))$$

and $\pi_1\varphi(x, y) \times Y\backslash\varphi(x, y)$ is open in $(X \times \pi_2\varphi(x, y)) \cup (\pi_1\varphi(x, y) \times Y)$. Similarly, $W_2\backslash\{y\}$ is open in V.

Let j be fixed and suppose that y had a neighborhood base in $W = W_j$ that consists of open-closed sets. Let $S, T \subseteq W$ be relatively open compact neighborhoods of y such that $S \cap (W\backslash T) \neq \varnothing$. Then $S \cap (W\backslash T)$ is compact, hence closed, in V. Since $y \notin S \cap (W\backslash T)$, $S \cap (W\backslash T)$ is open in W. Since $W\backslash\{y\}$ is open in Y, the set $S \cap (W\backslash T)$ is open in Y. But then $S \cap (W\backslash T)$ is both open and closed in Y, which contradicts the connectedness of Y. The Lemma is proved. □

11.7.9. Lemma. *If y is a corner of $x \times Y$, then x is a corner of $X \times y$.*

Proof. Let $U \times V$ be a neighborhood of (x, y) such that $x \times Y$ and $X \times y$ have at most one corner in U and V.

Since y is a corner of $x \times Y$, $\pi_j\varphi(x \times V) = T_j$ is not a neighborhood of $\pi_j\varphi(x, y)$, for $j = 1$ and 2. But by 11.7.7, $\pi_j\varphi((x \times V) \cup (U \times y))$ is a neighborhood of $\pi_j\varphi(x, y)$. Therefore $\pi_j\varphi(U \times y)$ clusters at $\pi_j\varphi(x, y)$, for $j = 1, 2$. Therefore x is a corner of $X \times y$. □

Proof of Theorem 11.7.1. We shall show that the existence of a corner contradicts the assumption that $X \times Y$ is connected. An application of 11.7.2 then completes the proof.

Suppose that y were a corner of $x \times Y$. Let $U \times V$ be a neighborhood of (x, y) such that V contains no corners of $x \times Y$ other than y, and such that U contains no corners of $X \times u$ other than x. For $j = 1, 2$, let U_j and V_j be defined by

$$U_j = \{u \in U: \pi_j\varphi(u, y) = \pi_j\varphi(x, y)\}$$

and

$$V_j = \{v \in V: \pi_j\varphi(x, v) = \pi_j\varphi(x, y)\}.$$

Then $\pi_2\varphi((U_1 \times y) \cup (x \times V_1))$ is a neighborhood of $q = \pi_2\varphi(x, y)$ by 11.7.7 and $\pi_1\varphi((U_2 \times y) \cup (x \times V_2))$ is a neighborhood of $p = \pi_1\varphi(x, y)$. We define four sets as follows (see Figure 1).

$$\mathrm{A} = [\pi_1\varphi(x \times V_2)] \times [\pi_2\varphi(x \times V_1)];$$
$$\mathrm{B} = [\pi_1\varphi(U_2 \times y)] \times [\pi_2\varphi(x \times V_1)];$$
$$\mathrm{C} = [\pi_1\varphi(U_2 \times y)] \times [\pi_2\varphi(U_1 \times y)]; \text{ and}$$
$$\mathrm{D} = [\pi_1\varphi(x \times V_2)] \times [\pi_2\varphi(U_1 \times y)].$$

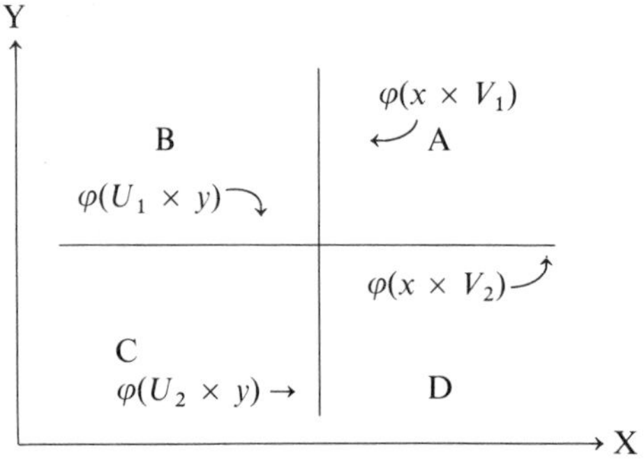

Figure 1

Let $U' \times V' \subseteq U \times V \cap (\varphi^{-1}(A \cup B \cup C \cup D))$ be a neighborhood of (x, y). Let W_j be the connected component of y in $V_j \cap V'$ and let T_j be the connected component of x in $U_j \cap U'$ for $j = 1$ and 2. Then $W_1 \cap W_2 = \{y\}$, $W_1 \neq \{y\} \neq W_2$, and similar relations hold for the T's and x.

Since the four sets U_1, V_1, U_2, and V_2 are infinite, the sets A, B, C, D are each distinct from their intersections with

$$\varphi(x \times (V_1 \cup V_2)) \cup \varphi((U_1 \cup U_2) \times y).$$

Let R be an open set in $(D \cup C)\backslash(A \cup B)$ that meets $\varphi(T_1 \cup T_2 \times y)$ and S an open set in $(B \cup C)\backslash(A \cup D)$ that meets $\varphi((W_1 \cup W_2) \times y)$.

Let $L = L(y') = \varphi((T_1 \cup T_2) \times y')$. If y' is sufficiently close to y, then $L \cap R$ and $L \cap S$ are both nonempty. Then

$$(1) \qquad L \cap (D \cup C) \neq \varnothing \quad \text{and} \quad L \cap (B \cup C) \neq \varnothing.$$

In what follows, we assume that $y' \neq y$ and that the intersections (1) are nonempty. Then

$$L \subseteq (A \cup B \cup C \cup D)\backslash\varphi((U_1 \cup U_2) \times y).$$

Therefore either $L \subseteq C$ or $L \subseteq A \cup B \cup D$, by the connectedness of L. Suppose first that $L \subseteq A \cup B \cup D$. Then $L \cap \varphi(x \times (V_1 \cup V_2)) = \{\varphi(x, y')\}$. If $\varphi(x, y') \in D$, then $L \cap A \cap B = \varnothing$, so $L \subseteq [B\backslash(A \cup D)] \cup [(A \cup D)\backslash B]$, and therefore L would not be connected, a contradiction. Therefore $L \subseteq C$. A similar argument shows that if x' is sufficiently close to x, and $x' \neq x$, then $M = \varphi(x' \times (W_1 \cup W_2)) \subseteq A$. Then of course $L \cup M \subseteq (A \cup C)\backslash(B \cup D)$. Since $[A\backslash(B \cup D)] \cap [C\backslash(B \cup D)] = \varnothing$, $L \subseteq C\backslash(B \cup D)$ and $M \subseteq A\backslash(B \cup D)$. Therefore $L \cap M = \varnothing$. That contradicts the fact that $L \cap M = \{\varphi(x', y')\}$. We therefore have a contradiction: $\varphi(x \times Y)$ and $\varphi(X \times y)$ cannot have corners. That ends the proof of 11.7.1. □

Remarks and Credits. Theorem 11.7.1 was proved by Saeki [8] after Graham [2] proved it for locally connected spaces. The proof given here is a combination of those by Saeki and Graham. The problem of determining the endomorphisms of $V(X, Y)$ remains open. Graham [2] contains examples. Varopoulos [18] has shown that if $f(x, y)$ has continuous second partial derivatives on $[0, 1] \times [0, 1]$ and $\|\exp(inf)\|_V \leq C$ for all n, then $f(x, y) = g(x) + h(y)$. If that could be extended to all continuous f with $\|\exp(inf)\|_V \leq C$, $n \in Z$, the endomorphism problem for $V([0, 1], [0, 1])$ would be solved.

Graham [2] also shows that if $\varphi: X \times Y \to X \times Y$ is a homeomorphism that maps V-Sidon sets onto V-Sidon sets, then φ induces an automorphism of $V(X, Y)$. The corresponding result for $A(T)$ is false; piecewise linear but non-linear maps provide the counterexample.

11.8. V-Sidon and V-Interpolation Sets

Let $X_1, \ldots, X_n$ be Hausdorff locally compact spaces, and let E be a closed subset of $X_1 \times \cdots \times X_n$. E is a *V-Helson set* if every element of $C_o(E)$ is the restriction to E of an element of $V_o(X_1, \ldots, X_n)$. (Here $V_o(X_1, \ldots, X_n) = C_o(X_1) \hat{\otimes} \cdots \hat{\otimes} C_o(X_n)$.) A V-Helson set that is either discrete or countable is called a *V-Sidon set*. If $X_1, \ldots, X_n$ are discrete and $E \subseteq X_1 \times \cdots \times X_n$ is such that every $f \in l^\infty(E)$ is the restriction of a function in $l^\infty(X_1) \hat{\otimes} \cdots \hat{\otimes} l^\infty(X_n)$, then E is a *V-interpolation* set. It is clear that every V-interpolation set is a V-Sidon set and that every V-Sidon set is a V-Helson set. We show that not all V-Sidon sets are V-interpolation sets, and, in fact, that the union of two V-interpolation sets is not necessarily a V-interpolation set. We first summarize results concerning V-Sidon and V-interpolation sets. Many of the results below do not yet have extensions for $n > 2$.

A subset $\{(x_j, y_j): 0 \le j \le 2n\}$ of $X_1 \times X_2$ is a *cycle* if $(x_0, y_0) = (x_{2n}, y_{2n})$ and $x_{2j} = x_{2j+1}$, $x_{2j} \ne x_{2j+2}$, $y_{2j} \ne y_{2j+1}$ and $y_{2j+1} = y_{2j+2}$ for $0 \le j < n$. A subset E of $X_1 \times X_2$ is *cycle-free* if E contains no cycles.

For $j = 1$ and 2, a subset E of $X_1 \times X_2$ is a *j-section* if the projection $\pi_j: X_1 \times X_2 \to X_j$ maps E one-to-one onto its image. A union of a 1-section and a 2-section is called a *bisection.* Sections are obviously cycle-free. Finite unions of sections are V-Sidon by 2.1.2. (Actually, in place of the union Theorem 2.1.2 for Sidon sets, all one needs is the weaker result that a finite union of independent sets is a Sidon set. That follows from the method of A.1.4; Rudin [1, p. 124] contains the details; Rider [1] carries the method to its logical conclusion.)

Let $\beta > 0$. A subset E of $X_1 \times X_2$ is an *R_β-set* if for all finite sets $F_1 \subseteq X_1$ and $F_2 \subseteq X_2$ with $\#F_1 = \#F_2$, $\#[E \cap (F_1 \times F_2)] \le \beta \# F_1$. It is obvious that a union of k sections is an R_k-set. For a proof of the next result, see Lindahl and Poulsen [1, p. 81].

11.8.1. Theorem. *Let X_1 and X_2 be locally compact Hausdorff spaces and E a discrete closed subset of $X_1 \times X_2$. Then the following are equivalent.*

(i) *E is V-Sidon.*
(ii) *E is a finite union of k cycle-free sets.*
(iii) *E is a finite union of k sections, for some integer $k \ge 1$.*
(iv) *For some $0 < \beta < \infty$, E is an R_β-set.*

Let βX and βY denote the Stone-Čech compactifications of the discrete spaces X and Y. Then $l^\infty(X) \hat{\otimes} l^\infty(Y) = V(\beta X, \beta Y)$, and, in that latter form, $l^\infty(X) \hat{\otimes} l^\infty(Y)$ is familiar. We shall write $V(X, Y)$ for $l^\infty(X) \hat{\otimes} l^\infty(Y)$. Thus, $V(X, Y) = V(\beta X, \beta Y)$. By $N(X, Y)$ we mean all $f \in l^\infty(X \times Y)$ such that $fg \in V_o(X, Y)$ for all $g \in V_o(X, Y)$. Clearly, N is a commutative Banach algebra with norm

$$\|f\|_N = \sup\{\|fg\|_{V_o}: g \in V_o, \|g\|_{V_o} \le 1\}.$$

For a proof of the next result, see Varopoulos [12].

11.8.2. Lemma. *Let Γ be a discrete abelian group. For each $\gamma \in \Gamma$ and each $f \in l^\infty(\Gamma \times \Gamma)$, let $\tau_\gamma f(x, y) = f(x - \gamma, y + \gamma)$. Then:*

(i) *τ_γ is an automorphism of each of the algebras $V_0(\Gamma, \Gamma)$, $V(\Gamma, \Gamma)$, and $N(\Gamma, \Gamma)$.*

(ii) $B^*(\Gamma) = \{f \in N(\Gamma, \Gamma): \tau_\gamma f = f$ for all $\gamma \in \Gamma\}$ *is isomorphic to* $B(\Gamma)$ *under the mapping* $M: B(\Gamma) \to B^*(\Gamma)$, *where* M *is defined by* $Mf(x, y) = f(x + y)$ *for all* $f \in B(\Gamma)$ *and* $x, y \in \Gamma$.

(iii) *The adjoint* M^* *of* M *maps the maximal ideal space* $\Delta N(\Gamma, \Gamma)$ *onto the maximal ideal space of* $B(\Gamma)$.

Let $\varepsilon > 0$. A subset E of the locally compact abelian group Γ is an ε-*Kronecker* set if for every continuous unimodular function f on E, there exists a continuous character γ, on Γ such that $|f - \gamma| < \varepsilon$ on E. We leave the proof of the following Lemma to the reader.

11.8.3. Lemma. *Let* Γ *be an infinite discrete abelian group. Then either* Γ *contains an infinite 1/3-Kronecker set or an infinite* K_p*-set for some prime* $p \geq 2$.

The proof of the next Theorem uses maps M and P, as in 11.1.1; P is obtained as a limit of averages; see Varopoulos [12] for the details.

11.8.4. Theorem. *Let* $0 < \varepsilon \leq 1/3$. *Let* X *and* Y *be countably infinite, disjoint subsets of the discrete abelian group* Γ. *Suppose that* $X \cup Y$ *is either* ε*-Kronecker or a* K_p*-set for some prime* p. *Then the following pairs of algebras are isomorphic. In each case the norm of the isomorphism depends only on* ε *or* p.

(i) $A(X + Y)$ *and* $V_o(X, Y)$
(ii) $B(X + Y)$ *and* $N(X, Y)$
(iii) $B_d(X + Y)$ *and* $V(X, Y)$
(iv) $AP(X + Y)$ *and* $APV(X, Y)$.

In (iv) $AP(X + Y)$ *denotes the restriction to* $X + Y$ *of the almost periodic functions on* Γ *and* $APV(X, Y)$ *denotes the uniform closure of* $V(X, Y)$ *in* $l^\infty(X \times Y)$.

Two subsets E and F of $X \times Y$ are *separated by n rectangles* if there exist subsets $X_1, \ldots, X_n$ of X and $Y_1, \ldots, Y_n$ of Y such that either $E \subseteq \bigcup [X_j \times Y_j]$ and $F \cap \bigcup [X_j \times Y_j] = \varnothing$ or $F \subseteq \bigcup [X_j \times Y_j]$ and $E \cap \bigcup [X_j \times Y_j] = \varnothing$. A subset E of $X \times Y$ has the *separation property* if any two disjoint subsets E_1 and E_2 of E may be separated by some number (depending on E_1 and E_2) of rectangles. A subset E has the *n-separation* property if any two disjoint subsets of E may be separated by at most n rectangles. A straightforward induction proves the following Lemma.

11.8.5. Lemma. *If* E *has the separation property, then* E *has the n-separation property for some integer* n.

For a set $F \subseteq X \times Y$, we let $r(F) = \pi_1 F \times \pi_2 F$. For $E \subseteq X \times Y$ and $x, y \in E$, we write $x \sim_E y$ if and only if $\#[r(\{x, y\}) \cap E] \geq 3$. Otherwise we write $x \nsim_E y$. The proof of the next Lemma is easy.

11.8.6. Lemma. *Suppose that $E \subseteq X \times Y$ and is such that for all $x, y \in E$, $x \not\sim_E y$ unless $x = y$. Then E is an R_2-set.*

We let $G(E)$ denote the graph whose vertices are the elements of E and whose edges connect $x, y \in E$ if and only if $x \sim_E y$. A *coloration* of $G(E)$ is a map $f: E \to N$ such that $x \sim_E y$ implies $f(x) \neq f(y)$. The smallest number of elements in the range of a coloration is denoted by $k(E)$ if it is finite, and otherwise we set $k(E) = \infty$. Each value of the coloration f will be called a *color*.

11.8.7. Theorem. *Let X and Y be discrete spaces and $E \subseteq X \times Y$. Then the following conditions are equivalent.*

(i) *E is a V-interpolation set.*
(ii) *E has the separation property.*
(iii) *$k(E) < \infty$.*

Proof. (i) $\Rightarrow$ (ii). Suppose that $E = E_1 \cup E_2$, and that $E_1 \cap E_2 = \varnothing$. Since E is a V-interpolation set, $\chi_{E_1} \in V(X, Y)|_E$. But χ_{E_1} can be approximated to within $\frac{1}{3}$, in V-norm, by finite linear combinations of functions $f_j \otimes g_j$, where f_j, g_j take on only the values zero and one for $1 \leq j \leq n$. Let φ be the function from $(X \times Y)$ to $\{0, 1\}^n$ given by $\varphi(x, y) = (f_1(x)g_1(y), \ldots, f_n(x)g_n(y))$, for $(x, y) \in E$. Then $\varphi(E_1) \cap \varphi(E_2) = \varnothing$. For each $z \in \{0, 1\}^n$, $\varphi^{-1}(z)$ is a rectangle, so some finite union of these contains E_1 and therefore separates E_1 and E_2. It follows that E has the separation property. The proof that (ii) $\Rightarrow$ (iii) $\Rightarrow$ (i) is left to the reader (see Varopoulos [12]). □

11.8.8. Theorem. *Let $\{(p_n, q_n): 1 \leq n < \infty\}$ be an enumeration of $N \times N$. Let $E_1 = \{(n, p_n): 1 \leq n < \infty\}$ and $E_2 = \{(n, q_n): 1 \leq n < \infty\}$. Then $E_1 \cup E_2$ is not a set of V-interpolation in $N \times N$.*

Proof. We argue by contradiction. Suppose that rectangles $X_1 \times Y_1, \ldots, X_n \times Y_n$ are given, with $E_1 \subseteq \bigcup_1^n X_k \times Y_k$. Without loss of generality, we may assume that either $Y_k = Y_j$ or $Y_k \cap Y_j = \varnothing$ for $1 \leq j, k \leq n$. (That follows by taking intersections and differences.) On replacing the collection $\{X_j\}$ by the collection $\{X'_j\}$ where $X'_j = \bigcup \{X_k: Y_j = Y_k\}$, we may assume that $Y_j \cap Y_k = \varnothing$ for $1 \leq j \neq k \leq n$.

Since $\{(p_m, q_m)\}_1^\infty = N \times N$ and at least one Y_j is infinite, there exists an integer m such that both p_m and q_m are elements of such a Y_j, say Y_{j_0}, and $p_m \neq q_m$. Since $Y_{j_0} \cap Y_k = \varnothing$ for all $k \neq j_0$, we must have $(m, p_m) \in X_{j_0} \times Y_{j_0}$. Therefore $(m, q_m) \in X_{j_0} \times Y_{j_0}$, so $(E_2 \backslash E_1) \cap (\bigcup X_k \times Y_k) \neq \varnothing$ and E_1 and E_2 cannot be separated by rectangles. By 11.8.7, the Theorem is proved. □

11.8.9. Theorem. *Let X and Y be infinite discrete spaces. Then:*

(i) *$N(X, Y) \cap C_o(X \times Y)$ is entire.*
(ii) *For every $n \geq 1$, there exists $f \in N(X, Y)$ such that $f^{n-1} \notin V_o(X, Y)$ and $f^n \in V_o(X, Y)$.*

Proof. (i) For $j \geq 1$, let G_j be a finite abelian group such that $A(G_j)$ contains a real-valued function f_j such that $\|f_j\| = 1$, $\|f_j^{j+1}\| < 2^{-j}$ and $\|\sum_1^j c_k f_j^k\| \geq (1 - 2^{-j}) \sum |c_k|$ whenever $\{c_1, \ldots, c_j\} \subseteq \mathbb{C}$. That such f_j, G_j exist follows from the method of 7.5. (We simply modify an appropriate Riesz product on $(Z_2)^k$, for example.)

Let $\{X_j \times Y_j\}$ be a sequence of bidisjoint rectangles with $\# X_j = \# Y_j =$ Card G_j for all j. Identifying X_j and Y_j with G_j, we have $Mf_j \in V(X_j, Y_j)$. Then $f = \sum Mf_j \in N(X, Y)$ by 11.1.4. Clearly $f \in C_o(X, Y)$, and the methods of 11.3 and 9.5 show that only entire functions operate.

(ii) For $j \geq 1$ let G_j be a finite abelian group such that $A(G_j)$ contains a function f_j such that $1 \leq \|f_j^k\| \leq 2$ for $1 \leq k \leq n - 1$ and $\|f_j^n\| < 2^{-j}$. Such G_j and f_j may be found by using the methods of 9.5. Then proceeding as in the second paragraph of (i), we have $f = \sum_1^\infty Mf_j \in N$, and $f^{n-1} \notin V_o$, while $f^n \in V_o$. We leave the details to the reader. □

Remarks and Credits. The results of this Section are due to Varopoulos [12], except for 11.8.9 (ii), which is a variant of 11.9.1 (ii). See also Lindahl and Poulsen [1, p. 76 ff]. Stegeman [1] shows that $k(E) \leq 4$ is always true for bisections and that that estimate is the best possible. In Stegeman [5], 11.8.7 is extended to V-interpolation sets for $V(X_1, \ldots, X_n)$. In case $n = 2$ Stegeman [5] obtains the result that E is of V-interpolation if and only if E is a union of m bisections $E_j = E_{1j} \cup E_{2j}$ where E_{kj} is a k-section and $\pi_k E_{kj} \cap \pi_k E_{kl} = \emptyset$ for $1 \leq k \leq 2$ and $1 \leq l \neq j \leq m$.

The union theorem for V-Sidon sets can be proved without using Drury's Lemma, using arguments simpler than those required in Chapter 2. See Galanis [3] for details.

11.9. Tilde Tensor Algebras

Let X and Y be compact Hausdorff spaces and $V = V(X, Y)$. We define a new algebra $\tilde{V} = \tilde{V}(X, Y)$ as follows. A function $f \in C(X \times Y)$ belongs to $\tilde{V}$ if and only if there exists a sequence $\{f_n\} \subseteq \tilde{V}$ such that

$$\lim \|f - f_n\|_\infty = 0 \tag{1}$$

and

$$\limsup \|f_n\|_V < \infty. \tag{2}$$

The norm $\|f\| = \|f\|_{\tilde{V}}$ is the infimum of the numbers (2) such that (1) holds. With that norm $\tilde{V}$ is a commutative Banach algebra of functions on $X \times Y$. It is the continuous analogue of the algebra $N(X, Y)$ of Section 11.8. For more general versions of tilde algebras, see Chapter 12.

11.9.1. Theorem. *Let X and Y be infinite compact Hausdorff spaces. Then:*

(i) *V is closed in $\tilde{V}$ and $\|f\|_{\tilde{V}} = \|f\|_V$ for all $f \in \tilde{V}$.*
(ii) *There exists $f \in \tilde{V}$ such that $f \notin V$ and $f^2 \in \tilde{V}$.*
(iii) *$\tilde{V}$ is entire.*

Proof. (i) is an immediate consequence of 11.1.4 and the definition of $\tilde{V}$.

(ii) For $j = 1, 2, \ldots$ let $\{X_j\}$ and $\{Y_j\}$ be sequences of compact pairwise disjoint subsets of X and Y. For each j, let $f_j \in V(X_j, Y_j)$ be such that $\|f_j\|_V = 1$, $f_j^{1/2} \in V$, and $\|f_j^{1/2}\|_V > 2^j$. Such f_j exists because square root does not operate in V. Let g_j be an extension of $f_j^{1/2}$ to an element of $V(X, Y)$ such that the g_j's have bidisjoint supports. Let $h_j = \|g_j\|_{\tilde{V}}^{-1} g_j$. Then $\|h_j\|_\infty$ tends to zero, so $f = \sum h_j \in \tilde{V}$, by 11.1.4 and the definition of $\tilde{V}$. But since $\|h_j^2\| < 2^{-j}$, we see that $f^2 \in \tilde{V}$.

What we have done so far does not prove that $f \notin V$. We leave the reduction to the case of metrizable X and Y to the reader. We return to our choice of $\{X_j\}$ and $\{Y_j\}$. We may assume that there exist points x and y such that for every neighborhood U of x and $V$of y, there exists a positive integer J such that $X_j \subseteq U$ and $Y_j \subseteq V$ whenever $j \geq J$. Then in every neighborhood of (x, y), f has restriction algebra norm at least one, while $f(x, y) = 0$. Since every singleton is a set of synthesis for V, we see that $f \notin V$.

(iii). This is proved exactly as Theorem 11.8.9 (i) is proved. □

Because V is a closed subalgebra of $\tilde{V}$, there exists a mapping $\pi: \Delta\tilde{V} \to X \times Y$, where $\pi(\psi) = (x, y)$ if $\hat{f}(\psi) = f(x, y)$ for all $f \in V$. Here, $\hat{f}$ denotes Gel'fand transform. The next theorem was suggested by the corresponding result for H^∞ (Hoffman [1, Chapter 10]).

11.9.2. Theorem. *Let X and Y be infinite compact Hausdorff spaces. Then the following hold.*

(i) *For each $z \in X \times Y$, $\pi^{-1}(z)$ is connected.*
(ii) *If $\{\psi_j\}$ is a sequence in $\Delta\tilde{V}$ that converges in the Gel'fand topology, then either $\lim \psi_j = \psi \in X \times Y$ or $\pi(\psi_j)$ is eventually constant.*
(iii) *If $\#\pi^{-1}(z) > 1$, then $\pi^{-1}(z)$ has non-empty interior in V.*

Proof. (i) We use the well-known Šilov Idempotent Theorem (see, for example, Gamelin [1, p. 88]) which states that if the maximal ideal space of a commutative Banach algebra contains a compact open subset, then there is an idempotent element of the algebra whose Gel'fand transform is the indicator function of that compact open subset.

Let I be the ideal of functions in $\tilde{V}$ whose Gel'fand transforms vanish on the fiber $\pi^{-1}(z)$. It is easily seen that the maximal ideal space of $\tilde{V}/I$ is $\pi^{-1}(z)$. If $\pi^{-1}(z)$ is disconnected, let $f + I$ be an idempotent given by Šilov's Theorem. We shall show that f has a representative g such that $g(y) = g(y)^2$ holds for all y in a neighborhood of x. It is easy to see that then g (and therefore f) is constant on $\pi^{-1}(z)$. Hence $\pi^{-1}(z)$ is connected. It remains to find the representative g.

Suppose S is any rectangular closed neighborhood of x. Consider the three sets:

$$\begin{aligned} &\{\psi: |\psi(f)| > \tfrac{1}{2}, \psi \in \pi^{-1}(S)\}, \\ (3) \qquad &\{\psi: |\psi(f)| < \tfrac{1}{2}, \psi \in \pi^{-1}(S)\}, \\ &\{\psi: |\psi(f)| = \tfrac{1}{2}, \psi \in \pi^{-1}(S)\}. \end{aligned}$$

It is easily seen that $\pi^{-1}(S)$ is the maximal ideal space of the restriction of $\tilde{V}$ to S, that is, the quotient $\tilde{V}/J(S)$, where $J(S)$ is the ideal of $f \in \tilde{V}$ such that $\hat{f}(\psi) = 0$ for all $\psi \in \pi^{-1}(S)$.

Suppose that for each S, the third set in (3) is never empty. Then choose a net $\{S(\beta)\}$ of closed rectangular neighborhoods of such that $\bigcap_\beta S(\beta) = \{z\}$. Let ψ_β be an element of $\pi^{-1}(S(\beta))$ such that $|\psi_\beta(f)| = \frac{1}{2}$, and let ψ be an accumulation point of the ψ_β. Now ψ must lie in the intersection of the cylinders $\pi^{-1}(S(\beta))$, that is in $\pi^{-1}(z)$. Hence we have a $\psi \in \pi^{-1}(z)$ such that $|\hat{f}(\psi)| = \frac{1}{2}$. That contradiction shows that for one closed rectangular neighborhood S of x, the maximal ideal space $\pi^{-1}(S)$ is the union of the first two sets in (3) above. Those two sets are open and compact, so there is an idempotent $g' + J(S)$ which is one on the first and zero on the second. Pick any representative g of g'. It is easy to see that $g = g^2$ in the neighborhood S. That completes the proof of (i).

(ii) We may assume that $\psi \notin X \times Y$ and that the ψ_j are all in different fibers $\pi^{-1}(z_j)$. Because π is continuous, z_j converges to $z = \pi(\psi)$. Because $\psi \notin X \times Y$ there is an $f \in \tilde{V}$ such that $f(z) = 0$ and $\hat{f}(\psi) = 1$.

We shall show that those hypotheses are contradictory by finding a subsequence $\{\psi_k\}$ of $\{\psi_j\}$ which does not converge. Since the π-projections $\pi\psi_j = z_j = (x_j, y_j)$ are distinct, there exists a subsequence $\{z_k = (x_k, y_k)\}$ of $\{z_j\}$ such that either the x_k are distinct in X, or the y_k are distinct in Y. We may assume that the x_k are distinct elements of X.

Let $g(v) = f(x, v)$ for all $v \in Y$. Then $f' = f - 1 \otimes g \in \tilde{V}$, $\hat{f}'(\psi) = 1$ and $f' = 0$ on $\{x\} \times Y$. We may therefore assume that $f = 0$ on $\{x\} \times Y$.

The x_k converge to x, the first coordinate of $z = \pi(\psi)$. For each j, let X_j be a closed neighborhood of x_j such that the X_j are pairwise disjoint, $x_j \in X_j$, and such that for each neighborhood U of x, there exists an integer N such that $j \geq N$ implies $X_j \subseteq U$. For each $j = 1, 2, \ldots$ choose $k_j \in C(X)$ such that $k_j(x_j) = 1$, the support of k_j is contained in X_j, and $\|k_j\|_\infty \leq 1$. Consider the finite sums

$$(4) \qquad F_n = \left(\sum_{j=1}^{n} (-1)^j k_j \otimes 1\right)(f).$$

Since the sum $\sum_1^n (-1)^j k_j$ has supremum one on X_1, the $\tilde{V}$-norm of F_n is bounded by $\|f\|$. For $n \geq j$, F_n agrees on a neighborhood of x_j with $[(-1)^j k_j \otimes 1]f$. Since $k_j \otimes 1 \in V$, it is constant on the fibers $\pi^{-1}(u)$ for all $u \in X \times Y$. Therefore

$$\hat{F}_n(\psi_j) = (-1)^j k_j(y_j)\hat{f}(\psi_j) = (-1)^j \hat{f}(\psi_j) \quad \text{for } 1 \leq j \leq n. \tag{5}$$

Since $f = 0$ on $\{x\} \times Y$, F_n converges uniformly on $X \times Y$, so $F = \lim F_n \in \tilde{V}$. But

$$\hat{F}(\psi_j) = (-1)^j \hat{f}(\psi_j) \quad \text{for } 1 \leq j < \infty. \tag{6}$$

Since ψ_j converges to ψ and $\hat{f}(\psi) = 1$, we have

$$\hat{F}(\psi_j) = (-1)^j \hat{f}(\psi_j) \approx (-1)^j,$$

for sufficiently large j, that is, ψ does not converge. That contradiction completes the proof of (ii).

(iii) Let $z = (x, y)$. If either x or y is isolated in its factor of $X \times Y$, then $V(K)$ restricted to the closed neighborhood $\{x\} \times X$ (or $X \times \{y\}$, as the case may be) of x is all of $C(\{y\} \times Y)$. Hence the fiber over x is one point. We may therefore assume that neither x nor y is isolated. Then, as in the proof of 11.8.9, we can find a sequence F_j of elements of $V(K)$ such that the following hold for $1 \leq j \neq k < \infty$. Supp $F_j \subseteq X_j \times Y_j$, where X_j and Y_j are closed;

$$Y_j \cap Y_k = \varnothing = X_j \cap X_j \quad \text{for } j \neq k < \infty; \tag{7}$$
$$\|F_j\|_\infty < 2^{-j-1};$$

$X_j \times Y_j$ is contained eventually in each neighborhood of z; and such that for any choice $c_1, \ldots, c_n$ of complex numbers

$$\sup\left[\left\| \sum_{k=1}^n c_k F_j^k \right\|_V : j = 1, 2, \ldots\right] = \sum_{k=1}^n |c_k|. \tag{8}$$

We set $F = \sum_{j=1}^\infty F_j$. By applying 11.1.4 repeatedly we see that $F \in \tilde{V}$. Formula (8) show that the spectral radius of $(F + 1)$ is two. Hence there exists a maximal ideal ψ of $\tilde{V}$ such that $\hat{F}(\psi) = 1$.

On the other hand, if φ is any maximal ideal of $\tilde{V}$ and $\varphi \notin \pi^{-1}(x)$, then $\hat{F}(\varphi)$ is given by $\sum_{j=1}^N \hat{F}_j(\varphi)$ for some finite N. That follows from the fact that the $Y_j \times Z_j$ are eventually in a neighborhood of x which does not contain $\pi(\varphi)$. By (7) and the fact that $F_j \in V$ are constant on fibers, we have $|\hat{F}(\varphi)| \leq \frac{1}{2}$.

Therefore, the non-void open set $\{\psi: |F(\varphi)| > \frac{3}{4}\}$ is contained in $\pi^{-1}(z)$. The reduction of the case of metrizable X, Y is left to the reader. □

Remarks and Credits. The results of this section are from Varopoulos [11] and Graham [1]. When $X = Y = G$, a compact group, then the group action on $G \times G$ maps the fibers $\pi^{-1}(z)$ homeomorphically onto each other. Thus, as sets, $\Delta\tilde{V}$ and $G \times G \times \pi^{-1}(z)$ are identical. Theorem 11.9.2(ii) shows that that set identification is not topological.

When X and Y are convergent sequences with limit points x and y, then the only nontrivial fiber is over (x, y).

By applying the methods of Chapter 12, one can construct sets E such that $V(E)$ is not closed in $V(E)^{\sim}$; See Saeki [11] for details.

Corollary 11.5.3 shows that smoothness conditions are of little obvious help in determining membership in $\tilde{V}(T, T)$.

The following are open questions. If X and Y are infinite compact Hausdorff spaces and $n \geq 3$, does there exist $f \in \tilde{V}(X, Y)$ such that $f^{n-1} \notin V(X, Y)$ and $f^n \in V(X, Y)$? Is the Šilov boundary of $\tilde{V}$ the entire maximal ideal space? Are the non-symmetric maximal ideals of $\tilde{V}$ dense in $\Delta\tilde{V}$?

Chapter 12

Tilde Algebras

12.1. Introduction

Let E be a closed subset of a locally compact abelian group Γ. We define $\tilde{A}(E) = \tilde{A}(E, \Gamma)$ to be the class of functions f in $C_o(E)$ such that the quantity

$$(1) \qquad \|f\|_{\tilde{A}(E)} = \sup\left\{\frac{|\langle f, \mu\rangle|}{\|\mu\|_{PM}} : \mu \in M(E), \mu \neq 0\right\}$$

is finite. Of course $N(E)$, the Banach space dual of $A(E)$, contains $M(E)$, and for $f \in A(E)$,

$$\|f\|_{A(E)} = \sup\left\{\frac{|\langle f, S\rangle|}{\|S\|_{PM}} : S \in N(E), S \neq 0\right\}.$$

Evidently $\|f\|_{\tilde{A}(E)} \leq \|f\|_{A(E)}$ for every $f \in A(E)$, and $A(E) \subset \tilde{A}(E) \subset C_o(E)$ for every E.

12.1.1. Proposition. *Let $f \in C_o(E)$. Then $f \in \tilde{A}(E)$ if and only if it is the uniform limit on E of a bounded sequence in $A(E)$. Moreover,*

$$(2) \qquad \|f\|_{\tilde{A}(E)} = \inf\left\{\sup_n \|f_n\|_{A(E)} : \|f - f_n\|_{C(E)} \to 0\right\}.$$

That result, whose proof appears below, makes it clear that with pointwise multiplication and the norm provided by (1), $\tilde{A}(E)$ is a Banach algebra. It is called the *tilde algebra* of E. Our purpose is to discuss the circumstances under which $\tilde{A}(E)$ is strictly larger than $A(E)$ and the properties that $\tilde{A}(E)$ enjoys when that happens. It is reasonable, up to a point, to treat the case of an arbitrary closed E in an arbitrary Γ, as we have been doing so far. However, the subject of tilde algebras has two rather different main parts, one for the case when E is countable (or else a subset of a discrete group), and one for the case when E is perfect. Section 1.8 treats a variant of the subject, involving sets E of positive Haar measure.

In Sections 12.2 and 12.3 we shall develop the theory for the case when E is a subset of a discrete group; then $\tilde{A}(E)$ is none other than $B(E) \cap C_o(E)$. As of this writing, there is no satisfactory characterization of the sets E for which $\tilde{A}(E) = A(E)$. For all we know, they may be precisely the Sidon sets. The best partial results are presented in Section 12.2.

In Section 12.4, turning to the tilde algebras of perfect sets, we shall describe the "Sigtuna set" $E \subset R$, for which $\tilde{A}(E)$ is non-separable (so that, in particular, $\tilde{A}(E) \neq A(E)$), and which has the further property that

$$\sup\left\{\frac{\|\mu\|_M}{\|\mu\|_{PM}} : \mu \in M_c(E), \mu \neq 0\right\} < \infty,$$

even though E is of course non-Helson. In Section 12.5 we shall present a delicate construction of sets $E \subset R$ such that $A(E)$ is a dense proper subspace of $\tilde{A}(E)$. The results presented have to do with perfect sets of zero Haar measure; the reader should compare the material of Section 1.8. The rest of the present introductory section is devoted to basic results and examples.

Proof of 12.1.1. Let $\|f\|'$ denote the right-hand side of (2). If $\{f_n\}$ is a bounded sequence in $A(E)$, $f_n \to f$ uniformly on E, and $\mu \in M(E)$, then

$$|\langle f, \mu\rangle| = \left|\lim_{n\to\infty} \langle f_n, \mu\rangle\right| \leq \left(\sup_n \|f_n\|_{A(E)}\right)\|\mu\|_{PM},$$

so that $f \in A(E)$ and $\|f\|_{A(E)} \leq \|f\|'$.

Let U be the unit ball in $A(E)$, but consider U as a subset of $C_o(E)$. To complete the proof of the proposition, it suffices to show that if $\|f\|_{A(E)} \leq 1$, then f is in the closure of U. If it were not, then there would exist $\mu \in M(E)$ such that $|\langle g, \mu\rangle| \leq 1$ for every $g \in U$, but $\langle f, \mu\rangle > 1$ (see Dunford and Schwartz [1, V.2.12]). These inequalities imply that $\|\mu\|_{PM} \leq 1$ and $\|f\|_{\tilde{A}(E)} > 1$, which is impossible. The Proposition is proved. □

Note that in 12.1.1 and its proof, norm convergence in $C_o(E)$ may be replaced, everywhere that it appears, by weak convergence, that is, convergence in the topology $\sigma(C_o(E), M(E))$.

Consider the quantity

$$\beta(E) = \sup\left\{\frac{\|f\|_{A(E)}}{\|f\|_{\tilde{A}(E)}} : f \in A(E), f \neq 0\right\}.$$

With $A(E)$, $M(E)$, and $\beta(E)$ in the roles of B, Γ, and s, respectively, Theorem 4.3.1 gives us a good bit of information, and in particular tells us that $\beta(E)$ is finite (and thus $A(E)$ is closed in $\tilde{A}(E)$) if and only if $M(E)$ is weak* sequentially dense in $N(E)$. When E is countable, $M(E)$ of course contains only discrete measures, and the inclusion of $A(E)$ in $\tilde{A}(E)$ is always an isometry, as we are about to show. On the other hand, among perfect sets E there are examples such that $A(E)$ is not even closed in $\tilde{A}(E)$, as we shall show in Section 12.5.

12.1.2. Proposition. *If E is countable, then $\beta(E) = 1$.*

Proof. Let $f \in A(E)$. For every $\varepsilon > 0$, there is a compact set $K \subset E$ such that $\|f\|_{A(K)} > \|f\|_{A(E)} \cdot (1 - \varepsilon)$, and thus a pseudomeasure $S \in PM(K)$ such that $\|S\|_{PM} = 1$ and $\langle f, S \rangle > \|f\|_{A(E)} \cdot (1 - \varepsilon)$. Since S is a pseudomeasure with countable norm spectrum, $\hat{S}$ is almost-periodic (by a theorem due to Loomis [1]). Therefore there is a sequence of discrete measures, whose supports are contained in the norm spectrum of S, that converges to S in PM-norm. Consequently $\|f\|_{\tilde{A}(E)} > \|f\|_{A(E)}(1 - \varepsilon)$. The Proposition is proved. □

For a treatment of distributions and almost-periodic functions that contains the basic results we have used here, see Katznelson [1, Chapter VI].

Tilde algebras of countable sets can be treated in terms of tilde algebras of subsets of discrete groups, about which we now state some useful facts.

12.1.3. Proposition. *Let E be a subset of a discrete group Γ, whose dual group is G. Then:*

(i) *$\tilde{A}(E) = C_o(E) \cap B(E)$, and*

$$\|f\|_{\tilde{A}(E)} = \inf\{\|\nu\|_M : \nu \in M(G) \quad \textit{and} \quad \hat{\nu} = f \textit{ on } E\}. \tag{3}$$

(ii) *If $f \in C_o(E)$, then $f \in A(E)$ if and only if for every $\varepsilon > 0$, there exists a finite set K such that $\|f\|_{\tilde{A}(E \setminus K)} < \varepsilon$.*

(iii) *If $\tilde{A}(E) \neq A(E)$, then $\tilde{A}(E)$ is non-separable.*

(iv) *If Γ is infinite, then $\tilde{A}(\Gamma) \neq A(\Gamma)$.*

Proof of (i). Let $B(E)_o$ denote $C_o(E) \cap B(E)$, and let $\|f\|_{B(E)_o}$ denote the right-hand side of (3). If $\nu \in M(G)$ and $\mu \in M(E)$, then $|\int_E \hat{\nu}\, d\mu| = |\int_G \hat{\mu}\, d\nu| \leq \|\mu\|_{PM}\|\nu\|_M$, so that $\|\hat{\nu}\|_{\tilde{A}(E)} \leq \|\nu\|_M$. It follows that $B(E)_o \subset \tilde{A}(E)$ and $\|f\|_{\tilde{A}(E)} \leq \|f\|_{B(E)_o}$ for $f \in B(E)_o$.

If $f \in \tilde{A}(E)$, then the mapping $\hat{\mu} \to \int f\, d\mu$ defines a bounded linear functional of norm $\|f\|_{\tilde{A}(E)}$ on the subspace $\{\hat{\mu} : \mu \in M(E)\}$ of $C(G)$. Therefore there exists $\nu \in M(G)$ such that $\|\nu\|_M \leq \|f\|_{\tilde{A}(E)}$ and $\int f\, d\mu = \int \hat{\mu}\, d\nu$ for all $\mu \in M(E)$. In particular, $\hat{\nu} = f$ on E. It follows that $\tilde{A}(E) \subset B(E)_o$ and $\|f\|_{B(E)_o} \leq \|f\|_{\tilde{A}(E)}$ for $f \in \tilde{A}(E)$. Part (i) is proved.

(ii). If $f \in A(\Gamma)$, and $\varepsilon > 0$, there exists $g \in A(\Gamma)$ with finite support K such that $\|f - g\|_A < \varepsilon$. Since $f - g = f$ on $\Gamma \setminus K$, it follows that $\|f - g\|_{A(E \setminus K)} < \varepsilon$. The "only if" part is proved. The other half of the proof goes as follows. Let $f \in C_o(E)$, let $\varepsilon > 0$, and let K be the corresponding finite set, as stated. Then there exists $\nu \in M(G)$ such that $\|\nu\|_M < \varepsilon$ and $\hat{\nu} = f$ on $E \setminus K$. There exists

$w \in A(\Gamma)$ such that $\|w\|_A < 2$, $w = 1$ on K, and $w = 0$ outside some finite set L. There exists $g \in L^1(G)$ such that $\hat{g} = f$ on L. Then $f = w\hat{g} + (1 - w)\hat{v}$ on E, so $f \in B(E)$ and $\|f - w\hat{g}\|_{B(E)} < 3\varepsilon$. Therefore the distance from f to $A(E)$ is zero, and hence $f \in A(E)$. Part (ii) is proved.

(iii). If E is uncountable, then $A(E)$ itself is non-separable.

Suppose that E is countable and that $\tilde{A}(E)$ contains a countable dense subset $\{g_j\}$. Let $f \in \tilde{A}(E)$, and let $\varepsilon > 0$. By 12.1.4 below, to prove that $f \in A(E)$, it suffices to show that for some neighborhood V of 0 in G, $\|f - f_x\|_{\tilde{A}(E)} < \varepsilon$ for all $x \in V$, where $f_x(\gamma) = \langle x, -\gamma\rangle f(\gamma)$ for $\gamma \in \Gamma$. It follows from (1) that for arbitrary $h \in \tilde{A}(E)$,

$$\|h\|_{\tilde{A}(E)} = \sup\{\|h\|_{\tilde{A}(F)}: F \text{ is a finite subset of } E\}.$$

Since E is countable, it follows that the mapping $x \to \|f_x - g_j\|_{A(E)}$ is measurable, and hence that each of the sets

$$G_j = \{x \in G: \|f_x - g_j\|_{\tilde{A}(E)} < \varepsilon\}$$

is measurable. Since $G = \bigcup_j G_j$, there exists k such that G_k has positive measure and hence, by Lemma 8.3.4, $G_k - G_k$ contains a neighborhood V of 0. If $x \in V$, then $x = x_1 - x_2$ where $x_i \in G_k$, so that $\|f - f_x\|_{\tilde{A}(E)} = \|f_{x_1} - f_{x_2}\| \leq \|f_{x_1} - g_k\| + \|g_k - f_{x_2}\| < 2$. Part (iii) is proved, using the Lemma below.

(iv). Since there exist singular measures μ such that $\hat{\mu} \in C_o(\Gamma)$ (see, for example, 4.7.1 or 6.8.1), (iv) follows from (i) and (ii) (note Lemma 9.5.2). □

12.1.4. Lemma. *Let G and Γ be dual locally compact abelian groups. Let $E \subset \Gamma$, and let $f \in B(E)$. For $\gamma \in \Gamma$ and $x \in G$, let $f_x(\gamma) = \langle x, -\gamma\rangle f(\gamma)$. If the mapping $x \to f_x$ is continuous from G into $B(E)$, then $f \in A(E)$.*

Proof. Let $\varepsilon > 0$. Let V be a neighborhood of 0 in G such that $\|f - f_x\|_{B(E)} < \varepsilon$ for $x \in V$. Let $g \in L^1(G)$, $\|g\|_1 = \int g = 1$, and $\operatorname{supp}(g) \subset V$. The product $f\hat{g}$ belongs to $A(E)$, and may be represented as the $B(E)$-valued integral $\int g(x) f_x\, dx$. Therefore

$$\operatorname{dist}(f, A(E)) \leq \|f - f\hat{g}\|_{B(E)} = \left\| \int g(x)(f - f_x)dx \right\|_{B(E)}$$

$$\leq \|g\|_{L^1(G)} \sup_{x \in V} \|f - f_x\|_{B(E)} < \varepsilon,$$

so $f \in A(E)$. The Lemma is proved. □

We shall now give an example of a countable set F of R_d (the discrete reals) such that $\tilde{A}(F) \neq A(F)$. This example was essentially the first to be discovered and has several striking features: $\tilde{A}(F)$ is not self-adjoint; F is not dense in the maximal ideal space; and the functions that operate in $A(F)$ are real-entire. It is not known whether (in discrete groups) those features follow from the property that $\tilde{A}(F) \neq A(F)$.

12.1.5. Proposition. *Let $\{K(j)\}$ be a sequence of integers tending to infinity. Let $\{r_j\}$ and $\{s_j\}$ be disjoint sequences of distinct real numbers chosen so that their union is independent over the rationals. Let $F_j = \{r_j + ks_j: k = 0, 1, \ldots, K(j)\}$. Let $F = \bigcup_{j=1}^{\infty} F_j$. Then $\tilde{A}(F, R_d) \neq A(F, R_d)$.*

Proof. By Kronecker's Theorem (Cassels [1, p. 53]), if a_j and b_j are arbitrary real numbers, for $1 \leq j \leq J$, and $\delta > 0$, then there exists an integer m such that

$$(4) \qquad |e^{imr_j} - e^{ia_j}| < \delta \quad \text{and} \quad |e^{ims_j} - e^{ib_j}| < \delta \quad \text{for } 1 \leq j \leq J.$$

We claim that if $\mu_j \in M(F_j)$ for $1 \leq j \leq J$, and if $\mu = \sum_{j=1}^{J} \mu_j$, then

$$(5) \qquad \|\mu\|_{PM} = \sum_{j=1}^{J} \|\mu_j\|_{PM}.$$

First, note that $\hat{\mu}_j$ has the form

$$\hat{\mu}_j(m) = \sum_{k=1}^{K(j)} c_{jk} e^{im(r_j + ks_j)} = e^{imr_j} \sum c_{jk}(e^{ims_j})^k.$$

Let $\varepsilon > 0$ and choose $m_j \in R$ such that

$$|\hat{\mu}_j(m_j)| \geq \|\mu_j\|_{PM}(1 - \varepsilon).$$

Let $\hat{\mu}_j(m_j) = p_j e^{i\theta_j}$ where $p_j \geq 0$ and θ_j is real. Let $b_j = m_j s_j$, $a_j = m_j r_j - \theta_j$. For a sufficiently small $\delta > 0$, if m is a solution of (4), then

$$\text{Re } \hat{\mu}_j(m) \geq |\hat{\mu}_j(m_j)|(1 - \varepsilon) \quad \text{for } 1 \leq j \leq J;$$

and therefore $\text{Re } \hat{\mu}_j(m) \geq (1 - \varepsilon)^2 \sum \|\mu_j\|_{PM}$. The conclusion (5) follows.

For $f \in C_o(F)$, $S \in PM(F)$, let f_j and S_j be the restrictions to F_j. If $\mu \in M(F)$, then

$$|\langle f, \mu \rangle| = \left| \int f \, d\mu \right| = \lim_{J \to \infty} \left| \sum_{j=1}^{J} \int f_j \, d\mu_j \right|$$

$$\leq \lim_{J \to \infty} \sum \|f_j\|_{A(F_j)} \|\mu_j\|_{PM}$$

$$\leq \sup_j \|f_j\|_{A(F_j)} \|\mu\|_{PM}.$$

Therefore $\|f\|_{\tilde{A}(F)} = \sup_j \|f_j\|_{A(F_j)}$, and $f \in \tilde{A}(F)$ if and only if this supremum is finite. On the other hand, if $f \in A(F)$, then $\|f_j\|_{A(F_j)} \to 0$ by 12.1.3(b). Since the Helson constant of F_j tends to infinity (see Section 1.6), there exist $g_j \in A(F_j)$ such that $\|g_j\|_{A(F_j)} = 1$ and $\|g_j\|_{C(F_j)} \to 0$. If g is the element of $C_o(F)$ that agrees with g_j on F_j, then evidently g belongs to $\tilde{A}(F)$ and not to $A(F)$. The Proposition is proved. □

Remark. One may see directly that $\tilde{A}(F)$ is non-separable. The sequences $\varepsilon = \{\varepsilon_j\}$, where each ε_j is 1 or -1, form an uncountable set. For each sequence ε, let g_ε be the element of $C_o(F)$ that agrees with $\varepsilon_j g_j$ on F_j. Then $g_\varepsilon \in \tilde{A}(F)$ and $\|g_\varepsilon\|_{\tilde{A}(F)} = 1$. If $\varepsilon \neq \varepsilon'$, then $\|g_\varepsilon - g_{\varepsilon'}\|_{\tilde{A}(F)} = 2$. Therefore $\tilde{A}(F)$ is non-separable.

Let us now point out some conditions, relevant to the study of perfect sets E, that suffice to imply that $\tilde{A}(E) = A(E)$. For the sake of simplicity, we shall deal only with subsets of the circle group. We define a class $A^*(E)$ as follows. A function $f \in C(E)$ belongs to $A^*(E)$ if there exists a sequence $\{f_n\} \subset A(Z)$ such that $\sup_n \|f_n\|_{A(Z)} < \infty$,

$$\|f - f_n\|_{C(E)} \to 0 \quad \text{as } n \to \infty, \tag{6}$$

and

$$\hat{f}_n(k) \to 0 \quad \text{as } n \to \infty \quad \text{for each } k \in Z. \tag{7}$$

With pointwise multiplication and the norm

$$\|f\|_{A^*(E)} = \inf\left\{\sup_n \|f_n\|_{A(Z)} \colon (6) \text{ and } (7) \text{ hold}\right\},$$

$A^*(E)$ is a Banach algebra. It may be defined equivalently as the class of functions $f \in C(E)$ such that for some constant c,

$$|\langle f, \mu\rangle| \leq c \limsup_{k \to \infty} |\hat{\mu}(k)| \quad \text{for every } \mu \in M(E). \tag{8}$$

Then $\|f\|_{A^*(E)} = \inf\{c\colon (8) \text{ holds}\}$. Evidently $A^*(E) = \{0\}$ if every portion of E is an M_o-set.

12.1.6. Proposition. *For every $f \in \tilde{A}(E)$, there exist $g \in A = A(T)$ and $h \in A^*(E)$ such that $f = g|_E + h$ and*

$$\|f\|_{\tilde{A}(E)} = \|g\|_A + \|h\|_{A^*(E)}. \tag{9}$$

Proof. Since $f \in \tilde{A}(E)$, there is a sequence $\{f_n\} \subset A$ such that $\|f_n - f\|_{C(E)} \to 0$ and $\lim_{n\to\infty} \|f_n\|_{A(E)} = \|f\|_{\tilde{A}(E)}$. Since we may replace $\{f_n\}$ by a subsequence, we may suppose that for each $k \in Z$, the numbers $\hat{f}_n(k)$ converge to some number, which we may call $\hat{g}(k)$. Let $h = f - g|_E$. It is easy to see that $g \in A$, $h \in A^*(E)$, and condition (9) holds. □

12.1.7. Proposition. *Let $f \in \tilde{A}(E)$, $f = g + h$ as above, and $\eta > 0$. If for every $\varepsilon > 0$, there exists $\mu \in M(E)$ such that*

$$\limsup_{k \to \infty} |\hat{\mu}(k)| < (1 - \eta)\|\mu\|_{PM} \quad \textit{and} \quad |\langle f, \mu \rangle| \geq (1 - \varepsilon)\|f\|_{A(E)}\|\mu\|_{PM},$$

then $h = 0$. Thus $\tilde{A}(E) = A(E)$.

Proof. For each $\varepsilon > 0$, there exists μ such that

$$\begin{aligned}(1 - \varepsilon)\|f\|_{\tilde{A}(E)}\|\mu\|_{PM} &\leq |\langle f, \mu\rangle| \leq |\langle g, \mu\rangle| + |\langle h, \mu\rangle| \\ &\leq \|\mu\|_{PM}(\|g\|_A + (1 - \eta)\|h\|_{A^*(E)}) \\ &= \|\mu\|_{PM}(\|f\|_{\tilde{A}(E)} - \eta\|h\|_{A^*(E)})\end{aligned}$$

(by (9)). It follows that $\|h\|_{A^*(E)}$ must be zero. □

Remarks and Credits. The study of tilde algebras was initiated by some questions posed by Professor Henry Helson in 1966. The algebras were studied systematically by Katznelson and McGehee [1, 2]. The propositions 12.1.1, 12.1.2, and parts (i) and (ii) of 12.1.3, are of course elementary and were understood in those two papers. The result 12.1.3(iii) is due to Blei, who had studied the problems in the setting of discrete groups (Blei [1, 3]).

Lemma 12.1.4 comes from Plessner [1], Raĭkov [1], and Wiener and Young [1]. Section 8.3 contains other consequences of the same idea.

Proposition 12.1.5 is from Katznelson and McGehee [1], where it has a slightly different form. Rosenthal [3] had the same idea. The idea of $A^*(E)$ comes from deLeeuw and Katznelson [2] and Katznelson and Körner [1].

One may show that 12.1.7 implies that $A(E(r)) = \tilde{A}(E(r))$ for every $r \in (2, \infty)$, where $E(r)$ is as defined in Section 4.1. Of course the result is easy if r is not a Pisot number, since then E is an M_o-set.

As a Banach space, $\tilde{A}(E)$ is an instance of what has been called the relative completion of one Banach space in another. J. R. Dorroh pointed out to us the following example of Banach spaces X and Y such that $X = \tilde{X} \subsetneq Y$ and such that the X and $\tilde{X}$ norms are not equal. Let $X = l^1(Z^+)$, with norm

$$\|x\|_X = \sum |x_k| + |\sum x_k|,$$

and let Y be the completion of $l^1(Z^+)$ with respect to the norm

$$\|x\|_Y = \left(\sup_{n>1} |x_n|\right) + \left|x_1 - \sum_{n>1} x_n\right|.$$

We consider the relative completion $\tilde{X}$ of X in Y. Let

$$x^{(n)} = (0, 2^{-1}, 2^{-2}, \ldots, 2^{-n}, \underbrace{-n^{-1}, \ldots, -n^{-1}}_{n \text{ times}}, 0, 0, \ldots),$$

$$x = (1, 2^{-1}, 2^{-2}, \ldots, 2^{-k}, \ldots).$$

Then $\|x\|_X = 4$, but $\|x\|_{\tilde{X}} \leq 2$, since $\|x^{(n)}\|_X = 2$ and $\|x^{(n)} - x\|_Y \to 0$. One may show easily that $X = \tilde{X}$.

12.2. Subsets of Discrete Groups

Let E be a subset of a discrete abelian group Γ. Then $A(E) \subset B_o(E) \subset \tilde{A}(E) \subset C_o(E)$, where $B_o(E)$ means $\{f|_E : f \in B(\Gamma) \cap C_o(\Gamma)\}$. As 12.1.3(i) points out, $\tilde{A}(E) = B(E) \cap C_o(E)$. As we shall show near the end of this section (12.2.13), $B_o(E)$ and $\tilde{A}(E)$ are not always the same. Therefore these two open questions are distinct:

(1) Is it true that if $\tilde{A}(E) = A(E)$, then E is a Sidon set?

(1′) Is it true that if $B_0(E) = A(E)$, then E is a Sidon set?

We are concerned with question (1). It is known that every non-Sidon set has a subset F such that $\tilde{A}(F) \neq A(F)$. That fact will be proved below, first for the case when $\Gamma = Z$ (12.2.4), and finally for the case of arbitrary Γ (12.2.12). Another partial result on question (1) is the following theorem, whose proof is the primary purpose of this section.

12.2.1. Theorem. *Let E be a non-Sidon subset of the discrete group G_d, where G is a non-discrete locally compact abelian group, such that E has precisely one cluster point in G. Then $\tilde{A}(E, G_d) \neq A(E, G_d)$.*

Before proving the theorem, we shall present two easier and less general results, 12.2.4 and 12.2.5. The first will exhibit a subset of the integer group Z analogous to the set F of 12.1.5. In preparation, we need to discuss conditions under which the norm of a sum of trigonometric polynomials is equivalent to the sum of their norms.

Let $F_1, \ldots, F_J$ be disjoint finite subsets of a locally compact abelian group Γ. Simply by virtue of the finite-dimensionality, there will be some constant c such that, whenever p_j is an F_j-polynomial and p denotes $\sum_{j=1}^{J} p_j$,

$$\sum_{j=1}^{J} \|p_j\| \leq c\|p\|, \tag{2}$$

where $\|f\|$ means $\|f\|_{C(G)} = \sup_{t \in G} |f(t)|$. We are concerned with conditions under which (2) holds for reasonably small values of c.

We shall say that the ranges of polynomials $p_1, \ldots, p_J$ are *ε-additive* if for every choice of $t_1, \ldots, t_J \in G$ there exists $t \in G$ such that

$$\left| p(t) - \sum_{j=1}^{J} p_j(t_j) \right| < \left(\frac{\varepsilon}{\pi}\right) \sum_{j=1}^{J} \|p_j\|.$$

12.2.2. Lemma. *If p_j is an F_j-polynomial, where $F_1, \ldots, F_J$ are disjoint finite subsets of Γ, and if the ranges of $p_1, \ldots, p_J$ are ε-additive, and if no F_j contains 0, then (2) holds with $c = \pi/(1 - \varepsilon)$.*

12.2.3. Lemma. *Every finite set X of complex numbers has a subset Y such that $\sum_{z \in X} |z| \le \pi |\sum_{z \in Y} z|$.*

Proof of 12.2.2 (using 12.2.3). Let $\delta > 0$. For each j, choose u_j such that $|p_j(u_j)| > (1 - \delta)\|p_j\|$. By 12.2.3 there is a set of indices $y \subset \{1, \ldots, J\}$ such that

$$\left| \sum_{j \in y} p_j(u_j) \right| > \frac{1 - \delta}{\pi} \sum_{j=1}^{J} \|p_j\|.$$

Let $\sum_{j \in y} p_j(u_j) = re^{ia}$, where $r \ge 0$ and a is real. For each j, $0 \notin F_j$, and therefore $e^{-ia}\hat{p}_j(0) = \hat{p}_j(0) = 0$. Therefore the average value of $e^{-ia}p_j$ on G (or, if G is not compact, on bG) is zero. Therefore there exists $v_j \in G$ such that

$$\mathrm{Re}(e^{-ia}p_j(v_j)) > -(\delta/\pi)\|p_j\|.$$

For $j \in y$, let $t_j = u_j$; for $j \notin y$, let $t_j = v_j$. Then

$$\left| \sum_{j=1}^{J} p_j(t_j) \right| > \frac{1 - 2\delta}{\pi} \sum_{j=1}^{J} \|p_j\|.$$

By hypothesis, there exists $t \in G$ such that

$$|p(t)| > \frac{1 - 2\delta - \varepsilon}{\pi} \sum \|p_j\|.$$

Since the choice of $\delta > 0$ was arbitrary, the Lemma is proved. □

Remark. One may show easily that if one of the sets F_j is allowed to contain zero, then (2) still holds, but with $c = 2\pi/(1 - \varepsilon)$.

Proof of 12.2.3. The obvious candidates for Y are the sets

$$Y_\theta = \{z \in X : |\theta - \arg(z)| \le \pi/2\}.$$

for $\theta \in [0, 2\pi)$. Let $f(\theta) = |\sum_{z \in Y_\theta} z|$. Then

$$f(\theta) = \left| \sum_{z \in Y_\theta} ze^{-i\theta} \right| \geq \sum_{z \in Y_\theta} |z| \cos(\theta - \arg(z)),$$

and

$$2\pi \max f(\theta) \geq \int_0^{2\pi} f(\theta) d\theta \geq \int_0^{2\pi} \sum_{z \in Y_\theta} |z| \cos(\theta - \arg(z)) d\theta$$

$$= \sum_{z \in X} |z| \int_{\arg(z)-(\pi/2)}^{\arg(z)+(\pi/2)} \cos(\theta - \arg(z)) d\theta = 2 \sum_{z \in X} |z|.$$

The Lemma is proved. □

Remarks. This proof is the one found in Bledsoe [1] and, in a more general version, in Kaufman and Rickert [1]. As shown in each of those references, the constant π is the smallest possible. The result has been known for a long time, and derives from the solution to the isoperimetric problem. For those and other connections and references, see Daykin *et al.* [1]. Incidentally, it is trivial to prove 12.2.3 with $4\sqrt{2}$ (say) insteadof π. Any finite constant at all would suffice for the purposes of this chapter.

Under the special circumstances of 12.1.5, it is easy to show that the ranges of the polynomials $\hat{\mu}_j$ are ε-additive for every $\varepsilon > 0$. Thus the range of the sum is dense in the sum of the ranges. Furthermore, the closure of the range of each $\hat{\mu}_j$ is circled, that is, closed under multiplication by every complex number of modulus one. That is why we could obtain (2) with $c = 1$ in the proof of 12.1.5.

12.2.4. Proposition. *Let $\{K_j\}$ and $\{m_j\}$ each be a sequence of positive integers tending to infinity, such that*

$$\frac{\pi K_{j-1} m_{j-1}}{m_j} < \frac{2^{-j}}{\pi}. \tag{3}$$

Let $F_j = \{km_j : k = 0, 1, \ldots, K_j\}$, $F = \bigcup_{j=1}^{\infty} F_j$. Then $\tilde{A}(F) \neq A(F)$.

Proof. We claim that it suffices to show that if p_j is an F_j-polynomial for $1 \leq j \leq J$, where J is arbitrary, then the ranges of the polynomials p_j are $\frac{1}{2}$-additive. If that is the case, then by Lemma 12.2.2,

$$\sum_{j=1}^{J} \|p_j\| \leq 2\pi \left\| \sum_{j=1}^{J} p_j \right\|. \tag{4}$$

Since the Helson constant of F_j is proportional to $\sqrt{K_j}$ (see Section 1.6), there exist $g_j \in A(F_j)$ such that $\|g_j\|_{A(F_j)} = 1$ and $\|g_j\|_{C(F_j)} \to 0$. Let g be the element of $C_o(F)$ that agrees with g_j on F_j. For $\mu \in M(F)$, let μ_j be the restriction of μ to F_j, so that $\hat{\mu}_j$ is an F_j-polynomial. Then

$$|\langle g, \mu\rangle| \leq \sum|\langle g_j, \mu_j\rangle| \leq \left(\sup_j \|g_j\|_{A(F_j)}\right)\left(\sum\|\mu_j\|_{PM}\right) \leq 2\pi\|\mu\|_{PM},$$

by (4). Therefore $g \in \tilde{A}(F)$ and $\|g\|_{\tilde{A}(F)} \leq 2\pi$, whereas $g \notin A(F)$. The claim is proved.

If p_k is an F_k-polynomial, then by Bernstein's Lemma A.4.1(b),

$$|p_k(u) - p_k(v)| \leq K_k m_k \|p_k\| \, |u - v|. \tag{5}$$

An F_k-polynomial p_k has period $2\pi/m_k$. Thus if $j < k$, then on every closed interval of radius π/m_k, p_k takes on every value that it ever takes on, while p_j varies by at most $(2^{-k}/\pi)\|p_k\|$, by virtue of (3) and (5). The proof of $\frac{1}{2}$-additivity should now be obvious in principle, and here it is in detail. Let $t_1, \ldots, t_J \in T$, and for each j let p_j be an F_j-polynomial. We shall choose a finite sequence $s_1, \ldots, s_J$, as follows. Let $s_1 = t_1$. When $s_1, \ldots, s_{k-1}$ have been chosen, choose s_k such that

$$|s_k - s_{k-1}| \leq \frac{\pi}{m_k} \quad \text{and} \quad p_k(s_k) = p_k(t_k).$$

In view of (3) and (5), it follows that for $1 \leq j < k$,

$$|p_j(s_k) - p_j(s_{k-1})| < \frac{\pi}{m_k} K_j m_j \|p_j\| < \frac{2^{-k}}{\pi} \|p_j\|.$$

Finally, $p_J(s_J) = p_J(t_J)$, and for $1 \leq j < J$,

$$\begin{aligned} |p_j(s_J) - p_j(t_j)| &\leq \sum_{k=j+1}^{J} |p_j(s_k) - p_j(s_{k-1})| \\ &\leq \left(\sum_{k=j+1}^{J} \frac{2^{-k}}{\pi}\right)\|p_j\| < \frac{1}{2\pi}\|p_j\|. \end{aligned}$$

Therefore, if $p = \sum_{j=1}^{J} p_j$,

$$|p(s_J) - \sum p_j(t_j)| < \frac{1}{2\pi} \sum\|p_j\|.$$

The Proposition is proved. □

In 12.1.5 and 12.2.4, our method of showing that $\tilde{A}(F) \neq A(F)$ is to exhibit a partition of F into parts F_j, such that $\alpha(F_j) \to \infty$ and

$$L_F^\infty \cong \oplus_{l^1} L_{F_j}^\infty.$$

Our next trick will be again to construct such a set F, by the same method, but subject to the constraint that F must be a subset of a prescribed non-Sidon set E. The reader might find it interesting to do this exercise on his own, and in any event, to try to see the difficulty in showing that $\tilde{A}(E) \neq A(E)$ whenever E is non-Sidon.

12.2.5. Proposition. *Let E be a non-Sidon subset of the integer group Z. Then there exists $F \subset E$ such that $\tilde{A}(F) \neq A(F)$.*

Proof. Let F_1 be a finite subset of E such that $\alpha(F_1) > 2$. Let $j \geq 1$ and suppose that disjoint finite sets $F_1, \ldots, F_{j-1}$ have been chosen. Let H_j be their union, and let

$$h_j = \max\{|n| : n \in H_j\},$$

so that if P is an H_j-polynomial, then by Bernstein's Lemma A.4.1(b),

$$|P(u) - P(v)| \leq h_j \|P\| \, |u - v|.$$

Let m_j be an even integer large enough so that

$$\frac{\pi h_j}{m_j} < \frac{2^{-j}}{2\pi}.$$

The set E is the union of the sets

$$E_n = E \cap (m_j Z + n), \; |n| \leq m_j/2.$$

For each n, there is an idempotent measure ν_n such that $\|\nu_n\|_M = 1$ and $\nu_n = \chi_{E_n}$, so that the mapping $S \to S|_{E_n}$ is a projection of norm one from $PM(E)$ onto $PM(E_n)$. It follows (without using the difficult union theorem of Chapter 2!) that at least one of the sets E_n is non-Sidon. Therefore we may select n_j and a finite set $F_j \subset m_j Z + n_j$ such that $\alpha(F_j) > 2^j$.

In the argument that follows, it is inconvenient to have $|n_j| > m_j/3$. If that holds, let $m'_j = m_j/2$ (we chose m_j to be even!). Then $m_j Z + n_j \subset m'_j Z + n'_j$, where $|n'_j| < (m_j/6) = (m'_j/3)$. So for those j such that $|n_j| > m_j/3$, we replace m_j, n_j by m'_j, n'_j, and then for every j, these conditions hold:

$$\frac{\pi h_j}{m_j} < \frac{2^{-j}}{\pi}, \qquad \alpha(F_j) > 2^j,$$

$$F_j \subset m_j Z + n_j, \qquad |n_j| \leq m_j/3.$$

Since we may replace $\{F_j\}$ by a subsequence, we may suppose that one of the following conditions holds.

(a) The integers n_j are all the same: $n_j \equiv n$.
(b) $n_j \to \infty$ and $n_j/m_j \to 0$.
(c) $n_j \to \infty$ and for some $a > 0$, $a < |n_j|/m_j \leq \frac{1}{3}$.

Let $F = \bigcup_{j=1}^{\infty} F_j$. In Case *a*, we may suppose that $n = 0$, since the properties that concern us are translation-invariant. The fact that $\tilde{A}(F) \neq A(F)$ then follows exactly as in the proof of 12.2.4. The only difference in the present situation is that F_j is not necessarily an arithmetic progression. The conditions needed for the argument are still present, namely that $F_j \subset m_j Z$ and $\alpha(F_j) \to \infty$ as $j \to \infty$.

Case b. Choose $\varepsilon > 0$ and $\varepsilon_j > 0$ with $\sum \varepsilon_j < \varepsilon/3$. Possibly at the cost of again replacing $\{F_j\}$ by a subsequence, we may suppose that

$$\frac{\pi h_j}{n_j} < \varepsilon_j \quad \text{and} \quad \frac{\pi n_j}{m_j} < \varepsilon_j \quad \text{for each } j \geq 1.$$

For each j, let p_j be an F_j-polynomial. Bypassing the Lemma 12.2.2, we shall prove by induction that for each J,

$$\left\| \sum_{j=1}^{J} p_j \right\| \geq (1 - \varepsilon) \sum_{j=1}^{J} \|p_j\|.$$

The case $J = 1$ is trivial. Let $P = \sum_{j=1}^{J-1} p_j$. We may suppose, as inductive hypothesis, that there exists $r \in T$ such that

$$|P(r)| > \left(1 - 3\sum_{j=1}^{J-1} \varepsilon_j\right) \sum_{j=1}^{J-1} \|p_j\|.$$

The polynomial p_J has the form

$$p_J(t) = e^{in_J t} q_J(t),$$

where $(2\pi/m_J)$ is a period of q_J. There exists $s' \in T$ such that $|q_J(s')| = \|p_J\|$. Let u and v be real numbers such that

$$|P(r)|e^{iu} = P(r), \qquad |q_J(s')|e^{iv} = q_J(s').$$

There exists $s'' \in T$ such that

$$|s'' - r| \leq \frac{\pi}{n_J} \quad \text{and} \quad e^{in_J s''} = e^{i(u-v)}.$$

Choose s so that

$$|s - s''| \leq \frac{\pi}{m_J} \quad \text{and} \quad q_J(s) = q_J(s') = \|p_J\| e^{iv}.$$

Then

$$\begin{aligned} |P(s) - P(r)| &\leq |P(s) - P(s'')| + |P(s'') - P(r)| \\ &\leq \varepsilon_J \|P\| + \varepsilon_J \|P\| \leq 2\varepsilon_J \sum_{j=1}^{J-1} \|p_j\|; \end{aligned}$$

$$|e^{in_J s} - e^{in_J s''}| < \varepsilon_J,$$

and

$$\begin{aligned} |e^{in_J s''} q_J(s) + P(r)| &= \left| e^{i(u-v)} \|p_J\| e^{iv} + e^{iu} |P(r)| \right| \\ &= \|p_J\| + |P(r)|. \end{aligned}$$

Therefore

$$\begin{aligned} \left| \sum_{j=1}^{J} p_j(s) \right| &= |e^{in_J s} q_J(s) + P(s)| \geq |e^{in_J s''} q_J(s) + P(r)| \\ &\quad - |(e^{in_J s} - e^{in_J s''}) q_J(s)| - |P(r) - P(s)| \\ &\geq \|p_J\| + \left(1 - 3 \sum_{j=1}^{J-1} \varepsilon_j\right) \sum_{j=1}^{J-1} \|p_j\| - \varepsilon_J \|p_J\| - 2\varepsilon_J \sum_{j=1}^{J-1} \|p_j\| \\ &\geq \left(1 - 3 \sum_{j=1}^{J} \varepsilon_j\right) \sum_{j=1}^{J} \|p_j\|. \end{aligned}$$

The proof in Case b is complete.

Case c. Choose $\varepsilon_j > 0$ with $\sum \varepsilon_j < \frac{1}{2} \sin \pi/6$. We may suppose that

$$\frac{\pi h_j}{m_j} < \frac{\pi h_j}{|n_j|} < \frac{\varepsilon_j}{1 + [a^{-1}]} \quad \text{for each } j \geq 1.$$

For each j, let p_j be an F_j-polynomial. It has the form

$$p_j(t) = e^{in_j t} q_j(t),$$

where $(2\pi/m_j)$ is a period of q_j.

Again we bypass the use of 12.2.2. We shall show by induction that for each J,

$$\left\| \sum_{j=1}^{J} p_j \right\| > \left(\frac{1}{2} \sin \frac{\pi}{6}\right) \sum_{j=1}^{J} \|p_j\|.$$

The case $J = 1$ is trivial. Letting $P = \sum_{j=1}^{J-1} p_j$, we may suppose, as inductive hypothesis, that there exists $r \in T$ such that

$$|P(r)| > \left(\left(\sin \frac{\pi}{6}\right) - \sum_{j=1}^{J-1} \varepsilon_j\right) \sum_{j=1}^{J-1} \|p_j\|.$$

If s belongs to the interval

$$I = \left\{s : |s - r| < \left(1 + \left[\frac{1}{a}\right]\right) \frac{\pi}{m_J}\right\},$$

then $|P(s) - P(r)| < \varepsilon_J \|P\|$. Since I contains $1 + [a^{-1}]$ complete periods of q_J, there are $1 + [a^{-1}]$ different values of s in I, distance $2\pi/m_J$ apart, such that $|q_J(s)| = \|p_J\|$. Let S be the set of those values. We would like to choose an $s \in S$ so that $e^{in_J s}$ will be as close as possible to $e^{i\theta}$, where

$$|P(s) + e^{i\theta} q_J(s)| = |P(s)| + |q_J(s)|.$$

As x moves from one value of $s \in S$ to the next, the number $e^{in_J x}$ moves along an arc of length $2\pi n_J/m_J$, which is bounded by $2\pi/3$; and the set $\{e^{in_J s} : s \in S\}$ intersects every arc of that length (on the unit circle). Therefore there exists $s \in S$ such that $|n_J s - \theta| \le \pi/3$ and hence

$$|P(s) + e^{in_J s} q_J(s)| > |P(s)| + \left(\sin \frac{\pi}{6}\right) |q_J(s)|$$

$$> |P(r)| - \varepsilon_J \|P\| - \left(\sin \frac{\pi}{6}\right) \|p_J\| \ge \left(\left(\sin \frac{\pi}{6}\right) - \sum_{j=1}^{J} \varepsilon_j\right) \sum_{j=1}^{J} \|p_j\|.$$

The proof of 12.2.5 is complete. □

We turn now to the task of proving the Theorem 12.2.1.

The next lemma asserts a relationship between a control on the support of a measure $\mu \in M(G)$, on the one hand, and a control on how fast $\hat{\mu}(\gamma)$ can change as γ moves about in the group Γ. It thus bears a resemblance to Bernstein's Lemma A.4.1.

12.2.6. Lemma. *Let U be a compact subset of a locally compact abelian group Γ, and let $\varepsilon > 0$. Then there exists a neighborhood W of 0 in the dual group G such that if $\mu \in M(W)$ and $u - v \in U$, then $|\hat{\mu}(u) - \hat{\mu}(v)| \le \varepsilon \|\mu\|_{PM}$.*

Proof. Let $W = \{x \in G: |\arg\langle\gamma, x\rangle| \leq \delta \text{ for all } \gamma \in U\}$. Then W is a closed symmetric neighborhood of 0 in G. By Lemma A.3.1. there exists $\delta > 0$ such that $\|1 - \langle\gamma, \cdot\rangle\|_{B(W)} \leq \varepsilon$ for $\gamma \in U$. If $\mu \in M(W)$ and $u - v \in U$, then

$$|\hat{\mu}(u) - \hat{\mu}(v)| = \left|\int_W (\langle u, -x\rangle - \langle v, -x\rangle) d\mu(x)\right|$$

$$\leq \|1 - \langle u - v, \cdot\rangle\|_{B(W)}\|\mu\|_{PM} \leq \varepsilon\|\mu\|_{PM}.$$

The Lemma is proved. □

In the proof of 12.2.4, we used the fact that an (mZ)-polynomial has period $2\pi/m$. We shall need the following allied fact, which comes from the theory of almost periodic functions.

12.2.7. Lemma. *Let F be a finite subset of a locally compact abelian group G, and let $\varepsilon > 0$. There exists a finite subset U of the dual group Γ such that if $t \in \Gamma$, then every translate of U contains an element s such that*

$$|\hat{\mu}(s) - \hat{\mu}(t)| \leq \varepsilon\|\mu\|_{PM} \quad \textit{for every } \mu \in M(F).$$

Proof. Let

$$V = \{w \in b\Gamma: |\arg\langle w, \gamma\rangle| < \delta \quad \text{for every } \gamma \in F\}.$$

Then V is a symmetric neighborhood of 0 in $b\Gamma$. By A.3.1, we may choose $\delta > 0$ such that $\|\langle w, \cdot\rangle - 1\|_{B(F)} < \varepsilon$ for each $w \in V$. If $\mu \in M(F)$, $u \in \Gamma$, and $w \in V$, then

$$|\hat{\mu}(u + w) - \hat{\mu}(u)| = \left|\int_F (\langle u + w, -\gamma\rangle - \langle u, -\gamma\rangle) d\mu(\gamma)\right|$$

$$\leq \|\langle w, \cdot\rangle - 1\|_{B(F)}\|\mu\|_{PM} < \varepsilon\|\mu\|_{PM}.$$

Since Γ is dense in $b\Gamma$ and $b\Gamma$ is compact, there is a finite set $U \subset \Gamma$ such that $b\Gamma = U + V$, and hence $\Gamma = U + (V \cap \Gamma)$. Let $t \in \Gamma$, and let $h + U$ be an arbitrary translate of U. Then $t - h = u + w$ for some $u \in U$ and $w \in V \cap \Gamma$. Let $s = h + u$. Then $t = s + w$, so that $|\hat{\mu}(t) - \hat{\mu}(s)| < \varepsilon\|\mu\|_{PM}$. The Lemma is proved. □

Remark. Lemma A.3.1 is not really needed for 12.2.7, since F is finite and for $w \in F$ we may write

$$|\hat{\mu}(u + w) - \hat{\mu}(u)| < \|\langle w, \cdot\rangle - 1\|_{C(F)}\|\mu\|_M < \varepsilon\alpha(F)\|\mu\|_{PM}.$$

12.2.8. Lemma. *Let V be a neighborhood of 0 in a locally compact abelian group G. Let $\varepsilon > 0$. Then there exists a smaller neighborhood K of 0 in G, and functions k and λ in $A(G)$, such that*

$$k + \lambda = 1 \quad on \quad K, \qquad k = \lambda = 0 \quad outside \quad V, \qquad k \geq 0,$$

$$\|k\|_{A(G)} = k(0) = 1, \ \|\hat{\lambda}\|_{L^1(\Gamma)} \leq \varepsilon, \ \|\hat{\lambda}\|_{L^\infty(\Gamma)} \leq \varepsilon.$$

Proof. Let Y be a compact symmetric neighborhood of 0 such that $Y + Y \subset V$, and let $k = \chi_Y * \chi_Y / m_G(Y)$. Then $k \in A(G)$, $k \geq 0$, k vanishes outside $Y + Y$ and hence outside V, and $\|k\|_{A(G)} = k(0) = 1$. Since $(1 - k)k^{-1}$ belongs to $A(G)$ locally at 0, and since $\{0\}$ obeys synthesis, there exists $g \in A(G)$ such that $g = (1 - k)k^{-1}$ on some neighborhood K of 0, and such that $\|g\|_{A(G)} < \varepsilon/\max(1, \|\hat{k}\|_\infty)$. Then $\|kg\|_{A(G)} < \varepsilon$ and $\|\hat{k} * \hat{g}\|_\infty \leq \|\hat{k}\|_\infty \|g\|_1 < \varepsilon$. Let $\lambda = kg$. The Lemma is proved. □

12.2.9. Lemma. *Let G be a locally compact abelian group with dual group Γ. Let V, ε, and K be as in 12.2.8. Let H be a finite subset of K. Then there exists a measure $\sigma \in M_d(V)$, with finite support, that assigns mass 1 to each point of the set $H \cup \{0\}$ and such that $\|\hat{\sigma}\|_{L^1(b\Gamma)} < (1 + \varepsilon)^2$. Furthermore, $\sigma = \tau + \eta$, where $\tau \geq 0$, $\|\hat{\tau}\|_1 = 1$, and $\|\hat{\eta}\|_1 \leq \|\hat{\eta}\|_\infty < 2\varepsilon + \varepsilon^2$.*

Proof. We claim that there exists a sequence of symmetric finite sets $Y_n \subset G$ such that

$$\frac{\#\{y \in Y_n : x - y \notin Y_n\}}{\# Y_n} \to 0 \quad \text{as} \quad n \to \infty \tag{6}$$

for each $x \in H$. Suppose for the moment that that has been done. Let $v_n = (\chi_{Y_n} * \chi_{Y_n}) / \# Y_n$. Then $\|\hat{v}_n\|_{L^1(b\Gamma)} = v_n(0) = 1$ and, by virtue of (6), $0 \leq v_n(x) \to 1$ for each $x \in H$. Let $\rho_n = \chi_H(1 - v_n)$. For each n, $v_n + \rho_n$ assigns mass 1 to each point of $H \cup \{0\}$. Choose n sufficiently large so that $\|\rho_n\|_M = \sum_{x \in H}(1 - v_n(x)) < \varepsilon$, and $v = v_n$, $\rho = \rho_n$. Note that $\|\hat{\rho}_n\|_{L^1(b\Gamma)} \leq \|\hat{\rho}_n\|_\infty \leq \|\rho_n\|_M < \varepsilon$. Let k and λ be the functions provided by 12.2.8, and let

$$\sigma = (k + \lambda)(v + \rho), \tag{7}$$

$$\tau = kv, \qquad \eta = \sigma - \tau = \lambda(v + \rho) + k\rho.$$

Then τ is a positive measure in $M(G_d)$ and $\|\hat{\tau}\|_{L^1(b\Gamma)} = \tau(0) = k(0)v(0) = 1$, while

$$\|\hat{\eta}\|_{L^1(\bar{\Gamma})} \leq \|\hat{\eta}\|_\infty \leq \|\hat{\lambda}\|_\infty \|\hat{v} + \hat{\rho}\|_1 + \|\hat{k}\|_1 \|\hat{\rho}\|_\infty$$
$$< \varepsilon(1 + \varepsilon) + \|\rho\|_M < 2\varepsilon + \varepsilon^2.$$

Thus 12.2.9 is proved provided suitable sets Y_n are found.

If G is not a torsion group, let $\{u_1, \ldots, u_k\}$ be an independent set in G such that the group generated contains H:

$$H \subset \mathrm{Gp}(u_1, \ldots, u_k), \tag{8}$$

and such that at least one of the elements u_j has infinite order. Condition (6) will be satisfied by the sets

$$Y_n = \left\{\sum_{j=1}^{k} a_j u_j \colon a_j \in Z, |a_j| \le n \quad \text{for each } j\right\}.$$

If G is a torsion group, let $\{u_j\}$ be an infinite independent set such that (8) is satisfied for some finite k, and let

$$Y_n = \left\{\sum_{j=1}^{\infty} a_j u_j \colon a_j \in Z \quad \text{and} \quad \sum |a_j| \le n\right\}.$$

The Lemma is proved. □

12.2.10. Lemma. *For $1 \le j \le k$, let ε_j, V_j, H_j, and σ_j be as in 12.2.8 and 12.2.9, let $H_j \subset F_j \subset \operatorname{supp} \sigma_j$, and let $E \subset G$. Suppose that the following two conditions hold.*

(i) *If $x = \sum_{j=1}^{k} x_j \in \bigcup_{j=1}^{k} H_j \cup \{0\}$, where $x_j \in \operatorname{supp} \sigma_j$, then one of the summands x_j equals x and each of the others equals 0.*
(ii) $(\sum_{j=1}^{k} \operatorname{supp} \sigma_j) \cap E = \bigcup_{j=1}^{k} F_j$.

Let $\rho = \bigstar_{j=1}^{k} \sigma_j$. Then

$$\rho(x) = 1 \quad \textit{for} \quad x \in \{0\} \cup \bigcup_{j=1}^{k} H_j, \tag{9}$$

$$\operatorname{supp}(\sigma) \cap E = \{0\} \cup \bigcup_{j=1}^{k} F_j;$$

and

$$\|\hat{\rho}\|_{L^1(\Gamma)} \le \prod_{j=1}^{k} (1 + \varepsilon_j)^2. \tag{10}$$

Proof. The hypotheses (i) and (ii) and the definition of convolution imply immediately the conditions (9).

We may write σ_j as $\tau_j + \eta_j$, as in 12.2.9. If $D \subset \{1, \ldots, k\}$ and $D \neq \emptyset$, then $\tau^{(D)} = \bigstar_{j \in D} \tau_j$ is positive and by virtue of (i), $\|\hat{\tau}^{(D)}\|_1 = \tau^{(D)}(0) = 1$. If D' is the complement of D in the set $\{1, \ldots, k\}$, then

$$\left\| \prod_{j \in D'} \hat{\eta}_j \prod_{j \in D} \hat{\tau}_j \right\|_1 \leq \left\| \prod_{j \in D'} \hat{\eta}_j \right\|_\infty \|\hat{\tau}^{(D)}\|_1$$
$$\leq \prod_{j \in D'} (2\varepsilon_j + \varepsilon_j^2).$$

The inequality (10) follows. The Lemma is proved. □

12.2.11. Lemma. *Let P be a non-Sidon set, $P \subset E$, and $\varepsilon > 0$. Then there exists $f \in A(E)$ such that $\|f\|_{C(E)} < \varepsilon \|f\|_{A(E)}$ and the support of f is a finite subset of P.*

Proof. Since functions with finite support are dense in $A(P)$, there exists $g \in A(P)$ with finite support $F \subset P$ such that $\|g\|_{C(P)} < \varepsilon \|g\|_{A(P)}$. Let h be an element of $A(E)$ that agrees with g on P. Let $f = \chi_F h$. Since $\chi_F \in A(E)$, $f \in A(E)$. Since $\|f\|_{A(E)} \geq \|f\|_{A(P)} > \varepsilon^{-1} \|f\|_{C(E)}$, the Lemma is proved. □

Proof of 12.2.1. We may suppose that 0 is the cluster point of E. We shall write $A(\cdot)$ and $\tilde{A}(\cdot)$ to designate $A(\cdot, G_d)$ and $\tilde{A}(\cdot, G_d)$, respectively.

This proof is analogous to that of Proposition 12.2.5. In that case, we moved along E toward infinity, selecting subsets F_j. In this case, we move along E toward 0, selecting subsets F_j. In both cases, we let $F = \bigcup_{j=1}^\infty F_j$ and show that $\tilde{A}(F) \neq A(F)$. But in this case, we take additional measures along the way in order to assure that $\tilde{A}(E) \neq A(E)$.

We shall select inductively a sequence of disjoint finite sets $F_j \subset E$, and a sequence of functions $f_j \in A(E)$ with $\operatorname{supp}(f_j) \subset F_j$, satisfying these conditions:

(i) If $\mu_j \in M(F_j)$ for $1 \leq j \leq J$, then the ranges of the transforms $\hat{\mu}_j$ are $\frac{1}{2}$-additive.
(ii) $\|f_j\|_{C(F_j)} \to 0$.
(iii) $1 \geq \|f_j\|_{A(F_j)} \to 1$.
(iv) There exists $\nu \in M(b\Gamma)$ such that $\hat{\nu} = 1$ on $\bigcup_{j=1}^\infty \operatorname{supp}(f_j)$, $\hat{\nu} = 0$ on $E \setminus F$, and $\|\nu\|_M < 2$.

Suppose that that selection has been made. Let $f = \sum_{j=1}^\infty f_j$. By (ii) and (iii), $f \in C_o(E)$ and $f \notin A(E)$. By (i), (iii), (iv), and 12.2.2, if $\mu \in M(E)$ and μ_j is the restriction of μ to F_j, then

$$|\langle f, \mu \rangle| = |\sum \langle f_j, \mu_j \rangle| = |\sum \langle f_j, \hat{\nu}\mu_j \rangle|$$
$$\leq \sup_j \|f_j\|_{A(F_j)} \sum_j \|\hat{\nu}\mu_j\|_{PM} \leq 1 \cdot 2\pi \|\hat{\nu}\mu\|_{PM} < 4\pi \|\mu\|_{PM}.$$

Therefore $f \in A(E)$ and $\|f\|_{A(E)} \leq 4\pi$. Thus 12.2.1 is proved, provided the selection of $\{F_j, f_j\}$ can be carried out as claimed.

The procedure uses a sequence of positive numbers ε_j, chosen so that $\prod_{j=1}^{\infty}(1 + \varepsilon_j)^2 < 2$, and a sequence $\{V_j\}$ of neighborhoods of 0 in G, beginning with $V_1 = G$.

When V_j has been selected, let K_j be a neighborhood of 0 corresponding to ε_j and V_j as in 12.2.8. Since $K_j \cap E\backslash\{0\}$ is a non-Sidon set, by 12.2.11 there is a function $f_j \in A(E)$ whose support is a finite set $H_j \subset K_j \cap E\backslash\{0\}$, such that $\|f_j\|_{A(E)} = 1$ and $\|f_j\|_\infty < 2^{-j}$. By 12.2.9 there exists a measure $\sigma_j \in M_d(V_j)$, with finite support, such that $\sigma_j(x) = 1$ for each $x \in H_j \cup \{0\}$ and $\|\hat{\sigma}_j\|_{L^1(b\Gamma)} < (1 + \varepsilon_j)^2$. Let $F_j = \operatorname{supp}(\sigma_j) \cap E\backslash\{0\}$. By 12.2.7, we may choose a finite set $U_j \subset \Gamma$ such that if $t \in \Gamma$, then every translate of U_j contains an element s such that

$$|\hat{\mu}(s) - \hat{\mu}(t)| \leq \frac{1}{4\pi}\|\mu\|_{PM} \quad \text{for every } \mu \in M(F_j).$$

We shall now choose the neighborhood V_{j+1}, imposing certain conditions. By 12.2.6, we may require that if $\mu \in M(V_{j+1})$ and $u - v \in U_1 + \cdots + U_j$, then $|\hat{\mu}(u) - \hat{\mu}(v)| \leq (1/4\pi)\|\mu\|_{PM}$. We may require also that

$$\text{(i')} \qquad (x + V_{j+1}) \cap \left(\bigcup_{i=1}^{j} H_i \cup V_{j+1}\right)\backslash\{x\} = \varnothing$$

$$\text{for each } x \in \left(\sum_{i=1}^{j} \operatorname{supp} \sigma_i\right)\backslash\{0\},$$

and, since 0 is the only cluster point of E, that

$$\text{(ii')} \qquad (x + V_{j+1}) \cap E\backslash\{x\} = \varnothing \quad \text{for each } x \in \left(\sum_{i=1}^{j} \operatorname{supp} \sigma_i\right)\backslash\{0\}.$$

The selection procedure is now completely described. Condition (ii) evidently holds, and it remains to verify (i), (iii) and (iv).

Condition (i). Let $\mu_j \in M(F_j)$ for $1 \leq j \leq J$, $\mu = \sum \mu_j$, and $t_1, \ldots, t_J \in \Gamma$. We shall select inductively a finite sequence $s_J, \ldots, s_1 \in \Gamma$ such that

$$\text{(11)} \qquad |\hat{\mu}(s_1) - \sum \hat{\mu}_j(t_j)| < \frac{1}{2\pi}\sum\|\mu_j\|_{PM}.$$

Let $s_J = t_J$. When $s_J, \ldots, s_{j+1}$ have been chosen, and $j \geq 1$, choose s_j so that $s_j - s_{j+1} \in U_j$ and $|\hat{\mu}_j(s) - \hat{\mu}_j(t_j)| < (1/4\pi)\|\mu_j\|_{PM}$.

Finally, since $\mu_j \in M(V_j)$, and $s_1 - s_j = \sum_{i=1}^{j-1} s_i - s_{i+1} \in \sum_{i=1}^{j-1} U_i$, it follows that $|\hat{\mu}_j(s_1) - \hat{\mu}_j(s_j)| < (1/4\pi)\|\mu_j\|_{PM}$. The inequality (11) follows, and (i) is proved.

Condition (iii). If $\mu \in M(E)$, then $\sigma_j \mu \in M(F_j)$, and $\|\sigma_j \mu\|_{PM} \leq (1 + \varepsilon_j)^2 \|\mu\|_{PM}$. If $\tau \in M(F_j)$, then $\tau = \sigma_j \mu$ for some $\mu \in M(E)$. Condition (iii) follows because

$$1 = \|f_j\|_{A(E)} \geq \|f_j\|_{A(F_j)} = \sup\left\{\frac{|\langle f_j, \tau\rangle|}{\|\tau\|_{PM}} : \tau \in M(F_j), \tau \neq 0\right\}$$

$$= \sup\left\{\frac{|\langle f_j, \mu\rangle|}{\|\sigma_j \mu\|_{PM}} : \mu \in M(E), \sigma_j \mu \neq 0\right\}$$

$$\geq (1 + \varepsilon_j)^{-2} \sup\left\{\frac{|\langle f_j, \mu\rangle|}{\|\mu\|_{PM}} : \mu \in M(E), \mu \neq 0\right\}$$

$$= (1 + \varepsilon_j)^{-2} \|f_j\|_{A(E)} = (1 + \varepsilon_j)^{-2}.$$

Condition (*iv*). Since (i′) and (ii′) are required to hold for every $j \geq 1$, conditions (i) and (ii) in the hypothesis of 12.2.1 hold for every $k \geq 1$. Therefore the measures $\rho_k = \bigstar_{j=1}^{k} \sigma_j$ converge pointwise to 1 on $\bigcup_{j=1}^{\infty} H_j$, and to 0 on $E \backslash F$. Since $\|\hat{\rho}_k\|_{L^1(b\Gamma)} \leq 2$, we may suppose that $\hat{\rho}_k$ converges weak $*$ in $M(b\Gamma)$ to some measure ν with the prescribed properties, so that (iv) is proved. The proof of 12.2.1 is complete. □

Remarks. The conclusion of the Theorem still follows whenever E is a closed subset of G and, for some open set U in G, $U \cap E$ is a non-Sidon set with only one cluster point.

Note that $A(E, G_d) = A(E, G) \cap C_o(E, G_d)$ and $\tilde{A}(E, G_d) = \tilde{A}(E, G) \cap C_o(E, G_d)$.

12.2.12. Proposition. *Let E be a non-Sidon subset of a discrete group Λ. Then E has a non-Sidon subset F such that $\tilde{A}(F) \neq A(F)$.*

Proof. By 12.2.1, it suffices to realize Λ as a subgroup of a compact group G, and thus as an open subgroup of G_d, and to find a non-Sidon set $F \subset E$ with only one cluster point in G.

The problem reduces to the case of countable Λ. For if Λ is uncountable, let E_1 be a countable non-Sidon subset of E, and let Λ_1 be the group generated by E_1; then $A(E \cap \Lambda_1, \Lambda_1)$ is identifiable with $A(E \cap \Lambda_1, \Lambda)$.

We may suppose, then, that Λ is countable, so that its dual group H is metrizable as well as compact and hence contains a countable dense subgroup Γ, which we consider as a discrete group. Then there is a natural isomorphism φ from Λ onto a subgroup of the dual group G of Γ:

$$\langle \varphi(s), \gamma\rangle = \langle s, \gamma\rangle \quad \text{for } \gamma \in \Gamma, s \in \Lambda.$$

The mapping $f \to f \circ \varphi^{-1}$ is an isometric isomorphism from $A(\Lambda)$ onto $A(\varphi(\Lambda), G_d)$, and from $A(E, \Lambda)$ onto $A(\varphi(E), G_d)$. Let us now write E for $\varphi(E)$ and regard it as a (non-Sidon) subset of G. We claim that there exists $x_0 \in G$

such that for every neighborhood N of x_0 in G, $N \cap E$ is a non-Sidon set. Then there is obviously a sequence $F = \{x_n\} \subset E$ such that F is a non-Sidon set and $x_n \to x_0$. If the claim were false, then for each $x \in G$ there would be a neighborhood N_x such that $N_x \cap E$ is a Sidon set. Since G is compact it would follow that E is a finite union of Sidon sets, and hence itself a Sidon set, which it is not. The claim, and the Proposition, are proved. □

12.2.13. Proposition. *There exists $E \subset Z$ such that $B_o(E) \neq \tilde{A}(E)$.*

Proof. Let μ be the measure in $M(T)$ given by the Riesz product

$$\prod_{j=1}^{\infty} (1 + \cos 3^j t).$$

Let $E_n = \{k \in Z: \hat{\mu}(k) < 2^{-n}\}$. Let $\{M_n\}$ be a sequence tending rapidly to infinity—it will suffice to require that $\log M_n \gg n^2$—and let

$$E = \bigcup_{n=1}^{\infty} (E_n \cap [-M_n, M_n]).$$

Clearly $\hat{\mu}$ belongs to $\tilde{A}(E)$. We shall show that if $\nu \in M(T)$ and $\hat{\nu} = \hat{\mu}$ on E, then

$$\liminf_{j \to \infty} |\hat{\nu}(\pm 3^j) - \tfrac{1}{2}| = 0,$$

and in particular, $\hat{\mu} \notin B_o(E)$.

Let $\nu \in M(T)$ such that $\hat{\nu} = \hat{\mu}$ on E, and suppose that

$$\liminf |\hat{\nu}(\pm 3^j) - \tfrac{1}{2}| > a > 0.$$

Since we may add to ν a trigonometric polynomial of the form $\sum a_\lambda e^{i\lambda t}$, $\lambda = \pm 3^j$, without changing values of $\hat{\nu}(k)$ for $k \in E$, we may suppose that

$$\inf |\hat{\nu}(\pm 3^j) - \hat{\mu}(\pm 3^j)| > a > 0,$$

noting that $\hat{\mu}(\pm 3^j) = \frac{1}{2}$. Let $p = (\nu - \mu) * (\nu - \mu)^{\sim}$ so that $\hat{p} = |\hat{\nu} - \hat{\mu}|^2$. Then $\hat{p} = 0$ on E and $\hat{p}(\pm 3^j) > a^2$. Let

$$\varphi_{k,x} = \prod_{j=1}^{k} \left(1 + \frac{ix}{\sqrt{k}} \cos 3^j t\right) \quad \text{for } -1 \leq x \leq 1,$$

$$g_k(x) = \int \varphi_{k,x}(t) dp(t) = \sum_{m \notin E} \hat{\varphi}_{k,x}(m) \hat{p}(m).$$

The function g_k is an algebraic polynomial whose degree n_k does not exceed the smallest n such that

$$[-3^{k+1}, 3^{k+1}] \cap E \supset [-3^{k+1}, 3^{k+1}] \cap E_n.$$

Hence $n_k = o(k^{1/2})$. Now $|g_k(x)| \le \|\varphi_{k,x}\|_{C(T)}\|p\|_{M(T)} \le e\|p\|$ uniformly for $-1 \le x \le 1$, and for all k. By Bernstein's Lemma A.4.1(b), $|g_k'(0)| = 0(n_k)$, and yet

$$g_k'(0) = \frac{i}{2k^{1/2}} \sum\{\hat{p}(\pm 3^j): j \le k\}$$

so that $|g_k'(0)| > a^2\sqrt{k}$. Thus $a^2\sqrt{k} = 0(n_k) = o(\sqrt{k})$, which cannot be true. The Proposition is proved. □

12.2.14. Proposition. *Let E be a subset of the integer group Z. Let $\{n_j\}$ be a sequence of integers such that for each k, the 3^k integers in the set*

$$\left\{\sum_{j=1}^{k} \varepsilon_j n_j: \varepsilon_j = -1, 0, \quad or \quad +1\right\}$$

are all distinct, and are all contained in E. Then $\tilde{A}(E) \neq A(E)$.

Proof. Let $j(k) = \sum_{n=0}^{k} n$. Let μ be the measure given by the Riesz product

$$\prod_{j=1}^{\infty} (1 + a_j \cos n_j x),$$

where $a_j = k^{-1/2}$ for $j(k-1) < j \le j(k)$. Let

$$p_k(x) = \left[\prod_{j(k-1)+1}^{j(k)} (1 + i a_j \cos n_j x)\right] - 1.$$

Then

$$\|p_k\|_{C(T)} \le (1 + k^{-1})^{k/2} + 1 < e^{1/2} + 1 < 3,$$

and the support of $\hat{p}_k$ is the set

$$F_k = \left\{\sum_{j(k-1)+1}^{j(k)} \varepsilon_j n_j: \varepsilon_j = -1, 0, \text{ or } +1\right\} \setminus \{0\}.$$

Note that $\inf\{|n|: n \in F_k\} \to \infty$ as $k \to \infty$, so that when we show that $\liminf_{k\to\infty} \|\hat{\mu}\|_{A(F_k)} > 0$, it will follow from 12.1.3(ii) that $\hat{\mu} \notin A(E)$.

$$\int p_k\, d\mu = \sum \hat{p}_k \hat{\mu} = \left[\sum_{q=0}^{k} \binom{k}{q} 2^q \left(\frac{1}{2\sqrt{k}}\right)^q \left(\frac{i}{2\sqrt{k}}\right)^q\right] - 1$$

$$= \left[\sum_{q=0}^{k} \binom{k}{q}\left(\frac{i}{2k}\right)^q\right] - 1 = \left(1 + \frac{i}{2k}\right)^k - 1 \to e^{i/2} - 1.$$

Since

$$\|\hat{\mu}\|_{A(F_k)} \geq \frac{|\int p_k\, d\mu|}{\|p_k\|_{C(T)}},$$

it follows that $\liminf_{k\to\infty} \|\hat{\mu}\|_{A(F_k)} \geq |e^{i/2} - 1|/3$. Therefore $\hat{\mu} \in \tilde{A}(E) \backslash A(E)$. □

Remarks. The lemmas in this section are all standard material. The Theorem 12.2.1 and its proof are due to Katznelson and McGehee [1], but they were formulated for the general setting by Blei [1, 3]. Proposition 12.2.4 is due to Rosenthal [3], who was primarily concerned with the condition

$$L_F^\infty(T) = C_F(T). \tag{12}$$

His objective was to show that (12) does not characterize Sidon sets, and the example makes that evident, since

$$L_F^\infty(T) \cong \oplus_{l^1} L_{F_j}^\infty(T) \tag{13}$$

and $\alpha(F_j) \to \infty$. Results related to property (12) are found in Hartman [1] and Pigno and Saeki [3]. The elegant Proposition 12.2.12 is due to Blei [1]. When Katznelson was told of Blei's result, he pointed out that the special case 12.2.5 has the straightforward proof given here, avoiding the use of Drury's union theorem. Blei [5] gave a much shorter proof, subsuming Katznelson's idea under his own and also avoiding Drury's theorem, but still not completely general. Blei [5] also gives more sets satisfying (12) and (13). Blei [7] produces sets satisfying (12) that cannot be partitioned into finite sets F_j satisfying (13). The Proposition 12.2.13 is from Katznelson and McGehee [3].

We presented 12.2.14 merely to illustrate the fact that when E is fairly large, $A(E) \neq \tilde{A}(E)$. That result was pointed out to us by L. T. Ramsey, but related results are found in Zygmund [1, V.7] and Zafran [1, Theorem 3.9].

12.3. The Connection with Synthesis

Let Γ be a discrete abelian group, and let G be its dual group. For a set $E \subset \Gamma$, we define $A_d(E) = A_d(E, \Gamma)$ to be the algebra $C_o(E) \cap B_d(E, \Gamma)$, with the quotient norm:

$$\|f\|_{A_d(E)} = \inf\{\|\mu\|_M : \mu \in M(G), \hat{\mu} = f \text{ on } E\}.$$

Since $M_d(G) \subset M(G)$, evidently $A_d(E) \subset \tilde{A}(E)$. Let E^c be the set of cluster points, and $\bar{E}$ the closure, of E in $b\Gamma$. Since $B_d(\Gamma)$ is identifiable with $A(b\Gamma)$, $A_d(E)$ is identifiable with $\{f \in A(\bar{E}, b\Gamma): f = 0 \text{ on } E^c\}$, and thus its dual space is $PM(\bar{E}, b\Gamma)$. Theorem 12.3.1 below characterizes the sets E for which $A(E, \Gamma) \subset A_d(E, \Gamma)$; evidently the condition that $E \cap E^c = \varnothing$ is necessary for this inclusion to hold. A corollary will state that if E is a countable set of real numbers whose closure with respect to R is compact, then $A(E, R_d) \subset A_d(E, R_d)$.

Let F be a set like that of Proposition 12.1.5, restricted so that its closure is compact. Thus if $E \subset F$, then

$$A(E, R_d) \subset A_d(E, R_d) \subset \tilde{A}(E, R_d) \subset C_o(E, R_d).$$

Recall that $F = \bigcup_{j=1}^{\infty} F_j$, where

$$F_j = \{r_j + ks_j : 0 \leq k \leq K(j)\},$$

$K(j) \to \infty$, and the set $\{r_j\} \cup \{s_j\}$ is independent over the rationals. The word "synthesis" in the title stands for the theory of harmonic synthesis, in which the salient fact is that the algebra $A(R)$ disobeys synthesis. We shall show (Theorem 12.3.3 below) that this fact implies the existence of sets $E \subset F$, obtained by thinning the arithmetic progressions F_j, which enjoy the interesting property that $A_d(E, R_d) \neq A(E, R_d)$. We shall prove also (Theorem 12.3.4) that the existence of sets with this property implies the failure of harmonic synthesis in $A(b(R_d))$. Theorem 12.3.5 has to do with the failure of this property, and shows in particular that $A_d(F, R_d) = A(F, R_d)$.

12.3.1. Theorem. *Let E be a subset of a discrete abelian group Γ with dual group G. Then $A(E) \subset A_d(E)$ if and only if $E \cap E^c = \varnothing$ and there exists $\mu \in M(G)$ such that $\hat{\mu} = 1$ on E and $\hat{\mu} = 0$ on $\Gamma \cap E^c$.*

Proof. Suppose that $A(E) \subset A_d(E)$. Since every element of $A_d(E)$ must vanish on E^c, it is evident that $E \cap E^c = \varnothing$. We need to show that $\chi_E \in B(F)$, where $F = \bar{E} \cap \Gamma$.

Let $f \in A(F)$. Then $f|_E \in A(E)$, so there exists $g \in A_d(\Gamma)$ such that $g = f$ on E. Then $g = \chi_E f$ on F. The mapping $f \to \chi_E f$ is easily shown to be continuous from $A(F)$ into $B(F)$, by the closed graph principle. Let k be the norm of this mapping.

Let p be an F-polynomial with $\|p\|_{C(G)} = 1$. There exists $h \in A(\Gamma)$ such that $h = 1$ on $\mathrm{supp}(\hat{p})$ and $\|h\|_A < 2$. Then

$$|\langle \chi_E, \hat{p} \rangle| = |\langle \chi_E h, \hat{p} \rangle| \leq \|\chi_E h\|_{B(F)} \|p\|_{C(G)} \leq 2k\|p\|_{B(F)}.$$

It follows that $\chi_E \in B(F)$. The only-if part is proved.

Suppose now that $E \cap E^c = \emptyset$ and $\chi_E \in B(F)$, and let $f \in A(E)$. Let $g \in A(F)$ such that $g = f$ on E, and let $h = \chi_E g$. Then $h \in A(F)$, $h = f$ on E, and $h = 0$ on $F \backslash E$. It suffices to show that $h \in A_d(F)$. Thus to complete the proof of 12.3.1, it suffices to prove the following result.

12.3.2. Lemma. *Let Γ be a discrete abelian group with dual group G. Let $E \subseteq \Gamma, F = \bar{E} \cap \Gamma, K = E^c \cap \Gamma$, where E^c is the set of cluster points of E in $b\Gamma$. Let $I(K, F) = \{f \in A(F)\colon f = 0$ on $K\}$. Then $I(K, F) \subseteq A_d(F)$.*

Proof. If P is a finite subset of Γ, and g is an arbitrary function defined on P, then $g = h|_P$ for some $h \in A_d(\Gamma)$. If $P \subseteq F \backslash K$ and $Q \subseteq K$, then P and $\bar{Q}$ are disjoint compact subsets of $b\Gamma$, and therefore there exists $k \in A_d(\Gamma)$ such that $k|_{P \cup Q} = \chi_P$. Thus hk, which agrees with g on P and vanishes on Q, belongs to $A_d(b\Gamma)$. It follows that in particular, the functions in $I(K, F)$ with finite support are contained in $A_d(F)$. Since those functions are dense in $I(K, F)$, to prove that $I(K, F) \subset A_d(F)$ if suffices to show that $\|f\|_{A_d(F)} = \|f\|_{A(F)}$ when f has finite support. Since $A(F)^* = PM(F, \Gamma)$ is contained isometrically in $A_d(F)^* = PM(\bar{F}, b\Gamma)$, it is obvious that $\|f\|_{A(F)} \le \|f\|_{A_d(F)}$. Let $S \in A_d(F)^*$, $\|S\|_{PM} \le 1$. If E_0 is the support of f and $\varepsilon > 0$, then there exists $\nu \in M(\Gamma)$ such that $\nu(x) = 1$ for $x \in E_0$, supp ν is finite, and $\|\hat{\nu}\|_{L^1(G)} < 1 + \varepsilon$. Then $\nu S \in PM(F)$, $\|\nu S\|_{PM} \le \|S\|_{PM}(1 + \varepsilon)$, and $\langle f, \nu S\rangle = \langle f, S\rangle$. It follows that for each $\varepsilon > 0$, $\|f\|_{A_d(F)} \le (1 + \varepsilon)\|f\|_{A(F)}$. The Lemma is proved. □

Let F be the set defined in the hypothesis of 12.1.5, and recall the notation established there. Let $\mathscr{H}$ be the family of all compact subsets of $(0, 1)$. For $H \in \mathscr{H}$, and for each $j = 1, 2, \ldots$, let H_j be the set of numbers $k/K(j)$, where k is an integer, whose distance from H is no greater than $1/K(j)$. Let $\varphi_j(x) = r_j + K(j)s_j x$ for $x \in R$. Let $E_j = \varphi_j(H_j)$, $E_H = \bigcup_{j=1}^{\infty} E_j$. Then $E_j \subset F_j$ and $E_H \subset F$. The mapping $H \to E_H$ is a one-to-one mapping from $\mathscr{H}$ into the family of subsets of F.

12.3.3. Theorem. *If $H \in \mathscr{H}$ and H disobeys harmonic synthesis, then $A_d(E_H, R_d) \neq A(E_H, R_d)$.*

Proof. Let $g \in I(H)$, $S \in PM(H)$ such that $\langle g, S\rangle \neq 0$. There is a sequence of measures $\mu_j \in M(H_j)$ such that $\mu_j \to S$ weak$*$; in particular, $\sup\|\mu_j\|_{PM} < \infty$ and $\langle g, \mu_j\rangle \to \langle g, S\rangle$, and hence $\lim_{j\to\infty}\|g\|_{A(H_j)} > 0$, whereas of course $\|g\|_{C(H_j)} \to 0$. (See Section 3.2. To apply that procedure to the present setting, one must use the following facts. If K is a compact set and $K \subseteq (-\pi, \pi)$, then $A(K, R)$ and $A(K, T)$ are isomorphic. If $m \neq 0$, then the mapping $f(x) \to f(mx + c)$ is an isometric automorphism of $A(R)$.)

Since $g \in A(R)$, g may be represented on each closed subinterval of $(-\pi, \pi)$ as the sum of an absolutely convergent Fourier series. Thus we may write,

$$g(x) = \sum c_n e^{inx} \quad \text{for} \quad -\pi/2 \le x \le \pi/2, \quad \text{where} \quad \sum|c_n| < \infty.$$

Since φ_j is an affine-linear map from R onto R, the mapping $f \to f \circ \varphi_j$ is an isometric automorphism of $A(R)$. In particular, $\|g \circ \varphi_j^{-1}\|_{A(E_j)} = \|g\|_{A(H_j)}$.

Let n be an integer. Because the set $\{r_j\} \cup \{s_j\}$ is independent, there is a character $\chi_n \in bR$ such that

$$\chi_n(r_j) = 1 \quad \text{and} \quad \chi_n(s_j) = e^{in/K(j)} \quad \text{for each } j.$$

Necessarily, $\chi_n(r_j + ks_j) = e^{ink/K(j)}$, and in particular, $\chi_n(x) = e^{in\varphi_j^{-1}(x)}$ for $x \in E_j$. Let $f = \sum c_n \chi_n$. Then $f \in B_d(R_d)$, since it is the transform of the discrete measure $\sum c_n \delta_{\chi_n}$. For $x \in E_j$, $f(x) = g \circ \varphi_j^{-1}(x)$. Evidently the restriction of f to E_H belongs to $A_d(E_H, R_d)$ but not to $A(E_H, R_d)$. The Theorem is proved. □

12.3.4. Theorem. *Let E be a sequence $\{x_n\}_{n=1}^{\infty}$ of nonzero real numbers. Considering E as a subset of the group $b(R_d)$, let $P = \bar{E}\backslash E$. If $A_d(E, R_d) \neq A(E, R_d)$, then P disobeys synthesis with respect to $A(b(R_d))$.*

Proof. The hypothesis implies the existence of a function $f \in A(b(R_d))$ such that $f \in C_o(E)$ and $\|f\|_{A(E_m)} > 1$ for each m, where $E_m = \{x_n : n \geq m\}$. Since f is of course continuous on $b(R_d)$, it vanishes on the closed set P, and hence belongs to the ideal $I(P, b(R_d))$. For every neighborhood V of P in $b(R_d)$, there exists m such that $V \supset E_m$, and hence $\|f\|_{A(V, b(R_d))} > 1$. The Theorem is proved. □

12.3.5. Theorem. *Let $f = \sum_{n=1}^{\infty} c_n \chi_n$ where $\sum |c_n| < \infty$ and $\chi_n \in b(R)$. For $j = 1, 2, \ldots$, let $F_j = \{r_j + ks_j : k = 0, 1, \ldots, K(j)\}$ where $s_j \neq 0$ and $K(j) \to \infty$. If $\|f\|_{C(F_j)} \to 0$, then $\|f\|_{A(F_j)} \to 0$.*

Proof. If $\chi \in b(R)$ and F is a finite set, then $\|\chi\|_{A(F)} = 1$. Therefore $\|f\|_{A(F_j)} \leq \sum |c_n|$ for every j. Let $L = \limsup_{j \to \infty} \|f\|_{A(F_j)}$, and let m be an arbitrary positive integer. We shall prove that in the definition of f, a subset of the set of coefficients $\{c_n\}$, including c_m, can be replaced by zeros without changing the value of L. The Theorem follows, since L must be zero.

For each j, consider a finite partition of the interval $[0, \pi]$:

$$P_j = \left\{\frac{k\pi}{K(j)} : k = 0, 1, \ldots, K(j)\right\},$$

and an affine-linear map from R onto R:

$$d_j(x) = r_j + s_j K(j)x/\pi.$$

Then d_j maps P_j onto F_j, and

$$\|f \circ d_j\|_{A(P_j)} = \|f\|_{A(F_j)},$$

while

$$\|f \circ d_j\|_{C(P_j)} = \|f\|_{C(F_j)} \to 0.$$

For each pair n, j, let $\theta(n, j)$ and $\lambda(n, j)$ be defined by these conditions:

$$e^{i\theta(n, j)} = \chi_n(r_j) = \chi_n \circ d_j(0), \qquad |\theta(n, j)| \le \pi,$$

$$e^{i\lambda(n, j)\pi/K(j)} = (\bar{\chi}_m \chi_n)(s_j) = (\bar{\chi}_m \chi_n) \circ d_j(\pi/K(j)), \qquad |\lambda(n, j)| \le K(j).$$

let

$$g_j(x) = \sum_n c_n e^{i[\theta(n, j) + \lambda(n, j)x]} \quad \text{for } x \in R.$$

For $x = \pi k/K(j) \in P_j$,

$$g_j(x) = g_j(\pi k/K(j)) = (\chi_m(s_j))^{-k} f \circ d_j(\pi k/K(j)).$$

Thus

$$\|g_j\|_{A(P_j)} = \|f\|_{A(F_j)}, \quad \text{while} \quad \|g_j\|_{C(P_j)} \to 0$$

and

$$L = \limsup_{j \to \infty} \|g_j\|_{A(P_j)}.$$

Since we may replace $\{g_j\}$ by a subsequence without changing L, we may suppose that for every n, (a) $\theta(n, j)$ converges to a finite limit $\theta(n)$, and (b) either $|\lambda(n, j)| \to \infty$ or $\lambda(n, j)$ converges to a finite limit $\lambda(n)$. Let

$$X_1 = \{n : |\lambda(n, j)| \to \infty\},$$
$$X_2 = \{n : \lambda(n, j) \to \lambda(n) \in R\}.$$

For $i = 1$ and 2, let

$$g_{ij}(x) = \sum_{n \in X_i} c_n e^{i[\theta(n, j) + \lambda(n, j)x]},$$
$$f_i = \sum_{n \in X_i} c_n \chi_n,$$
$$h(x) = \sum_{n \in X_2} c_n e^{i[\theta(n) + \lambda(n)x]}.$$

Since $\|e^{iux} - e^{ivx}\|_{A(E)} = 0(\|e^{iux} - e^{ivx}\|_{C(E)})$ as $u - v \to 0$, for any closed set E, it is evident that

$$\|g_{2j} - h\|_{A[0, \pi]} \to 0.$$

We shall show that $h = 0$ on $[0, \pi]$ by showing that $\int_a^b h = 0$ whenever $0 \le a < b \le \pi$. Using Riemann sums, one finds that

$$\int_a^b h = \lim_{j \to \infty} \frac{\pi}{K(j)} \sum\{h(x): x \in P_j \cap [a, b]\}.$$

Since $\|h + g_{1j}\|_{C(P_j)} \le \|h - g_{2j}\|_{C(P_j)} + \|g_j\|_{C(P_j)} \to 0$, it follows that

$$\begin{aligned} \int_a^b h &= -\pi \lim_{j \to \infty} K(j)^{-1} \sum\{g_{1j}(x): x \in P_j \cap [a, b]\} \\ &= -\pi \lim_{j \to \infty} \sum_{n \in X_1} c_n b(n, j), \end{aligned}$$

where $b(n, j) = K(j)^{-1} \sum\{e^{i\lambda(n, j)x}: x \in P_j \cap [a, b]\}$. In every case, $|b(n, j)| \le 2(b - a)$. If $\lambda(n, j) \ne 0$, then $b(n, j)$ involves a sum of powers of $e^{i\lambda(n, j)\pi/K(j)}$ and one may estimate it as follows.

$$\begin{aligned} |b(n, j)| &\le \frac{2}{K(j)|1 - e^{i\lambda(n, j)\pi/K(j)}|} \\ &= \frac{1}{K(j)|\sin(\pi\lambda(n, j)/2K(j))|} \\ &\le \frac{1}{|\lambda(n, j)|} \quad (\text{since } |\sin x| \ge 2x/\pi \quad \text{for } |x| \le \pi/2). \end{aligned}$$

Therefore

$$\left| \int_a^b h \right| \le \pi \sum_{n \in X_1} |c_n| \min(2(b - a), |\lambda(n, j)|^{-1}),$$

which tends to zero as $j \to \infty$. Therefore $h = 0$ on $[0, \pi]$, so that $\|f_2\|_{A(F_j)} = \|g_{2j}\|_{A(P_j)} \le \|g_{2j}\|_{A[0, \pi]} \to 0$, and thus L remains the same when c_n is replaced by 0 for every $n \in X_2$. The Theorem is proved. □

12.3.6. Corollary. $A_d(F, R_d) = A(F, R_d)$.

Remarks. Theorem 12.3.1 is from Pigno and Saeki [1]. Other conditions that suffice to imply that $A(E) \subset A_d(E)$ are developed in Blei [4]. The connections between Blei's results and those of Pigno and Saeki are not clear. Theorems 12.3.3 and 12.3.5 are from Katznelson and McGehee [2]. Theorem 12.3.4 is due to Graham [6].

12.4. Sigtuna Sets

12.4.1. Theorem. *There exists a perfect set $E \subset R$ such that for every portion U of E, $A(U) \neq \tilde{A}(U)$; and such that the quantity*

$$\alpha_c(E) = \sup\left\{\frac{\|\mu\|_M}{\|\mu\|_{PM}}: \mu \in M_c(E), \mu \neq 0\right\}$$

is finite (even though, of course, $\alpha(E) = \alpha_d(E) = \infty$).

Proof. We shall construct a set E with the stated properties. It will be the closure of the union of inductively selected arithmetic progressions

$$F_j = \{r_j + ms_j: 0 \leq m \leq 10^j\}, j = 1, 2, \ldots.$$

Suppose that $k \geq 1$ and that $F_1, \ldots, F_k$ have been selected so that the set $\{r_j, s_j: 1 \leq j \leq k\}$ is independent over the rationals. Let g be a function defined on the set

$$E_k = \bigcup_{j=1}^{k} F_j$$

such that on each F_j, g is constant and equals 0 or 1. Then there exists a measure $\omega = \omega_{k,g} \in M_d(R)$ such that $\omega(x) = g(x)$ for each $x \in E_k$ and $\|\omega_{k,g}\|_{A(R_d)} < \frac{4}{3}$. (This is standard business. One may apply the proof of 12.2.9 ignoring V and K; letting $\varepsilon = \frac{1}{3}$, $H = g^{-1}(1)$, and $\{u_j\} = \{r_j, s_j: g(F_j) = 1\}$; and taking ω to be $v + \rho$). Let $\{V_\lambda: \lambda > 0\}$ be the familiar de la Vallée Poussin kernel, defined in Appendix A.5. There exists $\lambda_g > 0$ such that if $\lambda \leq \lambda_g$, then

$$\|\omega_g * V_\lambda\|_A \leq 4,$$

and

$$(\omega_g * V_\lambda)(t) = g(x) \quad \text{if} \quad x \in E_k \quad \text{and} \quad |t - x| \leq \lambda.$$

Let $d(k)$ be the minimum value of λ_g, considering all the 2^k possible functions g. We now select F_{k+1}, subject to two conditions: first,

$$F_{k+1} \subset (x_0 - \tfrac{1}{2}d(k), x_0 + \tfrac{1}{2}d(k)),$$

where x_0 is a point of E_k chosen so that

$$\operatorname{dist}(x_0, E_k \backslash \{x_0\}) = \max\{\operatorname{dist}(x, E_k \backslash \{x\}): x \in E_k\};$$

second, r_{k+1} and s_{k+1} are chosen so that the set $\{r_1, \ldots, r_{k+1}, s_1, \ldots, s_{k+1}\}$ is independent over the rationals. The first stipulation ensures that the set $E = (\bigcup_{j=1}^{\infty} F_j)^-$ is a perfect set, and that

$$E \subset E_k + (-d(k), d(k)) \quad \text{for } k = 1, 2, \ldots.$$

Evidently E is a totally disconnected set with Lebesgue measure zero.

Let $S \in N(E)$, $\|S\|_{PM} = 1$. We shall show that

(1) $$S = \mu + \sum_{j=1}^{\infty} S_j, \quad \text{where} \quad S_j \in M(F_j),$$

$$\sum \|S_j\|_{PM} \le 4, \; \mu \in M(E), \quad \text{and} \quad \|\mu\|_M \le 16;$$

so that, in particular, $\alpha_c(E) \le 16$. For each k, let v_k be a measure that assigns mass 1 to each point of E_k and annihilates each point of $\bigcup_{j=k+1}^{\infty} F_j$, and such that

$$\|v_k * V_{d(p)}\|_A \le 4 \quad \text{for } p \ge k.$$

($\omega_{k,1}$ will serve as v_k if it is obtained as in the parenthetical remark above). The sequence $\{(v_k * V_{d(p)})S : p = 1, 2, \ldots\}$ is bounded in PM-norm by 4 and hence has a subsequence that converges weak$*$ to an element of $M(E_k)$. Therefore by a diagonal process we may find $S_j \in M(F_j)$ for $j \ge 1$ and a sequence $\{p(m) : m = 1, 2, \ldots\}$ such that for each k,

$$\text{weak}*\lim_{m \to \infty} (v_k * V_{d(p(m))})S = \sum_{j=1}^{k} S_j.$$

By Kronecker's Theorem (as in the proof of 12.1.5),

$$\sum_{j=1}^{k} \|S_j\|_{PM} = \left\| \sum_{j=1}^{k} S_j \right\|_{PM} \le 4 \quad \text{for } k \ge 1.$$

Therefore the series $\sum S_j$ converges in norm to a pseudomeasure S, whose transform is almost-periodic. To prove that the remainder

$$\mu = S - \sum_{j=1}^{\infty} S_j$$

is a measure with $\|\mu\|_M \le 16$, it suffices to show that

(2) $$|\langle f, \mu \rangle| \le 4 \quad \text{for all } f \text{ in } C(E) \text{ with range } \{0, 1\}.$$

Considering such an f, let $\varepsilon > 0$. Fix k large enough so that f is constant on each F_j for $j > k$, and so that

$$\left(\sum_{j=k+1}^{\infty} \|S_j\|_{PM}\right)\|f\|_{A(E)} < \varepsilon$$

and hence

$$\left|\left\langle f, \mu - \left(S - \sum_{j=1}^{k} S_j\right)\right\rangle\right| < \varepsilon.$$

Now fix $p = p(m) \geq k$ large enough so that

$$\left|\left\langle f, \sum_{j=1}^{k} S_j - (v_k * V_d)S\right\rangle\right| < \varepsilon,$$

where $d = d(p(m))$. Then

$$|\langle f, \mu - (1 - v_k * V_d)S\rangle| < 2\varepsilon.$$

Let g vanish on E_k and agree with f on $E_p \backslash E_k$. Let $\omega = \omega_{p,g}$. Note that

$$E \subset E_p + (-d, d);$$

$$\omega * V_d \quad \text{and} \quad 1 - (v_k * V_d) \quad \text{both vanish on} \quad E_k + (-d, d);$$

and

$$\omega * V_d \quad \text{agrees with } f \text{ and } v_k * V_d \text{ vanishes on } (E_p \backslash E_k) + (-d, d).$$

Therefore

$$|\langle f, (1 - v_k * V_d)S\rangle| = |\langle \omega * V_d, S\rangle| \leq 4,$$

and hence $|\langle f, \mu\rangle| \leq 4 + 2\varepsilon$. The inequality (2) follows, and the decomposition property (1) is established.

For each j, let f_j be a function in $A(R)$ such that f_j is constant on each of the 10^j intervals $\{x + [-d(j), d(j)]: x \in F_j\}$ and zero on every other portion of E, and such that $\|f_j\|_{A(F_j)} = 1$ and $\sum \|f_j\|_{C(E)} < 1$. Let $f = \sum_{j=1}^{\infty} f_j$. It is easy to show that by virtue of (1), $\|f\|_{\tilde{A}(E)} \leq 20$.

Let $x \in E$. Since $\{x\}$ obeys synthesis, if f were in $A(E)$, then

(3) $$\|f\|_{A(E \cap (x-\varepsilon, x+\varepsilon))} \to |f(x)| \quad \text{as} \quad \varepsilon \to 0.$$

But every neighborhood of x contains F_j for infinitely many j, so that the left-hand side of (3) is always at least one, whereas $|f(x)| < 1$. Thus f is not even locally in $A(E)$ at any $x \in E$. In particular, if U is a portion of E, then $f \notin A(U)$. But of course $\|f\|_{\tilde{A}(U)} \leq 20$. The Theorem is proved. □

Remarks. In 1966, at the meeting in Sigtuna, Sweden, Katznelson was asked whether there exists a set whose every subset obeys synthesis (a "set of resolution") and whose every portion is non-Helson. In response he concocted the construction given above. Later in 1966, he realized that the Sigtuna set was an example of a set E with $\tilde{A}(E) \neq A(E)$, and was led immediately to the simpler example, the set F of 12.1.5.

Rudin [11] had constructed a perfect M_o-set F whose points are independent over the rationals. Such an F is not a Helson set, even though $\|\mu\|_M = \|\mu\|_{PM}$ for all $\mu \in M_d(F)$, so that $\alpha_d(F) = 1$ while $\alpha_c(F) = \alpha(F) = \infty$. Note that the Sigtuna set E, on the other hand, has the property that $\alpha_c(E) < \infty$ while $\alpha_d(E) = \alpha(E) = \infty$. (See Section 4.7 for Körner's method of constructing Rudin's type of set).

If $S \in PM(R)$ and $x \in R$, there may be more than one cluster point of $(V_\lambda * \delta_x)S$ as $\lambda \to 0$. In other words, there is no well-defined restriction of a pseudomeasure to a point (see Katznelson [1, Chapter VI]). That is why the measures S_j in the above proof must be approached with care. The set E is special in that (as it turns out) for $S \in PM(E)$ and $x \in E$, weak$*$ $\lim_{\lambda \to 0}(V_\lambda * \delta_x)S$ always exists. Sets with this property were named *ergodic sets* and studied at length by G. S. Woodward [1, 3]. The Sigtuna sets are important examples of ergodic sets.

12.5. An Example in which $A(E)$ Is a Dense Proper Subspace of $\tilde{A}(E)$

The quantity $\beta(E)$ and the algebra $A^*(E)$, defined and discussed in Section 12.1, will be used here.

12.5.1. Theorem. *There exists a perfect set $E \subseteq T$ such that $A(E)$ is a dense proper subspace of $\tilde{A}(E)$.*

Proof. Suppose that $E = \{0\} \cup \bigcup_{n=1}^{\infty} E_n$, where E_n is a perfect set contained in the middle third of the interval $(1/(n+1), 1/n)$ such that

$$A(E_n) = \tilde{A}(E_n) \quad \text{but} \quad \beta(E_n) > 2^n. \tag{1}$$

The spacing of the sets E_n ensures the existence of functions $w_n(x)$ (equal to $V_\lambda(x - a)$ for appropriate λ and a) such that

$$\|w_n\|_{A(T)} \leq 3, \; w_n|_E = \chi_{E_n}.$$

By virtue of (1), there exist $f_n \in A(T)$ such that

$$\|f_n\|_{\tilde{A}(E_n)} < 2^{-n} \quad \text{and} \quad \|f_n\|_{A(E_n)} = 1.$$

It follows that

$$\|w_n f_n\|_{\tilde{A}(E)} < 3 \cdot 2^{-n} \quad \text{and} \quad \|w_n f_n\|_{A(E)} \geq 1,$$

so that $\beta(E) = \infty$. Thus $A(E)$ is indeed a proper subspace of $\tilde{A}(E)$, if (1) holds and the sets E_n are positioned as stated.

Theorem 12.5.3 allows the selection of such a sequence $\{E_n\}$, and a sequence $\{\Lambda_n\}$ of disjoint infinite subsets of Z such that

(2) for each $h_n \in A(E_n)$, there exists $\mu_n \in M(E_n)$ such that $\|\hat{\mu}_n\|_{PM} \leq 21$, $\hat{\mu}_n \in C_o(Z \backslash \Lambda_n)$, and $\int h_n \, d\mu_n \geq \|h_n\|_{A^*(E_n)}$.

(The sets E_n are obtained as follows. First we choose sequences $Q^{(n)}$ so that the sets $\Lambda_n = Y_{Q^{(n)}}$, obtained as in 12.5.2. are disjoint. Then we apply 12.5.3 with $Q = Q^{(n)}$, $K = 2^n$, and $\varepsilon < \frac{1}{21}$, to obtain E_n, which we may require to lie in the desired interval.)

It will follow that $A(E)$ is dense in $\tilde{A}(E)$, for we may let $f \in \tilde{A}(E)$ and argue as follows. Let $f = g|_E + h$, where $g \in A(T)$ and $h \in A^*(E)$, as in 12.1.6. Let $h_n = w_n h$, which belongs to $A(E) \cap A^*(E)$. Let μ_n be the measure provided by condition (2), and for $N \geq 1$ let $\mu = \sum_{n=1}^N \mu_n$. Then

$$\limsup_{k \to \infty} |\hat{\mu}(k)| \leq 21,$$

and

$$\sum_{n=1}^N \|h_n\|_{A^*(E)} \leq 3 \sum_{n=1}^N \|h_n\|_{A^*(E_n)}$$

$$\leq 3 \sum \int h_n \, d\mu_n = 3 \int h \, d\mu \leq 63 \|h\|_{A^*(E)}.$$

It follows that f is the limit, in $\tilde{A}(E)$-norm, of the sequence $\{g + \sum_{n=1}^m h_n\}_{m=1}^\infty$, which lies in $A(E)$. The Theorem is proved. □

12.5.2. Lemma. *Let $Q = \{q_k\}$ be a sequence of positive integers such that if*

$$b_k = \frac{q_k}{1000 q_{k-1}} \quad \textit{for } k > 1,$$

then b_k is an integer and $b_k \geq k$. Let

$$\tau_k = \frac{1}{b_k}\sum_{j=0}^{b_k-1}\delta_{(2\pi j/q_k)}, \qquad \sigma_n = \mathop{\bigstar}_{k=n+1}^{\infty}\tau_k.$$

Then

(3) $$\hat{\sigma}_n \in C_o(Z\setminus Y_Q), \quad \textit{where } Y_Q = \bigcup_{k=1}^{\infty}(q_k - kq_{k-1}, (k+1)q_k)$$

and

(4) $$\limsup_{j\to\infty}|\hat{\sigma}_n(j)| \leq 1 - \eta,$$

where η is a positive number independent of n and of Q.

Proof. For $b = 2, 3, \ldots,$ consider the functions

$$f_b(x) = \frac{1 - e^{ibx}}{b(1 - e^{ix})}, \; f_b(0) = 1.$$

It is elementary to show that

$$\sup\left\{|f_b(x)|: b \geq 2, \; \frac{s}{b} \leq x \leq 2\pi - \frac{s}{b}\right\} \to 0 \quad \text{as } s \to \infty,$$

and that for each $\varepsilon > 0$ there exists $\eta > 0$ such that

$$\sup\left\{|f_b(x)|: \frac{\varepsilon}{b} \leq x \leq 2\pi - \frac{\varepsilon}{b}\right\} \leq 1 - \eta \quad \text{for all } b.$$

For $m \neq 0$,

$$\hat{\tau}_k(m) = \frac{1}{b_k}\sum_{j=0}^{b_k-1}e^{2\pi ijm/q_k} = \frac{1 - e^{2\pi ib_k m/q_k}}{b_k(1 - e^{2\pi im/q_k})} = f_{b_k}(2\pi m/q_k).$$

It follows that

$$\sup\{|\hat{\tau}_k(m)|: kq_{k-1} \leq m \bmod(q_k) \leq q_k - kq_{k-1}\} \to 0 \quad \text{as } k \to \infty,$$

and that for some $\eta > 0$,

$$|\hat{\tau}_k(m)| \leq 1 - \eta \quad \text{for} \quad \frac{q_{k-1}}{2} \leq m \bmod(q_k) \leq q_k - \frac{q_{k-1}}{2}, \quad \text{for all } k.$$

Conditions (3) and (4) follow, since $\hat{\sigma}_n = \prod_{k=n+1}^{\infty}\hat{\tau}_k$. □

12.5.3. Theorem. *Let $Q = \{q_k\}$ be a sequence as in 12.5.2, and let $K > 0$. There exists a perfect set E such that*

(5)
$$\tilde{A}(E) = A(E),$$

(6)
$$\beta(E) > K,$$

and

(7) *for every $h \in A(E)$ and $\varepsilon > 0$, there exists $\rho \in M(E)$ such that $\|\rho\|_{PM} \le 20$, $\hat{\rho} \in C_o(Z \setminus Y_Q)$, and $\int h\, d\rho \ge \|h\|_{A^*(E)}(1 - \varepsilon)$.*

Proof. It follows from Theorem 4.6.2 that there exists a perfect M-set P_0 such that

(8)
$$\|f_r - 1\|_{C(P_0)} \to 0 \quad \text{as } r \to \infty, \quad \text{where} \quad f_r(x) = (p_{r+1} - p_r)^{-1} \sum_{j=p_r+1}^{p_{r+1}} e^{iq_j x}$$

and $\{p_r\}$ is some increasing sequence of integers. It follows that for every $\mu \in N(P_0)$, $\|\mu\|_{PM} = \limsup_{j\to\infty} |\hat{\mu}(j)|$. Since P_0 is an M-set, there exists a nonzero $S \in PM(P_0)\setminus N(P_0)$ such that $\hat{S} \in C_o(Z)$.

Beginning with P_0, we define a sequence of sets P_n. For $n \ge 1$, let

$$H_n = \left\{x = \frac{2\pi j}{q_n} : j \quad \text{is an integer and } \operatorname{dist}(x, P_{n-1}) \le \frac{2\pi}{q_n}\right\},$$

and let $P_n = P_{n-1} \cup (H_n + \operatorname{supp}(\sigma_{n+1}))$, where σ_{n+1} is as defined in 12.5.2. The set $P = (\bigcup_{n=0}^{\infty} P_n)^-$ enjoys three interesting properties, which we are about to explain.

(a) *P obeys synthesis, so that, in particular, $S \in N(P)$.* In fact, if $x \in P\setminus(P_{n-1} \cup H_n)$, then

$$\operatorname{dist}(x, H_n) \le \sum_{p>n+1} \frac{2\pi b_p(p - n - 1)}{q_p} + \sum_{p>n} \frac{2\pi}{q_p} < 2\pi 10^{-3} \sum_{p>n} \frac{p}{q_p}$$
$$= \frac{2\pi}{q_n} 10^{-3} \sum_{p>n} \frac{pq_n}{q_p} < \frac{2\pi}{q_n} 10^{-3} \sum_{p>n} 10^{-3(p-n)}\left(\frac{n!}{p!}\right).$$

It follows that if $u = 2\pi j/q_n \notin P$, then

$$\operatorname{dist}(u, P) \ge \frac{2\pi(1 - \varepsilon_n)}{q_n} \quad \text{where } \varepsilon_n \to 0.$$

By 3.2.2, P obeys harmonic synthesis. In fact, we now make use of the procedure in the proof of that theorem. For arbitrary $\mu \in PM(P)$ let v_{q_n} be obtained from μ as was v_q in formula (14) of Section 3.2. Let

$$\mu_n = v_{q_n} * \sigma_{n+1}. \tag{9}$$

Then $\mu_n \in M(H_n + \operatorname{supp}(\sigma_{n+1})) \subset M(P)$, $\|\mu_n\|_{PM} \leq \|\mu\|_{PM}$, and $\mu_n \to \mu$ weak* (since $\sigma_n \to \delta_0$ weak*).

(b) *$A(P) = \tilde{A}(P)$, and $\beta(P) = 1$.* Let $f \in \tilde{A}(P)$, $\varepsilon > 0$. Choose $\mu \in M(P)$ such that $\|\mu\|_{PM} = 1$ and $\langle f, \mu\rangle > \|f\|_{\tilde{A}(P)}(1 - \varepsilon)$. Let μ_n be as in (9). Then $\langle f, \mu_n\rangle \to \langle f, \mu\rangle$, and for each n,

$$\limsup_{k\to\infty} |\mu_n(k)| \leq (1 - \eta).$$

By 12.1.7, $f \in A(P)$ and $\|f\|_{A(P)} = \|f\|_{\tilde{A}(P)}$.

(c) *P enjoys a variant of the approximation property* (8) *of the set* P_0. Let φ be the function defined on $[-\pi, \pi]$ as follows:

$$\varphi(x) = \begin{cases} 1 & \text{for } |x| \leq \dfrac{\pi}{4}, \\ -1 & \text{for } |x| \geq \dfrac{3\pi}{4}, \\ 2 - \dfrac{4}{\pi}\left(|x| - \dfrac{\pi}{4}\right) & \text{for } \dfrac{\pi}{4} \leq |x| \leq \dfrac{3\pi}{4}. \end{cases}$$

Then $\varphi \in A(T)$, $\hat{\varphi}(0) = 0$, and $\|\varphi\|_A \leq 4$. Let

$$g_r(x) = (p_{r+1} - p_r)^{-1} \sum_{j=p_r+1}^{p_{r+1}} \varphi(q_j x).$$

Then $g_r \in A(T)$, $\|g_r\|_A \leq 4$, and, since $\hat{\varphi}(0) = 0$,

$$\inf\{|n| : \hat{g}_r(n) \neq 0\} \to \infty \quad \text{as} \quad r \to \infty. \tag{10}$$

The meaning of our statement (c) is that

$$\|g_r - 1\|_{C(P)} \to 0 \quad \text{as} \quad r \to \infty. \tag{11}$$

Incidentally, (10) and (11) imply that P is a U_0-set, since for $\mu \in M(P)$ and $k \in Z$,

$$\begin{aligned} |\hat{\mu}(k)| &= \left| \lim_{r\to\infty} \int g_r(x) e^{-ikx}\, d\mu(x) \right| = \left| \lim_{r\to\infty} \sum_n \hat{g}_r(n)\hat{\mu}(k-n) \right| \\ &\leq \lim_{r\to\infty} 4 \sup\{|\hat{\mu}(k-n)| : \hat{g}_r(n) \neq 0\}, \end{aligned}$$

and it follows that, in fact,

$$\|\mu\|_{PM} \le 4 \limsup_{|n|\to\infty} |\hat{\mu}(n)| \quad \text{for every } \mu \in M(P).$$

The reader should be advised that the fact that φ equals one near 0 is the only property of φ that is involved in proving (11), which we shall now do. It follows from (8) that

$$\max_{x\in P_0} \operatorname{card}\{j: p_r < j \le p_{r+1}, |e^{iq_j x} - 1| \ge \tfrac{1}{10}\} \le a_r(p_{r+1} - p_r)$$

where $a_r \to 0$ as $r \to \infty$. To prove (11), it suffices to show that

(12) $$\max_{y\in \bigcup_n P_n} \operatorname{card}\{j: p_r < j \le p_{r+1}, |e^{iq_j y} - 1| \ge \tfrac{1}{5}\} \le a_r(p_{r+1} - p_r).$$

To prove (12), it suffices to show that for every $j \ge 1$, and for every

$$y \in \bigcup_{n=1}^{\infty} P_n \backslash P_0,$$

either

(13) $$|e^{iq_j y} - 1| < \tfrac{1}{10}$$

or

(14) there exists $u \in P_0$ such that $|e^{iq_j y} - e^{iq_j u}| < \tfrac{1}{10}$.

Fix j and y. Then $y \in P_n \backslash P_{n-1}$ for some $n \ge 1$, so that $y = (2\pi m/q_n) + z$, where m is an integer and $z \in \operatorname{supp}(\sigma_{n+1})$. The number z has the form

$$z = 2\pi \sum_{k>n+1} \frac{j_k}{q_k}, \quad \text{where } 0 \le j_k < b_k.$$

If $j \le n + 1$, then

$$|e^{iq_j z} - 1| \le 2\pi q_j \sum_{k>n+1} \frac{b_k}{q_k} < 2\pi 10^{-3} \sum_{k>n} 10^{-3(k-j)} < 10^{-2-3(n+1-j)}.$$

If $j > n + 1$, then

$$|e^{iq_j z} - 1| \le 2\pi q_j \sum_{k>j} \frac{b_k}{q_k} < 2\pi 10^{-3} \sum_{k\ge j} 10^{-3(k-j)} < 10^{-2}.$$

If $j \geq n$, then $|e^{iq_j y} - 1| = |e^{iq_j z} - 1| < 10^{-2}$, so that (13) is satisfied. If $j < n$, then there exists $x \in P_{n-1}$ such that

$$\left| x - \frac{2\pi m}{q_n} \right| \leq \frac{2\pi}{q_n}.$$

Then

$$\left| q_j \left(x - \frac{2\pi m}{q_n} \right) \right| < 2\pi 10^{-3(n-j)},$$

and

$$|e^{iq_j x} - e^{iq_j y}| < 10^{1-3(n-j)}.$$

If $x \in P_0$, then (14) is satisfied with $u = x$. If not, then $x \in P_{n'} \setminus P_{n'-1}$ for some n', $0 < n' < n$, and we deal with the pair x, n' the way we dealt with y, n. Eventually we reach either (13) or (14). Condition (11) follows.

Now we shall make use of the fact that there exists $S \in PF \cap N(P)$, $S \neq 0$. We may suppose that $\hat{S}(0) = \|S\|_{PM} = 1$. Choose m_0 so that

$$|m| \geq m_0 \Rightarrow |\hat{S}(m)| < \frac{1}{4K}.$$

Then choose k_0 so that if $t = \pi \sum_{k > k_0} q_k^{-1}$, then

$$m_0 t < \frac{1}{2K}.$$

Since

$$|(S - S * \delta_t)^\wedge(m)| = |\hat{S}(m)(1 - e^{imt})| \leq \min(2|\hat{S}(m)|, |m|t) \quad \text{for all } m \in Z,$$

it follows that $\|S - S * \delta_t\|_{PM} < 1/2K$.

By virtue of (11), and since $e^{iq_j t} \to -1$ as $j \to \infty$,

$$\|g_r + 1\|_{C(P+t)} \to 0 \quad \text{as } r \to \infty. \tag{15}$$

Clearly $P \cap (P + t) = \varnothing$. Let $E = P \cup (P + t)$. Since $A(P) = \tilde{A}(P)$, and since that property of a set is preserved under translation and finite disjoint unions, it is also the case that $A(E) = \tilde{A}(E)$. Let

$$g(x) = \begin{cases} 1 & \text{for } x \in P, \\ -1 & \text{for } x \in P + t. \end{cases}$$

By (11) and (15), $\|g_r - g\|_{C(E)} \to 0$. Therefore $g \in \tilde{A}(E)$ and $\|g\|_{\tilde{A}(E)} \leq 4$, whereas, since $S - S * \delta_t \in N(E)$,

$$\|g\|_{A(E)} \geq \frac{|\langle g, S - S * \delta_t \rangle|}{\|S - S * \delta_t\|_{PM}} > \frac{2}{1/2K} = 4K.$$

Thus $\beta(E) > K$. We have proved (5) and (6). It remains to prove (7).

We need the fact that

$$\|\mu\|_{PM} \leq 4 \limsup_{n \to \infty} |\hat{\mu}(n)| \quad \text{for every } \mu \in M(E). \tag{16}$$

Let $\tilde{g}_r(x) = g_r(2x)$. Then $\|\tilde{g}_r - 1\|_{C(E)} \to 0$ and $\|\tilde{g}_r\|_A \leq 4$. For $\mu \in M(E)$ and $k \in Z$,

$$\begin{aligned} |\hat{\mu}(k)| &= \left| \lim_{r \to \infty} \int \tilde{g}_r(x) e^{-ikx}\, d\mu(x) \right| \\ &\leq \lim_{r \to \infty} (4 \sup\{|\hat{\mu}(k - n)| : \hat{\tilde{g}}_r(n) \neq 0\}). \end{aligned}$$

Condition (16) follows.

Let $h \in A(E)$, $\varepsilon > 0$. Choose $\mu \in M(E)$ such that $\limsup_{|n| \to \infty} |\hat{\mu}(n)| = 1$ and $\int h\, d\mu \geq (1 - \varepsilon/2)\|h\|_{A^*(E)}$. In view of (16), we know that $\|\mu\|_{PM} \leq 4$. Let $E_1 = P$, $E_2 = P + t$. Let $\mu = \mu^1 + \mu^2$ such that $\mu^i \in M(E_i)$. For $i = 1$ and 2, there is a sequence $\{\mu_n^i\} \subset M(E_i)$, obtained as in (9), such that

$$\mu_n^i \to \mu^i \text{ weak}*, \qquad \hat{\mu}_n^i \in C_o(Z \setminus Y_Q),$$

$$\|\mu_n^i\|_{PM} \leq \|\mu^i\|_{PM}.$$

For each $k \in Z$, $|\hat{\mu}^1(k) - \hat{\mu}^2(k)| = |\langle \mu, g e^{ikx} | \leq 4\|\mu\|_{PM} \leq 16$. Thus

$$\left\| \sum_{i=1}^{2} \mu_n^i \right\|_{PM} \leq \sum_{i=1}^{2} \|\mu^i\|_{PM} \leq \|\mu^1 - \mu^2\|_{PM} + \|\mu^1 + \mu^2\|_{PM} \leq 20.$$

If $\rho = \mu_n^1 + \mu_n^2$ for n sufficiently large, then $|\int h\, d\rho| \geq (1 - \varepsilon)\|h\|_{A^*(E)}$, and of course $\|\rho\|_{PM} \leq 20$. Condition (7) and Theorem 12.5.3 are proved. □

Remarks. The construction is due to Katznelson and Körner [1], and our presentation is a revision of theirs. The fundamental ingredient is the idea for producing sets E with arbitrarily small $\beta(E)$. This idea is due to Varopoulos [13, Theorem 3] (see Lindahl and Poulsen [1, Chapters 10 and 11] or Katznelson and McGehee [2, Section 4]).

Chapter 13

Unsolved Problems

13.1. Dichotomy

The reader will recall that the Banach algebra $A(G)$ is analytic whenever G is infinite. That result is Theorem 9.3.1 (or 9.3.3), and the concept of an analytic algebra is discussed again in Sections 10.1 and 10.3. The reader will recall also what it means for a set $E \subseteq G$ to be a Helson set, and how the Helson constant $\alpha(E)$ is defined.

When E is a non-Helson set, it is "like G" in the sense that not every continuous function on it is a Fourier transform. Speaking loosely, we say that E then "participates in the harmonic analysis of G," and we ask the extent of that participation. We ask, if $A(E) \neq C_o(E)$, then how much is $A(E)$ "like $A(G)$"? In particular, is $A(E)$ an analytic algebra? That is the *Dichotomy Problem.* The *Dichotomy Conjecture* is as follows.

Either $A(E) = C_o(E)$, or else $A(E)$ is an analytic algebra.

The reputation of the Problem earns it first place in this chapter. The truth or falsehood of the Conjecture is not known for any infinite G. Nor is it known for the special case of countable E. The question is not "central"; there is no great column of other problems lined up like dominoes behind it. Still, the techniques that would solve it might prove very interesting and useful.

An equivalent formulation of the Conjecture is as follows.

(1) If E is non-Helson, then there exist $a > 0$, $c > 0$ such that $N_E(u) > ce^{au}$ for every $u > 0$, where

$$N_E(u) = \sup\{\|e^{iuf}\|_{A(E)}: f \in A(E), f \text{ is real-valued, and } \|f\|_{A(E)} \leq 1\}.$$

The Problem was introduced and studied at length by Katznelson [3]. Most of the known sufficient conditions for $A(E)$ to be analytic are developed efficiently in Kahane [2, Sections VI.8 and VIII.8]. We shall summarize them here.

If there exists a constant $K = K(E)$ such that for every $N > 0$ there exists $\mu \in N(E)$ and an integer $\lambda > 0$ such that

$$\sup_{p \in Z} \sum_{|m| \leq N} |\hat{\mu}(p - m\lambda)| \leq K |\hat{\mu}(0)|,$$

then $A(E)$ is analytic. In particular, if E is an M_o-set, or if E contains arbitrarily large arithmetic progressions or meshes, then $A(E)$ is analytic.

Varopoulos proved that if for every integer n, E contains the sum of two sets of n points, then $A(E)$ is analytic; see 11.3.4. Salinger and Varopoulos [1] showed that that sum condition holds whenever G is metrizable and there exist positive continuous measures $\mu, \nu \in M(G)$ such that $\mu * \nu(E) > 0$. Saeki [16] removed the metrizability condition.

In short, just about every convenient condition that will make E non-Helson will also make $A(E)$ analytic. The search for a counterexample to the Conjecture is further narrowed by the results obtained probabilistically by Katznelson and Malliavin [1, 2]. We state them now in the form of two theorems.

A subset V of the unit ball of $C(E)$ is a *set of majorization for E* if

$$\sup_{\gamma \in \Gamma} |\hat{\mu}(\gamma)| \leq 2 \sup_{f \in V} \left| \int f \, d\mu \right| \quad \text{for every } \mu \in M(E).$$

The *arithmetic diameter* of E, denoted by $d(E)$, is the smallest cardinality found among the sets of majorization for E. It is easy to show that if E is a subset of an arithmetic progression $Q = \{x + ky : k = 1, 2, \ldots, n\}$, then $d(E)$ is bounded by cn, for some universal constant $c > 0$. Similarly, if E is independent, then $d(E) \geq c_1 \exp(c_2 \# E)$, for some universal constants $c_1, c_2 > 0$, by 11.6.11. A sequence $\{Q_j\}$ of arithmetic progressions is *independent* if there exists $C > 0$ such that whenever $\mu_j \in M(Q_j)$ for $1 \leq j \leq J < \infty$,

$$\sum \|\mu_j\|_{PM} \leq C \|\sum \mu_j\|_{PM}.$$

Such a sequence is used in 12.1.5.

13.1.1. Theorem. *Let E be a closed subset of a locally compact abelian group G. Suppose that for some $s > 0$, and some sequence $\{E_j\}$ of finite subsets of E, $\# E_j \to \infty$ as $j \to \infty$, and $d(E_j) \leq (\# E_j)^s$ for every j. Then $A(E)$ is analytic.*

13.1.2. Theorem. *Let $\{Q_j\}$ be a sequence of arithmetic progressions contained in a locally compact abelian group G. Let $\{p_j\}$ be a sequence of positive integers, and for each j let E_j be a set of p_j points chosen at random from Q_j. Let the choices of the sets E_j be independent. Let E be the closure of $\bigcup_{j=1}^{\infty} E_j$, so that E is a set-valued random variable on a probability space.*

(A) If $p_j/\log(\#Q_j) \to \infty$, then $A(E)$ is almost surely analytic (and of course never a Helson set, by Theorem 11.6.6).

(B) Suppose that the sequence $\{Q_j\}$ is independent, and that $\sum(\#Q_j)^{\varepsilon-1} < \infty$ for some $\varepsilon > 0$. If the sequence $\{p_j/\log(\#Q_j)\}$ is bounded, then E is almost surely a Helson set.

The probabilistic idea used in proving (*B*) reappears in Katznelson [6, 7], where a probability space is constructed whose points are almost all Sidon sets.

For a given set E, let $a_E = \limsup_{u\to\infty} u^{-1} \log N_E(u)$. (Since $\log N_E$ is subadditive we could as well write "lim inf".) For $0 < \eta \leq 1$, let

$$N_{E,\eta}(u) = \sup\{\|e^{iuf}\|_{A(E)}: f \in A(E),\ f \text{ real-valued},$$

$$\|f\|_{A(E)} \leq 1, \quad \text{and} \quad \|f\|_{C(E)} \leq \eta\}.$$

Let $a_{E,\eta} = \limsup_{u\to\infty} u^{-1} \log N_{E,\eta}(u)$. One may consider the following five conjectures, of which the first is equivalent to (1).

(I) $$a_E > 0 \quad \text{for every non-Helson } E.$$

(II) $$\inf\{a_E: E \text{ is non-Helson}\} > 0.$$

(III) $$a_{E,\eta} > 0 \quad \text{for every non-Helson } E \text{ and every } \eta > 0.$$

(IV) $$b_E \equiv \inf\{a_{E,\eta}: \eta > 0\} > 0 \quad \text{for every non-Helson } E.$$

(V) $$\inf\{b_E: E \text{ is non-Helson}\} > 0.$$

Evidently (II) $\Rightarrow$ (I) $\Leftrightarrow$ (III) $\Leftarrow$ (IV) $\Leftarrow$ (V). We do not know whether any of the other possible implications hold. Conditions (I)–(IV) are considered in the study by Katznelson and McGehee [2, Section 3] of the Dichotomy Problem for countable E.

Zafran [4, 5] discovered strongly homogeneous algebras B, with $A(T) \subsetneq B \subsetneq C(T)$, such that many non-analytic functions operate in B. Pisier [3] produces such a B in which all Lipschitz-one functions operate—an example that cannot be improved. Katznelson and Malliavin [3] had obtained a related but less interesting result, in which many non-analytic functions operate from $A(T)$ into such a B. Their paper also explained how for every E the functions that operate in $A(E)$ are the same as those that operate in a certain associated homogeneous or strongly homogeneous algebra on T.

Graham [6] pointed out that if a set $E \subseteq G$ has the property that $B(E) = B_d(E) \neq l^\infty(E)$, then the closure of E in bG is a counterexample to the Conjecture.

Salinger [1] studied the also-unsolved dichotomy problem for tensor algebras and proved an analogue of 13.1.2.

See Burckel [2, Chapters IV, VII and VIII] and deLeeuw and Katznelson [1] for other interesting results in the general area of functions that operate in Banach algebras.

See also the general Banach algebra question raised at the end of Section 10.2.

13.2. Finite Sets

An *arithmetic relation* in a set $E \subseteq G$ is an equation $\sum_{j=1}^{k} u_j x_j = 0$, where k is finite, $\{u_j\} \subseteq Z$, and $\{x_j\} \subseteq E$. If $u_j x_j = 0$ for every j, the relation is *trivial*. If there are no non-trivial relations in E, then E is *independent*.

To have a list of all the relations in E is to know a great deal about (1) how E participates in the algebraic structure of G. When G is discrete or E is countable, that surely ought to tell us (2) how E participates in the harmonic analysis on G. Yet the connection between (1) and (2) is poorly understood.

Let E be finite. If E is independent, then $\alpha(E) \leq \sqrt{2}$. If E is very rich in relations, then $\alpha(E)$ is roughly $\sqrt{\#E}$ (see Sections 1.6 and 11.6). Except for those facts about the two extreme cases, we know little about how to estimate $\alpha(E)$ by looking at the list of relations in E. Given a function f on E, we do not know how to compute its $A(E)$-norm. If we knew more, perhaps we could settle the Dichotomy Conjecture in the case of countable E, in view of its equivalent formulation 13.1(1).

Here are two modest and specific questions.

1. If G is an n-element abelian group, then $\alpha(E)$ is exactly $\sqrt{n}$ (11.6.1). If $\alpha(E) = (\#E)^{1/2}$, must E be a coset?
2. Is it true that in Z, $\alpha(\{1, 2, \ldots, n\})n^{-1/2} \to 1$ as $n \to \infty$? See D. J. Newman [1] for a discussion.

See also the next Section.

13.3. Isomorphisms between Quotients of $A(G)$

We are concerned with the occurrence of isomorphisms

$$H: A(F) \to A(E)$$

where E and F are closed subsets of T. Even when H is merely a homomorphism with dense range, we know from elementary Banach algebra theory (see Katznelson [1, Section VIII.4]) that it has the form

$$Hf(x) = f(h(x)) \quad \text{for all } x \in E \quad \text{and all } f \in A(F),$$

where h is a homeomorphism from E into F. Evidently such an h defines a homomorphism H if and only if

$$\sup_{n \in Z} \|e^{inh}\|_{A(E)} < \infty.$$

Whenever E and F are finite sets with the same cardinality k, every one-to-one correspondence h induces an isomorphism H with norm no greater than $\sqrt{k}$. Let $E = \{x_j: 1 \le j \le k\}$ and $F = \{y_j = h(x_j): 1 \le j \le k\}$, and suppose that h preserves all arithmetic relations:

$$\{u_j\} \subseteq Z, \sum u_j x_j = 0 \Rightarrow \sum u_j y_j = 0.$$

Evidently then, $\|H\| \le 1$. Perhaps a much weaker hypothesis on h would suffice to place a bound on $\|H\|$ independent of k. We formulate the question specifically as follows.

Fix $C > 1$. Let $\{x_j\}_{j=1}^{\infty}$ and $\{y_y\}_{j=1}^{\infty}$ be two sequences in T converging to zero. Let E and F denote their respective closures. Suppose that for every vector of integers $(u_1, \ldots, u_n)$ with n finite and $|u_k| \le C$ for each k,

$$\sum u_k x_k = 0 \Leftrightarrow \sum u_k y_k = 0.$$

Let $h(x_j) = y_j$, $h(0) = 0$. Is the induced mapping H then an isomorphism from $A(F)$ onto $A(E)$?

When thickness conditions are imposed on E and F, one finds limitations on the kinds of h that will induce homomorphisms. Thus if E is an interval, h must be linear. But with many perfect sets of measure zero, more complicated h will work, and the picture is far from being simple or complete. One open question is: When $A(E)$ and $A(F)$ are isomorphic and E is a U_0-set, must F also be a U_0-set?

We offer now a selective survey of the related literature, beginning with the characterization of homomorphisms H from $A(\Gamma)$ into $B(\Gamma')$, where Γ and Γ' are arbitrary locally compact abelian groups. Given such an H, let $Y = \{\gamma' \in \Gamma': Hf(\gamma') \text{ is nonzero for some } f \in A(\Gamma)\}$. Then Y is the disjoint union of sets $S_1, \ldots, S_n$ belonging to the coset ring of Γ', and H is induced by a mapping $h: Y \to \Gamma$ that for each j is affine on some open coset that contains S_j. (A function h is *affine on* a set E if $h(x + y - z) = h(x) + h(y) - h(z)$ whenever $x, y, z \in E$.) Conversely, every such mapping $h: Y \to \Gamma$, for such a set Y, induces a homomorphism H. That characterization, for the general case, is due to Cohen [2], and follows from the characterization of the idempotents in a measure algebra. It is presented in Rudin [1, Chapter 4].

In particular, if H is an automorphism of $A(R)$, then $Hf(x) = f(ax + b)$ for some real a and b, $a \neq 0$; and every automorphism of $A(T)$ has the form $f(x) \to f(x + b)$ or $f(x) \to f(-x + b)$. The result for R was proved first by

Beurling and Helson [1], and their technique is of great interest for our purposes. For two somewhat different expositions, see Katznelson [1, VIII.4.5] and Kahane [2, Section VI.9], who uses the technique to get the result for T. (Leĭbenson [1] approached the problem for T by another method.) The technique works also for T^n and R^n and leads to other related results; see Brenner [3, 1, 2, in that order] and Self [1]. The Beurling-Helson proof seems most relevant to the following problem, posed by R. S. Pierce. If $h: T \to T$ is a homeomorphism such that E is a Helson set if and only if $h(E)$ is a Helson set, what can be said about h? The only functions that we know to have that property are the piecewise linear ones. Compare the tensor algebra case, treated in Graham [2].

Leblanc [2] showed that the endomorphisms of the algebras $A_\alpha = \{f(x) = \sum c_n e^{inx}: \|f\|_\alpha = \sum |c_n|(1 + |n|)^\alpha < \infty\}$ for $\alpha > 0$ are all of the form $f(x) \to f(nx + b)$. Domar [4] gave a new proof and obtained more general results.

If an isomorphism exists between $A(\Gamma)$ and $A(\Gamma')$ whose norm is less than $\sqrt{2}$, then Γ and Γ' are isomorphic; and that is the best possible constant. See Kalton and Wood [1].

For the structure of homomorphisms from $L^1(G)$ into $M(G')$, where G and G' are arbitrary locally compact groups, see Greenleaf [1].

We return now to the cases involving subsets of T in which h is not even locally linear, and thus when E and F are relatively thin. Early negative results, respecting the case $\|H\| = 1$, are due to de Leeuw and Katznelson [2]; see also the later McGehee [4]. McGehee [2] produced an example in which e^{ih} is the restriction to a perfect set E of a discontinuous character and yet belongs to $A(E)$ and has bounded powers. Many more examples as well as interesting negative results were discovered by Varopoulos, Schneider [1, 2], and Meyer [5, 6]. For expositions of most of those results, see our Theorem 11.7.1, Section 11.4, and Kahane [2, Chapter IX and Section VIII.5]. Finally, we recommend the masterful study by Leblanc [3] of homomorphisms occurring when E is a symmetric set.

The growth of powers. We take this occasion to survey selected studies of how powers of certain functions grow in norm, in $B(\Gamma)$ and in various quotients.

For a compact set $E \subseteq \Gamma$ that is the closure of its interior, Andersson [1] has characterized the set of functions $f \in B(E)$ of constant modulus one such that $\sup_{n \geq 0} \|f^n\|_{B(E)} < \infty$.

For estimates of $\|e^{inf}\|_{A(T)}$ as $n \to \infty$ when f is real-valued and either piecewise linear or C^2, see Kahane [2, Sections VI.2 and VI.3]. Kahane has posed the question: When f is a real-valued function in $A(T)$, not of the form $ax + b$, must it be true for some $K > 0$ that $\|e^{inf}\|_{A(T)} > K \log n$ for all $n \geq 1$?

For estimates of how $\|f^n\|_{A(T)}$ grows when $|f| \leq 1$ and f satisfies various conditions, see Hedstrom [1], where the motivation was to understand the behavior of solutions of certain difference equations. For the case when $\{t: |f(t)| = 1\}$ is a finite set, the asymptotic behavior of $\|f^n\|_{A(T)}$ as $n \to \infty$ is precisely described by Girard [1]. The conditions for the powers of such an f to be bounded had been identified by Clunie and Vermes [1] and Bajsanski [1], and discovered independently by D. J. Newman [2], who provides a readable treatment under the title "Homomorphisms of l^+". Girard [2] describes the asymptotic behavior of $\|f^n\|_{A(T)}$ when f is an automorphism of the unit disc:

$$f(z) = e^{i\zeta}\frac{z - \alpha}{1 - \bar{\alpha}z} \qquad (0 < |\alpha| < 1, \zeta \text{ real}).$$

Studies of how $\|f^n\|_{A(T^k)}$ behaves as $n \to \infty$, for $k > 1$, include Hedstrom [2] and Heiberg [1, 2, 3, 4].

Domar [3, 6] studies the growth of $\|e^{inf}\|_{A(E)}$ as $n \to \infty$ for smooth f and smooth manifolds $E \subseteq R^n$.

Frisch [1] refers to the growth of norms in certain quotients in his study of how solutions $f(x, t)$: $T^n \times R \to \mathbb{C}$ of the wave equation behave.

13.4. The Rearrangements Question of N. N. Lusin

Is it true that for every $f \in C(T)$, there exists a homeomorphism h: $T \to T$ such that $f \circ h \in A(T)$? The answer is not known even for real-valued f. With the dyadic group D in place of T, the answer is Yes, due to Wells [1]. See also 1.3.15, and Bary [2, 3].

13.5. Continuity of Linear Operators on $L^1(R)$

Johnson [2] shows that if U is a linear map from $L^1(R)$ to $L^1(R)$ such that for some $x \neq 0$, $U(\delta_x * f) = \delta_x * Uf$ for all $f \in L^1(R)$, then U is continuous. Can δ_x be replaced in the hypothesis by any measure $\mu \in M(R)$ such that $\hat{\mu}$ is not constant on any open interval?

13.6. p-Helson Sets

Gregory [1] defines a set $E \subseteq T$ to be a *p-Helson set* (for $1 < p < 2$) if $C(E) = A^p(E)$, where $A^p = \{f \in C(T): \hat{f} \in l^p\}$. For $q = p/(p - 1)$, let $M(q, E) = \{\mu \in M(E): \hat{\mu} \in l^q\}$. He shows (among other results) that if E is a p-Helson set, then $M(q, E) = \{0\}$. Is the converse true?

13.7. Questions of Atzmon on Translation Invariant Subspaces of L^p and C_o

Is every closed, translation-invariant subspace of $C_o(R)$ or $L^p(R)$ (for $2 < p < \infty$) generated by the translates of some one element? One may ask the same question about $L^\infty(R)$ with the weak* topology.

For $1 < p < 2$ or $2 < p < \infty$, is every closed translation-invariant subspace of $L^p(R)$ generated by the translates of some finite number of elements? See the related results in Atzmon [5, 6].

If $2 < p < \infty$ and f is a nonzero element of $L^p(R)$ or of $l^p(Z)$ that vanishes on $(-\infty, 0)$, do the translates of f span the space? The answer is easily Yes for $p = 2$. It is also Yes for $L^\infty(R)$ or for $l^\infty(Z)$ with the weak* topology; that result for R is proved in a dual form in D. J. Newman [3]. The answer for $1 < p < 2$ is No (unpublished work of Andrei Hilpher).

13.8. Questions on Subsets E of the Integer Group

The following inclusions always hold:

$$A(E) \subseteq B_o(E) \subseteq \tilde{A}(E) \subseteq C_o(E).$$

Each of them is a proper inclusion for some E. That $\tilde{A}(E)$ can be strictly larger than $B(E)$ is due to Katznelson and McGehee [3]. It is not known whether $A(E) = C_o(E)$ whenever $A(E) = \tilde{A}(E)$; that is the "tilde problem" (see Chapter 12). Perhaps $A(E) = C_o(E)$ whenever $A(E) = B_o(E)$.

A set E is an *I-set* if $AP(E) = l^\infty(E)$. Ryll-Nardzewski [1] showed that if E is an I-set and $n \in Z \setminus E$, then $E \cup \{n\}$ is an I-set; there is an easier proof due to L. T. Ramsay [3]. The accumulation points of an I-set in the Bohr group are all outside Z. It is easy to see that $\{3^n\} \cup \{n + 3^n\}$ is a Sidon set but not an I-set. See 11.8.8 for another example. An I-set is always a *Riesz set*, that is, one such that $M_E(T) = L_E^1(T)$; that is one of the interesting results in Meyer [4]. For more on I-sets and related types of sets see Kahane [2, p. 129], Hartman, Kahane, and Ryll-Nardzewski [1], Kahane [9], and Hartman and Ryll-Nardzewski [1, 2, 3].

Pigno and Saeki [1] have a curious characterization of the sets E such that $A(E) \subseteq B_d(E)$.

Let $\bar{E}$ denote the closure of E in bZ.

If $A(E) = C_o(E)$, must $\bar{E}$ have m_{bZ}-measure zero? Must $\bar{E}$ be a Helson set in bZ? Must E and $\bar{E} \cap Z$ both be Riesz sets? See the related results in Section 7.6.

If E is dense in the maximal ideal space of the algebra $\tilde{A}(E)$, must $A(E) = C_o(E)$?

13.9. Questions on Sets of Synthesis

Is the union of two sets of synthesis also a set of synthesis? To answer that question affirmatively, it would suffice to answer one of the following two affirmatively.

Is it true whenever E_1 and E_2 are closed subsets of G that $I_0(E_1 \cup E_2) = I_0(E_1) \cap I_0(E_2)$?

Is it true whenever E is a closed subset of G and $f \in I_0(E)$, that there is a sequence of functions $u_n \in A(G)$ such that $u_n^{-1}(0)$ is a neighborhood of E and $\|u_n f - f\|_A \to 0$?

Let $E(r) = \{\sum_{j=1}^{\infty} \varepsilon_j r^{-j}: \varepsilon_j = 0 \text{ or } 1\}$ where $r > 2$. If r is a Pisot number, then $E(r)$ is a set of bounded synthesis (see Meyer [1, Chapter VII]). Is the same true for other values of r? Are all the sets $E(r)$ at least sets of synthesis? See also Saeki [17].

If $f \in A$, $S \in PM$, $f^{-1}(0) \supseteq \text{supp } S$, and $fS \neq 0$, then must the support of fS disobey synthesis? If the boundary of a closed set E obeys synthesis, must E obey synthesis?

Does there exist a continuous curve in R^2 that is a Helson set and does not obey synthesis?

Is the class of so-called ergodic sets (see Woodward [3, 1, 2]) closed under finite unions?

13.10. Characterizing Sidon Sets in Certain Groups

Malliavin and Malliavin-Brameret [1] proved that if p is a prime, then every Sidon set in the product $(Z_p)^\infty$ is the finite union of independent sets (see Lindahl and Poulsen [1, p. 75]). Does that result hold for other product groups with bounded order?

13.11. Subalgebras of L^1

A closed subalgebra L of $L^1(G)$, where G is compact, induces an equivalence $\sim$ on Γ: $m \sim n \Leftrightarrow \hat{f}(m) = \hat{f}(n)$ for all $f \in L$. Let $\{E_\lambda\}$ be an indexing of the classes and distinguish the class $E_0 = \bigcap_{f \in L} (\hat{f})^{-1}(0)$; the other classes must be finite. Let L_0 and L_1 be the smallest and largest closed subalgebras that induce $\sim$. Kahane [8] showed that L_0 may be strictly smaller than L_1. Friedberg [1] showed that with $\Gamma = Z$, if there is a constant M such that for each $\lambda \neq 0$, $|m - n| \leq M(\min|m|, |n|)^{1/2}$ whenever $m, n \in E_\lambda$, then $L_0 = L_1$. We recommend the comprehensive treatment by Rider [3], who shows (among other results) that $\frac{1}{2}$ cannot be replaced by 1 in Friedberg's condition. That gap raises one obvious question, and one may ask others.

Recently, Kahane and Katznelson [5] have found an example in which $L_0 \neq L_1$, and each E_λ contains at most two elements. Bachelis and Gilbert [1] have somewhat generalized Rider's construction methods.

13.12. $\Lambda(p)$-Sets and Multipliers

A set $E \subseteq Z$ is a $\Lambda(p)$-set if $L_E^1(T) = L_E^p(T)$; see Lòpez and Ross [1, Chapter 5] or the early paper by Rudin [10] for an introduction. For which values of p and r is it true that every $\Lambda(p)$-set is a $\Lambda(r)$-set? It is true when $r < p$; it is untrue when

$4 \leq p < r$ and p is an even integer, as proved in Rudin [10]. Bachelis and Ebenstein [1], using work of Rosenthal [6], showed that for every E, the set $\{r \in (1, 2)$: E is a $\Lambda(r)$-set$\}$ is open.

Let C_p be the collection of sets $E \subseteq Z$ such that $\chi_E \in M_p(Z)$. Then C_1 is the coset ring and C_2 contains all subsets of Z. Does C_p contain C_r property whenever $1 \leq r < p \leq 2$? All we know is that when q is an even integer, $q \geq 4$, and $p = q/(q-1)$, then the result of Rudin mentioned above provides a set $E \in C_p$ that is not in C_r if $r < p$.

13.13. Identifying the Maximal Ideal Spaces of Certain L-Subalgebras of $M(G)$

The maximal ideal space of the L-subalgebra N generated by $M_c(E) \cup M_d(G)$, where E is algebraically scattered (or dissociate) is identified in 6.2.9 (or 6.3.9). Brown and Moran [7] identify ΔN and the structure semigroup of N when N is the L-algebra generated by the measure $\bigstar_{n=1}^{\infty}[\frac{1}{2}\delta(0) + \frac{1}{2}\delta(1/a_1, \ldots, a_n)]$ where the integers $a_j \geq 2$ are such that lim sup $a_j = \infty$. It remains to do the analogous work when N is generated by the measure $\bigstar_{n=1}^{\infty}[\frac{1}{2}\delta(0) + \frac{1}{2}\delta(x^n)]$, where $0 < x < 1$. Another open problem is to identify ΔA when A is the algebra of measures concentrated on the sets belonging to the Raĭkov system generated by an algebraically scattered σ-compact set (see Section 5.2).

13.14. A Question of Katznelson on Measures with Real Spectra

If $\mu \in M(G)$ and the range of the Gel'fand transform on $\Delta M(G)$ is real, does it follow that $\hat{\mu}(\Gamma)$ is dense in $\hat{\mu}(\Delta M(G))$?

13.15. The Support Group of a Tame Measure

Let μ be a tame measure in $M(T)$. Let $\mu(E)$ be nonzero, where E is a Borel subset of G. Does it follow that $m_T(E + E) > 0$? The answer is Yes in the case $\mu = \bigstar_{n=1}^{\infty}[\frac{1}{2}\delta(0) + \frac{1}{2}\delta(1/a_1 \cdots a_n)]$, where each a_j is an integer, $a_j \geq 2$, and sup $a_j < \infty$. See Theorem 6.2.15 and Proposition 6.6.6.

13.16. The Šilov Boundary of $M(G)$

If $\psi \in \Delta M(G)$ and $|\psi_\mu|^2 = |\psi_\mu|$ a.e. $d\mu$ for all $\mu \in M(G)$, does it follow that $\psi \in \partial M(G)$? See Section 8.5.

13.17. Taylor's Theorems

Are there elementary proofs of Theorems 5.3.3 and 5.3.7?

13.18. Two Factorization Questions

Is every element f of $L^1(G)$ of the form $f = \mu * \nu$ where μ and ν are singular? The answer is Yes for G compact and f a trigonometric polynomial; see the Remarks and Credits of Section 7.1.

Does every Riesz product ω have the form $\omega = \mu * \nu$ where μ and ν are continuous?

13.19. Questions about Tensor Algebras

Suppose that $f \in V(X, Y)$ where X and Y are compact connected spaces, and that $|f| > 0$ on $X \times Y$. If $\sup\{\|f^n\|_V : n \in Z\}$ is finite, must f have the form $f = g \otimes h$? The answer is Yes if $X = Y = [0, 1]$ and $f \in C^2$ (Varopoulos [18]).

Is a cycle-free compact subset of $X \times Y$ necessarily a V-Helson set? See 11.8.1.

Let $\tilde{V} = \tilde{V}(X, Y)$ be the algebra defined in 11.9. Is it true that $\partial\tilde{V} \neq \Delta\tilde{V}$?

13.20. Other Question Lists

Kahane [2, p. 89] gives an array of open questions. The first three have been answered by Körner's construction of a Helson set disobeying synthesis (Section 4.6). The others are still open.

Körner [2, pp. 209–225] also gives a list of open questions. Number 14 was answered by Katznelson and Körner [1] (see Section 12.5). Number 16 was answered by Saeki [15]. The others are still open.

Appendix

A.1. Riesz Products in Brief

Two useful types of products are

(1) $$f_k(x) = \prod_{j=1}^{k} (1 + r_j \cos(n_j x + \theta_j))$$

and

(2) $$g_k(x) = \prod_{j=1}^{k} (1 + ir_j \sin(n_j x + \theta_j)),$$

where the numbers r_j are non-negative, and the numbers θ_j are real. Note that

(3) $$\|g_k\|_{C(T)} \leq \prod_{j=1}^{k} (1 + r_j^2)^{1/2}.$$

It is usual to require that

(4) the set $\left\{ \sum_{j=1}^{k} \varepsilon_j n_j : \varepsilon_j = 0, 1 \text{ or } -1 \right\}$ contains 3^k distinct integers,

which would be the case if, for example, $n_{j+1} \geq 3n_j$ for every j. Since the transform of a product is the convolution of the transforms, the values of $\hat{f}_k$ are readily written down, in view of (4):

$$\hat{f}_k\left(\sum \varepsilon_j n_j\right) = 2^{-\Sigma|\varepsilon_j|} e^{i\Sigma_j \theta_j} \prod r_j^{|\varepsilon_j|};$$

in particular, $\hat{f}_k(0) = 1$ and $\hat{f}_k(n_j) = \frac{1}{2} r_j e^{i\theta_j}$; and similarly for $\hat{g}_k$. We present now an elementary but representative collection of applications. Recall that a set $E \subseteq Z$ is a *Sidon set* if $A(E) = C_o(E)$; equivalently, if $B(E) = l^\infty(E)$.

A.1.1. Proposition. A *set* $E = \{n_j\}_1^\infty$ *such that* (4) *is satisfied for every* k *is a Sidon set.*

Proof. For an arbitrary sequence of norm one in $l^\infty(E)$, say $\{r_j e^{i\theta_j}\}$ with $0 \le r_j \le 1$, the functions f_k are each non-negative, so that $\|f_k\|_1 = \hat{f}_k(0) = 1$. It is easy to see that f_k converges weak$*$ in $M(T)$ to a measure $\mu \in M(T)$, with $\|\mu\| = 1$. Then $(2\mu)\hat{}(n_j) = r_j e^{i\theta_j}$ for all j. It follows that the natural mapping of $B(E)$ into $l^\infty(E)$ is surjective, and that the norm of its inverse is at most 2. Therefore E is a Sidon set and $\alpha(E) \le 2$. □

A.1.2. Proposition. *A set* $E = \{n_j\}_1^k$ *such that* (4) *is satisfied has the property that if* $f \in L^1(T)$ *and* $\hat{f}(n) = 0$ *for all* $n \notin E$, *then* $\|f\|_1 \ge (4ek)^{-1/2} \sum |\hat{f}(n)|$.

Proof. Write $\hat{f}(n_j)$ as $|f(n_j)|e^{-i\theta_j}$. Use g_k as in (2) with $r_j = k^{-1/2}$. Then

$$\frac{1}{2k^{1/2}} \sum |\hat{f}(n_j)| = \int f g_k \le \|f\|_1 \|g_k\|_{C(T)}.$$

The result now follows from (3). □

A.1.3. Proposition. *A set* $E = \{n_j\}_1^\infty$ *such that* (4) *is satisfied for every k has the property that if* $\sum |\beta_j|^2 < \infty$, *then there exists* $g \in L^\infty(T)$ *such that* $\hat{g}(n_j) = \beta_j$ *for each j.*

Proof. The sequence $\{g_k\}$ given by (2), where we define r_j and θ_j by requiring that $r_j = 2|\beta_j|$ and $\beta_j = \frac{1}{2} r_j e^{i\theta_j}$, converges weak$*$ in $L^\infty(T)$ to a suitable g. In fact $\|g\|_\infty \le \prod (1 + 4|\beta_j|^2)^{1/2}$. □

A.1.4. Proposition. *Let E be a set as in A.1.1. Let F be an arbitrary Sidon set. Then* $E \cup F$ *is a Sidon set.*

Proof. We may suppose that $E \cap F = \varnothing$ and that $0 \notin F$. It suffices to show that for some constants $K > 0$ and $\varepsilon \in (0, 1)$, and for every $h \in l^\infty(E \cup F)$, there exists $g \in B(Z)$ such that $\|g\|_B \le K\|h\|_\infty$ and $|g - h| < \varepsilon \|h\|_\infty$ on $E \cup F$; because then an iterative procedure will yield an $f \in B(Z)$ such that $f = h$ on $E \cup F$ and $\|f\|_B < (K/(1 - \varepsilon))\|h\|_\infty$.

It suffices to deal with $h \in l^\infty(E \cup F)$ of norm one. There exists $g_1 \in B(Z)$ such that $g_1 = h$ on F and $\|g_1\|_B < \alpha(F) + 1$. Choose an integer $m \ge 0$ such that $2^{-m}(\alpha(F) + 2) < \frac{1}{2}$. Let $(h - g_1)^{1/m}$ denote some choice of a function whose mth power is $h - g_1$. The proof of A.1.1 provides a function g_2 such that $g_2 = (h - g_1)^{1/m}$ on E, $|g_2| \le \frac{1}{2}\|h - g_1\|_\infty{}^{1/m} \le \frac{1}{2}(\alpha(F) + 2)^{1/m}$ on F, and $\|g_2\|_{B(Z)} \le 2\|h - g_1\|_\infty^{1/m} \le 2(\alpha(F) + 2)^{1/m}$. Let $g = g_1 + g_2^m$. Then $g = h$ on E, $g - h = |g_2|^m < \frac{1}{2}$ on F, and $\|g\|_B \le \alpha(F) + 1 + 2^m(\alpha(F) + 2)$. The Proposition is proved. □

Remarks. A thoughtful study of the foregoing proofs will prepare the neophyte for most Riesz product applications. Katznelson [1, Section V.1] provides a good list of exercises.

The method of A.1.4 is carried to its logical conclusion in Rider [1] and also in Lòpez and Ross [1, pp. 24–30].

Condition (4) is not present in all applications; see the proof of 1.3.12.

A.2. Norbert Wiener's Theorem on the Average Value of $|\hat{\mu}|^2$

We give first a version for the circle group, and then a version for arbitrary G. See 7.5.4 and 8.3.7 for two variants of this result.

A.2.1. Theorem. *Let $\mu \in M(T)$, and let the discrete part of μ be $\mu_d = \sum_{k=1}^{\infty} m_k \delta(x_k)$ where the points x_k are distinct. Then for each $n_o \in Z$,*

$$\lim_{N\to\infty} \frac{1}{2N+1} \sum_{n=-N}^{N} |\hat{\mu}(n - n_o)|^2 = \sum_{k=1}^{\infty} |m_k|^2.$$

In fact, the limit is uniform in n_o.

Proof. Let $d\nu(x) = e^{in_o x}\, d\mu(x)$. Then $\hat{\nu}(n) = \hat{\mu}(n - n_o)$, $d(\nu * \tilde{\nu})(x) = e^{in_o x}\, d(\mu * \tilde{\mu})(x)$, and $(\nu * \tilde{\nu})(\{0\}) = \sum |m_k|^2$. Let $D_N(x) = \sum_{n=-N}^{N} e^{inx}$ ($\{D_N\}$ is the familiar Dirichlet kernel). Then $\sum_{n=-N}^{N} |\hat{\mu}(n - n_o)|^2 = \sum_{n=-N}^{N} |\hat{\nu}(n)|^2 = \int_{-\pi}^{\pi} D_N(x) d(\nu * \tilde{\nu})(x)$. For each $h \in (0, \pi)$, and all integers N,

$$\left|(2N+1)^{-1} \int_{-h}^{h} D_N(x) d(\nu * \tilde{\nu})(x) - D_N(0)(\nu * \tilde{\nu})(\{0\})\right| \leq |\nu * \tilde{\nu}|([-h, 0) \cup (0, h]),$$

which tends to zero as $h \to 0$; and

$$\left|\left(\int_{-\pi}^{-h} + \int_{h}^{\pi}\right) D_N(x) d(\nu * \tilde{\nu})(x)\right| \leq \sup\{|D_N(x)| : h \leq |x| \leq \pi\} \|\nu\|^2$$

which tends to zero as $N \to \infty$ (for each h). The result follows. □

A.2.2. Theorem. *Let $\{V_\alpha\}$ be a neighborhood base at 0 in a locally compact abelian group G. For each α, let f_α be a continuous positive-definite function whose support is a compact subset of V_α, such that $f_\alpha(0) = 1 = \|\hat{f}_\alpha\|_{L^1(\Gamma)}$. Let $\mu \in M(G)$. Then for each $\gamma_o \in \Gamma$,*

$$\lim_{\alpha} \int \hat{f}_\alpha(\gamma) |\hat{\mu}(\gamma - \gamma_o)|^2\, d\gamma = \sum_{x \in G} |\mu(\{x\})|^2.$$

In fact, the limit is uniform in γ_o.

Proof. Let $\nu = \gamma_o\mu$ and $\sigma = \nu * \tilde{\nu}$. Then $\hat{\sigma}(\gamma) = |\hat{\mu}(\gamma - \gamma_o)|^2$ and $\sigma(\{0\}) = \sum_{x \in G}|\mu(\{x\})|^2$. Let $\varepsilon > 0$, and choose a symmetric neighborhood U of 0 in G such that $|\sigma|(U \setminus \{0\}) < \varepsilon$. The choice of U is independent of γ_o. If $V_\alpha \subseteq U$, then

$$\left|\sigma(\{0\}) - \int_\Gamma \hat{f}_\alpha(\gamma)\hat{\sigma}(\gamma)d\gamma\right| = \left|\sigma(\{0\}) - \int_G f_\alpha(x)d\sigma(x)\right| < \varepsilon.$$

The Theorem follows. □

Remark. The measures $f_\alpha(\gamma)d\gamma$ converge weak$*$ in $M(b\Gamma)$ to $m_{b\Gamma}$.

A.3. A Proof That Singletons Obey Synthesis, and How

The fact that a one-point subset of G obeys harmonic synthesis with respect to $A(G)$ is of course an instance of a more general proposition that applies to many Banach algebras. Our purpose here is to give an *ad hoc* proof that yields some useful information about the special case. It suffices to deal with the singleton $\{0\}$, and for the moment we consider only $G = T$. To say that $\{0\}$ is a set of synthesis for $A = A(T)$ is to say that

(1) if $f \in A$ and $f(0) = 0$, then $\|f\|_{A([-h, h])} \to 0$ as $h \to 0$.

It suffices to prove that

(2) $$\|1 - e^{ix}\|_{A([-h, h])} - h + 0(h^2) \quad \text{as} \quad h \to 0,$$

for then, if $f \in A$ and $f(0) = 0$,

$$f = \sum_n \hat{f}(n)e^{inx} = \sum_n \hat{f}(n)(1 - e^{inx}),$$

so

$$\begin{aligned}\|f\|_{A([-h, h])} &\leq \sum|\hat{f}(n)| \max_{|n| \leq N} \|1 - e^{inx}\|_{A([-h, h])} + 2 \sum_{|n| > N} |\hat{f}(n)| \\ &\leq \|f\|_A N(h + 0(h^2)) + \sum_{|n| > N} |\hat{f}(n)|,\end{aligned}$$

for every N, and (1) follows.

To prove (2), consider the function defined on T by

$$g(x) = 1 - \frac{|x|}{\pi} \quad \text{for } |x| \leq \pi.$$

Then $g(0) = \sum \hat{g}(n) = \sum |\hat{g}(n)| = 1$ and $\hat{g}(0) = \frac{1}{2}$. Let $G(x) = \pi(g(x - \pi) - \frac{1}{2})$. Then $G(x) = x$ for $|x| \leq \pi/2$, and $\|G\|_A = \pi/2$. For each k, let $G_k(x) = k^{-1}G(kx)$. Then

$$G_k(x) = x \quad \text{for } |x| \leq \pi/2k,$$

and

$$\|G_k\|_A = \pi/2k.$$

Therefore

$$\|x\|_{A([-\pi/2k,\, \pi/2k])} = \pi/2k \quad \text{for } k = 1, 2, \ldots,$$

and hence

$$\|x\|_{A([-h,\, h])} = h + 0(h^2) \quad \text{as} \quad h \to 0.$$

Since

$$\|1 - e^{ix}\|_{A([-h,\, h])} = \sup\{|\langle S, 1 - e^{ix}\rangle| : S \in PM([-h, h]),\ \|S\|_{PM} \leq 1\},$$

it suffices to prove that $|\langle S, 1 - e^{ix}\rangle| \leq \|x\|_{A([-h,\, h])}\|S\|_{PM}$ for S with support in $[-h, h]$. Now $|\langle S, 1 - e^{ix}\rangle| = |\hat{S}(0) - \hat{S}(1)| = |\hat{S}'(t)|$ for some $t \in (0, 1)$, and $\hat{S}' = (ixS)\hat{}$, so that for such S,

$$|\hat{S}'(t)| \leq \|\hat{S}'\|_\infty = \|ixS\|_{PM} \leq \|x\|_{A([-h,\, h])}\|S\|_{PM},$$

and (2) is proved.

The result (2) has the following corollary, which we use in Chapter 12.

A.3.1. Lemma. *(a) If $n \in Z$, $t \in R$, $F \subseteq T$ and $\min_{k \in Z}|nx - t - 2\pi k| < h$ for $x \in F$, then $\|e^{inx} - e^{it}\|_{A(F)} \leq h + 0(h^2)$ as $h \to 0$.*

(b) Let G and Γ be dual locally compact abelian groups, $\gamma \in \Gamma$, $F \subset G$. Let θ be defined on G so that $e^{i\theta(x)} = \langle x, \gamma\rangle$ and $|\theta(x) - t| < h$ for $x \in F$. Then $\|\gamma - e^{it}\|_{B(F)} = h + 0(h^2)$.

Proof. It suffices to prove the more general part (b). We may write

$$1 - e^{ix} = \sum_{n \in Z} c_n e^{inx} \quad \text{for } |x| \leq h,$$

where $\sum |c_n| = h + 0(h^2)$. Then

$$e^{it} - e^{iu} = \sum (c_n e^{i(1-n)t}) e^{inu} \quad \text{for } |u - t| \leq h.$$

Let $\mu = \sum(c_n e^{i(1-n)t})\delta_{-n\gamma}$. Then

$$\hat{\mu}(x) = \sum(c_n e^{i(1-n)t})\langle x, \gamma\rangle^n = e^{it} - \langle x, \gamma\rangle \quad \text{for } x \in F.$$

Therefore $\|e^{it} - \gamma\|_{B(F)} \leq \|\mu\|_M = h + 0(h^2)$. The Lemma is proved. □

Remark. In (2), and hence in A.3.1, we can replace "$h + 0(h^2)$" with "$h + h^2 + \cdots + h^n + 0(h^{n+1})$" for any $n > 1$.

A.4. S. Bernstein's Inequality

A.4.1. Proposition. (*a*) *If* $S \in PM(R)$ *and the support of* S *is contained in* $[-a, a]$, *then* $\|\hat{S}'\|_\infty \leq a\|\hat{S}\|_\infty$.

(*b*) *If* $P(x) = \sum_{k=-n}^{n} c_k e^{ikx}$, *then* $\|P'\|_\infty \leq n\|P\|_\infty$.

Proof. It suffices to prove the more general part (a). Let G be the function of norm $\pi/2$ in $A(T)$ constructed in A.3. We may regard G as an element of norm $\pi/2$ in $B(R)$. Let $G_a(x) = (2a/\pi)G(\pi x/2a)$. Then $\|G_a\|_{B(R)} = a$, and $G_a(x) = x$ for $x \in \operatorname{supp} S$. Therefore $\hat{S}'$ is the transform of $iG_a S$, whose PM norm is bounded by $\|G_a\|_{B(R)}\|S\|_{PM} = a\|S\|_\infty$. The result is proved. □

For a fuller discussion of Bernstein's inequality, see Edwards and Gaudry [1, pp. 197–201].

A.5. Triangles and Trapezoids

The declared prerequisites for the book include the basic facts about approximate identities and functions that separate sets. Nevertheless we offer, as *lagniappe*, this quick review.

When we treat $L^1(R)$ and its transform space $A(R)$, there are two copies of R involved, which we distinguish as an x-axis and a y-axis with Haar measures dx and $dy/2\pi$, respectively. For $\lambda > 0$, let

(1) $$\hat{K}_\lambda(y) = \max(0, 1 - |y|/\lambda).$$

Its graph reminds one of an isosceles triangle of height 1 and base 2λ. It is the transform of

(2) $$K_\lambda(x) = \frac{\lambda}{2\pi}\left(\frac{\sin(\lambda x/2)}{\lambda x/2}\right)^2.$$

Then $K_\lambda(x) = \lambda K_1(\lambda x)$; and $\|K_\lambda\|_1 = 1$. The net $\{K_\lambda : \lambda \to \infty\}$ is the Fejér kernel, an approximate identity for $L^1(R)$. Thus $K_\lambda * f \to f$ in $L^1(R)$, and of course $\hat{K}_\lambda \hat{f} \to \hat{f}$ in $A(R)$, for every $f \in L^1(\mathrm{R})$.

(To make the acquaintance of a relative of the Fejér kernel, momentarily change Haar measure on the y-axis to dy and let

$$\Delta_\lambda(y) = \lambda^{-1}\hat{K}_\lambda(y) = \lambda^{-1}\max(0, 1 - |y|/\lambda). \tag{3}$$

Then $\|\Delta_\lambda\|_1 = 1$; and in $M(R)$, the net $\{\Delta_\lambda : \lambda \to 0^+\}$ converges weak$*$ to the point mass δ_o.)

It is often desirable to use an approximate identity in $A(R)$ whose elements equal 1 on large compact sets. Such trapezoids are combinations of triangles; for $\lambda > 0$ and $a > 1$, let

$$V_{\lambda, a} = \frac{a}{a-1} K_{a\lambda} - \frac{1}{a-1} K_\lambda. \tag{4}$$

Then

$$V_{\lambda, a}(y) = \begin{cases} 1 \quad \text{for } |y| \le \lambda. \\ \dfrac{1}{a-1}(a - |y|/\lambda) \quad \text{for } \lambda \le |y| \le a\lambda, \\ 0 \quad \text{for } |y| \ge a\lambda; \end{cases} \tag{5}$$

$$\|V_{\lambda, a}\|_1 \le \frac{a+1}{a-1};$$

and the net $\{V_{\lambda, a} : \lambda \to \infty\}$ is an approximate identity for $L^1(R)$. A popular choice for a is 2; let

$$V_\lambda = V_{\lambda, 2} = 2K_{2\lambda} - K_\lambda. \tag{6}$$

The net $\{V_\lambda : \lambda \to \infty\}$ is the *de la Vallée Poussin kernel*. Note that $\|V_\lambda\|_1 \le 3$ for all λ.

The functions $\hat{K}_\lambda$ and $\hat{V}_{\lambda, a}$ may be obtained as convolutions, as follows. If $h_t = \chi_{[-t, t]}$, then

$$\hat{K}_\lambda = \frac{2\pi}{\lambda}(h_{\lambda/2} * h_{\lambda/2}) \quad \text{and} \quad \hat{V}_{\lambda, a} = \frac{2\pi}{(a-1)\lambda}(h_{(a+1)\lambda/2} * h_{(a-1)\lambda/2}).$$

That observation is a good hint toward finding the kernel's counterpart in the general setting. For such results we are glad to refer to Rudin [1, Chapter 2], but for the following result, which Rudin calls Theorem 2.6.8, we provide the easier proof pointed out by Bachelis, Parker, and Ross [1].

A.5.1. Proposition. *Let F be a compact subset of Γ, and let $\varepsilon > 0$. Then there exists $k \in L^1(G)$ such that $\hat{k} = 1$ on F, $\hat{k}$ has compact support, and $\|k\|_1 < 1 + \varepsilon$.*

Proof. Let $v \in L^1(G)$ be such that $\hat{v} = 1$ on F and $\hat{v}$ has compact support (to obtain v, take $\hat{v}$ to be an appropriate multiple of $\chi_F * \chi_U$ where U is a suitable compact neighborhood of 0 in Γ). Let $g \in L^1(G)$ be such that $\|g\|_1 = 1$, $\hat{g}$ has compact support, and $\|v - v * g\|_1 < \varepsilon$ (g may be selected from a Fejér-type kernel in $L^1(G)$). Let $k = v + g - v * g$. Then $\|k\|_1 \leq \|g\|_1 + \|v - v * g\|_1 < 1 + \varepsilon$, and $\hat{k} = \hat{v} + \hat{g} - \hat{v}\hat{g}$, which equals 1 on F and has compact support. □

A.6. Convolution and Relative Absolute Continuity

The following result, for $S = R$, is at least as old as Šreĭder [1, Lemma 2].

A.6.1. Proposition. *Let S be a locally compact semitopological commutative semigroup. Let μ and $\nu \in M(S)$ be non-negative. If $\mu_1 \ll \mu$ and $\nu_1 \ll \nu$, then $\mu_1 * \nu_1 \ll \mu * \nu$.*

Proof. Consider the special case that $\mu_1 = f\mu$ and $\nu_1 = g\nu$, where f and g are indicator functions of Borel sets. If $h \in C_o(S)$ is non-negative, then

$$(1) \qquad \int h \, d(\mu_1 * \nu_1) = \iint h(xy) f(x) g(y) d\mu(x) d\nu(y) \leq \int h \, d(\mu * \nu).$$

The regularity of $\mu_1 * \nu_1$ and $\mu * \nu$ and formula (1) show that $\mu_1 * \nu_1(E) \leq \mu * \nu(E)$ for every Borel set E. Therefore $\mu_1 * \nu_1 \ll \mu * \nu$.

Let $\varepsilon \in (0, 1)$. Let $f = \sum_{j=1}^m c_j f_j$ and $g = \sum_{k=1}^n d_k g_k$ be simple functions such that $\|f\mu - \mu_1\| < \min(1, \varepsilon/\|\nu_1\|)$ and $\|g\nu - \nu_1\| < \min(1, \varepsilon/\|\mu_1\|)$. Then $\|g\nu\| \leq 1 + \|\nu_1\|$ and

$$\begin{aligned}\|\mu_1 * \nu_1 - (f\mu) * (g\nu)\| &= \|\mu_1 * (\nu_1 - g\nu) + (\mu_1 - f\mu) * (g\nu)\| \\ &\leq \|\mu_1\|\varepsilon/\|\mu_1\| + \varepsilon\|g\nu\|/(1 + \|\nu_1\|) < 2\varepsilon.\end{aligned}$$

Of course $(f\mu) * (g\nu)$ is a sum of terms, each of which is absolutely continuous with respect to $\mu * \nu$, by the special case. Hence, $\mu_1 * \nu_1$ is a limit in $M(S)$ of measures that are absolutely continuous with respect to $\mu * \nu$. The Lemma is proved. □

A.7. An Extension Theorem for Fourier–Stieltjes Transforms

Various versions of the following result have been in the folklore for many years. The one that we present is from Pigno and Saeki [4, Lemma 1], and Graham and A. MacLean [1, Lemma 6]. For the terms "tame" and "monotrochic," see Section 6.1.

A.7.1. Theorem. *Let G be a locally compact abelian group with dual group Γ. Let H be a closed subgroup of G such that G/H is compact. Let $\Lambda = \{\lambda \in \Gamma: \langle x, \lambda \rangle = 1$ for all $x \in H\}$. Let τ be a probability measure in $M(G)$ and $S = \{\gamma \in \Gamma: \hat{\tau}(\gamma) \neq 0\}$. Suppose that $\Lambda \cap S = \{0\}$ and that $\Lambda \cap (S - \gamma)$ is finite for all $\gamma \in \Gamma$. Then there exists a unique linear mapping $J = J_\tau: M(G/H) \to M(G)$ such that*

$$(1) \qquad (J\mu)\hat{}(\gamma) = \sum_{\lambda \in \Lambda} \hat{\mu}(\lambda)\hat{\tau}(\gamma - \lambda) \quad \text{for all } \gamma \in \Gamma \quad \text{and all } \mu \in M(G/H).$$

Furthermore the following hold.

(i) *J is an isometry.*
(ii) *$J\mu \geq 0$ if and only if $\mu \geq 0$.*
(iii) *$J\mu \in L^1(G)$ if $\tau \in L^1(G)$ and $\mu \in L^1(G/H)$.*
(iv) *$J\mu \notin L^1(G)$ if $\mu \notin L^1(G/H)$.*
(v) *If $\sup_{\gamma \in \Gamma} \#[\Lambda \cap (S - \gamma)] < \infty$ and $\tau \in M_o(G)$, then $J\mu \in M_o(G)$ if and only if $\mu \in M_o(G/H)$.*
(vi) *If H is a discrete subgroup of G, then $J\mu$ is tame (or monotrochic) if and only if μ is tame (or monotrochic).*

Remarks. (i) The usual application of A.7.1 is to $G = R^n \times K$, where K is a compact abelian group and $H = Z^n \times 0$. In conjunction with the structure theorem, A.7.1 permits the extension of results known for compact abelian groups to all locally compact abelian groups.

(ii) When $G = R$ and $H = Z$ and $\hat{\tau}$ is a triangle of base 2, then A.7.1 says that elements of $B(Z)$ can be extended linearly in the gaps to elements of $B(R)$, with no increase in norm.

(iii) Some restriction on τ or H is needed in A.7.1 (iii) and (v). To see that, consider the case $G = T^2$, $H = T \times \{0\}$, $\Lambda = \{0\} \times Z$ and $\tau = \delta_0 \times m_T$. It is easy to see that $J\mu = \delta_0 \times \mu$, for all $\mu \in M(G/H) = M(T)$. From that it follows that the conclusions of (iii) and (iv) fail for Jm_T.

(iv) The condition $\sup_{\gamma \in \Gamma} \#[\Lambda \cap (S - \gamma)] < \infty$ follows from

$$\#[\Lambda \cap (S - S)] < \infty,$$

and the latter condition will usually be imposed in practice.

(v) Whether the restriction on H in A.7.1 (vi) can be dropped appears to be unknown.

Proof of A.7.1. That J is unique is immediate from (1). Note that the hypothesis on S implies that each sum in (1) is finite, so $\mu \mapsto (J\mu)^\wedge$ defines a mapping from $M(G/H)$ to the bounded continuous functions on Γ. We shall use a double weak* limit argument to show that J maps into $M(G)$.

Let p denote the projection mapping from G to G/H and $\check{p}$ the induced mapping of measures. We may assume that Haar measures on G and G/H are normalized so that

$$(2) \qquad \int_G f\, dm_G = \int_{G/H} \int_H f(x+t)dm_H(t)dm_{G/H}(px)$$

for all $f \in C_o(G)$.

We first suppose that τ is absolutely continuous with $w = d\tau/dm_G \in C_o(G)$. (The most important case arises when $\hat{\tau}$ has compact support). Let $g \in C(G/H)$ and $(Jg)(x) = g(px)w(x)$. If g is a trigonometric polynomial, then (1) obviously holds. (As customary, we identify g with $g\, dm_{G/H}$.) Also, by (2) and the easily proved fact that $p\tau = m_{G/H}$,

$$\|Jg\|_1 = \int |g(px)|w(x)dm_G(x) = \int_{G/H} \int_H |g(px)|w(x+t)dm_H(t)dm_{G/H}(px)$$
$$= \int_{G/H} |g(y)|dm_{G/H}(y) = \|g\|_1.$$

It now follows that (1) defines an element $J\mu$ of $M(G)$ whenever $\mu \in M(G/H)$ is absolutely continuous with $d\mu/dm_{G/H} \in C(G)$, and in that case $\|J\mu\| = \|\mu\|$. We now take weak* limits, concluding that (1) defines an element of $M(G)$ for each $\mu \in M(G/H)$ and that $\|J\mu\| \le \|\mu\|$.

To life the restriction on τ, we take weak* limits again, say $\tau =$ weak* lim τ_α, where $|\tau_\alpha(\gamma)| \le |\tau(\gamma)|$ for all α and all $\gamma \in \Gamma$. Of course, the τ_α are to be probability measures in $L^1(G)$ with $d\tau_\alpha/dm_G \in C_o(G)$. It is obvious from (1) that $J_{\tau_\alpha}\mu$ converges weak* to a measure $J_\tau\mu$. Of course, $\|J_\tau\mu\| \le \|\mu\|$. That ends the proof of the existence and uniqueness of J. We now establish (i)–(vi).

(i) is easy: $\check{p}J\mu = \mu$ for all $\mu \in M(G/H)$. Since $\|J\| \le 1$, J is an isometry.

(ii) Since J is an isometry and $(J\mu)^\wedge(0) = \hat{\mu}(0)$, $J\mu \ge 0$ if and only if $\mu \ge 0$.

(iii) We may assume that $\hat{\mu}$ has compact support in Λ. Let $g = d\mu/dm_{G/H}$. Let τ_n be a sequence of elements of $L^1(G)$ such that lim $\tau_n = \tau$ and such that each τ_n is a probability measure with supp $\hat{\tau}_n$ compact and $|\hat{\tau}_n(\gamma)| \le |\hat{\tau}(\gamma)|$ for all $\gamma \in \Gamma$. Then

$$\|J_{\tau_n} g - J_\tau g\| \le \|g\|_\infty \|\tau_n - \tau\|_1.$$

But $(J_{\tau_n} g)^\wedge$ is supported in supp $\hat{g}$ + supp $\hat{\tau}_n$, which is compact. Therefore $J_\tau g$ is a limit in $L^1(G)$-norm of elements of $L^1(G)$, so $J_\tau g \in L^1(G)$.

(iv) follows easily from the observations that $\mu = PJ_\mu \in L^1(G/H)$ if $J\mu \in L^1(G)$.

(v) Suppose that $\sup \#[\Lambda \cap (S - \gamma)] = N < \infty$ and that $\tau \in M_o(G)$. It is clear that if $\mu \notin M_o(G/H)$, then $J\mu \notin M_o(G)$: $(J\mu)\hat{}$ does not vanish along Λ. We may assume that $\mu \in M_o(G/H)$. Let $\varepsilon > 0$, and let $E \subseteq \Gamma$, $F \subseteq \Lambda$ be compact sets such that $|\hat{\tau}| \leq \varepsilon^{1/2}/N$ on $\Gamma \backslash E$ and $|\hat{\mu}| < \varepsilon^{1/2}$ on $\Lambda \backslash F$. Then if $\gamma \in \Gamma \backslash (E + F)$, $|(J\mu)\hat{}(\gamma)| < \varepsilon$ as a simple calculation shows.

(vi) Suppose that μ is tame and $\psi \in \Delta M(G)$. Then the restriction of ψ to $M(H)$ is given by integration against a continuous character γ of G. (That is because we assume that H is discrete as well as closed.) Then the kernel of $\bar{\gamma}\psi$ contains the kernel of $\check{p}$, so $\bar{\gamma}\psi$ has the form $\rho \circ \check{p}$, where $\rho \in \Delta M(G/H)$. Since μ is tame, $\rho_\mu = a\lambda$, where $\lambda \in \Lambda$ and a is a constant of modulus at most one. It now follows that $\psi_{J\mu} = a\lambda\bar{\gamma}$ a.e. $d(J\mu)$. Therefore $J\mu$ is tame. A similar argument shows that $J\mu$ is monotrochic if μ is monotrochic. The reverse implications are trivial. That ends the proof of A.7.1. □

References

N. I. Achieser (Ahiezer)

[1] Theory of Approximation. English translation by C. J. Hyman. New York: Ungar 1956. Augmented 2nd Russian edition, Moscow: Izdat, "Nauka" 1965.

S. Agmon and S. Mandelbrojt

[1] Une généralisation du théorème Tauberien. Acta Sci. Math. Szeged **12**, 167–176 (1950).

G. Akst

[1] L^1-algebras on semigroups. Illinois J. Math. **19**, 454–466 (1975).

I. Amemiya and I. Ito

[1] A simple proof of the theorem of P. J. Cohen. Bull. Amer. Math. Soc. **70**, 774–776 (1964).

R. Andersson

[1] Power bounded restrictions of Fourier-Stieltjes transforms. Thesis, University of Göteborg, Sweden, c. 1977

A. Atzmon

[1] Non-singly-generated closed ideals in group algebras. Israel J. Math. **7**, 303–310 (1969).

[2] Spectral synthesis in regular Banach algebras. Israel J. Math. **8**, 197–212 (1970).

[3] Sur les sous-algèbres fermées de A(G). C. R. Acad. Sci. Paris Sér. A-B **270**, 946–948 (1970).

[4] Non-finitely-generated closed ideals in group algebras. J. Functional Analysis **11**, 444–462 (1972).

[5] Translation invariant subspaces of $L^p(G)$. Studia Math. **48**, 245–250 (1973).

[6] Addition to the paper: Translation invariant subspaces of $L^p(G)$. Studia Math. **52**, 291–292 (1975).

[7] Spectral synthesis of functions of bounded variation. Proc. Amer. Math. Soc. **47**, 417–422 (1975).

[8] Spectral synthesis in some spaces of bounded continuous functions. Pacific J. Math. **74**, 277–284 (1978).

A. P. Baartz

[1] The measure algebra of a locally compact semigroup. Pacific J. Math. **21**, 199–214 (1967).

G. F. Bachelis and S. E. Ebenstein

[1] On $\Lambda(p)$ sets. Pacific J. Math. **54**, 35–38 (1974).

G. F. Bachelis and J. E. Gilbert

[1] Banach algebras with Rider subalgebras (preprint).

G. F. Bachelis, W. A. Parker, and K. A. Ross

[1] Local units in $L^1(G)$. Proc. Amer. Math. Soc. **31**, 312–313 (1972).

W. G. Bade and P. C. Curtis, Jr.

[1] The Wedderburn decomposition of commutative Banach algebras. Amer. J. Math. **82**, 851–866 (1960).

W. J. Bailey, G. Brown, and W. Moran

[1] Spectra of independent power measures. Proc. Camb. Philos. Soc. **72**, 27–35 (1972).

B. Bajsanski

[1] Sur une classe générale de procédés de sommations du type d'Euler-Borel. Acad. Serbe Sci. Publi. Inst. Math. **10**, 131–152 (1956).

A. C. Baker and J. W. Baker

[1] Algebras of measures on a locally compact semigroup I, II, III. J. London Math. Soc. (2) **1**, 249–259 (1969); ibid. **2**, 651–659 (1970); ibid **4**, 685–695 (1972).

J. W. Baker

[1] Convolution measure algebras with discrete spectra. J. London Math. Soc. (2) **5**, 193–201 (1972).

[2] A characterization of a certain class of convolution measure algebras. Proc. Edinburgh Math. Soc. (2) **18**, 199–205 (1973).

S. Banach

[1] Théorie des opérations linéaires. New York: Chelsea, 1955.

N. Bary

[1] A Treatise on Trigonometric Series. Translated from the Russian by M. F. Mullins. Two volumes. New York: Macmillan, 1964.

[2] Memoire sur la répresentation fine des fonctions continues I. Math. Ann. **103**, 185–248 (1930).

[3] Memoire sur la répresentation fine des fonctions continues II. Math. Ann. **103**, 598–653 (1930).

J. Benedetto

[1] Harmonic Analysis on Totally Disconnected Sets. Berlin-Heidelberg-New York: Springer-Verlag 1971.

G. Benke

[1] Sidon sets and the growth of L^p norms. Thesis, University of Maryland, 1971.

[2] Arithmetic structure and lacunary Fourier series. Proc. Amer. Math. Soc. **34**, 128–132 (1972).

A. Bernard and N. Th. Varopoulos

[1] Groupes de fonctions continues sur un compact. Studia Math. **35**, 199–205 (1970).

A. Beurling

[1] Local harmonic analysis with some applications to differential operators. In: Annual Science Conference Proceedings I (1962–1964), Belfer Graduate School of Science, Yeshiva University, New York, 1966, pp. 109–125.

A. Beurling and H. Helson

[1] Fourier-Stieltjes transforms with bounded powers. Math. Scand. **1**, 120–126 (1953).

W. W. Bledsoe

[1] An inequality about complex numbers. Amer. Math. Monthly **77**, 180–183 (1970).

R. C. Blei

[1] On trigonometric series associated with separable, translation invariant subspaces of $L^1(G)$. Trans. Amer. Math. Soc. **173**, 491–499 (1972).

[2] On subsets with associated compacta in discrete abelian groups. Proc. Amer. Math. Soc. **37**, 453–455 (1973).

[3] Some thin sets in discrete abelian groups. Trans. Amer. Math. Soc. **193**, 55–65 (1974).

[4] On Fourier-Stieltjes transforms of discrete measures. Math. Scand. **35**, 211–214 (1974).

[5] A simple diophantine condition in harmonic analysis. Studia Math. **52**, 195–202 (1975). Erratum and addendum: to appear.

[6] Sidon partitions and p-Sidon sets. Pacific J. Math. **65**, 307–313 (1976).

[7] Rosenthal sets that cannot be sup-norm partitioned and an application to tensor products. Colloq. Math. (to appear).

W. R. Bloom

[1] Sets of p-spectral synthesis. Pacific J. Math. **60**, 7–19 (1975).

J. R. Blum, B. Eisenberg, and L. S. Hahn

[1] Ergodic theory and the measure of sets in the Bohr group. Acta Sci. Math. (Szeged) **34**, 17–24 (1973).

J. R. Blum and B. Epstein

[1] On the Fourier transform of an interesting class of measures. Israel J. Math. **10**, 302–305 (1971).

[2] On the Fourier-Stieltjes coefficients of Cantor-type distributions. Israel J. Math. **17**, 35–45 (1974).

S. Bochner
[1] Monotone Functionen, Stieltjessche Integrale, und harmonische Analyse. Math. Ann. **108**, 378–410 (1933).
[2] A theorem on Fourier-Stieltjes integrals. Bull. Amer. Math. Soc. **40**, 271–276 (1934).

H. Bohr
[1] Über fastperiodische ebene bewegungen. Comment. Math. Helv. **4**, 51–64 (1934).

A. Bonami
[1] Étude des coefficients de Fourier des fonctions de $L^p(G)$. Ann. Inst. Fourier (Grenoble) **20**, fasc. 2, 335–402 (1970).

P. Brenner
[1] Power bounded matrices of Fourier-Stieltjes transforms. Math. Scand. **22**, 115–129 (1968).
[2] Power-bounded matrices of Fourier-Stieltjes transforms II. Math. Scand. **25**, 39–48 (1969).
[3] Corrections to the papers "Power bounded matrices of Fourier-Stieltjes transforms I, II." Math. Scand. **30**, 150–151 (1972).

G. Brown (listed also with W. J. Bailey and W. Moran)
[1] On convolution measure algebras. Proc. London Math. Soc. (3) **20**, 490–506 (1970).
[2] $M_0(G)$ has a symmetric maximal ideal off the Šilov boundary. Proc. London Math. Soc. (3) **27**, 484–504 (1973).
[3] Riesz products and generalized characters. Proc. London Math. Soc. (3) **30**, 209–238 (1975).
[4] Singular infinitely divisible distribution functions whose characteristic functions vanish at infinity. Math. Proc. Camb. Phil. Soc. **82**, 277–287 (1977).
[5] Spectral extension and power independence in measure algebras. Studia Math. (to appear).

G. Brown, C. C. Graham, and W. Moran
[1] Translation and symmetry in M(G). Symposia Mathematica **22**, 371–392 (1977).

G. Brown and E. Hewitt
[1] Continuous singular measures equivalent to their convolution squares. Math. Proc. Camb. Phil. Soc. **80**, 249–268 (1976).

G. Brown and W. Moran
[1] On the Šilov boundary of a measure algebra. Bull. London Math. Soc. **3**, 197–203 (1971).
[2] In general, Bernoulli convolutions have independent powers. Studia Math. **47**, 141–152 (1973). Corrigendum and addendum **61**, 25–28 (1977).
[3] Translation and power independence for Bernoulli convolutions. Colloq. Math. **27**, 301–313 (1973).
[4] A dichotomy for infinite convolutions of discrete measures. Proc. Cambridge Philos. Soc. **73**, 307–316 (1973).
[5] L-ideals of M(G) determined by continuity of translation. Proc. Edinburgh Math. Soc. (2) **18**, 307–316 (1972–73).
[6] $L^{\frac{1}{2}}(G)$ is the kernel of the asymmetric maximal ideals of M(G). Bull. London Math. Soc. **5**, 179–186 (1973).
[7] Bernoulli measure algebras. Acta Math. **132**, 77–109 (1974).
[8] On orthogonality for Riesz products. Proc. Cambridge Philos. Soc. **76**, 173–181 (1974).
[9] Sums of random variables in groups and the purity law. Z. Wahrscheinlichkeitstheorie und Verw. Gebiete **30**, 227–234.
[10] $M_0(G)$-boundaries are M(G)-boundaries. J. Functional Analysis **18**, 350–368 (1975).
[11] Products of random variables and Kakutani's criterion for orthogonality of product measures. J. London Math. Soc. (2) **10**, 401–405 (1975).
[12] Coin tossing and powers of singular measures. Math. Proc. Camb. Philos. Soc. **77**, 349–364 (1975).
[13] Gleason parts for measure algebras. Math. Proc. Camb. Phil. Soc. **79**, 321–327 (1976).
[14] Point derivations on M(G). Bull. London Math. Soc. **8**, 57–64 (1976).
[15] Analytic discs in the maximal ideal space of the measure algebra. Pacific J. Math. **75**, 45–57 (1978).
[16] Maximal elements in the maximal ideal space of a measure algebra (preprint).

R. B. Burckel

[1] Weakly Almost Periodic Functions on Semigroups. New York-London-Paris: Gordon and Breach, 1970.

[2] Characterizations of C(X) among Its Subalgebras. New York: Marcel Dekker, 1972.

A. P. Calderòn

[1] Intermediate spaces and interpolation, the complex method. Studia Math. **24**, 113–190 (1964).

G. Cantor

[1] Über die Ausdehnung eines Satzes aus der Theorie der trigonometrischen Reihen. Math. Ann. **5**, 123–132 (1872).

L. Carleson and P. Sjölin

[1] Oscillatory integrals and a multiplier theorem for the disk. Studia Math. **44**, 287–299 (1972).

J. W. S. Cassels

[1] An Introduction to Diophantine Approximation. Cambridge: The University Press, 1957.

P. S. Chow and A. J. White

[1] The structure semigroup of some convolution measure algebras. Quart. J. Math. Oxford (2) **22**, 221–229 (1971).

J. B. R. Christensen and P. Ressel

[1] Functions operating on positive definite matrices and a theorem of Schoenberg. Trans. Amer. Math. Soc. **243**, 89–95 (1978).

J. Clunie and P. Vermes

[1] Regular Sonnenschein type summability methods. Acad. Roy. Belg. Bull. Cl. Sci. (5) **45**, 930–954 (1959).

P. J. Cohen

[1] On a conjecture of Littlewood and idempotent measures. Amer. J. Math. **82**, 191–212 (1960).

[2] On homomorphisms of group algebras. Amer. J. Math. **82**, 213–226 (1960).

R. R. Coifman and G. Weiss

[1] Representations of compact groups and spherical harmonics. Ens. Math. **14**, 121–173 (1968).

[2] (Review of: Edwards and Gaudry [1]) Bull. Amer. Math. Soc. **84**, 242–250 (1978).

B. Connes

[1] Mesure de Hausdorff des complementaires des U_ε de Zygmund. Israel J. Math. **23**, 1–7 (1976).

D. M. Connolly and J. H. Williamson

[1] An application of a theorem of Singer. Proc. Edinburgh Math. Soc. (2) **19**, 119–123 (1974–75).

M. G. Cowling

[1] Functions which are restrictions of L^p-multipliers. Trans. Amer. Math. Soc. **213**, 35–51 (1975).

M. G. Cowling and J. J. F. Fournier

[1] Inclusions and noninclusions of spaces of convolution operators. Amer. Math. Soc. Trans. **221**, 59–95 (1976).

P. C. Curtis, Jr.: listed with W. G. Bade

H. Davenport

[1] On a theorem of P. J. Cohen. Mathematika **7**, 93–97 (1960).

A. M. Davie

[1] The approximation problem for Banach spaces. Bull. London Math Soc. **5**, 261–266 (1973).

[2] The Banach approximation problem. J. Approx. Theory **13**, 395–412 (1975).

D. E. Daykin *et al*

[1] An analytic inequality. Amer. Math. Monthly **81**, 787–788 (1974).

M. Déchamps-Gondim

[1] Ensembles de Sidon topologiques. Ann. Ins. Fourier (Grenoble) **22**, fasc. 3, 51–79 (1972).

[2] Interpolation linéaire approchée des fonctions bornées définies sur un ensemble de Sidon. Colloq. Math. **28**, 255–259 (1973).

K. deLeeuw
[1] On L^p multipliers. Ann. Math. **81**, 364–379 (1965).

K. deLeeuw and I. Glicksberg
[1] The decomposition of certain group representations. J. Analyse Math. **15**, 135–192 (1965).

K. deLeeuw and C. S. Herz
[1] An invariance property of spectral synthesis. Illinois. J. Math. **9**, 220–229 (1965).

K. deLeeuw, J.-P. Kahane, and Y. Katznelson
[1] Sur les coefficients de Fourier des fonctions continues. C. R. Acad. Sci. Paris Sér. A-B. **285**, A1001–A1002 (1977).

K. deLeeuw and Y. Katznelson
[1] Functions that operate on non-self-adjoint algebras. J. Analyse Math. **11**, 207–219 (1963).
[2] On certain homomorphisms of quotients of group algebras. Israel J. Math. **2**, 120–126 (1964).
[3] The two sides of a Fourier-Stieltjes transforms and almost idempotent measures. Israel J. Math. **8**, 213–229 (1970).

L. deMichele and P. Soardi
[1] Symbolic calculus in $A_p(G)$-I. Atti Accad. Naz. Lincei Rend. Cl. Sci. Fis. Mat. Natur. (8) **57**, 24–30 (1974).
[2] Symbolic calculus in $A_p(G)$-II. Atti Accad. Naz. Lincei Rend. Cl. Sci. Fis. Mat. Natur. (8) **57**, 31–35 (1974).
[3] Functions which operate in the Fourier algebra of a discrete group. Proc. Amer. Math. Soc. **45**, 389–392 (1974).
[4] A remark on sets of uniqueness of l^p. Boll. Un. Mat. Ital. (4) **11**, 64–65 (1975).

J. Dixmier
[1] Sur un théorème de Banach. Duke Math. J. **15**, 1057–1071 (1948).
[2] Quelques exemples concernant la synthèse spectrale. C. R. Acad. Sci. Paris Sér. A-B **247**, A24–A26 (1958).

P. G. Dixon
[1] A lower bound for the L^1-norm of certain exponential sums. Mathematika **24**, 182–188 (1977).

Y. Domar
[1] A property of Fourier-Stieltjes transforms on the discrete group of real numbers. Ann. Inst. Fourier (Grenoble) **20**, fasc. 2, 325–334 (1970).
[2] On the spectral synthesis problem for (n-1)-dimensional subsets of R^n, $n \geqslant 2$. Ark. Mat. **9**, 23–37 (1971).
[3] Estimates of $\|e^{itf}\|_{A(\Gamma)}$, when $\Gamma \subseteq R^n$ is a curve and f is a real-valued function. Israel J. Math. **12**, 184–189 (1972).
[4] A theorem of Buerling-Helson type. Math. Scand. **33**, 139–144 (1973).
[5] On spectral synthesis for curves in R^3. Math. Scand. **39**, 282–294 (1976).
[6] On the Banach algebra $A(\Gamma)$ for smooth sets $\Gamma \subseteq R^n$. Comment. Math. Helvetici **52**, 357–371 (1977).
[7] A C^∞ curve of spectral non-synthesis. Mathematika **24**, 189–192 (1977).

R. Doss
[1] Approximation and representations for Fourier transforms. Trans. Amer. Math. Soc. **153**, 211–221 (1971).
[2] Elementary proof of a theorem of Helson. Proc. Amer. Math. Soc. **27**, 418–420 (1971).
[3] Convolution of singular measures. Studia Math. **45**, 111–117 (1973).
[4] Representations of continuous functions of several variables. Amer. J. Math. **98**, 375–378 (1976).

R. G. Douglas and J. L. Taylor
[1] Wiener-Hopf operators with measure kernels. Colloq. Math. Soc. Janos Bolyai 5 (Hilbert Space Operators), 1970. Amsterdam: North Holland 1972.

S. W. Drury
[1] Properties of certain measures on a topological group. Proc. Camb. Philos. Soc. **64**, 1011–1013 (1968).
[2] Studies in regular algebras. Doctoral dissertation, Cambridge University, 1969.
[3] On non-triangular sets in tensor algebras. Studia Math. **34**, 253–263 (1970).

[4] Sur les ensembles de Sidon. C. R. Acad. Sci. Paris Sér. A-B **271**, A162–A163 (1970).
[5] Unions of sets of interpolation. In: Conference on Harmonic Analysis, College Park, Maryland, 1971, pp. 23–33. Berlin-Heidelberg-New York: Springer-Verlag 1972.
[6] The Fatou-Zygmund property for Sidon sets. Bull. Amer. Math. Soc. **80**, 535–538 (1974).
[7] Birelations and Sidon sets. Proc. Amer. Math. Soc. **53**, 123–128 (1975).

N. Dunford and J. T. Schwartz
[1] Linear Operators, Part I: General Theory. New York: Interscience 1958.

C. F. Dunkl
[1] The measure algebra of a locally compact hypergroup. Trans. Amer. Math. Soc. **179**, 331–348 (1973).
[2] Structure hypergroups for measure algebras. Pacific J. Math. **47**, 413–425 (1973).

P. L. Duren
[1] Theory of H^p Spaces. New York: Academic Press 1970.

H. A. M. Dzinotyiweyi
[1] On the analogue of the group algebra for locally compact semigroups. J. London Math. Soc. (2) **17**, 489–506 (1978).

S. E. Ebenstein: listed with G. F. Bachelis

W. F. Eberlein
[1] Characterizations of Fourier-Stieltjes transforms. Duke Math. J. **22**, 465–468 (1955).

D. A. Edwards
[1] On translates of L^∞ functions. J. London Math. Soc. **36**, 431–432 (1961).

R. E. Edwards
[1] Functional Analysis. New York: Holt, Rinehart, and Winston, 1965.
[2] Translates of L^∞ functions of bounded measures. J. Austral. Math. Soc. **4**, 403–409 (1964).

R. E. Edwards and G. I. Gaudry
[1] Littlewood-Paley and Multiplier Theory. Berlin-Heidelberg-New York: Springer Verlag 1977.

B. Eisenberg: listed with J. R. Blum and L. S. Hahn

P. Enflo
[1] A counterexample to the approximation problem in Banach spaces. Acta Math. **13**, 309–317 (1973).

B. Epstein: listed with J. R. Blum

P. Eymard
[1] Algèbres A_p et convoluteurs de L^p. In: Séminaire Bourbaki 1969/70, Exposés 364–381, pp. 55–72. Berlin-Heidelberg-New York: Springer-Verlag 1971.

C. Fefferman
[1] The multiplier problem for the ball. Ann. of Math. (2) **94**, 330–336 (1971).

C. Fefferman and H. S. Shapiro
[1] A planar face on the unit sphere of the multiplier space M_p, $1 < p < \infty$. Proc. Amer Math. Soc. **36**, 435–439 (1972).

W. Feller
[1] An introduction to Probability Theory and its Applications, Volume I, 3rd edition. New York-London-Sydney: Wiley 1968.

A. Figà-Talamanca
[1] Translation invariant operators in L^p. Duke Math. J. **32**, 495–501 (1965).

A. Figà-Talamanca and G. I. Gaudry
[1] Multipliers and sets of uniqueness of L^p. Michigan Math. J. **17**, 179–191 (1970).
[2] Multipliers of L^p which vanish at infinity. J. Functional Analysis **7**, 475–486 (1971).

M. Filippi
[1] Variété de non-synthèse spectrale sur un groupe abélien localement compact. Israel J. Math. **3**, 43–60 (1965).

J. J. F. Fournier (listed also with M. G. Cowling)
[1] Fourier coefficients after gaps. J. Math. Anal. Appl. **42**, 255–270 (1973).
[2] On a theorem of Paley and the Littlewood conjecture. Ark. Mat. (to appear).

S. H. Friedberg
[1] Closed subalgebras of group algebras. Trans. Amer. Math. Soc. **147**, 117–125 (1970).
[2] The Fourier transform is onto only when the group is finite. Proc. Amer. Math. Soc. **27**, 421–422 (1971).

S. H. Friedberg and L. E. Spence
[1] Local-belonging sets and multiplier-induced ideals in group algebras. Illinois J. Math. **21**, 293–299 (1977).

M. Frisch
[1] Propriété asymptotiques des vibrations du tore. J. Math. Pures Appl. **54**, 285–303 (1975).

L. Fuchs
[1] Infinite Abelian Groups, Volumes I, II. New York: Academic Press, 1970 and 1973.

E. Galanis
[1] A problem on the union of two Helson sets. Ark. Math. **9**, 193–196 (1971).
[2] Some results on the union of Helson sets. Proc. Cambridge Philos. Soc. **70**, 235–241 (1971).
[3] Separating function for V-Sidon sets. Proc. Cambridge Philos. Soc. **71**, 39–42 (1972).
[4] The Helson constant of the union of independent countable Helson sets. Prakt. Akad. Athénon **48**, 332–337 (1973).

T. W. Gamelin
[1] Uniform Algebras. Cleveland Cliffs, New Jersey: Prentice-Hall 1969.

M. Gatesoupe
[1] Characterisation locale de la sous-algèbre fermée des fonctions radiales de $L^1(R_n)$. Ann. Inst. Fourier (Grenoble) **17**, fasc. 1, 93–107 (1967).

G. I. Gaudry (listed also with: R. E. Edwards; A. Figà-Talamanca)
[1] Bad behavior and inclusion results for multipliers of type (p,q). Pacific J. Math. **35**, 83–94 (1970).
[2] Restrictions of multipliers to closed subgroups. Math. Ann. **197**, 171–179 (1972).

I. M. Gel'fand, D. A. Raĭkov, and G. E. Šilov
[1] Commutative Normed Rings. New York: Chelsea 1964. (Previous version: Uspehi Mat. Nauk 1, No. 2 (**12**), 48–146 (1946) [Russian]; Amer. Math. Soc. Transl. (2) **5**, 115–220 (1957).

J. E. Gilbert (listed also with G. F. Bachelis)
[1] On a strong form of spectral synthesis. Ark. Mat. **7**, 571–575 (1969).
[2] On projections of $L^\infty(G)$ onto translation-invariant subspaces. Proc. London Math. Soc. (3) **19**, 69–88 (1969).

D. M. Girard
[1] The asymptotic behavior of norms of powers of absolutely convergent Fourier series. Pacific J. Math. **37**, 357–381 (1971).
[2] The behavior of the norm of an automorphism of the unit disk. Pacific J. Math. **47**, 443–456 (1973).

I. Glicksberg (listed also with K. deLeeuw)
[1] Fourier-Stieltjes transforms with an isolated value. In: Conference on Harmonic Analysis, College Park, Maryland, 1971, pp. 59–77. Berlin-Heidelberg-New York: Springer-Verlag 1972.
[2] Some remarks on absolute continuity in groups. Proc. Amer. Math. Soc. **40**, 135–139 (1973).
[3] Two remarks on Fourier-Stieltjes transforms (preprint).

I. Glicksberg and I. Wik
[1] Multipliers of quotients of L_1. Pacific J. Math. **38**, 619–624 (1971).

R. R. Goldberg and R. B. Simon
[1] Characterization of some classes of measures. Acta Sci. Math. (Szeged) **27**, 157–161 (1966).

C. C. Graham (listed also with G. Brown and W. Moran)
[1] On a Banach algebra of Varopoulos. J. Functional Analysis **4**, 317–328 (1969).
[2] Automorphisms of tensor algebras. J. Math. Anal. Appl. **30**, 385–397 (1970).
[3] Symbolic calculus for positive-definite functions. J. Functional Analysis **11**, 465–478 (1972).

[4] Compact independent sets and Haar measure. Proc. Amer. Math. Soc. **36**, 578–582 (1972).
[5] Interpolation sets for convolution measure algebras. Proc. Amer. Math. Soc. **38**, 512–522 (1973).
[6] Sur un théorème de Katznelson et McGehee. C. R. Acad. Sci. Paris Sér. A-B **276**, A37–A40 (1973).
[7] Measures vanishing off the symmetric maximal ideals of M(G). Proc. Cambridge Philos. Soc. **75**, 51–61 (1974).
[8] Maximal ideals and discontinuous characters. Acta Sci. Math. (Szeged) **36**, 233–238 (1974).
[9] Idempotent maximal ideals and independent sets. Proc. Amer. Math. Soc. **54**, 133–137 (1976).
[10] Non-Sidon sets in the support of a Fourier-Stieltjes transform. Colloq. Math. **36**, 269–273 (1976).
[11] Functional calculus and positive-definite functions. Trans. Amer. Math. Soc. **231**, 215–231 (1977).
[12] The Sidon constant of a finite abelian group. Proc. Amer. Math. Soc. **68**, 83–84 (1978).

C. C. Graham, B. Host, and F. Parreau
[1] Sur les supports des transformées de Fourier-Stieltjes. Colloq. Math. (to appear).

C. C. Graham and A. MacLean
[1] A multiplier theorem for continuous measures. Studia Math. (to appear).

F. P. Greenleaf
[1] Norm decreasing homomorphisms of group algebras. Pacific J. Math. **15**, 1187–1219 (1965).

M. B. Gregory
[1] p-Helson sets, $1 < p < 2$. Israel J. Math. **12**, 356–368 (1972).

A. Grothendieck
[1] Sur les applications linéaires faiblement compactes d'espaces du type C(K), Canadian J. Math **5**, 129–173 (1953).
[2] Produits tensorielles topologiques et espaces nucléaires. Mem. Amer. Math. Soc. **16** (1955).

R. C. Gunning and H. Rossi
[1] Analytic Functions of Several Complex Variables. Englewood Cliffs, N. J.: Prentice-Hall 1965.

L. S. Hahn: listed with J. R. Blum and B. Eisenberg

J. A. Haight
[1] Difference covers which have small k-sums for any k. Mathematica **20**, 109–118 (1973).

H. Halberstam and K. F. Roth
[1] Sequences, Volume I. Oxford: Clarendon Press 1966.

P. Halmos
[1] Measure Theory. Princeton, N.J.: D. Van Nostrand 1950.

G. H. Hardy and J. E. Littlewood
[1] A new proof of a theorem on rearrangements. J. London Math. Soc. **23**, 163–168 (1948).

S. Hartman
[1] Rosenthal sets on the line. Math. Nachr. **76**, 153–158 (1977).

S. Hartman, J.-P. Kahane, and C. Ryll-Nardzewski
[1] Sur les ensembles d'interpolation. Bull. Acad. Polon. Sci. Sér. Sci. Math. Astronom. Phys. **13**, 83-86 (1965).

S. Hartman and C. Ryll-Nardzewski
[1] Almost periodic extensions of functions I. Colloq. Math. **12**, 23–29 (1964).
[2] Almost periodic extensions of functions II. Colloq. Math. **15**, 79–86 (1966).
[3] Almost periodic extensions of functions III. Colloq. Math. **16**, 223–224 (1967).
[4] Quelques résultats et problèmes en algèbre des mesures continues. Colloq. Math. **22**, 271–277 (1971).

G. W. Hedstrom
[1] Norms of powers of absolutely convergent Fourier series. Michigan Math. J. **13**, 249–259 (1966).

[2] Norms of powers of absolutely convergent Fourier series in several variables. Michigan Math. J. **14**, 493–495 (1967).

C. H. Heiberg

[1] Norms of powers of absolutely convergent Fourier series of several variables. J. Functional Analysis **14**, 382–400 (1973).

[2] Norms of powers of absolutely convergent Fourier series: an example. Pacific J. Math. **66**, 131–152 (1967).

[3] Fourier series in several variables with bounded convolution powers. Pacific J. Math. (to appear).

[4] Limits of norms of powers of absolutely convergent Fourier series in several variables (preprint).

H. Helson (listed also with A. Beurling)

[1] Note on harmonic functions. Proc. Amer. Math. Soc. **4**, 686–691 (1953).

[2] Proof of a conjecture of Steinhaus. Proc. Nat. Acad. Sci. U.S. **40**, 205–206 (1954).

[3] Fourier transforms on perfect sets. Studia Math. **14**, 209–213 (1954).

[4] On a theorem of Szegö. Proc. Amer. Math. Soc. **6**, 235–242 (1955).

H. Helson and J.-P. Kahane

[1] Sur les fonctions opérant dans les algèbras de transformées de suites ou de fonctions sommable. C. R. Acad. Sci. Paris. Sér. A-B **247**, A626–A628 (1958).

[2] A Fourier method in Diophantine problems. J. Analyse Math. **15**, 245–262 (1965).

H. Helson, J.-P. Kahane, Y. Katznelson, and W. Rudin

[1] The functions which operate on Fourier transforms. Acta Math. **102**, 135–157 (1959).

C. S. Herz (listed also with K. deLeeuw)

[1] Spectral synthesis for the Cantor set. Proc. Nat. Acad. Sci. U.S.A. **42**, 42–43 (1956).

[2] Spectral synthesis for the circle. Ann. of Math. **68**, 709–712 (1958).

[3] The spectral theory of bounded functions. Trans. Amer. Math. Soc. **94**, 181–232 (1960).

[4] Fourier transforms related to convex sets. Ann. of Math. **75**, 81–92 (1962).

[5] On the number of lattice points in a convex set. Amer. J. Math. **84**, 126–133 (1962).

[6] Fonctions opérant sur les fonctions définies-positives. Ann. Inst. Fourier (Grenoble) **13**, fasc. 1, 161–180 (1963).

[7] Remarques sur la note précédente de Varopoulos. C. R. Acad. Sci. Paris Sér. A-B **260**, A6001–A6004 (1965).

[8] (Review of Varopoulos [5, 6, 7]). Math. Reviews 31 # 2567, pp. 462–463 (1966).

[9] Drury's lemma and Helson sets. Studia Math. **42**, 205–219 (1972).

[10] Harmonic synthesis for subgroups. Ann. Inst. Fourier (Grenoble) **23**, fasc. 3, 91–123 (1973).

[11] Une géneralization de la notion de transformée de Fourier-Stieltjes. Ann. Inst. Fourier (Grenoble) **24**, fasc. 3, 145–157 (1974).

E. Hewitt (listed also with G. Brown)

[1] The asymmetry of certain algebras of Fourier-Stieltjes transforms. Michigan Math. J. **5**, 149–158 (1958).

E. Hewitt and S. Kakutani

[1] A class of multiplicative linear functionals on the measure algebra of a locally compact abelian group. Illinois J. Math. **4**, 553–574 (1960).

[2] Some multiplicative linear functionals on M(G). Ann. of Math. **79**, 489–505 (1964).

E. Hewitt and K. A. Ross

[1] Abstract Harmonic Analysis, Volumes I and II. New York-Heidelberg-Berlin: Springer-Verlag 1963 and 1970.

E. Hewitt and K. Stromberg

[1] Real and Abstract Analysis. Berlin-Heidelberg-New York: Springer-Verlag 1965.

[2] A remark on Fourier-Stieltjes transforms. An. Acad. Brasil Cien. **34**, 175–180 (1962).

E. Hewitt and H. S. Zuckerman

[1] On a theorem of P. J. Cohen and H. Davenport. Proc. Amer. Math. Soc. **14**, 847–855 (1963).

[2] Singular measures with absolutely continuous convolution squares. Proc. Cambridge Philos. Soc. **62**, 399–420, (1966) Corrigendum, ibid. **63**, 367–368 (1967).

[3] Some singular Fourier-Stieltjes series. Proc. London Math. Soc. (3) **19**, 310–326 (1969).

[4] Structure theory for a class of convolution algebras. Pacific J. Math. **7**, 913–941 (1957).

E. Hille

[1] Analytic Function Theory. Two volumes. New York: Chelsea 1973.

E. Hille and R. S. Phillips

[1] Functional Analysis and Semi-groups, revised edition. Providence, R.I.: American Mathematical Society 1957.

E. Hille and J. D. Tamarkin

[1] Remarks on a known example of a monotone continuous function. Amer. Math. Monthly **36**, 255–264 (1929).

E. Hlawka

[1] Integrale auf konvexen Körpern. Monat. Math. **54**, 1–36 and 81–99 (1950).

K. Hoffman

[1] Banach Spaces of Analytic Functions. Englewood Cliffs, N.J.: Prentice-Hall 1962.

K. H. Hofmann and P. Mostert

[1] Elements of Compact Semigroups. Columbus: Charles E. Merrill, 1966.

J. A. R. Holbrook

[1] The iterates of a contraction and its adjoint. Proc. Amer. Math. Soc. **29**, 543–546 (1971).

L. Hörmander

[1] Estimates for translation invariant operators in L^p spaces. Acta Math. **104**, 93–140 (1960).

A. Horn

[1] A characterization of unions of linearily independent sets. J. London Math. Soc. **30**, 494–496 (1955).

R. A. Horn

[1] On inequalities between Hermitian and symmetric forms. Linear Algebra and Appl. **11**, 189–218 (1975).

[2] Quadratic forms in harmonic analysis and the Bochner-Eberlein theorem. Proc. Amer. Math. Soc. **52**, 263–270 (1975).

B. Host: listed also with C. C. Graham and F. Parreau

B. Host and F. Parreau

[1] Sur un problème de I. Glicksberg: les idéaux fermés de type fini de M(G). Ann. Inst. Fourier (Grenoble) **28**, fasc. 3, 143–164 (1978).

[2] Sur les mesures dont la transformée de Fourier-Stieltjes ne tend pas vers zero a l'infini. Colloq. Math. (to appear).

[3] Produits de Riesz generalisés. Séminaire Math. Univ. de Paris-Nord, 1977–78.

S. Igari

[1] Functions of L^p-multipliers. Tôhoku Math. J. (2) **21**, 304–320 (1969).

[2] Functions of L^p-multipliers II. Tôhoku Math. J. (2) **26**, 555–561 (1974).

I. Ito: listed with I. Amemiya

K. Ito and Y. Kawada

[1] On the probability distribution on a compact group (I). Proc. Phys.-Math. Soc. Japan (3) **22**, 977–998 (1940).

T. Ito and B. M. Schreiber

[1] A space of sequences given by pairs of unitary operators. Proc. Japan Acad. **46**, 637–641 (1970).

O. S. Ivašëv-Musatov

[1] On the Fourier-Stieltjes coefficients of singular functions. Dokl. Akad. Nauk. SSSR (N.S.) **82**, 9–11 (1952) [Russian].

[2] On Fourier-Stieltjes coefficients of singular functions. Izv. Akad. Nauk SSSR Ser. Mat. **20**, 179–196 [Russian]. English translation in Amer. Math. Soc. Transl. (2) **10**, 107–124 (1958).

[3] On the coefficients of trigonometric null series. Izv. Akad. Nauk SSSR Ser. Mat. **21**, 559–578 (1957) [Russian]. English translation in Amer. Math. Soc. Transl. (2) **14**, 289–310 (1960).

K. Izuchi

[1] On a problem of J. L. Taylor. Proc. Amer. Math. Soc. **53**, 347–352 (1975).

B. Jessen and A. Wintner

[1] Distribution functions and the Riemann zeta function. Trans. Amer. Math. Soc. **38**, 48–88 (1935).

R. I. Jewitt
[1] Spaces with an abstract convolution. Adv. Math. **18**, 1–101 (1975).

M. Jodeit, Jr.
[1] Restrictions and extensions of Fourier multipliers. Studia Math. **34**, 215–226 (1970).
[2] A note on Fourier multipliers. Proc. Amer. Math. Soc. **27**, 423–424 (1971).

B. E. Johnson
[1] Symmetric maximal ideals in M(G). Proc. Amer. Math. Soc. **18**, 1040–1044 (1967).
[2] Continuity of linear operators commuting with continuous linear operators. Trans. Amer. Math. Soc. **128**, 88–102 (1967).
[3] The Šilov boundary of M(G). Trans. Amer. Math. Soc. **134**, 289–296 (1968).
[4] Separate continuity and measurability. Proc. Amer. Math. Soc. **20**, 420–422 (1969).

G. W. Johnson and G. S. Woodward
[1] On p-Sidon sets. Indiana Univ. Math. J. **24**, 161–167 (1974).

J.-P. Kahane (listed also with K. deLeeuw and Y. Katznelson; S. Hartman and C. Ryll-Nardzewski; H. Helson, Y. Katznelson, and W. Rudin)
[1] Some Random Series of Functions. Lexington, Mass.: D. C. Heath 1968.
[2] Séries de Fourier Absolument Convergentes. Berlin-Heidelberg-New York: Springer-Verlag 1970.
[3] Sur les fonctions sommes des séries trigonométriques absolument convergentes. C. R. Acad. Sci. Paris Sér. A-B **240**, A36–A37 (1955).
[4] Sur certaines classes de séries de Fourier absolument convergentes. J. Math. Pures et Appl. **35**, 249–259 (1956).
[5] Sur un problème de Littlewood. Nederl. Acad. Wetensch. Indag. Math. **19**, 268–271 (1957).
[6] Sur un théorème de Wiener-Lévy. C. R. Acad. Sci. Paris Sér. A-B **246**, A1949–A1951 (1958).
[7] Algèbres tensorielles et analyse harmonique. Seminaire Bourbaki **17**, 291-1–291-10 (1964/65).
[8] Idempotents and closed subalgebras of A(Z). In: Function Algebras, Proc. Intern. Symp. on Function Algebras, New Orleans, 1965, Frank Birtel editor, 198-207. Chicago: Scott, Foresman 1966.
[9] Ensembles de Ryll-Nardzewski et ensembles de Helson. Colloq. Math **15**, 87–92 (1966).
[10] Sur les réarrangements de suites de coefficients de Fourier-Lebesgue. C. R. Acad. Sci. Paris Sér. A-B **265**, A310–A312 (1967).
[11] Sur les réarrangements de fonctions de la classe A. Studia Math. **31**, 287–293 (1968).
[12] A metric condition for a closed circular set to be a set of uniqueness. J. Approx. Theory **2**, 233–236 (1969).

J.-P. Kahane and Y. Katznelson
[1] Sur la réciproque du Théorème de Wiener-Lévy. C. R. Acad. Sci. Paris Sér. A-B **248**, A1279–A1281 (1959).
[2] Contribution à deux problèmes, concernant les fonctions de la classe A. Israel J. Math. **1**, 110–131 (1963).
[3] Lignes de niveau et séries de Fourier absolument convergentes. Israel J. Math **6**, 346–353 (1968).
[4] Sur les ensembles d'unicité U(ε) de Zygmund. C. R. Acad. Sci. Paris Sér. A-B **277**, A893–A895 (1973).
[5] Mauvaise approximation par polynômes trigonométriques dans certaines algèbres de convolution.

J.-P. Kahane and W. Rudin
[1] Caractérisation des fonctions qui opèrent sur les coefficients de Fourier-Stieltjes. C. R. Acad. Sci. Paris Sér. A-B **247** A773–A775 (1958).

J.-P. Kahane and R. Salem
[1] Ensembles parfaits et séries trigonométriques. Paris: Hermann 1963.

J.-P. Kahane, M. Weiss, and G. Weiss
[1] On lacunary power series. Ark. Mat. **5**, 1–26 (1965).

S. Kaijser
[1] Representations of tensor algebras as quotients of group algebras. Ark. Mat. **10**, 107–141 (1972).

[2] An application of Grothendieck's inequality to a problem in harmonic analysis. Seminaire Maurey-Schwartz 1975–1976, exposé V, 25 nov.
[3] Some results in the metric theory of tensor products. Uppsala University Dept. of Math. Report 2, 1973.

S. Kakutani (listed also with E. Hewitt)
[1] On the equivalence of infinite product measures. Ann. of Math. **49**, 214–224 (1948).

N. J. Kalton and G. V. Wood
[1] Homomorphisms of group algebras with norm less than $\sqrt{2}$. Pacific J. Math. **62**, 439–460 (1976).

Y. Katznelson (listed also with K. deLeeuw; K. deLeeuw and J.-P. Kahane; H. Helson, J.-P. Kahane and W. Rudin; J.-P. Kahane)
[1] An introduction to Harmonic Analysis (Second Corrected Edition). New York: Dover 1976.
[2] Sur les fonctions opérant sur l'algèbre des séries de Fourier absolument convergentes. C. R. Acad. Sci. Paris Sér. A-B **247**, A404–A406 (1958).
[3] Sur le calcul symbolique dans quelques algèbres de Banach. Ann. Sci. Ec. Norm. Sup. **76**, 83–123 (1959).
[4] Sets of uniqueness for some classes of trigonometrical series. Bull. Amer. Math. Soc. **70**, 722–723 (1964).
[5] Trigonometric series with positive partial sums. Bull. Amer. Math. Soc. **71**, 718–719 (1965).
[6] Sequences of integers dense in the Bohr group. Proc. Royal. Inst. of Tech. (Stockholm), 76–86 (June 1973).
[7] Suites aléatoire d'entier. In: L'Analyse Harmonique dans le Domaine Complexe, E. J. Akutowicz editor, pp. 148–152. Berlin-Heidelberg-New York: Springer-Verlag 1975.
[8] On a theorem of Menchoff. Proc. Amer. Math. Soc. **53**, 396–398 (1975).

Y. Katznelson and T. W. Körner
[1] An algebra dense in its tilde algebra. Israel J. Math. **17**, 248–260 (1974).

Y. Katznelson and P. Malliavin
[1] Un critère d'analyticité pour les algèbres de restriction. C. R. Acad. Sci. Paris Sér. A-B **261**, A4964–A4967 (1965).
[2] Vérification statistique de la conjecture de la dichotomie sur une classe d'algèbres de restriction. C. R. Acad. Sci. Paris Sér. A-B **262**, A490–A492 (1966).
[3] Analyse harmonique dans quelques algèbres homogènes. Israel J. Math. **5**, 107–117 (1967).

Y. Katznelson and O. C. McGehee
[1] Measures and pseudomeasures on compact subsets of the line. Math. Scand. **23**, 57–68 (1968).
[2] Some Banach algebras associated with quotients of $L^1(R)$. Indiana Univ. Math. J. **21**, 419–436 (1971).
[3] Un ensemble d'entiers E tel que $B_0(E) \neq B(E)_0$. C. R. Acad. Sci. Paris Sér. A-B **280**, A31–A32 (1975).
[4] Some sets obeying harmonic synthesis. Israel J. Math. **23**, 88–93 (1976).

R. Kaufman
[1] The spectrum of an infinite product measure. Studia Math. **29**, 59–62 (1967).
[2] A functional method for linear sets, I. Israel J. Math **5**, 185–187 (1967).
[3] Measures in independent sets. Studia Math. **30**, 17–20 (1968).
[4] On certain singular measures, Colloq. Math. **19**, 265–269 (1968).
[5] Some measures determined by mappings of the Cantor set. Colloq. Math. **19**, 77–83 (1968).
[6] A functional method for linear sets, II. Israel J. Math. **7**, 293–298 (1969).
[7] Gap series and an example to Malliavin's theorem. Pacific J. Math. **28**, 117–119 (1969).
[8] Remark on Fourier-Stieltjes transforms of continuous measures. Colloq. Math. **22**, 279–280 (1971).
[9] Kronecker sets and metric properties of M_0-sets. Proc. Amer. Math. Soc. **36**, 519–524 (1972).
[10] Sets of multiplicity and differentiable functions. Proc. Amer. Math. Soc. **32**, 472–476 (1972).
[11] Lacunary series and Hausdorff measures. J. Analyse Math. **25**, 163–167 (1972).

[12] M-sets and distributions. Astérisque (Soc. Math. France) **5**, 225–230 (1973).
[13] On Menchoff's set of multiplicity. Math. Scand. **34**, 235–240 (1974).
[14] Sets of multiplicity and differentiable functions II. Trans. Amer. Math. Soc. **200**, 427–435 (1974).
[15] Bernoulli convolutions and differentiable functions. Trans. Amer. Math. Soc. **217**, 99–104 (1976).

R. Kaufman and N. W. Rickert
[1] An inequality concerning measures. Bull. Amer. Math. Soc. **72**, 672–676 (1966).

Y. Kawada: listed with K. Ito

T. W. Körner (listed also with Y. Katznelson)
[1] Some results on Kronecker, Dirichlet, and Helson sets. Ann. Inst. Fourier (Grenoble) **20**, fasc. 2, 219–324 (1970).
[2] A pseudofunction on a Helson set, I and II. Astérisque **5**, 3–224 and 231–239 (1973).
[3] Some results on Kronecker, Dirichlet, and Helson sets, II. J. Analyse Math. **27**, 260–388 (1974).
[4] On the theorem of Ivašëv-Musatov I. Ann. Inst. Fourier (Grenoble) **27**, fasc. 3, 97–115 (1977).
[5] On the theorem of Ivašëv-Musatov II. Ann. Inst. Fourier (Grenoble) **28**, fasc. 3, 123–142 (1978).
[6] A Rudin Shapiro type theorem. Illinois J. Math. (to appear).
[7] Fourier transforms and Hausdorff measure I (preprint).
[8] Fourier transforms and Hausdorff measure II (preprint).

M. G. Kreĭn
[1] On the representation of a function through Fourier-Stieltjes integrals. Kuybishov 1943.

D. Kuhlmann
[1] Measure algebras on sums and products of intervals. Thesis, Northwestern University, Evanston, Illinois 1978.

A. Kurosh
[1] The Theory of Groups, Volume I. New York: Chelsea 1960.

J. D. Lawson, J. R. Liukkonen, and M. W. Mislove
[1] Measure algebras of semilattices with finite breadth. Pacific J. Math. **69**, 125–140 (1977).

N. Leblanc
[1] Les fonctions qui opèrent dans certaines algèbres à poids. Math. Scand. **23**, 190–194 (1969).
[2] Les endomorphismes d'algèbres à poids. Bull. Soc. Math. France **99**, 387–396 (1971).
[3] Sur les isomorphismes d'algèbres de restriction. Studia Math. **49**, 7–34 (1973).

Z. L. Leĭbenson
[1] On the ring of functions with absolutely convergent Fourier series. Usp. Mat. Nauk **9**, 157–162 (1954) [Russian].

P. Lévy
[1] Sur la convergence absolue des séries de Fourier. Compositio Math. **1**, 1–14 (1935).

C. Lin and S. Saeki
[1] Bernoulli convolutions in LCA groups. Studia Math. **58**, 165–177 (1976).

L.-Å. Lindahl and F. Poulsen (editors)
[1] Thin Sets in Harmonic Analysis. New York: Marcel Dekker 1971.

J. E. Littlewood: listed with G. H. Hardy

W. Littman
[1] Fourier transforms of surface-carried measures and differentiability of surface averages. Bull. Amer. Math. Soc. **69**, 766–770 (1963).

J. R. Liukkonen: listed with J. D. Lawson and M. W. Mislove.

N. Lohoué
[1] Sur certains ensembles de synthèse dans les algèbres $A_p(G)$. C. R. Acad. Sci. Paris Sér. A-B **270**, A589–A591 (1970).
[2] Une condition d'appartenance à $A_p(T)$. C. R. Acad. Sci. Paris Sér. A-B **270**, A736–A738 (1970).

[3] Ensemble de non-synthèse uniforme dans les algèbres $A_p(G)$. Studia Math. **36**, 125–129 (1970).

[4] Algèbres $A_p(G)$ et convoluteurs de $L^p(G)$. Thèse, Université Paris-Sud 1971.

[5] La synthèse des convoluteurs sur un groupe abélien localement compact. C. R. Acad. Sci. Paris. Sér. A-B **272**, A27–A29 (1971).

[6] Sur certaines propriétés remarquables des algèbres $A_p(G)$, C. R. Acad. Sci. Paris Sér. A-B **273**, A893–A896 (1971).

L. Loomis

[1] The spectral characterization of a class of almost periodic functions. Ann. of Math. **72**, 362–368 (1960).

J. M. Lòpez and K. A. Ross

[1] Sidon Sets. New York: Marcel Dekker 1975.

P. Ludvik

[1] Discontinuous translation invariant linear functionals on $L^1(G)$. Studia Math. **56**, 21–30 (1976).

[2] On a singular measure of Connolly and Williamson. Proc. Edinburgh Math. Soc. (2) **20**, 59–61 (1967).

G. Lumer

[1] Bochner's theorem, states, and the Fourier transforms of measures. Studia Math. **46**, 135–140 (1973).

F. Lust (publishing now under the name F. Lust-Piquard)

[1] Sur la réunion de deux ensembles de Helson. C. R. Acad. Sci. Paris Sér. A-B **272**, A720–A723 (1971).

[2] Étude d'ensembles de Ditkin fort dans les algèbres tensorielles et les algèbres de groupe. Math. Scand. **28**, 317–324 (1971).

A. M. Macbeath

[1] On measures of sum sets, II: the sum-theorem for the torus. Proc. Camb. Philos. Soc. **49**, 40–43 (1953).

A. MacLean: listed with C. C. Graham

B. M. Makarov

[1] An example of a singular measure equivalent to its convolution square. Vestnik Leningrad Univ. **23**, No. 2, 51–54 (1968).

P. Malliavin (listed also with Y. Katznelson)

[1] Sur l'impossibilité de la synthèse spectrale dans une algèbre de fonctions presque périodiques. C. R. Acad. Sci., Sér. A-B **248**, A1756–A1759 (1959).

[2] Sur l'impossibilité de la synthèse spectrale sur la droite. C. R. Acad. Sci., Sér. A-B, **248**, A2155–A2157 (1959).

[3] Impossibilité de la synthèse spectrale sur les groupes abéliens non compacts. Publ. Math. Inst. Hautes Études Sci. Paris, 61–68 (1959).

[4] Calcul symbolique et sous-algèbres de $L^1(G)$. Bull. Soc. Math. France **87**, 181–186, 187–190 (1959).

[5] Ensembles de résolution spectrales. In: Proc. Internat. Congr. Mathematicians, pp. 368–378. Stockholm: 1962.

P. Malliavin and M.-P. Malliavin-Brameret

[1] Caractérisation arithmétique d'une classe d'ensembles de Helson. C. R. Acad. Sci. Paris Sér. A-B, **264**, A192–A193 (1967).

S. Mandelbrojt: listed with S. Agmon

J. Marcinkiewicz

[1] Sur la convergence absolue des séries de Fourier. Mathematica Cluj. **16**, 66–73 (1940).

D. K. Mason

[1] The functions which operate on Fourier-Stieltjes transforms of certain sub-vector spaces of M[T]. Thesis, Northwestern Univ., Evanston, IL 1972; Notices Amer. Math. Soc. **20**, A-381 (1973).

O. C. McGehee (listed also with Y. Katznelson)

[1] Sets of uniqueness and sets of multiplicity. Israel J. Math. **4**, 83–96 (1966).

[2] Certain isomorphisms between quotients of a group algebra. Pacific J. Math. **21**, 133–152 (1967).

[3] A proof of a statement of Banach about the weak$*$ topology. Mich. Math. J. **15**, 135–140 (1968).
[4] Sur un théorème de Noël Leblanc. C. R. Acad. Sci. Paris **273A**, 1226–1227 (1971).
[5] Helson sets in T^n. Conf. Harmonic Analysis, College Park, Maryland, (1971), Lecture Notes in Math. **266**, 229–237 (1972).
[6] Fourier transforms and measure-preserving transformations, Proc. Amer. Math. Soc. **44**, 71–77 (1974),

O. C. McGehee and G. S. Woodward
[1] Continuous manifolds in R^n that are sets of interpolation for the Fourier algebra (preprint).

S. A. McKilligan and A. J. White
[1] Representations of L-algebras. Proc. London Math. Soc. (3) **25**, 655–674 (1972).

D. E. Men'shov
[1] Sur l'unicité du développment trigonométrique. C. R. Acad. Sci. Paris Sér. A-B **163**, A433–A436 (1916).

Y. Meyer
[1] Algebraic Numbers and Harmonic Analysis. Amsterdam: North Holland 1972.
[2] Idéaux fermés de L^1 dans lesquels une suite approche l'identité. Math. Scand. **19**, 219–222 (1966).
[3] Endomorphisms des idéaux fermés de $L^1(G)$, classes de Hardy et séries de Fourier lacunaires. Ann. Sci. École Norm. Sup. (4) **1**, 499–580 (1968).
[4] Spectres des mesures et mesures absolument continues. Studia Math. **30**, 87–99 (1968).
[5] Isomorphismes entre certaines algèbres de restrictions. Ann. Inst. Fourier (Grenoble) **18**, 73–86 (1969).
[6] Algèbres de restriction non isomorphes. Ann. Inst. Fourier (Grenoble) **19**, 117–124 (1969).
[7] Trois problèmes sur les sommes trigonometriques. Astérisque **1**, Soc. Math. France, 1973.

Y. Meyer and H. P. Rosenthal
[1] Convexité et ensembles de Ditkin forts. C. R. Acad. Sci. Paris Sér. A-B **262**, A1404–A1406 (1966).

H. Milicer-Gruzewska
[1] Sur la continuité de la variation. C. R. Soc. Sci. de Varsovie, **21**, 164–177 (1928).

R. B. Miller
[1] Gleason parts and Choquet boundary points in convolution measure algebras. Pacific J. Math. **31**, 755–771 (1969).

C. B. Miller and M. Rajagopalan
[1] Topologies in locally compact groups, III. Proc. London Math. Soc. (3) **31**, 55–78 (1975).

M. W. Mislove: listed with J. D. Lawson and J. R. Liukkonen

W. Moran (listed also with W. J. Bailey; G. Brown; C. C. Graham and G. Brown)
[1] Separate continuity and supports of measures. J. London Math. Soc. **44**, 320–324 (1969).
[2] The Šilov boundary of $M_0(G)$. Trans. Amer. Math. Soc. **179**, 455–464 (1973).
[3] The individual symbolic calculus for measures. Proc. London Math. Soc. (3) **31**, 385–417 (1975) and (3) **38**, 481–496 (1979).

P. Mostert: listed with K. H. Hofmann

D. J. Newman
[1] An L^1-extremal problem for polynomials. Proc. Amer. Math. Soc. **16**, 1287–1290 (1965).
[2] Homomorphisms of l_+. Amer. J. Math. **91**, 37–46 (1969).
[3] Translates are always dense on the half line. Proc. Amer. Math. Soc. **21**, 511–512 (1969).

S. E. Newman
[1] Measure algebras on idempotent semigroups. Pacific J. Math. **31**, 161–169 (1969).
[2] Measure algebras and functions of bounded variation on idempotent semigroups. Trans. Amer. Math. Soc. **163**, 189–205 (1972).

D. M. Oberlin
[1] $M_p(G) \neq M_q(G)(p^{-1} + q^{-1} = 1)$. Israel J. Math. **22**, 175–179 (1975).

A. B. Paalman-de Miranda

[1] Topological Semigroups. Amsterdam: Mathematish Centrum, 1970.

O. Padé

[1] Sur le spectre d'une classe de produits de Riesz. C. R. Acad. Sci. Paris Sér. A-B **276**, A1453–A1455 (1973).

W. A. Parker: listed with G. F. Bachelis and K. A. Ross.

F. Parreau: listed with C. C. Graham and B. Host; B. Host.

J. Peyrière

[1] Sur les produits de Riesz. C. R. Acad. Sci. Paris Sér. A-B, **276**, A1417–A1419 (1973).

[2] Étude de quelques propriétés des produits de Riesz (French summary). Ann. Inst. Fourier (Grenoble) **25**, fasc. 2, xii, 127–169 (1975).

R. S. Phillips (listed also with E. Hille)

[1] On Fourier-Stieltjes integrals. Trans. Amer. Math. Soc. **69**, 312–323 (1950).

S. K. Pichorides

[1] A lower bound for the L^1 norm of exponential sums. Mathematika **21**, 155–159 (1974).

[2] A remark on exponential sums. Bull. Amer. Math. Soc. **83**, 283–285 (1977).

[3] Norms of exponential sums. Publications Math. d'Orsay no. 77–73, 65 pp. 1977.

[4] On a conjecture of Littlewood concerning exponential sums. Bull. Greek Math. Soc. (to appear).

[5] On a conjecture of Littlewood concerning exponential sums (II) (preprint).

L. Pigno and S. Saeki

[1] Interpolation by transforms of discrete measures. Proc. Amer. Math. Soc. **52**, 156–158 (1975).

[2] Fourier-Stieltjes transforms which vanish at infinity. Math. Zeitsch. **141**, 83–92 (1975).

[3] On the spectra of almost periodic functions. Indiana Univ. Math. J. **25**, 191–194 (1976).

[4] Constructions of singular measures with remarkable convolution properties (preprint).

L. Pigno and B. P. Smith

[1] Almost idempotent measures on compact abelian groups. J. Functional Analysis (to appear).

G. Pisier

[1] Ensembles de Sidon et processus Gaussiens. C. R. Acad. Sci. Paris Sér. A-B, **286**, A671–A674 (1978).

[2] Lacunarité et processus Gaussiens. C. R. Acad. Sci. Paris Sér. A-B, **286**, A1003–A1006 (1978).

[3] A remarkable homogeneous Banach algebra (preprint).

H. R. Pitt: listed with N. Wiener

A. Plessner

[1] Eine Kennzeichnung der totstetigen Funktionen. J. Reine und Angew. Math. **60**, 26–32 (1929).

H. Pollard

[1] The harmonic analysis of bounded functions. Duke Math. J. **20**, 499–512 (1953).

F. Poulsen: listed with L.-Å. Lindahl

J. F. Price

[1] Some strict inclusions between spaces of L^p-multipliers. Trans. Amer. Math. Soc. **152**, 321–330 (1970).

I. I. Pyateckiĭ-Šapiro

[1] Supplement to the work "On the problem of uniqueness of expansion of a function in a trigonometric series" (in Russian). Moscov. Gos. Univ. Uč. Zap. 165 Mat. **7**, 79–97 (1954).

J. Rago

[1] Convolutions of continuous measures and sums of an independent set. Proc. Amer. Math. Soc. **44**, 123–128 (1974).

David L. Ragozin

[1] Rotation-invariant measure algebras on Euclidean space. Indiana Univ. Math. J. **23**, 1139–1154 (1974).

D. A. Raĭkov (listed also with I. M. Gel'fand and G. E. Šilov)

[1] On absolutely continuous set functions. Doklady **34**, 239–241 (1942).

M. Rajagapolan: listed with C. B. Miller

A. Rajchman

[1] Une classe de séries trigonométriques qui convergent presque partout vers zéro. Math. Ann. **101**, 686–700 (1929).

L. T. Ramsey

[1] Fourier-Stieltjes transforms of measures with a certain continuity property. J. Functional Analysis **25**, 306–316 (1977).

L. T. Ramsey and B. B. Wells

[1] Fourier-Stieltjes transforms of strongly continuous measures. Mich. Math. J. **24**, 13–19 (1977).

J. F. Rennison

[1] Arens products and measure algebras. J. London Math. Soc. **44**, 369–377 (1969).

[2] A supplement to "Arens products and measure algebras." J. London Math. Soc. (2) **1**, 232–236 (1969).

P. Ressel: listed with J. B. R. Christensen

I. Richards

[1] On the disproof of spectral synthesis. J. Comb. Theory **2**, 61–70 (1967).

L. F. Richardson

[1] A class of idempotent measures on compact nilmanifolds. Acta Math. **135**, 129–154 (1975).

C. E. Rickart

[1] The General Theory of Banach Algebras. Princeton: Van Nostrand, 1960.

N. W. Rickert (listed also with R. Kaufman)

[1] Locally compact topologies for groups. Trans. Amer. Math. Soc. **126**, 225–235 (1967).

D. Rider

[1] Gap series on groups and spheres. Canadian J. Math. **18**, 389–397 (1966).

[2] Transformations of Fourier coefficients. Pacific J. Math. **19**, 347–355 (1966).

[3] Closed subalgebras of $L^1(T)$. Duke Math. J. **36**, 105–116 (1969).

[4] Functions which operate on positive-definite functions. Proc. Cambridge Philos. Soc. **69**, 87–97 (1971).

[5] Functions which operate in the Fourier algebra of a compact group. Proc. Amer. Math. Soc. **28**, 525–530, (1971).

[6] Functions which operate on $\mathscr{F}L^p(T)$, $1 < p < 2$. Pacific J. Math. **40**, 681–693 (1972).

[7] Randomly continuous functions and Sidon sets. Duke Math. J. **42**, 752–764 (1975).

F. Riesz

[1] Über die Fourierkoeffizienten einer stetigen Funktion von beschränkter Schwankung. Math. Zeitschr. **18**, 312–315 (1918).

G. Ritter

[1] Unendliche Produkte unkorrelierte Funktionen auf kompakten Abelschen Gruppen. Math. Scand. **42**, 251–270 (1978).

[2] On dichotomy of Riesz products. Math. Proc. Cambridge Philos. Soc. **85**, 79–90 (1978).

[3] On Kakutani's theorem for infinite products of not necessarily independent functions. Math. Ann. **239**, 35–53 (1979).

N. M. Rivière and Y. Sagher

[1] The converse of the Wiener-Lévy-Marcinkiewicz Theorem. Studia Math. **28**, 133–138 (1966).

H. P. Rosenthal (listed also with Y. Meyer)

[1] Projections onto Translation-Invariant Subspaces of $L^p(G)$. Mem. Amer. Math. Soc., **63**, 1966.

[2] Caracterization d'ensembles de Helson, par l'existence de certains projecteurs. C. R. Acad. Sci. Paris Sér. A-B **262**, A286–A288 (1966).

[3] On trigonometric series associated with weak$*$ closed subspaces of continuous functions. J. Math. Mech. **17**, 485–490 (1967).

[4] On the existence of approximate identities in ideals of group algebras. Ark. Mat. **7**, 185–191 (1967).

[5] A characterization of restrictions of Fourier-Stieltjes transforms. Pacific J. Math. **23**, 403–418 (1967).
[6] A characterization of the linear sets satisfying Herz's criterion. Pacific J. Math. **28**, 663–668 (1969).
[7] On subspaces of L^p. Ann. Math. **97**, 344–373 (1973).

K. A. Ross (listed also with: G. F. Bachelis and W. A. Parker; E. Hewitt; J. M. Lòpez)
[1] Sur les compacts associés à un ensemble de Sidon. C. R. Acad. Sci. Paris Sér. A-B **275**, A183–A185 (1972).
[2] The structure of certain measure algebras. Pacific J. Math. **11**, 723–736 (1961).

H. Rossi (listed with R. C. Gunning)
[1] The local maximum modulus principle. Ann. Math. **72**, 1–11 (1960).

K. F. Roth (listed also with H. Halberstam)
[1] On cosine polynomials corresponding to sets of integers. Acta Arithmetica **24**, 87–98 (1973).

H. L. Royden
[1] Real Analysis. New York: Macmillan, 1963.

W. Rudin (listed also with: H. Helson, J.-P. Kahane and Y. Katznelson; J.-P. Kahane)
[1] Fourier Analysis on Groups. Interscience Tract No. 12. New York: Wiley, 1962.
[2] Real and Complex Analysis (second edition). New York: McGraw-Hill, 1974.
[3] Functional Analysis. New York: McGraw-Hill, 1973.
[4] Non-analytic functions of absolutely convergent Fourier series. Proc. Nat. Acad. Sci. (U.S.) **41**, 238–240 (1955).
[5] Transformations des coefficients de Fourier. C. R. Acad. Sci. Paris Sér. A-B **243**, A638–A640 (1956).
[6] Independent perfect sets in groups. Mich. Math. J. **5**, 159–161 (1958).
[7] Positive definite sequences and absolutely monotonic functions. Duke Math. J. **26**, 617–622 (1959).
[8] Measure algebras on abelian groups. Bull. Amer. Math. Soc. **65**, 227–247 (1959).
[9] Idempotent measures on abelian groups. Pacific J. Math. **9**, 195–209 (1959).
[10] Trigonometric series with gaps. J. Math. Mech. **9**, 203–227 (1960).
[11] Fourier-Stieltjes transforms of measures on independent sets. Bull. Amer. Math. Soc. **66**, 199–202 (1960).
[12] Some theorems on Fourier coefficients. Proc. Amer. Math. Soc. **66**, 199–202 (1960).
[13] A strong converse of the Wiener-Lévy Theorem. Canad. J. Math. **14**, 694–701 (1962).

C. Ryll-Nardzewski (listed also with: S. Hartman and J.-P. Kahane; S. Hartman)
[1] Concerning almost periodic extensions of functions. Colloq. Math. **12**, 235–237 (1964).

S. Saeki (listed also with: C. Lin; L. Pigno)
[1] On norms of idempotent measures. Proc. Amer. Math. Soc. **19**, 600–602 (1968).
[2] On norms of idempotent measures II. Proc. Math. Soc. **19**, 367–371 (1968).
[3] Operating functions on $B_0(\hat{G})$ in plane regions. Tôhoku Math. J. (2) **21**, 112–116 (1969).
[4] Spectral synthesis for the Kronecker sets. J. Math. Soc. Japan **21**, 549–563 (1969).
[5] Translation invariant operators on groups. Tôhoku Math. J. **22**, 409–419 (1970).
[6] The ranges of certain isometries of tensor products of Banach spaces. J. Math. Soc. Japan **23**, 27–39 (1971).
[7] On the union of two Helson sets. J. Math. Soc. Japan **23**, 636–648 (1971).
[8] Homomorphisms of tensor algebras Tôhoku Math. J. (2) **23**, 173–199 (1971).
[9] Tensor products of Banach algebras and harmonic analysis. Tôhoku Math. J. (2) **24**, 281–299 (1972).
[10] On strong Ditkin sets. Arkiv Mat. **10**, 1–7 (1972).
[11] On restriction algebras of tensor algebras. J. Math. Soc. Japan **25**, 506–522 (1973).
[12] Symmetric maximal ideals in M(G). Pacific J. Math. **54**, 229–243 (1974).
[13] Helson sets which disobey spectral synthesis. Proc. Amer. Math. Soc. **47**, 371–377 (1975).
[14] Infinite tensor products in Fourier algebras. Tôhoku Math. J. **27**, 355–379 (1975).
[15] On the sum of two Kronecker sets. Illinois J. Math. **19**, 127–130 (1975).
[16] Convolutions of continuous measures and sets of non-synthesis. Proc. Amer. Math. Soc. **60**, 215–220 (1976).
[17] Most symmetric sets are of synthesis. Illinois J. Math. **20**, 171–176 (1976).
[18] Asymmetric maximal ideals in M(G). Trans. Amer. Math. Soc. **222**, 241–253 (1976).

[19] Singular measures having absolutely continuous convolution powers. Illinois J. Math. **21**, 395–412 (1977).
[20] On infinite convolution products of discrete probability measures. J. London Math. Soc. (2) **16**, 172–176 (1977).
[21] Bohr compactification and continuous measures (preprint).
[22] On convolution squares of singular measures (preprint).

S. Saeki and E. Sato
[1] Critical points and point derivations on M(G). Mich. Math. J. **25**, 147–161 (1978).

Y. Sagher: listed with N. M. Rivière

K. Saka
[1] On a characterization of some L-subalgebras in measure algebras. J. London Math. Soc. (2) **9**, 261–271 (1974).

R. Salem (listed also with J.-P. Kahane)
[1] Algebraic Numbers and Fourier Analysis. Boston: D. C. Heath, 1963.
[2] On a problem of Littlewood. Amer. J. Math. **77**, 535–540 (1955).

D. L. Salinger
[1] Large squares and sets of analyticity in tensor algebras. Studia Math. **36**, 259–267 (1970).

D. L. Salinger and N. Th. Varopoulos
[1] Convolutions of measures and sets of analyticity. Math. Scand. **25**, 5–18 (1969).

D. Sarason
[1] Weak-star generators of H^∞. Pacific J. Math. **17**, 519–528 (1966).
[2] On the order of a simply connected domain. Mich. Math. J. **15**, 129–133 (1968).

E. Sato (listed also with S. Saeki)
[1] Convolutions of measures on some thin sets. Hokkaido Math. J. **7**, 49–57 (1978).

R. Schneider
[1] On certain homomorphisms of restriction algebras of symmetric sets. Israel J. Math. **6**, 223–232 (1968).
[2] Some theorems in Fourier analysis on symmetric sets. Pacific J. Math. **31**, 175–195 (1969).

I. J. Schoenberg
[1] A remark on the preceding note by Bochner. Bull. Amer. Math. Soc. **40**, 277–278 (1934).
[2] Metric spaces and completely monotone functions. Ann. Math. (2) **39**, 811–841 (1938).

B. M. Schreiber (listed also with T. Ito)
[1] On the coset ring and strong Ditkin sets. Pacific J. Math. **32**, 805–812 (1970).

A. Schwartz
[1] On the ideal structure of the algebra of radial functions. Proc. Amer. Math. Soc. **26**, 621–624 (1970).

J. T. Schwartz: listed with N. Dunford

J. Schwartz
[1] A remark on inequalities of Calderòn-Zygmund type for vector-valued functions. Comm. Pure Appl. Math. **14**, 785–799 (1961).

L. Schwartz
[1] Sur une propriété de synthèse spectrale dans les groupes non compacts. C. R. Acad. Sci. Sér. A-B **227**, A424–A426 (1948).

William M. Self
[1] Some consequences of the Beurling-Helson theorem. Rocky Mountain J. Math. **6**, 177–180 (1976).

G. S. Shapiro
[1] Linear projections which implement balayage in Fourier transforms. Proc. Amer. Math. Soc. **61**, 295–299 (1976).

H. S. Shapiro (listed also with C. Fefferman)
[1] Extremal problems for polynomials and power series. M. S. Thesis, Mass. Inst. Tech., Cambridge, Mass. 1971.
[2] Fourier multipliers whose multiplier norm is an attained value. In: Linear operators and approximation (Proc. Conf., Oberwolfach, 1971), pp. 338–347. Basel. Birkhaüser 1972.

R. Shatten

[1] A theory of cross spaces. Ann. Math. Studies, no. 26. Princeton: Princeton Univ. Press, 1950.

G. E. Šilov: listed with I. M. Gel'fand and D. A. Raĭkov

S. M. Simmons

[1] A converse Steinhaus theorem for locally compact groups. Proc. Amer. Math. Soc. **49**, 383–386 (1975).

A. B. Simon (listed also with R. R. Goldberg)

[1] Symmetry in measure algebras. Bull. Amer. Math. Soc. **66**, 399–400 (1960).

[2] Homomorphisms of measure algebras. Illinois J. Math. **5**, 398–408 (1961).

[3] The ideal space and Šilov boundary of a subalgebra of measures on a group. J. Math. Anal. App. **6**, 266–276 (1963).

P. Sjölin: listed with L. Carleson

G. L. G. Sleijpen

[1] Convolution measure algebras on semigroups. Thesis, Catholic University of Nijmegen, Nijmegen, The Netherlands, 1976.

B. P. Smith

[1] Helson sets not containing the identity are uniform Fatou-Zygmund sets. Indiana Univ. Math. J. **27**, 331–347 (1978).

P. M. Soardi: listed with L. deMichele

R. Spector

[1] Apercu de la théorie des hypergroupes. In: Analyse Harmonique sur les Groupes de Lie, pp. 643–672. Berlin-Heidelberg-New York: Springer-Verlag, 1975.

[2] Mesures invariantes sur les hypergroupes. Trans. Amer. Math. Soc. **239**, 147–165 (1978).

L. E. Spence: listed with S. H. Friedberg

Y. A. Šreĭder

[1] The structure of maximal ideals in rings of measures with involution. Mat. Sb. (N.S.) **27** (69), 297–318 (1950); Amer. Math. Soc. Transl. (1st ser.) No. 81 (1953), 28 pp.

[2] On an example of a generalized character. (Russian). Mat. Sb. (N.S.) **29**, (71), 419–426 (1951).

J. D. Stafney

[1] Arens multiplication and convolution. Pacific J. Math. **4**, 1423–1447 (1964).

J. D. Stegeman

[1] Méthodes combinatoires en analyse harmonique. Séminaire d'Analyse Harmonique, No. 6, 1968–9, Orsay, France.

[2] On a theorem of N. Th. Varopoulos. Math. Scand. **27**, 50–52 (1970).

[3] On unions of Helson sets. Indag. Math. **32**, 456–462 (1970).

[4] Studies in Fourier and Tensor Algebras. Utrecht: Pressa Trajectina 1971.

[5] A criterion for sets of V-interpolation. J. Functional Analysis **8**, 189–196 (1971).

E. M. Stein

[1] Singular Integrals and Differentiability Properties of Functions, Princeton, N.J.: Princeton Univ. Press, 1970.

[2] Topics in Harmonic Analysis Related to the Littlewood-Paley Theory. Princeton, N.J.: Princeton Univ. Press, 1970.

E. M. Stein and G. Weiss

[1] Introduction to Fourier Analysis on Euclidean Spaces. Princeton, N.J.: Princeton Univ. Press, 1971.

H. Steinhaus

[1] Sur les distances des points des ensembles de mesure positive. Fund. Math. **1**, 93–104 (1920).

J. Stewart

[1] Positive definite functions and generalizations, an historical survey. Rocky Mountain J. **6**, 409–434 (1976).

R. Strichartz

[1] Isometric isomorphisms of measure algebras. Pacific J. Math. **15**, 315–317 (1965).

K. Stromberg (listed also with E. Hewitt)
[1] An elementary proof of Steinhaus's Theorem. Proc. Amer. Math, Soc. **36**, 308 (1972).
[2] Large families of singular measures having absolutely continuous convolution squares. Proc. Camb. Philos. Soc. **65**, 1015–1022 (1968).

L. J. Sulley
[1] On countable inductive limits of locally compact abelian groups. J. London Math. Soc. (2) **5**, 629–637 (1972).

T. Suzuki
[1] The singularity of infinite product measures. Osaka J. Math. **11**, 653–661 (1971).

Z. Takeda
[1] Notes on Fourier-Stieltjes Integral II. Kōdai Mathematical Seminar Reports, Tokyo Institute of Technology, 33–36 (1953).

M. Talagrand
[1] Solution d'un problème de R. Haydon. Pub. de Dept. Math. Université de Lyon 12–2, 43–46 (1975).
[2] Somme vectorielles d'ensembles de mesure nulle. Ann. Inst. Fourier (Grenoble) **26**, fasc. 3, 137–172 (1976).

J. D. Tamarkin: listed with E. Hille

J. L. Taylor (listed also with R. G. Douglas)
[1] Measure Algebras. Regional Conference Series in Math., No. 16, Providence, R. I.: American Math. Soc., 1972.
[2] The Šilov boundary of the algebra of measures on a group. Proc. Amer. Math. Soc. **16**, 941–945 (1965).
[3] The structure on convolution measure algebras. Trans. Amer. Math. Soc. **119**, 150–166 (1965).
[4] Ideal theory and Laplace transforms for a class of measure algebras on a group. Acta Math. **121**, 251–292 (1968).
[5] The cohomology of the spectrum of a measure algebra. Acta Math. **126**, 195–225 (1971).
[6] Inverses, logarithms, and idempotents in M(G). Rocky Mountain J. Math. **2**, 183–206 (1972).
[7] On the spectrum of a measure. Adv. Math. **12**, 451–462 (1974).

Y. Uno
[1] Lipschitz functions and convolution. Proc. Japan, Acad. **50**, 785–788 (1974).

E. R. Van Kampen
[1] Infinite product measures and infinite convolutions. Amer. J. Math. **62**, 417–448 (1940).

N. Th. Varopoulos (listed also with: A. Bernard; D. L. Salinger)
[1] Sur les mesures de Radon d'un groupe localement compact abélien. C. R. Acad. Sci. Paris. Sér. A-B, **258**, A3805–A3808 (1964).
[2] Measure algebras of a locally compact abelian group. In: Séminaire Bourbaki No. 282 (1964–65), pp. 1–10.
[3] The functions that operate on $B_0(\Gamma)$ of a discrete group. Bull. Soc. Math. France **93**, 301–321 (1965).
[4] Sur les ensembles parfaits et les séries trigonométriques. C. R. Acad. Sci. Paris Sér. A-B **260**, A3831–A3834 (1965).
[5] Sur les ensembles parfaits et les séries trigonométriques. C. R. Acad. Sci. Sér. A-B **260**, A4668–A4670 (1965).
[6] Sur les ensembles parfaits et les séries trigonométriques. C. R. Acad. Sci. Sér. A-B **260**, A5165–A5168 (1965).
[7] Sur les ensembles parfaits et les séries trigonométriques. C. R. Acad. Sci. Paris Sér. A-B **260**, A5997–A6000 (1965).
[8] Spectral synthesis on spheres. Proc. Camb. Philos. Soc. **62**, 379–387 (1966).
[9] Sets of multiplicity in locally compact abelian groups. Ann. Inst. Fourier (Grenoble) **16**, fasc. 2, 123–158 (1966).
[10] Tensor algebras and harmonic analysis. Acta Math. **119**, 51–112 (1968).
[11] On a problem of A. Beurling. J. Functional Analysis **2**, 24–30 (1968).
[12] Tensor algebras over discrete spaces. J. Functional Analysis **3**, 321–335 (1969).
[13] Groups of continuous functions in harmonic analysis. Acta Math. **125**, 109–152 (1970).
[14] Sur la réunion de deux ensembles de Helson. C. R. Acad. Sci. Paris Sér. A-B **271**, A251–A253 (1970).

[15] Sidon sets in R^n. Math. Scand. **27**, 39–49 (1970).
[16] Sur les ensembles de Helson. C. R. Acad. Sci. Paris Sér. A-B **272**, A592–A593 (1971).
[17] Ensembles pics et ensembles d'interpolation pour les algèbres uniformes. C. R. Acad. Sci. Paris Sér. A-B **272**, A866–A867 (1971).
[18] On the endomorphism problem for tensor algebras. Prakt. Akad. Athénón **47**, 193–197 (1972).
[19] Sur la réunion de deux ensembles d'interpolations d'une algèbre uniforme. C. R. Acad. Sci. Paris Sér. A-B **272**, A950–A952 (1971).
[20] Une remarque sur les ensembles de Helson (preprint).

P. Vermes: listed with J. Clunie.

J. von Neumann
[1] Ein System algebraisch unabhängiger Zahlen. Math. Ann. **99**, 134–141 (1928).

L. Waelbroeck
[1] The holomorphic functional calculus. In: Algebras in Analysis, edited by J. H. Williamson, pp. 187–249. New York: Academic Press, 1975.

A. Weil
[1] L'intégration dans les Groupes Topologiques et ses Applications. Paris: Gautiers-Villars, 1938.

G. Weiss: listed with R. R. Coifman; J.-P. Kahane and M. Weiss; E. M. Stein

M. Weiss (listed also with J.-P. Kahane and G. Weiss)
[1] On a problem of Littlewood. J. London Math. Soc. **34**, 217–221. (1959).

B. B. Wells, Jr. (listed also with L. T. Ramsey)
[1] Rearrangements of functions on the ring of integers of a p-series field. Pacific J. Math. **65**, 253–259 (1976).

J. G. Wendel
[1] Left centralizers and isomorphisms of group algebras. Pacific J. Math. **2**, 251–261 (1952).

A. J. White (listed also with: P. S. Chow; S. A. McKilligan)
[1] Convolution of vector measures. Proc. Royal Soc. Edinburgh (73A) **7**, 117–135 (1974/75).

N. Wiener.
[1] Tauberian Theorems. Ann. Math. **33**, 1–100 (1932).
[2] The Fourier Integral and Certain of Its Applications. New York: Dover, 1958.

N. Wiener and H. E. Pitt
[1] On absolutely convergent Fourier-Stieltjes transforms. Duke Math. J. **4**, 420–436 (1938).

N. Wiener and A. Wintner
[1] Fourier-Stieltjes transforms and singular infinite convolutions. Amer. J. Math. **60**, 513–522 (1938).

N. Wiener and R. C. Young
[1] The total variation of $g(x + h) - g(x)$. Trans. Amer. Math. Soc. **33**, 327–340. (1935).

I. Wik (listed also with I. Glicksberg)
[1] A strong form of spectral synthesis. Ark. Mat. **6**, 55–64 (1965).

J. H. Williamson (listed also with D. M. Connolly)
[1] A theorem on algebras of measures on topological groups. Proc. Edinburgh Math. Soc. **11**, 795–806 (1958/59).
[2] Raikov systems and the pathology of M(R). Studia Math. **31**, 399–409 (1968).
[3] Banach algebra elements with independent powers. In: Proc. Int. Symp. on Function Algebras, F. Birtel Ed., pp. 186–197. Chicago: Scott-Foresman, 1966.

A. Wintner (listed also with: B. Jessen; N. Wiener)
[1] Asymptotic distributions and infinite convolutions. Princeton, N.J.: The Institute for Advanced Study, 1938.

J. C. S. Wong
[1] Convolution and separate continuity. Pacific J. Math. **75**, 601–611 (1978).

G. V. Wood (listed also with N. J. Kalton)
[1] Distance between group algebras (preprint).

G. S. Woodward (listed also with: G. W. Johnson; O. C. McGehee)
[1] Sur une classe d'ensembles épars. C. R. Acad. Sci. Paris. Sér. A-B **274**, A221–A223 (1972).

[2] The generalized almost periodic part of an ergodic function. Studia Math. **50**, 103–116 1974.

[3] Invariant means and ergodic sets in Fourier analysis. Pacific J. Math. **54**, 281–299 (1974).

K. Ylinen

[1] Tensor products of complex L-spaces and convolution measure algebras. Ann. Acad. Sci. Fenn. I. Mathematica, No. 558, Ser. A, 1–10 (1973).

R. C. Young: listed with N. Wiener

W. H. Young

[1] A note on trigonometric series. Mess. for Math. **38**, 44–48 (1909).

M. Zafran

[1] On the spectra of multipliers. Pacific J. Math. **47**, 609–626 (1973).

[2] The spectra of multiplier transformations on the L^p spaces. Ann. Math. **103**, 355–374 (1976).

[3] The functions operating on multiplier algebras. J. Functional Analysis **26**, 289–314 (1977).

[4] The dichotomy problem for homogeneous Banach algebras. Ann. Math. (to appear).

[5] On the symbolic calculus in homogeneous Banach algebras (preprint).

H. S. Zuckerman: listed with E. Hewitt

A. Zygmund

[1] Trigonometric Series, two volumes. Cambridge: The University Press, 1959.

Index

Numbers such as [1] after an author's name refer to items in the References (pp. 425–445). There are no citations here to the section Symbols, Conventions, and Terminology (p. xv).

Grundlehren der mathematischen Wissenschaften

A Series of Comprehensive Studies in Mathematics

A Selection

114. Mac Lane: Homology
131. Hirzebruch: Topological Methods in Algebraic Geometry
144. Weil: Basic Number Theory
145. Butzer/Berens: Semi-Groups of Operators and Approximation
146. Treves: Locally Convex Spaces and Linear Partial Differential Equations
152. Hewitt/Ross: Abstract Harmonic Analysis. Vol. 2: Structure and Analysis for Compact Groups. Analysis on Locally Compact Abelian Groups
153. Federer: Geometric Measure Theory
154. Singer: Bases in Banach Spaces I
155. Müller: Foundations of the Mathematical Theory of Electromagnetic Waves
156. van der Waerden: Mathematical Statistics
157. Prohorov/Rozanov: Probability Theory. Basic Concepts. Limit Theorems. Random Processes
158. Constantinescu/Cornea: Potential Theory on Harmonic Spaces
159. Köthe: Topological Vector Spaces I
160. Agrest/Maksimov: Theory of Incomplete Cylindrical Functions and their Applications
161. Bhatia/Szegö: Stability of Dynamical Systems
162. Nevanlinna: Analytic Functions
163. Stoer/Witzgall: Convexity and Optimization in Finite Dimensions I
164. Sario/Nakai: Classification Theory of Riemann Surfaces
165. Mitrinovic/Vasic: Analytic Inequalities
166. Grothendieck/Dieudonné: Eléments de Géometrie Algébrique I
167. Chandrasekharan: Arithmetical Functions
168. Palamodov: Linear Differential Operators with Constant Coefficients
169. Rademacher: Topics in Analytic Number Theory
170. Lions: Optimal Control of Systems Governed by Partial Differential Equations
171. Singer: Best Approximation in Normed Linear Spaces by Elements of Linear Subspaces
172. Bühlmann: Mathematical Methods in Risk Theory
173. Maeda/Maeda: Theory of Symmetric Lattices
174. Stiefel/Scheifele: Linear and Regular Celestial Mechanic. Perturbed Two-body Motion–Numerical Methods–Canonical Theory
175. Larsen: An Introduction to the Theory of Multipliers
176. Grauert/Remmert: Analytische Stellenalgebren
177. Flügge: Practical Quantum Mechanics I
178. Flügge: Practical Quantum Mechanics II
179. Giraud: Cohomologie non abélienne
180. Landkof: Foundations of Modern Potential Theory
181. Lions/Magenes: Non-Homogeneous Boundary Value Problems and Applications I
182. Lions/Magenes: Non-Homogeneous Boundary Value Problems and Applications II
183. Lions/Magenes: Non-Homogeneous Boundary Value Problems and Applications III
184. Rosenblatt: Markov Processes. Structure and Asymptotic Behavior
185. Rubinowicz: Sommerfeldsche Polynommethode
186. Handbook for Automatic Computation. Vol. 2. Wilkinson/Reinsch: Linear Algebra

187. Siegel/Moser: Lectures on Celestial Mechanics
188. Warner: Harmonic Analysis on Semi-Simple Lie Groups I
189. Warner: Harmonic Analysis on Semi-Simple Lie Groups II
190. Faith: Algebra: Rings, Modules, and Categories I
191. Faith: Algebra II, Ring Theory
192. Mallcev: Algebraic Systems
193. Pólya/Szegö: Problems and Theorems in Analysis I
194. Igusa: Theta Functions
195. Berberian: Baer*-Rings
196. Athreya/Ney: Branching Processes
197. Benz: Vorlesungen über Geometric der Algebren
198. Gaal: Linear Analysis and Representation Theory
199. Nitsche: Vorlesungen über Minimalflächen
200. Dold: Lectures on Algebraic Topology
201. Beck: Continuous Flows in the Plane
202. Schmetterer: Introduction to Mathematical Statistics
203. Schoeneberg: Elliptic Modular Functions
204. Popov: Hyperstability of Control Systems
205. Nikollskii: Approximation of Functions of Several Variables and Imbedding Theorems
206. André: Homologie des Algèbres Commutatives
207. Donoghue: Monotone Matrix Functions and Analytic Continuation
208. Lacey: The Isometric Theory of Classical Banach Spaces
209. Ringel: Map Color Theorem
210. Gihman/Skorohod: The Theory of Stochastic Processes I
211. Comfort/Negrepontis: The Theory of Ultrafilters
212. Switzer: Algebraic Topology—Homotopy and Homology
213. Shafarevich: Basic Algebraic Geometry
214. van der Waerden: Group Theory and Quantum Mechanics
215. Schaefer: Banach Lattices and Positive Operators
216. Pólya/Szegö: Problems and Theorems in Analysis II
217. Stenström: Rings of Quotients
218. Gihman/Skorohod: The Theory of Stochastic Processes II
219. Duvaut/Lions: Inequalities in Mechanics and Physics
220. Kirillov: Elements of the Theory of Representations
221. Mumford: Algebraic Geometry I: Complex Projective Varieties
222. Lang: Introduction to Modular Forms
223. Bergh/Löfström: Interpolation Spaces. An Introduction
224. Gilbarg/Trudinger: Elliptic Partial Differential Equations of Second Order
225. Schütte: Proof Theory
226. Karoubi: K-Theory. An Introduction
227. Grauert/Remmert: Theorie der Steinschen Räume
228. Segal/Kunze: Integrals and Operators
229. Hasse: Number Theory
230. Klingenberg: Lectures on Closed Geodesics
231. Lang: Elliptic Curves: Diophantine Analysis
232. Gihman/Skorohod: The Theory of Stochastic Processes III
233. Stroock/Varadhan: Multi-dimensional Diffusion Processes
234. Aigner: Combinatorial Theory
235. Dynkin/Yushkevich: Markov Control Processes and Their Applications
236. Grauert/Remmert: Theory of Stein Spaces
237. Köthe: Topological Vector Spaces II
238. Graham/McGehee: Essays in Commutative Harmonic Analysis
239. Elliott: Probabilistic Number Theory I
240. Elliott: Probabilistic Number Theory II